HIGH-THROUGHPUT ANALYSIS IN THE PHARMACEUTICAL INDUSTRY

T0136194

CRITICAL REVIEWS IN COMBINATORIAL CHEMISTRY

Series Editors

BING YAN
School of Pharmaceutical Sciences
Shandong University, China

ANTHONY W. CZARNIK
Department of Chemistry
University of Nevada–Reno, U.S.A.

A series of monographs in molecular diversity and combinatorial chemistry, high-throughput discovery, and associated technologies.

Combinatorial and High-Throughput Discovery and Optimization of Catalysts and Materials
Edited by Radislav A. Potyrailo and Wilhelm F. Maier

Combinatorial Synthesis of Natural Product-Based Libraries
Edited by Armen M. Boldi

High-Throughput Lead Optimization in Drug Discovery
Edited by Tushar Kshirsagar

High-Throughput Analysis in the Pharmaceutical Industry
Edited by Perry G. Wang

HIGH-THROUGHPUT ANALYSIS IN THE PHARMACEUTICAL INDUSTRY

Edited by

Perry G. Wang

CRC Press
Taylor & Francis Group
Boca Raton London New York

CRC Press is an imprint of the
Taylor & Francis Group, an **informa** business

CRC Press
Taylor & Francis Group
6000 Broken Sound Parkway NW, Suite 300
Boca Raton, FL 33487-2742

First issued in paperback 2019

© 2009 by Taylor & Francis Group, LLC
CRC Press is an imprint of Taylor & Francis Group, an Informa business

No claim to original U.S. Government works

ISBN-13: 978-1-4200-5953-3 (hbk)
ISBN-13: 978-0-367-38700-6 (pbk)

This book contains information obtained from authentic and highly regarded sources Reasonable efforts have been made to publish reliable data and information, but the author and publisher cannot assume responsibility for the validity of all materials or the consequences of their use. The Authors and Publishers have attempted to trace the copyright holders of all material reproduced in this publication and apologize to copyright holders if permission to publish in this form has not been obtained. If any copyright material has not been acknowledged please write and let us know so we may rectify in any future reprint

Except as permitted under U.S. Copyright Law, no part of this book may be reprinted, reproduced, transmitted, or utilized in any form by any electronic, mechanical, or other means, now known or hereafter invented, including photocopying, microfilming, and recording, or in any information storage or retrieval system, without written permission from the publishers.

For permission to photocopy or use material electronically from this work, please access www.copyright.com (http://www.copyright.com/) or contact the Copyright Clearance Center, Inc. (CCC) 222 Rosewood Drive, Danvers, MA 01923, 978-750-8400. CCC is a not-for-profit organization that provides licenses and registration for a variety of users. For organizations that have been granted a photocopy license by the CCC, a separate system of payment has been arranged.

Trademark Notice: Product or corporate names may be trademarks or registered trademarks, and are used only for identification and explanation without intent to infringe.

Library of Congress Cataloging-in-Publication Data

High-throughput analysis in the pharmaceutical industry / edited by Perry G. Wang.
    p. ; cm. -- (Critical reviews in combinatorial chemistry)
  Includes bibliographical references and index.
  ISBN-13: 978-1-4200-5953-3 (hardcover : alk. paper)
  ISBN-10: 1-4200-5953-X (hardcover : alk. paper)
  1. High throughput screening (Drug development) 2. Combinatorial chemistry. I. Wang, Perry G. II. Title. III. Series.
  [DNLM: 1. Combinatorial Chemistry Techniques--methods. 2. Drug Design. 3. Pharmaceutical Preparations--analysis. 4. Technology, Pharmaceutical--methods. QV 744 H6366 2009]

RS419.5.H536 2009
615'.19--dc22
                                              2008003895

Visit the Taylor & Francis Web site at
http://www.taylorandfrancis.com

and the CRC Press Web site at
http://www.crcpress.com

Contents

Preface

I had the pleasure of developing and exploiting the high-throughput techniques used for drug analysis in the pharmaceutical industry at Abbott Laboratories. My major duties as project leader involved bioanalytical method development and validation by liquid chromatography with tandem mass spectrometry (LC/MS/MS). While organizing a symposium titled "High-Throughput Analyses of Drugs and Metabolites in Biological Matrices Using Mass Spectrometry" for the 2006 Pittsburgh Conference, it became my dream to edit a book called *High-Throughput Analysis in the Pharmaceutical Industry*.

It is well known that high-throughput, selective and sensitive analytical methods are essential for reducing timelines in the course of drug discovery and development in the pharmaceutical industry. Traditionally, an experienced organic chemist could synthesize and finalize approximately 50 compounds each year. However, since the introduction of combinatorial chemistry technology to the pharmaceutical industry, more than 2000 compounds can be easily generated yearly with certain automation. Conventional analytical approaches can no longer keep pace with the new breakthroughs and they now constitute bottlenecks to drug discovery. In order to break the bottlenecks, a revolutionary improvement of conventional methodology is needed. Therefore, new tools and approaches for analysis combined with the technologies such as combinatorial chemistry, genomics, and biomolecular screening must be developed. Fortunately, liquid chromatography/mass spectrometry (LC/MS)-based techniques provide unique capabilities for the pharmaceutical industry. These techniques have become very widely accepted at every stage from drug discovery to development.

This book discusses the most recent and significant advances of high-throughput analysis in the pharmaceutical industry. It mainly focuses on automated sample preparation and high-throughput analysis by high-performance liquid chromatography (HPLC) and mass spectrometry (MS). The application of high-performance liquid chromatography combined with mass spectrometry (HPLC-MS) and the use of tandem mass spectrometry (HPLC/MS-MS) have proven to be the most important analytical techniques for both drug discovery and development. The strategies for optimizing the application of these techniques for high-throughput analysis are also discussed. Microparallel liquid chromatography, ADME/PK high-throughput assays, MS-based proteomics, and advances in capillary and nano-HPLC technology are also introduced in this book.

I sincerely hope that readers—ranging from college students to professionals and academics in the fields of pharmaceutics and biotechnology—will find the chapters in this book to be helpful and valuable resources for their current projects and recommend this volume to their colleagues.

I would like to note my appreciation to all the contributors who found time in their busy schedules to provide the chapters herein. Many thanks to my previous colleagues, Shimin Wei, Min S. Chang, and Tawakol El-Shourbagy for their friendship and support. I would like to take this opportunity to acknowledge and thank the late Dr. Raymond Wieboldt for his priceless mentoring, without which I could not have been so successful in establishing my career in the pharmaceutical industry. I would also like to thank Bing Yan, Lindsey Hofmeister, Pat Roberson, Marsha Hecht, and Hilary Rowe for their much valued assistance throughout the preparation of this book. My thanks and gratitude go also to my family, whose support and encouragement greatly assisted me in editing this book.

Perry G. Wang
Wyomissing, Pennsylvania

Editor

Dr. Perry G. Wang is currently a principal scientist at Teleflex Medical. His interests include analytical method development and validation, medicated device products, and environmental engineering. His expertise focuses on high-throughput analysis of drugs and their metabolites in biological matrices with LC/MS/MS.

Dr. Wang received a President's Award for Extraordinary Performance and Commitment in 2005 for his dedication in leading the Kaletra® reformulation project at Abbott Laboratories. He was presented with a President's Award for Excellence while he worked in the U.S. Environmental Protection Agency's research laboratories.

Dr. Wang is an author of more than 20 scientific papers and presentations. He organized and presided over symposia for the Pittsburgh Conference in 2006 and 2008, respectively. He has been an invited speaker and presided over several international meetings including the Pittsburgh Conference and the Federation of Analytical Chemistry and Spectroscopy Societies (FACSS). His current research focuses on developing new medicated-device products applied to critical care medicine and testing drug release kinetics and impurities released from drug-device combination products. He earned a B.S. in chemistry from Shandong University and an M.S. and Ph.D. in environmental engineering from Oregon State University.

Contributors

Min Shuan Chang
Abbott Laboratories
Abbott Park, Illinois, USA

Kojo S. J. Elenitoba-Johnson
Department of Pathology
University of Michigan Medical School
Ann Arbor, Michigan, USA

Tawakol El-Shourbagy
Abbott Laboratories
Abbott Park, Illinois, USA

Michael G. Frank
Agilent Technologies
Waldbronn, Germany

Roger K. Gilpin
Brehm Research Laboratory
Wright State University
Fairborn, Ohio, USA

Sergio A. Guazzotti
Engineering Division
Nanostream, Inc.
Pasadena, California, USA

Krishna Kallury
Phenomenex
Torrance, California, USA

Walter Korfmacher
Exploratory Drug Metabolism
Schering-Plough Research Institute
Kenilworth, New Jersey, USA

Yau Yi Lau
Abbott Laboratories
Abbott Park, Illinois, USA

Zhong Li
Merck Research Laboratories
West Point, Pennsylvania, USA

Timothy L. Madden
Pharmaceutical Development Center
MD Anderson Cancer Center
The University of Texas
Houston, Texas, USA

Douglas E. McIntyre
Agilent Technologies
Santa Clara, California, USA

Patrick J. Rudewicz
Genentech
South San Francisco, California, USA

Young Shin
Genentech
South San Francisco, California, USA

Katty X. Wan
Abbott Laboratories
Abbott Park, Illinois, USA

Dong Wei
Biogen Idec, Inc.
Cambridge, Massachusetts, USA

Quanyun A. Xu
Pharmaceutical Development Center
MD Anderson Cancer Center
The University of Texas
Houston, Texas, USA

Richard Xu
Micro-Tech Scientific
Vista, California, USA

Xiaohui Xu
Pharmaceutical Research Institute
Bristol-Myers Squibb
Princeton, New Jersey, USA

Frank J. Yang
Micro-Tech Scientific
Vista, California, USA

Liyu Yang
Biogen Idec, Inc.
Cambridge, Massachusetts, USA

Qin Yue
Genentech
South San Francisco, California, USA

Wanlong Zhou
Brehm Research Laboratory
Wright State University
Fairborn, Ohio, USA

1 High-Throughput Sample Preparation Techniques and Their Application to Bioanalytical Protocols and Purification of Combinatorial Libraries

Krishna Kallury

CONTENTS

1

1.1 NEED FOR HIGH-THROUGHPUT SAMPLE PURIFICATION AND CLEAN-UP IN DRUG DISCOVERY

The drug discovery process took a revolutionary turn in the early 1990s through the adaptation of combinatorial chemistry for generating large volumes of small organic molecules (generally having molecular weights below 750 Daltons) so that the products of all possible combinations of a given set of starting materials (building blocks) can be obtained at once. The collection of these end products is called a combinatorial library.

Production of such libraries can be achieved through either solid phase synthesis or solution chemistry. This newly acquired capability of synthetic chemists to produce a large number of compounds with a wide range of structural diversity in a short time, when combined with high-throughput screening, computational chemistry, and automation of laboratory procedures, led to a significantly accelerated drug discovery process compared to the traditional one-compound-at-a-time approach. During the high-throughput biological screening of combinatorial compounds, initial sample purification to remove assay-interfering components is required to ensure "true hits" and prevent false positive responses. This created needs for rapid purification of combinatorial synthesis products along with rapid evaluation of the purities of these large numbers of synthetic products. In addition, screening biological activities of combinatorial libraries at the preclinical and clinical (phases I through III) trial stages generates drug and metabolite samples in blood, plasma, and tissue matrices. Because these biological matrices carry many other constituents (proteins, peptides, charged inorganic and organic species) that can interfere with the quantitation of the analytes and also damage the analytical instrumentation (especially mass spectrometers and liquid chromatographic columns), rapid clean-up methods are required to render the samples amenable for analysis by fast instrumental techniques. This chapter addresses the progress made during the past decade in the areas of rapid purification of combinatorial libraries and sample preparation and clean-up for high-throughput HPLC and/or LC/MS/MS analysis.

In addition to the large volume synthesis of small molecules, combinatorial approaches are also used to generate catalysts, oligonucleotides, peptides, and oligosaccharides. High-throughput purification has also found applicability for the isolation and clean-up of natural products investigated for biological activity. Several reviews and monographs are available on various topics related to the synthetic and biological screening aspects of the drug discovery process. Since the focus of this chapter is on the purification of combinatorial libraries and clean-up of drugs and their metabolites in biological matrices, it is suggested that the readers refer to the latest literature available on solid phase[1–10] and solution phase[11–17] combinatorial synthesis, ADME studies,[18–26] rapid instrumental analysis techniques,[27–33] and high-throughput methods in natural products chemistry[34–40] for more detailed insights into these areas of relevance to combinatorial synthesis and high-throughput screening.

1.2 RAPID PURIFICATION TECHNIQUES FOR DRUGS AND METABOLITES IN BIOLOGICAL MATRICES

1.2.1 State-of-Art Sample Preparation Protocols

A number of advances have been made during the past decade to convert sample preparation techniques used for about 30 years for the clean-up of drugs in biological matrices into formats that are amenable for high volume processing with or without automation. Detailed accounts about the fundamentals of these techniques can be found in the literature.[41–50] Therefore, only brief descriptions of the principles of these methods are presented. For isolating drugs and metabolites from biological matrices, several approaches have been reported, which consist of:

- Solid phase extraction (SPE)
- Liquid—liquid extraction (LLE)
- Protein precipitation (PPT)
- Affinity separations (MIP)
- Membrane separations
- Preparative high performance liquid chromatography (HPLC)
- Solid phase microextraction (SPME)
- Ultrafiltration and microdialysis

SPE, LLE, and PPT are the most commonly used sample preparation techniques and hence most of the discussion will be devoted to them. The others will be dealt with briefly. All of these methods have certain ultimate goals:

- Concentrate analyte(s) to improve limits of detection and/or quantitation
- Exchange analyte from a non-compatible environment into one that is compatible with chromatography and mass spectrometric detection
- Remove unwanted matrix components that may interfere with the analysis of the desired compound
- Perform selective separation of individual components from complex mixtures, if desired
- Detect toxins in human system or in environment (air, drinking water, soil)
- Identify stereochemical effects in drug activity and/or potency
- Follow drug binding to proteins
- Determine stability and/or absorption of drugs and follow their metabolism in human body

1.2.2 Matrix Components and Endogenous Materials in Biological Matrices

Biological matrices include plasma, serum, cerebrospinal fluid, bile, urine, tissue homogenates, saliva, seminal fluid, and frequently whole blood. Quantitative analysis of drugs and metabolites containing abundant amounts of proteins and large numbers of endogenous compounds within these matrices is very complicated. Direct injection of a drug sample in a biological matrix into a chromatographic column would result in the precipitation or absorption of proteins on the column packing material, resulting in an immediate loss of column performance (changes in retention times, losses of efficiency and capacity). Similar damage can occur to different components of the ESI/MS/MS system commonly utilized for analyzing drugs. Matrix components identified by different analytical techniques are shown in Table 1.1. Major classes encountered in plasma consist of inorganic ions, proteins and/or macromolecules, small organic molecules, and endogenous materials.[51–56]

Mass spectrometry is the most preferred technique employed during high-throughput screening. It provides specificity based on its capability to monitor selected mass ions, sensitivity because it affords enhanced signal-to-noise ratio, and speed due to very short analysis times that allow analysis

TABLE 1.1

Interferences Identified in Human Plasma

Classification	Components	Concentration (mg/L)	Reference
Inorganic ions	Sodium [Na^+]	3.2×10^3 to 3.4×10^3	51
	Potassium [K^+]	148.6 to 199.4	
	Calcium [Ca^{2+}]	92.2 to 112.2	
	Magnesium [Mg^{2+}]	19.5 to 31.6	
	Chloride [Cl^-]	3.5×10^3 to 3.8×10^3	
	Hydrogencarbonate [HCO_3^-]	1.5×10^3 to 2.1×10^3	
	Inorganic phosphorus [P], total	21.7 to 41.6	
	Iron [Fe] in men	1.0 to 1.4	
	Iron [Fe] in women	0.9 to 1.2	
	Iodine [I], total	34.9×10^{-3} to 79.9×10^{-3}	
	Copper [Cu] in men	0.7 to 1.4	
	Copper [Cu] in women	0.9 to 1.5	
Proteins/Macromolecules (g/L)	Prealbumin	0.1 to 0.4	51
	Albumin	42.0	
	Acid-α_1-glycoprotein	0.2 to 0.4	
	Apolipoproteins (globulins)	4.0 to 9.0	
	Haptoglobin (α_2-globulin)	1.0	
	Hemopexin (β_1-globulin)	0.7	
	Transferin (β_2-globulin)	2.9	
	Ceruloplasmin (α_2-globulin)	0.4	
	Transcortin (α_1-globulin)	0.04	
	Transcobalamin	94.0×10^{-8}	
	α_2-Macroglobulin	2.5	
	α_1-Antitrypsin	2.5	
	Protein-binding metal (α_1-globulin)	0.06	
	Antithrombin III (α_2-globulin)	0.2	
	Fibrinogen	4.0	
	Immunoglobulins (γ-globulins)	15.0 to 16.0	
Endogenous components (small organic molecules)	**Amino Acids**		52
	Alanine	NA	
	Valine	NA	
	Leucine	NA	
	Serine	NA	
	Threonine	NA	
	Methionine	28.8 μM	
	Aspartate	NA	
	Glutamate	43.5 μM	
	Phenyl alanine	55.8 μM	
	Glycine	NA	
	Lysine	127.3 μM	
	Tyrosine	NA	
	Proline	289.1 μM	
	Cystine	NA	
	Tryptophan	55.7 μM	

TABLE 1.1 (CONTINUED)
Interferences Identified in Human Plasma

Classification	Components	Concentration (mg/L)	Reference
	Fatty Acid Derivatives		
	2-Hydroxybutyrate	NA	
	3-Hydroxybutyrate	NA	
	3-Methyl-2-hydroxybutyrate	NA	
	Palmitate	125.8 μM	
	Oleate	NA	
	Stearate	NA	
	Laurate	NA	
	Linoleate	NA	
	Other Small Organics		
	Urea	NA	
	Glycerate	NA	
	Creatinine	106.5 μM	
	Glycerol phosphate isomer	NA	
	Citrate	318.6 μM	
	Ascorbic acid	NA	
	Carbohydrate Derivatives		
	Glucose	NA	
	Myoinositol	24.5 μM	
	Inositol phosphates	NA	
	Purine Derivatives		
	Urate	331.5 μM	
	Nucleosides	NA	
	Steroids		
	Cholesterol	2109.7 μM	
Endogenous phospholipids	Phosphatidylcholine	NA	53
	Lysophosphatidylcholines (18:2, 16:0 and 18:0)		
Prostaglandins	Prostaglandin D_2 and F_2	NA	54
Hormones	Melatonin	NA	55
Polysaccharides	Glycosaminoglycans	NA	56
Unseen endogenous matrix components (dosing excipients)	Hydroxypropyl-β-cyclodextrin	NA	
	Polyethyleneglycol 400	NA	
	Propyleneglycol	NA	
	Tween 80	NA	

of dozens of samples per hour. One important factor affecting the performance of a mass detector is ion suppression, with the sample matrix, coeluting compounds, and cross talk contributing to this effect. Operating conditions and parameters also play a role in inducing matrix effects that result in suppression of the signal, although enhancement is also observed occasionally. The main cause is a change in the spray droplet solution properties caused by the presence of nonvolatile or less volatile solutes. These nonvolatile materials (salts, ion-pairing agents, endogenous compounds, drugs, metabolites) change the efficiency of droplet formation or droplet evaporation, affecting the concentrations of charged ions in the gas phase reaching the detector.

The literature clearly reviews how plasma constituents and endogenous materials adversely affect the quantitation of drugs and their metabolites in these matrices.[57–70] It follows that when drugs or metabolites in biological matrices are analyzed, a thorough purification step must be invoked to eliminate (or at least minimize) these adverse effects. In the context of high-throughput screening of ADME (or DMPK) samples, the following discussion elaborates on protocols popularly employed for the high-throughput clean-up of biological matrix components and/or endogenous materials.

1.2.3 SOLID PHASE EXTRACTION (SPE)

Application of SPE to sample clean-up started in 1977 with the introduction of disposable cartridges packed with silica-based bonded phase sorbents. The *solid phase extraction* term was devised in 1982. The most commonly cited advantages of SPE over liquid–liquid extraction (LLE) as practiced on a macroscale include the reduced time and labor requirements, use of much lower volumes of solvents, minimal risk of emulsion formation, selectivity achievable when desired, wide choices of sorbents, and amenability to automation. The principle of operation consists of four steps: (1) conditioning of the sorbent with a solvent and water or buffer, (2) loading of the sample in an aqueous or aqueous low organic medium, (3) washing away unwanted components with a suitable combination of solvents, and (4) elution of the desired compound with an appropriate organic solvent.

With increasing popularity of the SPE technique in the 1980s and early 1990s, polymeric sorbents started to appear to offset the two major disadvantages of silica based sorbents, i.e., smaller surface area resulting in lower capacities and instability to strongly acidic or basic media. Around the mid-1990s, functionalized polymers were introduced to overcome the shortcomings of the first generation polymers such as lower retention of polar compounds and loss of performance when the solvent wetting them accidentally dried. Tables 1.2 and 1.3 list some of the popular polar functionalized neutral and ion exchange polymeric SPE sorbents, respectively, along with structure and

---

TABLE 1.2
Functionalized Neutral Polymeric Sorbents

Source	Sorbent	Chemistry	Mode of Interaction	Examples from Literature (Plasma Samples Only)
Waters (see 2006–2007 Catalog, SPE products)	Oasis HLB	Divinylbenzene-N-vinyl-pyrrolidone copolymer	Reversed phase with some hydrogen bond acceptor and dipolar reactivity	Rosuvastatin (71); NSAIDs (72); fexofenadine (73); catechins (74); valproic acid (75)
Phenomenex (see 2006 Catalog, SPE products)	Strata-X	Polar functionalized styrene-divinylbenzene polymer	Reversed phase with weakly acidic, hydrogen bond donor, acceptor, and dipolar interactions	Cetirizine (76); pyridoxine (77); omeprazole (78); mycophenolic acid (79); 25-hydroxy-vitamin D_3 (80)
Varian (see Catalog, SPE products)	Focus	Polar functionalized styrene-divinylbenzene polymer	Reversed phase with strong hydrogen bond donor, acceptor, and dipolar character	Fluoxetine, verapamil, olanzapine, tramadol, loratidine, and sumatriptane (81); verdanafil (82)
Varian (see Catalog, SPE products)	Bond Elut Plexa	Highly cross-linked polymer with hydroxylated surface	Hydrophobic retention of small molecules and hydrophilic exclusion of proteins	See catalog

TABLE 1.3
Functionalized Ion Exchange Polymeric Sorbents

Source	Sorbent	Chemistry	Mode of Interaction	Examples from Literature (Plasma Samples Only)
Waters	Oasis MCX	Sulfonated divinylbenzene-N-vinylpyrrolidone	Mixed mode with strong cation exchange and reversed phase activities	Alkaloids (83); illicit drugs (84); general screening of therapeutic and toxicological drugs (85)
	Oasis MAX	Quarternary amine functionalized divinylbenzene-N-vinylpyrrolidone	Mixed mode with strong anion exchange and reversed phase activities	NSAIDs (86); glutathione (87)
	Oasis WCX	Carboxy functionalized divinylbenzene-N-vinylpyrrolidone	Mixed mode with weak cation exchange and reversed phase activities	Basic drugs (88)
	Oasis WAX	Cyclic secondary/tertiary amine functionalized divinylbenzene-N-vinylpyrrolidone	Mixed mode with weak anion exchange and reversed phase activities	NSAIDs (86)
Phenomenex	Strata-X-C	Sulfonated styrene-divinylbenzene polymer with polar surface modification	Mixed mode with both strong cation exchange and reversed phase interactions	Stanazolol (89); antidepressant drugs (90); sulfonamides (91); acrylamide (92)
	Strata-X-CW	Carboxylated styrene-divinylbenzene polymer	Mixed mode with weak cation exchange, hydrogen bond donor and acceptor, and reversed phase activities	Phenothiazine drugs (93); basic drugs (94)
	Strata-X-AW	Primary and secondary amine-functionalized styrene-divinylbenzene polymer	Weak anion exchange and reversed phase interactions	Nucleotide phosphates (95)

manufacturer information, modes of interaction, and references on representative applications in sample preparation. Other known hydrophilic polymeric materials are summarized in a recent review.[96]

1.2.3.1 Interactions of Sorbent and Analyte in SPE and Selective Extractions Based on Sorbent Chemistry

The interactions of a sorbent and an analyte fall into three classes: hydrophobic (also called dispersive or van der Waals interactions with associated energy of 1 to 5 kJ/mol), polar, and ionic. Polar interactions are further divided into dipole-induced dipole (2 to 7 kJ/mol), dipole–dipole (5 to 10 kJ/mol), hydrogen bonding (10 to 25 kJ/mol), and ion–dipole (10 to 50 kJ/mol). Ionic interactions are electrostatic with the highest associated energy levels of 50 to 500 kJ/mol. These energy values reflect the fact that when analytes interact with neutral sorbents only through hydrophobic interactions, a thorough organic wash (with 100% solvent) cannot be carried out and hence extracts may contain some contaminants or interference. On the other hand, sorbents possessing ion exchange functionalities can retain ionizable analytes via ionic mechanisms and are amenable to 100% organic solvent washes, thereby furnishing much cleaner extracts.[58]

Ion exchange resins based on poly(styrene-divinylbenzene) backbones display mixed mode retention mechanisms. The ion exchange functionality (sulfonic acid or carboxylic acid for cation exchangers and quarternary or primary, secondary, or tertiary amines for anion exchangers) contributes to the ionic mechanism and the backbone polymer to hydrophobic retention. This is exemplified

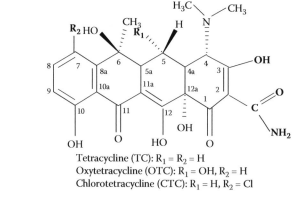

FIGURE 1.1 Structures of THC and THC-COOH, its main metabolite from urine.

by a recent report demonstrating the retention of a hydrophobic molecule like THC carboxylic acid (a metabolite of THC, the major constituent of marijuana; see Figure 1.1) on a strong cation exchanger like strata-X-C even when subjected to a 30 to 40% acetonitrile wash without breakthrough.[97]

The mechanisms of retention of apparently basic analytes on either strong or weak cation exchanger resins depend upon the structures of these analytes and the intra-molecular interactions of the functional groups on these analytes. Thus, tetracycline and its analogs are not eluted from the sulfonic acid-functionalized strata-X-C resin with methanol containing 5% ammonium hydroxide or with acetonitrile containing 0.1M oxalic acid. However, these antibiotics are eluted from strata-X-C with acetonitrile containing 1.0M oxalic acid. On the other hand, they could be easily eluted from the carboxy functionalized weak cation exchanger strata-X-CW with methanol containing formic acid.

The differences in the elution patterns for the two sorbents have been explained[98] by invoking the zwitterionic structures for the antibiotics under the basic pH conditions employed for strata-X-C

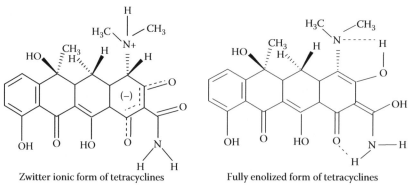

FIGURE 1.2 Neutral, zwitterionic, and fully enolic forms of tetracyclines.

(see Figure 1.2). At acidic pH, the antibiotics exist in their non-ionized enol forms and can be eluted from the weaker carboxylic resin with formic acid, which is stronger than the surface carboxylic acid moieties. However, the sulfonic acid is a stronger acid than either formic or even 0.1M oxalic acid (pH 2.0) and hence the basic dimethylamino moieties on the antibiotics preferentially stay with the sulfonic acid; these antibiotics can be eluted with 1.0M oxalic acid (pH 0.8) from these strong ion exchange resins. Neutral polar functionalized polymers like Oasis HLB or strata-X do not retain the tetracyclines even when a 5% methanol wash is used.

Another interesting selectivity difference was observed[99] during the extraction of benzodiazepine drugs from plasma employing different sorbents. With silica-based strata-C18E, the neutral polymeric strata-X sorbent, or the strata-X-CW weak cation exchanger, diazepam, nordiazepam, oxazepam, lorazepam, and temazepam could all be eluted in excellent yields (Table 1.4) with methanol. On the other hand, with the strong strata-Screen C (silica-based sulfonic acid) and strata-X-C cation exchangers, methanol eluted oxazepam, lorazepam, and temazepam, while methanol containing 5% ammonia was needed to elute diazepam and nordiazepam.

The differential elution with strong cation exchangers does not stem from differences in pH (see Figure 1.3 for structures and pH values). On the contrary, oxazepam, lorazepam, and temazepam possess a hydroxyl at the C-3 position of the diazepine ring system that can stabilize their enolic forms while simultaneously promoting hydrogen bonding with the basic N-4 nitrogen, resulting in the

TABLE 1.4
Results of SPE of Benzodiazepines from Plasma

Sorbent	Main Mode of Interaction	Benzodiazepine	% Recovery with Methanol	% Recovery with Methanol/5% Ammonia
strata-C18-E (silica based)	Reversed phase	Nordiazepam	104	Not applicable
		Diazepam	101	
		Oxazepam	97	
		Lorazepam	95	
		Temazepam	95	
strata-Screen C	Cation exchange	Nordiazepam	9	90
		Diazepam	24	93
		Oxazepam	63	
		Lorazepam	104	
		Temazepam	87	
strata-X	Reversed phase	Nordiazepam	103	Not applicable
		Diazepam	98	
		Oxazepam	94	
		Lorazepam	95	
		Temazepam	92	
strata-X-CW	Weak cation exchanger	Nordiazepam	94	Not applicable
		Diazepam	97	
		Oxazepam	96	
		Lorazepam	100	
		Temazepam	98	
strata-X-C	Strong cation exchanger	Nordiazepam	14	96
		Diazepam	18	95
		Oxazepam	65	
		Lorazepam	88	
		Temazepam	87	

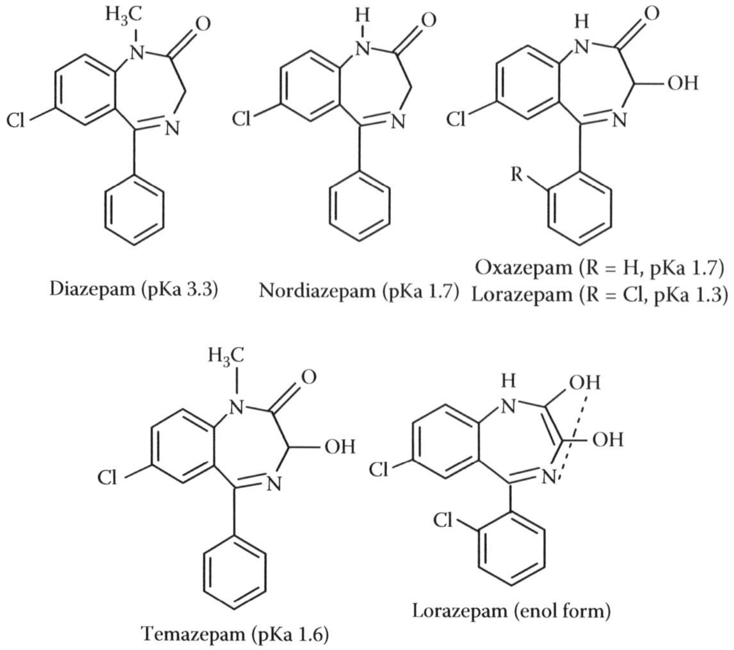

Diazepam (pKa 3.3)   Nordiazepam (pKa 1.7)

Oxazepam (R = H, pKa 1.7)
Lorazepam (R = Cl, pKa 1.3)

Temazepam (pKa 1.6)

Lorazepam (enol form)

FIGURE 1.3 Structures and pKa values of benzodiazepines.

failure of this nitrogen to interact with the sulfonic acid moieties of strata-X-C and the silica-based strata-Screen C through ionic mechanism. Consequently, these drugs are eluted off in methanol.

Selective extraction of the flavonoid components from the ginkgolide and bilabalide terpenoids (see Figure 1.4) of health supplement extracts from *Ginkgo biloba* leaves has been demonstrated[100] by solid phase extraction with the weak strata-X-CW cation exchanger. While the terpenoids could be eluted with 60:40 methanol:water, the flavonoids required a strong organic (methanol:acetonitrile: water, 40:40:20 or acetonitrile:dichloromethane, 50:50) for elution. In comparison, the silica-based strata-C18E and the neutral strata-X polymer did not exhibit this kind of selectivity (see Table 1.5 for recovery data), the former eluting all components with 60:40 methanol:water, while the latter eluted the terpenoid partially in this solvent and partially with the stronger organic.

This protocol was modified to enable automation. In a later publication,[107] 20 mg of plant material (*Arabidopsis thaliana*) was extracted with 1 mL of methanol, water, and formic acid. The extract was transferred to glass tubes in an Aspec XL4 robot. After an initial clean-up with a C18 cartridge, the extract was evaporated and the residue reconstituted in formic acid and transferred to the robot. SPE purification was carried out with Oasis MCX. After buffering and methanol wash, the cytokinins were eluted with methanol and aqueous ammonium hydroxide (see Figure 1.5). After evaporation, the residue was derivatized with either propionic anhydride or benzoic anhydride. The derivatives were analyzed by LC/MS/MS using a 10 × 1 mm BetaMax Neutral Guard cartridge as the LC column. Lower detection limits in the femtomole to attomole range were obtained. The protocol was also successfully applied to non-cytokinin compounds such as adenosine mono-, di-, and tri-phosphates, adenosine, uridinophosphoglucose, and flavin mononucleotide with the same limits of detection. The ESI sensitivity of the derivatives was found to be far superior compared to underivatized cytokinins and nucleotides. The procedure can be applied to strongly hydrophilic molecules from any biological matrix and serves as an example of high-throughput automated solid phase extraction.

The propensity of mixed mode cation exchange resins to retain highly water-soluble compounds like gamma-aminobutyric acid (GABA) was demonstrated in a recent publication.[108] Animal tissue

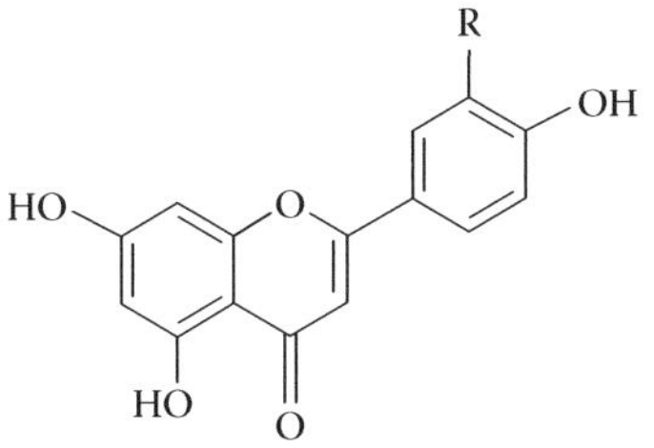

Flavonoid aglycones (isolated by acid hydrolysis of the corresponding flavonoid glycosides)
Kaempferol (R = H)
Quercetin (R = OH)
Isorhamnetin (R = OCH_3)

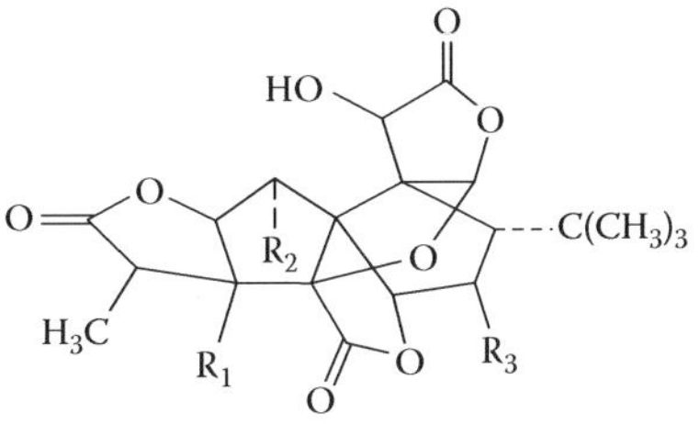

Ginkgolides A (R_1 = OH; R_2 = R_3 = H); B (R_1 = R_2 = OH; R_3 = OH); C (R_1 = R_2 = R_3 = OH)

Bilabalide

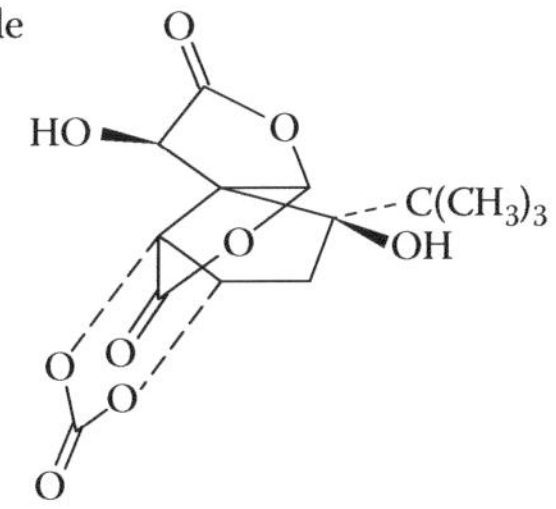

FIGURE 1.4 Structures of flavonoids and terpenoids from *Gingko biloba*.

TABLE 1.5
Selective Elution of Flavonoids from Terpenoids in *Ginkgo biloba* Leaf Extracts

Analyte (MW)	Strata C18-E		Strata-X		Strata-X-CW	
	Elut 1	Elut 2	Elut 1	Elut 2	Elut 1	Elut 2
Quercetin (302)	83	8	0	89	25	76
Kaempferol (286)	84	13	0	85	0	90
Isorhamnetin (316)	90	13	0	80	0	103
Bilobalide (326)	113	0	9	102	86	2
Ginkgolide A (408)	82	0	51	45	78	0

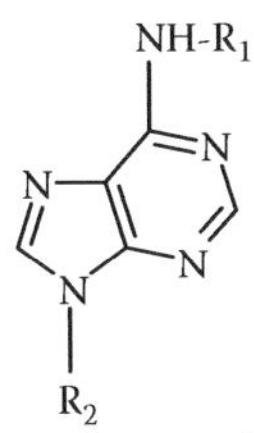

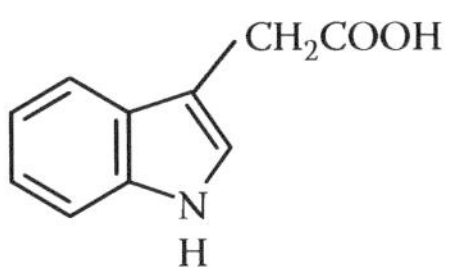

Auxin (Indole-3-acetic acid)

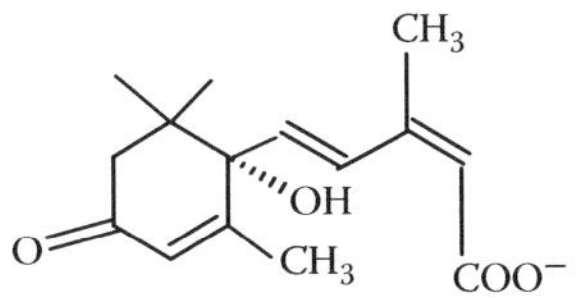

Abscisic Acid

Cytokinins and nucleosides
R_1 = -CH_2CH = $CHCH_2OH$
    CH_2CH = $CHCH_2O$ Ribofuranoside (or glucopyranoside)
    -$CH_2CH_2CH(CH_3)CH_2OH$
    -CH_2Phenyl
Cytokinin nucleotides
R_2 = H or beta-D-ribofuranosyl-5'-monophosphate

FIGURE 1.5 Structures of auxin and abscisic acid derivatives of cytokinin.

was extracted with simultaneous protein precipitation using 2% sodium dodecylsulfate in 0.1M potassium dihydrogen phosphate buffer (pH 2). After centrifugation, the supernatant was loaded onto an Oasis MCX cartridge. Washing with methanol and formic acid in acetonitrile (10:90) selectively eluted gamma-hydroxybutyric acid and 1,4-butanediol. GABA was then eluted with water: methanol:ammonia (94.5:5:05 v/v). All the analytes were derivatized with N-(t-butyldimethylsilyl)-N-methyl trifluoroacetamide (MTBSTFA) and analyzed by GC/MS. This procedure is potentially suitable for evaluating PMI (postmortem interval) in humans because the amount of GABA in blood increases after death and the increase may be correlated to time of death.

Relative extraction efficiencies of polar polymeric neutral, cation, and anion exchange sorbents (HLB, MCX, and MAX) for 11 beta antagonists and 6 beta agonists in human whole blood were probed.[109] Initial characterization of MCX and MAX for acidic and basic load conditions, respectively, showed that both the agonists and antagonists were well retained on MCX, while they were recovered from MAX in the wash with either methanol or 2% ammonia in methanol (see Table 1.6). Blood samples were treated with ethanol containing 10% zinc sulfate to precipitate proteins and the supernatants loaded in 2% aqueous ammonium hydroxide onto the sorbents. After a 30% methanol and 2% aqueous ammonia wash, the analytes were eluted with methanol (HLB), 2% ammonia in methanol (MCX), or 2% formic acid in methanol (MAX). The best recoveries were observed with MCX under aqueous conditions or blood supernatant (after protein precipitation) spiked sample load conditions (see Table 1.7). Ion suppression studies by post-column infusion showed no suppression for propranolol and terbutaline with MCX, while HLB and MAX exhibited suppression (see Figure 1.6).

TABLE 1.6
Comparison of MCX and MAX for SPE of β-Agonists and Antagonists

SPE Column Equilibration and Loading Washing Elution	MCX 2% HCOOH aq MeOH 2% NH₄OH in MeOH		2% HCOOH in MeOH		MAX 2% NH₄ OH aq MeOH 2% NH₄OH in MeOH		2% HCOOH in MeOH	
Collected Fractions	Washing	Elution	Washing	Elution	Washing	Elution	Washing	Elution
β-Antagonists:								
Atenolol	0	100	0	102	99	5	98	4
Sotalol	0	93	0	94	0	113	0	111
Carteolol	0	96	0	97	91	4	91	2
Pindolol	Trace	58	0	76	86	15	87	16
Timolol	Trace	90	0	85	99	4	97	3
Metoprolol	Trace	92	0	88	91	17	91	16
Bisoprolol	0	93	0	90	90	18	86	18
Labetalol	Trace	85	0	83	0	108	0	114
Betaxolol	Trace	90	0	93	86	23	86	22
Propranolol	Trace	84	0	91	73	28	70	29
Carvediol	1.5	83	0	82	65	23	63	22
β-Agonists:								
Salbutamol	0	98	0	99	27	53	38	47
Terbutaline	0	101	Trace	95	0	80	0	77
Fenoterol	0	104	0	54	0	94	0	87
Formoterol[a]	Trace	53	0	32	26	41	18	42
Clenbuterol	Trace	94	Trace	93	84	6	77	5
Bambuterol	0	97	0	96	91	3	92	4

Source: From M. Joseffson and A. Sabanovic, *J. Chromatogr. A*, 2006, 1120, 1. With permission from Elsevier.

TABLE 1.7

Comparison of HLB, MCX and MAX, under Optimized Sorbent Conditions for Recovery of β-Agonists and Antagonists

SPE Column	HLB 2% NH$_4$OH aq / 2% NH$_4$OH aq / 30% MeOH in 2% NH$_4$OH aq / MeOH	Diluted Supernatant*	MCX 2% HCOOH aq / 2% HCOOH aq / 30% MeOH in 2% HCOOH aq / 2% NH$_4$OH in MeOH	Diluted Supernatant*	MAX 2% NH$_4$OH aq / 2% NH$_4$OH aq / 30% MeOH in 2% NH$_4$OH aq / 2% HCOOH in MeOH	Diluted Supernatant*
β-Antagonists:						
Atenolol	98	0	95	91	22	0
Sotalol	2	0	92	84	97	0
Carteolol	99	1	100	105	107	1
Pindolol	82	3	74	85	86	2
Timolo	96	2	87	90	100	2
Metoprolol	95	3	86	88	98	2
Bisoprolol	98	42	93	88	98	2
Labetalol	80	71	76	75	93	62
Betaxolol	88	88	91	94	103	16
Propramolol	73	79	80	81	96	48
Carvediol	69	77	70	70	106	116
β-Agonists:						
Salbutarmol	5	0	88	75	85	0
Terbutaline	2	0	86	85	74	0
Fenoterol	9	1	59	58	67	2
Formoteral	47	11	48	46	65	3
Clenbuterol	74	18	80	76	87	4
Bambuterol	95	6	88	77	92	1

Source: From M. Joseffson and A. Sabanovic, *J. Chromatogr. A*, 2006, 1120, 1. With permission from Elsevier.

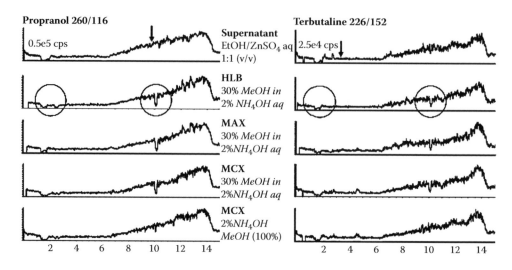

FIGURE 1.6 Comparison of ion suppression data for propranolol and terbutaline after solid phase extraction with HLB, MAX, and MCX.[109] (Reproduced with permission from Elsevier.)

1.2.3.2 Elimination of Proteinaceous and Endogenous Contaminants from Biological Matrices to Minimize Ion Suppression during SPE: Comparison of Ion Exchange and Mixed Mode Sorbents

In a recent communication, Wille and coworkers[110] compared the efficacies of ion exchange sorbents and mixed mode, neutral polymeric and silica-based reverse phase sorbents. The analytes consisted of 13 new generation antidepressants with pKa values ranging from 6.7 to 10.5 and log P values ranging from 0.04 to 7.10. The authors utilized derivatization with hexafluorobutyryl imidazole to facilitate GC/MS quantitation of SPE recoveries and to assess the purity of the extracts. They also utilized HPLC for optimizing SPE and for investigating the protein binding of these antidepressant drugs to plasma proteins. Since water absorption and/or retention by the sorbents is not compatible with derivatization and GC/MS analysis, although Oasis HLB and Focus sorbents retained all the drugs well, they were excluded after initial screening because they are very hydrophilic. The inability to withstand 100% methanol wash for all the drugs tested eliminated silica-based nonpolar sorbents and neutral polymer strata-X and Certify.

Although Certify is a mixed mode sorbent with C8 and sulfonic acid moieties, the authors rationalized that the hydrophobic retention on this sorbent is more dominant and caused the nonretention of certain drugs during methanol wash. The weak WCX ion exchanger was also excluded for similar reasons. Both the mixed mode strata-X-C and the ion exchange sorbent SCX were found to be most amenable for the derivatization-based GC/MS analysis and both yielded pure extracts. However, the yields were consistently lower with strata-XC than SCX and the authors hypothesized that this was due to the inability of the 5% ammonia/methanol eluent to completely disrupt the hydrophobic and dipolar interactions between the analytes and XC.

Wille et al.[110] made interesting observations on the protein binding of the 13 antidepressant drugs investigated. These drugs were divided into two groups—one consisting of desmethylmirtazapine, O-desmethylvenlafaxine, desmethylcitalopram, didesmethylcitaloporam, reboxetine, paroxetine, maprotiline, fluoxetine, norfluoxetine, and m-chlorophenylpiperazine. The other group included mirtazapine, viloxazine, desmethylmianserin, citalopram, mianserin, fluvoxamine, desmethylsertraline, sertraline, melitraen, venlafaxine, and trazodone. They tested protein precipitation by four methods: dilution with (1) pH 2.5 or pH 6.5 phosphate buffer, (2) glycine hydrochloride, (3) 2% phosphoric acid, and (4) organic solvents (methanol and acetonitrile). Since the sorbents used for SPE were cation exchangers, Willie's group did not investigate inorganic salts.

After equilibration at 4°C overnight, the samples were vortexed and centrifuged and the super-natants subjected to SPE on SCX and strata-X-C. HPLC analysis of eluates indicated that glycine HCl and dilution with acidic phosphate buffer yielded 89 and 88% recoveries, which result was interpreted on the basis of negligible drug binding by α-1-acid glycoprotein of the plasma (isoelectric point 3.0) at pH 2.5 for both reagents. On the other hand, the lower recoveries for acetonitrile (62%) and methanol (78%) were interpreted as arising from the hydrophobic binding of the drugs to albumin and lower solubility of the drugs in acetonitrile. Phosphoric acid gave 73% recovery. The importance of load pH and disruption of hydrophobic interactions while using ion exchange and mixed mode sorbents is thus emphasized.

Of particular interest is the comparison of the performance of cation exchange and mixed mode sorbents for their efficacy in cleaning up endogenous phospholipids. Unlike the protein-related materials that are eluted in the very early stages of HPLC, these phospholipids elute in the hydrophobic region and interfere with drug peaks which also elute around the same time.

Shen and coworkers[111] compared SPEC SCX disks with SPEC MP1 disks and Oasis MCX. SPEC-SCX is a phenylsulfonic acid, while MP1 is a mixed mode C8/sulfonic acid and MCX is a polymeric sulfonic acid on a divinylbenzene–vinylpyrrolidone polymer backbone. The sorbents were conditioned with methanol and then with 2% formic acid. The sample was loaded in 2% formic acid solution and washing was done with 2% formic acid, followed by acetonitrile:methanol (70:30). Analytes were eluted with two aliquots of methanol:acetonitrile:water:ammonia (45:45:10:4% v/v/v/v). The eluent was dried under nitrogen and the residue reconstituted in the mobile phase (80% 10mM ammonium formate containing 0.2% formic acid and 20% 10mM ammonium formate in methanol with 0.2% formic acid). Their data on desloratadine and its 3-hydroxy analog (see Figure 1.7), along with data on phosphatidylcholine indicates that MCX retains about seven times as much phospholipids as SCX does and MP1 retains around 60 times more than SCX (see Figure 1.8). Post-column infusion experiments with blank plasma extracts showed ion suppression in the hydrophobic region for MP1 and MCX, but not for SCX (see Figures 1.9 through 1.11). The observations were rationalized through hydrophobic retention of the phospholipids by the mixed mode sorbents; SCX did not exhibit such retention mechanisms.

1.2.3.3   Formats for Rapid and/or High-Throughput Solid Phase Extraction of Drugs in Biological Matrices

With the advent of fast analytical techniques such as LC/MS/MS, 96-well plate formats gained preference around 1995 to cater to the high-throughput sample preparation needs of bioanalysis. The historical development of these 96-well plate formats was well documented by Wells[42] and will not be detailed here. In this well format, the sorbent is packed at the bottom of the plate with popular bed mass sizes ranging from 10 to 500 mg. Further refinements of this 96-well flow-through system include miniaturization of the plate and well geometry to accommodate as little as 2 mg of sorbent in a particle bed or disk (laminar, sintered, glass fiber, or particle-loaded membranes) that allows the use of very small elution volumes (e.g., 25 μL). Other modifications consist of modular geometries

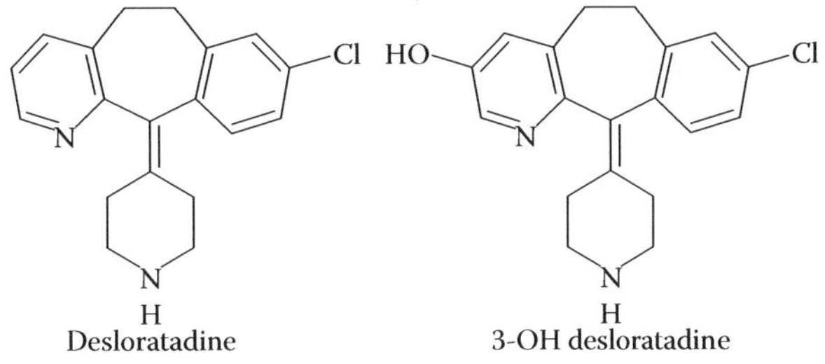

FIGURE 1.7 Structures of desloratadine and 3-hydroxydesloratadine.[111] (Reproduced with permission from Elsevier.)

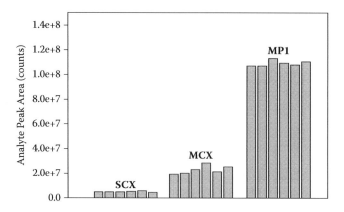

FIGURE 1.8 Comparison of SCX, MCX, and MP1 for retention of phosphatidylcholine.[111] (Reproduced with permission from Elsevier.)

with removable well plates containing 10 to 100 mg of sorbent. Increased well plate formats (384 to 1536) are also available for very high volume turnarounds in sample processing.

Liquid handling systems such as the Tomtec Quadra Model 320 or Packard Multiprobe II EX (HT) are used to automate the solid phase extraction process. The former carries 96 pipette tips for simultaneous delivery of liquid into all 96 wells, while the latter is designed with 8 tips. Processing of a well plate using the Tomtec takes 10 min. Multiprobe processing requires 30 to 60 min to complete SPE on a 96-well plate. This compares favorably in terms of time and labor to manual SPE of a 96-well plate that requires more than 5 hr for completion of one extraction. Other popular liquid handling systems include the Sciclone Advanced Liquid Handler Workstation (Zymark), Cyberlab (Gilson, Inc.), Multimek (Beckman Coulter), and Personal Pipettor (Apricot Designs), all of which use 96-tip channel pipettors. Among the 4- to 8-channel pipettors, SPE 215 (Gilson), Genesis (Tecan), Biomek 2000 (Beckman Coulter), and Microlab (Hamilton) are widely used. Wells[42] contains detailed accounts of these automated liquid stations; they are not discussed here due to spatial considerations.

A few examples from the latest literature will be presented to illustrate the use of the 96- and higher well formats and pipette tip formats for high-throughput sample preparation. An interesting example of orthogonal extraction chromatography and ultra-pressure liquid chromatography (UPLC) of plasma samples of desloratadine and 3-hydroxy-desloratadine (see Figure 1.7 for structures) was recently reported.[112] Sample clean-up was achieved in a 96-well plate containing 10 mg of the MCX mixed mode polymeric sorbent. After conditioning with 400 μL of methanol and then 400 μL of 2% formic acid, a sample solution of the analytes (250 μL of plasma spiked with the metabolites diluted with 500 μL of 2% formic acid) was loaded. Washing was done with 400 μL of 2% formic acid and then 400 μL of methanol:acetonitrile (1:1 volume %). Elution with two 200 μL aliquots of methanol:acetonitrile:water:ammonia (45:45:10:4% v/v/v/v%) solution followed. After concentration under nitrogen, the eluate was reconstituted in the mobile phase (A = 10mM ammonium formate/0.2% formic acid; B = 10mM ammonium formate in methanol with 0.2% formic acid, A:B = 80:20). Experiments to evaluate extraction efficiencies showed that 4% ammonium hydroxide was optimal (5% ammonia reduced extraction yield by about 15%, while 2 or 3% ammonia showed about 6% lower recoveries). It was hypothesized that higher concentration of ammonia in the eluent co-eluted the phospholipids or the excess ammonium ions caused ion suppression.

The presence of 10% water in the eluent minimized variations. Two different LC modes were used for the analysis of the extracts: a smaller (50 × 2.1 mm) Atlantis C18 column (5 μm particle size) on a Shimadzu liquid chromatograph and a UPLC column (Acquity C18, 1.7 μm particle size, 50 × 2.1 mm) on a Waters Acquity system. In the Shimadzu experiment, a gradient from 0.5 min

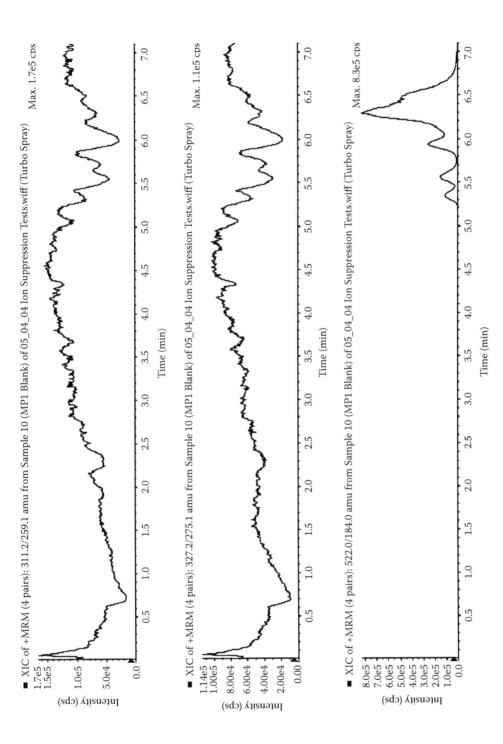

FIGURE 1.9 Multiple reaction monitored ion chromatograms for desloratadine (top), 3-hydroxydesloratadine (middle), and phosphatidylcholine monoester (bottom) during post-column infusion and subsequent injection of a SPEC(R) MP1-extracted control blank plasma sample.[111] (Reproduced with permission from Elsevier.)

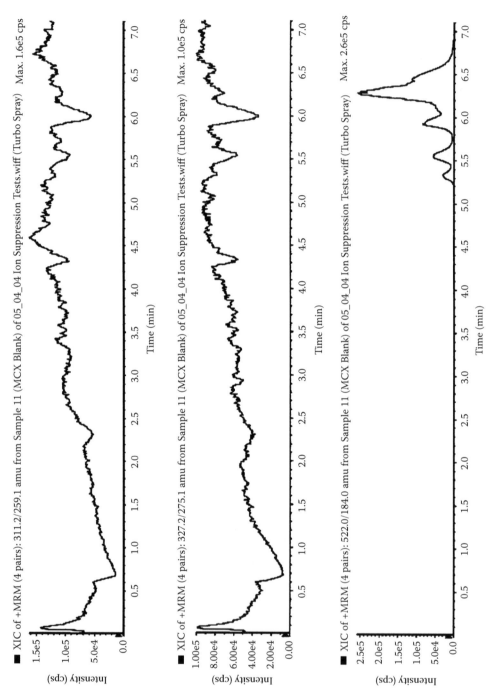

FIGURE 1.10 Multiple reaction monitored ion chromatograms for desloratadine (top), 3-hydroxydesloratadine (middle), and phosphatidyl-choline monoester (bottom) during post-column infusion and subsequent injection of an Oasis[R] HLB MCX-extracted control blank plasma sample.[111] (Reproduced with permission from Elsevier.)

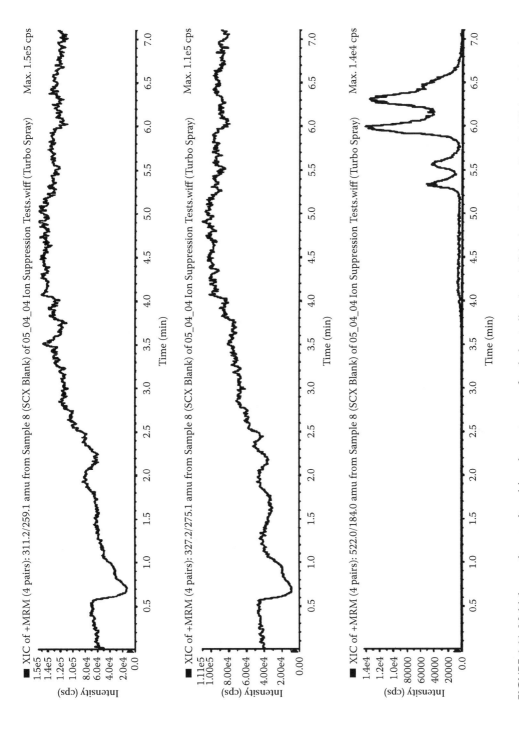

FIGURE 1.11 Multiple reaction monitored ion chromatograms for desloratadine (trace), 3-hydroxydesloratadine (middle), and phosphatidylcholine monoester (bottom) during post-column infusion and subsequent injection of a SPEC® SCX-extracted control blank plasma sample.[111] (Reproduced with permission from Elsevier.)

(20% B) to 90% B in 3.80 min was used, then returned to 20% B at 3.81 min with equilibration until 4.2 min. In the Acquity experiment, the gradient was from 0.25 min (20% B) to 90% B in 1.65 min, maintained at 90% B through 1.90 min, then returned to 20% at 1.91 min with equilibration until 2.10 min. When an injection volume of 30 μL (from a reconstituted solution volume of 150 μL) was used, the Shimadzu experiment yielded a typical response for 0.955 pg of desloratadine and 1.05 pg of its 3-hydroxy analog. In the Acquity experiment, 15 μL of injection volume was necessary to achieve significantly better signal-to-noise ratio compared to the Shimadzu experiment representing an LLOQ of 0.478 pg for desloratadine and 0.525 pg for the 3-hydroxy metabolite.

In a control experiment, the peak widths with UPLC were found at 0.15 min and at 0.16 min for desloratadine and its 3-hydroxy analog, respectively. The corresponding values from the Shimadzu experiment were 0.37 min and 0.32 min, respectively. Nevertheless, only a marginal improvement in sensitivity (peak height) was found under UPLC conditions. The accuracy and precision values for the two drugs under the two sets of LC conditions were very similar.

In another application of the 96-well plate-based sample preparation, Qi Song et al.[113] reported the extraction of cetirizine (Zyrtec) on strata-X (a neutral surface hydrophilic functionalized styrene–divinylbenzene polymer) using a combination of Packard MultiProbe II and Tomtec Quadra 96-320. Plasma samples (100 μL) were spiked with cetirizine and its d-8 labeled analog (50 μL in 1:1 acetonitrile:water) used as internal standard and diluted with 1% TFA (150 μL) in a 96-well collection plate. The solution was aspirated and dispensed 10 times for thorough mixing on a MultiProbe. The plate was transferred to the Quadra manually and automated SPE was carried out. The strata-X 96-well plate (10 mg sorbent/well) was conditioned with 500 μL of methanol, followed by 500 μL of water, after which 300 μL of the sample solution was added to each well and eluted under gravity over 10 min. Subsequently, low vacuum (0.5 scfh) was applied and washing was done with 2% formic acid followed by 5% methanol in water. After application of low vacuum, the drug was eluted with acetonitrile (200 μL) under gravity and another aliquot of acetonitrile (200 μL) was passed through the plate under low vacuum. The eluate was evaporated under nitrogen (15 min) and reconstituted in acetonitrile. The addition of TFA during sample dilution was aimed at keeping the carboxylic groups of cetirizine in the protonated form and the piperazine ring nitrogens protonated. The reconstituted eluates were analyzed in the HILIC mode on a Betasil silica column (50 × 3 mm, 5 μm, Keystone Scientific) using a mobile phase of acetonitrile:water:TFA:acetic acid (93:7.0:0.025:1 v/v/v/v) under isocratic conditions and MRM detection on an API 3000 or 4000 mass spectrometer using a run time of 2 min. The transitions monitored were 389 $\rightarrow$ 201 for cetirizine and 397 $\rightarrow$ 201 for the internal standard. A minimum detection limit of 1.0 ng/mL was achieved. Matrix lot-to-lot reproducibility tests revealed an RSD of 5%; the RSD for precision and accuracy was less than 3%. For LLOQ (n = 6), the RSD of measured concentration was 7%. The authors recommend the HILIC mode of LC for rapid analysis based on their analyses of several drugs and analyses by other laboratories.

An automated solid phase extraction method for human biomonitoring of urinary polycyclic aromatic hydrocarbon (PAH) metabolites using the RapidTrace SPE workstation was recently reported.[114] PAHs are formed during incomplete combustion of organic materials such as coal, gas, wood, and tobacco. Exposure is primarily through inhaling polluted air or tobacco smoke, and by ingestion of contaminated and processed food and water. Dermal exposure may also be a major pathway. Following absorption in the human body, PAHs are rapidly biotransformed into hydroxylated metabolites by cytochrome P450 mono-oxygenases and these are further converted into glucuronide or sulfate conjugates to enhance their polarity and consequently aid in their urinary excretion. Romanoff and coworkers[114] compared four polymeric sorbents in their studies and found that Focus (a Varian product) yielded the best results (recoveries of 69 to 93%).

The protocol consisted of preconditioning with methanol (1 mL) followed by water (1 mL). Urine samples (3 mL) were deconjugated by treatment with β-glucuronidase and arylsulfatase (10 μL and 200 μg/μL) in 0.1M sodium acetate (pH 5.5) and then loaded onto conditioned cartridges. After washing with water (1 mL) and methanol:sodium acetate (3 mL, 4:6, pH 5.5), the PAH metabolites were eluted with dichloromethane (3 mL). The eluate was spiked with dodecane (used

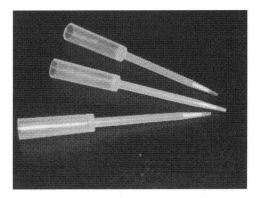

FIGURE 1.12 OMIX tips for Tomtec Quadra.[115] (Reproduced with permission from Elsevier.)

for minimizing loss of volatile metabolites) and concentrated to about 5 μL. Toluene (20 μL) was added and the metabolites were derivatized with MSTFA. The silyl derivatives were analyzed by GC/MS using a DB-5 column. Appropriate C13 labeled metabolites were used as standards and the molecular ions and their [M-15] fragment ions were monitored. Variations were within stipulated limits for N >157 samples and ranged from 5 to 17%. The method was used to quantify 23 PAH metabolites in the fifth NHANES study, consisting of nearly 3000 samples.

A novel variation from the 96-well formats for high-throughput sample purification is the micro-pipette design called the μ-SPE tip. These devices are constructed from automated pipette tips. One example is Varian's OMIX Tip for the Tomtec Quadra (see Figure 1.12), which is constructed by inserting a plug of OMIX C18 SPE material into the tip section of a Tomtec 450-μL pipette. This material is based on a monolithic glass fiber sorbent bed functionalized with the octadecyl chains through silanization and provides superior flow characteristics in comparison with a packed bed. No additional filters or frits are present in this prototype of the μ-SPE tip. A mixed phase cation exchange SPE sorbent (MP1) in this tip format is also available from Varian.

When these tips are used for extraction, a sample solution is aspirated and then dispensed using an automated liquid handler like Tomtec Quadra and circulates across the solid phase media. The use of a monolithic glass fiber results in a design that has less sorbent density than that used in a traditional plate format and enables free flow of a liquid across the media without assistance from a vacuum. A recent publication by Shen et al.[115] demonstrates an application of this tip format for posaconazole, a potent selective inhibitor of the 14α-demethylase (CYP 51) enzyme, the structure of which is shown in Figure 1.13. SCH 56984, a closely related compound, was used as an internal standard. The extraction procedure was the same as in a typical SPE procedure: conditioning, application of the sample solution, wash, and elution. The wash and elution solutions were pre-aliquoted into individual wells of a 96-well block before placement on the Tomtec. Prior to aspiration, a 50- to 150-μL air gap was drawn into the μ-SPE tips followed by an aliquot of the sample solution and then another 5 to 10 μL of air gap. The entire tip contents were dispensed in a single step with the top air gap acting as a pump to expel all liquid from the μ-tip. After each dispense cycle, 25 μL of air from the system air compressor was blown into the tips to dislodge remaining liquid. No vacuum application or manual operator intervention was needed.

In a typical extraction, 50 μL of plasma was treated with 25 μL of internal standard solution and then diluted with 200 μL of 3% ammonium hydroxide. The C18 μ-SPE tips were conditioned with 150 μL of methanol and then 300 μL of 3% ammonium hydroxide. The sample was exhaustively extracted by aspirating and dispensing the plasma samples seven times from the dilution tube. For the wash, 90 μL of 3% ammonium hydroxide followed by 100 μL of methanol:water (20:80 v/v) was used. Elution was achieved with 50 μL of methanol:water (90:10 v/v). After evaporation of the collected eluate in the 96-well block, the residues were reconstituted in 200 μL of mobile phase A:B

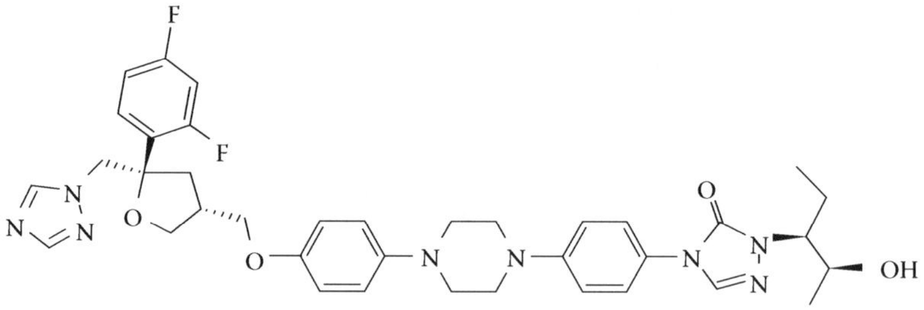

Posaconazole

Formula Weight: 700.8

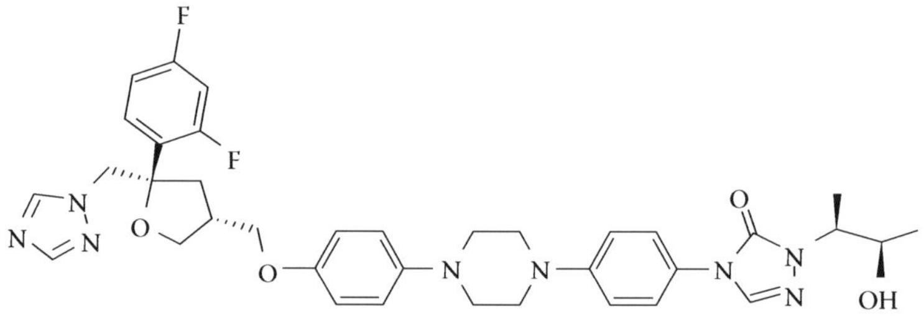

SCH 56984

Formula Weight: 686.8

FIGURE 1.13 Structures of posaconazole and SCH 56984.[115] (Reproduced with permission from Elsevier.)

(1:1) and a 10-μL portion was injected into the LC/MS/MS system. A Varian Polaris-C18A column (50 × 2 mm) was used; mobile phase A was water:methanol:formic acid (90:10:0.1 v/v/v); mobile phase B was acetonitrile:methanol:formic acid (90:10:0.1 v/v/v) with a gradient from 10% A at 0.3 min to 75% in 1.3 min, held until 2.5 min, then to 100% B at 2.6 min and back to 10% at 3.6 min and equilibrated until 4 min. Posaconazole and the internal standard had retention times of 2.0 and 2.1 min, respectively.

Wash–elution and aspiration–dispensing cycle optimization experimental results are shown in Figures 1.14 and 1.15, respectively. A comparison of recovery yields between the tip experiment and a 96-well plate containing 15 mg of Varian SPEC C18 under the same extraction conditions gave a value of 70% for the latter, a figure obtained from three aspiration–dispensing cycles for the former. For intra-run accuracy of calibration standards, a %CV range from −3.6% to 3.5% was recorded, while for the QC samples, 7.7%, 1.3%, and 0% were obtained for QCL, QCM, and QCH (n = 18), respectively. Run precisions were 1.1 to 9.2% and 5.1 to 5.7%, respectively, for calibration and QC samples. An LLOQ of 10 ng/mL was established.

An analogous pipette tip-based solid phase extraction of ten antihistamine drugs from human plasma was reported by Hasegawa and coworkers.[116] A MonoTip C18 tip (GL Sciences, 200-μL pipette tip volume, C18-bonded monolithic silica gel with a diameter of 2.8 mm and thickness of 1 mm) was utilized. The monolithic silica with a continuous mesoporous (pore size ~20 nm) silica skeleton ~10 μm in size and ~10 to 20 μm through-pores was fixed in the point of the 200-μL pipette tip and chemically modified with the C18 phase. The advantages of this sorbent include ease of extraction coupled with rapidity compared to conventional SPE cartridges. The small bed volume and the sorbent mass within the MonoTip C18 permit use of a small volume of solvent, smaller elution volumes, and reduced evaporation times, leading to higher throughput. A plasma sample containing ten antihistamines (100 μL) was diluted with 400 μL of water and 25 μL of a 1M potassium phosphate (pH 8.0) buffer. After centrifugation, the supernatant was decanted into

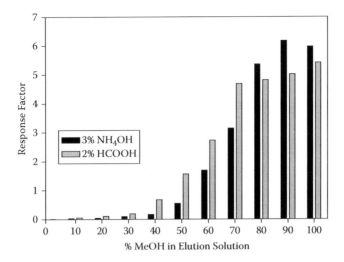

FIGURE 1.14 Wash–elution optimization of posaconazole on C18 OMIX tips.[115] (Reproduced with permission from Elsevier.)

a sample tube and 200 μL aspirated into the conditioned MonoTip C18. The same volume of sample was then dispensed back into the sample tube. Twenty-five such cycles were performed. The tips were washed with 200 μL of water, dried for 3 min by drawing air continuously and then the drugs were eluted with 100 μL of methanol by five repeated aspirating–dispensing cycles. A 2-μL aliquot of eluate was subjected to GC/MS analysis. Quantitation was performed on the base peaks of the respective antihistamines by SIM. The total time for extraction was 8 min, as opposed to >20 min by cartridge extraction. Recoveries ranged from 73.8 to 105% and detection limits of 0.2 to 2.0 ng/0.1 mL plasma. Within-day and day-to-day CVs were less than 8.8 and 9.9%, respectively. The method was applied successfully to dosed human plasma samples after oral administration of diphenhydramine and chlorpheneramine to healthy volunteers. Their respective concentrations immediately after administration were 18.0 and 15.1 ng/0.1 mL for diphenhydramine and 1.65 and 1.07 ng/0.1 mL for chlorpheneramine 3 and 4 hr after administration.

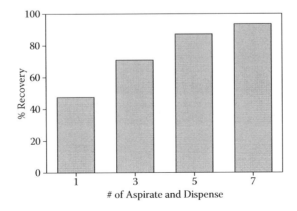

FIGURE 1.15 Optimization of aspiration and dispensing cycles of posaconazole on C18 OMIX tips.[115] (Reproduced with permission from Elsevier.)

1.2.3.4 Online Solid Phase Extraction as Tool for High-Throughput Applications

Features that make online SPE more attractive compared to off-line SPE consist of:

- Direct elution of analyte from extraction cartridge into LC system
- Elimination of time-consuming evaporation, reconstitution, and preparation for injection
- Achievement of maximum sensitivity for detection because the entire volume of eluate is utilized for analysis
- Processing of samples and SPE cartridges in a completely enclosed system, allowing protection for light- and air-sensitive compounds and preventing of exposure of operator to solvents
- Less handling and manipulation; no loss of analyte

Examples of automated systems developed for online SPE include Prospekt and Prospekt-2 (Spark Holland, The Netherlands) for cartridges and SPE Twin Pal (LEAP Technologies, Carrboro, North Carolina) for traditional 96-well plates. Although several alternative formats for online sample clean up such as dual and multiple columns, turbulent flow chromatography, restricted access media, immuno-affinity and multiple column extraction with multiple column separation have been used,[42] the most promising and viable mode is use of online SPE cartridges based on economic considerations and the ability of the cartridges to withstand high pressures and pH extremes.

Generally, an online SPE LC/MS/MS system consists of three major hardware components: an online SPE module, a separation (LC) module, and a detection (MS) module. A multicomponent LC pumping assembly with two individual HPLC pumping units connected by one or two switching valves (six- or ten-port type) is used in most of the online applications reported in the literature. One pump is used for plasma sample loading and washing of the SPE cartridge; the other is used for analytical separation of compounds eluted from the SPE cartridge after removal of plasma proteins.

In a recent publication by Zang and coworkers,[117] an online SPE LC/MS/MS assay was reported for verapamil, indiplon, and six investigative drug compounds, using a strata-X online extraction cartridge (2.1 × 20 mm) and a monolithic Chromolith Speed ROD RP-18e (4.6 × 50 mm) column as the analytical column. This combination permits exploitation of the speed of the monolithic columns and provides the advantages of polymeric online SPE that also include the ability to utilize hydrophobic and hydrophilic interactions simultaneously in addition to the favorable features cited above. The flow rate for achieving optimal removal of proteins was established initially, by comparing 2, 3, and 4 mL/min flow rates using 90:10 water:acetonitrile with 0.1% formic acid as the mobile phase; 4 mL/min was determined to yield the cleanest profile.

A single six-port switching valve was used in two settings. In position A, the autosampler (HTC Pal, LEAP Technologies) loads the plasma sample onto the strata-X, followed by a 30-sec washing using the same mobile phase as above at 4 mL/min; in position B, the drugs are back-eluted off strata-X (after the 30-sec wash, the six-port valve switches to connect the monolithic column) into the monolithic column that effectively provides baseline separations for all eight drug compounds. The autosampler syringe depth was adjusted such that the syringe needle only slightly penetrated the top layer of the diluted plasma solution in the autosampler vial. This avoided clogging from the diluted plasma sample. The linear range was validated from 1.95 to 1000 ng/mL of each drug and greater than 0.997 correlation coefficient values were obtained. The set-up enabled the analysis of 300 samples on one strata-X cartridge without any noticeable changes in HPLC pump back pressure, chromatographic retention time, baseline noise level, or peak shape for each analyte. The method proved to be rugged and comparison of off-line LLE data with results from this online method for pharmacokinetic screening samples for 0.25 to 12 hr time periods showed that the online SPE method was as efficient as the LLE method.

Alnouti et al.[118] recently reported a novel method for online introduction of internal standard (IS) for quantitative analysis of drugs from biological matrices using LC/MS/MS. Using the

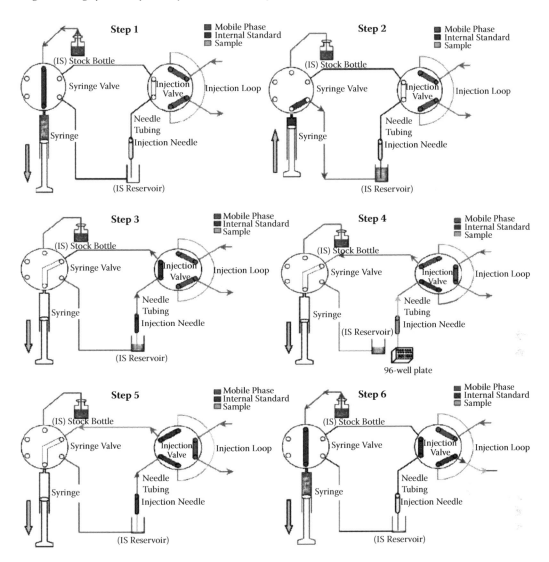

FIGURE 1.16 Configuration and operational details of online introduction of internal standard for quantitative analysis of drugs from biological matrices.[118] (Reproduced with permission from the American Chemical Society and the authors.)

autosampler as a device to measure and introduce both sample (analyte) and IS, two off-line (manual) sample preparation steps (measuring fixed amounts of samples and spiking with fixed amount of IS) can be eliminated. The applicability of this method for propranolol and diclofenac using ketoconazole and ibuprofen as ISs, respectively, was demonstrated on the Symbiosis system (Spark Holland) with C18 HD cartridges (2 × 10 mm, Spark Holland) as online SPE columns and a Luna C18 (2.1 × 50 mm, 5 μm particles) as the analytical column. The operational details of introduction of sample and IS are illustrated in Figure 1.16. The IS may be injected from a vial via autosampler or directly into the injection loop (using one of the injection modes of the Symbiosis autosampler), the latter avoiding cross contamination possible with the former. The variation (RSD) in IS peak areas of samples spiked with IS off-line were 10.1% for ketoconazole and 2.1% for ibuprofen. For online introduction, the values were 6.8 and 3.1% for ketoconazole and ibuprofen, respectively. Individual absolute recoveries for the two analytes and their respective ISs exceeded 90%. Precision ranged from 2 to 12% for off-line and 2 to 16% for online introductions.

1.2.3.5 Utility of 384-Well Plates for High-Throughput Applications and In-Process Monitoring of Cross Contamination

An application involving the use of monoclonal antibody fragments for selective extraction of the d-enantiomer of an experimental drug belonging to the diarylalkyl triazole system was reported recently.[119] The antibody fragment was immobilized on chelating Sepharose and dispensed as a 50% suspension in PBS (pH 7.4, 25 μL) into a 384-well plate (Whatman). A plasma sample of a mixture of the d- and a-enantiomers of the triazole drug was loaded in PBS after conditioning the sorbent in the 384-well plate with 60 μL of 100mM imidazole-PBS, pH 7.4. After sample load, washing was done with ammonium acetate (10mM, pH 5.0). The bound d-enantiomer was eluted with 25mM ammonium acetate, pH 2.5 (3 × 60 μL).

Analysis was performed on an ES-Ovomucoid column for stereoselectivity assessment, and for MS/MS, an X-Terra MS C18 column (2.1 × 100mm, 5 μm) was used. Figure 1.17 shows the wash and elution fractions from the SPE in a 384-well plate. The SPE conditions evaluated are listed in the table below the figure. The binding of the drug to the affinity sorbent in a 96-well plate was less efficient than the 384-well plate because the sorbent formed a disk on the former and a column on the latter. The efficiency is reflected in the >95% recoveries achieved with the 384-well format.

Progress with the 384-well plate solid phase extraction has been slow since the first examples of bioanalytical applications were published[120,121] in 2001. Reasons for this observation include increased cross contamination, lack of appropriate supplies and tools, lack of demand and interest, presence of other upstream and downstream bottlenecks, and sample volume and sensitivity limits. Min Chang and coworkers[122] evaluated the 384 technology by developing assays for lopinavir and ritonavir, the active ingredients of the Kaletra anti-HIV drug. Samples in individual vials were

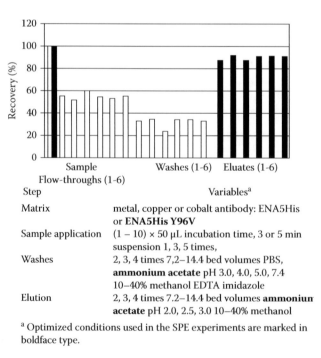

Step	Variables[a]
Matrix	metal, copper or cobalt antibody: ENA5His or **ENA5His Y96V**
Sample application	(1 – 10) × 50 μL incubation time, 3 or 5 min suspension 1, 3, 5 times,
Washes	2, 3, 4 times 7,2–14.4 bed volumes PBS, **ammonium acetate** pH 3.0, 4.0, 5.0, 7.4 10–40% methanol EDTA imidazole
Elution	2, 3, 4 times 7.2–14.4 bed volumes **ammonium acetate** pH 2.0, 2.5, 3.0 10–40% methanol

[a] Optimized conditions used in the SPE experiments are marked in boldface type.

FIGURE 1.17 Load–wash–elution profile of SPE extracts of d-enantiomer of a traizole drug studied on a 384-well plate.[119] (Reproduced with permission from the American Chemical Society.)

transferred to 96-well plates using MicroLab AT plus II and then to the 384-format by a multichannel pipettor or MicroLab Starlet. Two procedures were used for SPE on these 384-formatted well plates (Orochem, Lombard, Illinois): loading and washing using a vacuum box and use of a centrifuge for loading, washing, and eluting. Eluates were analyzed by LC/MS/MS using a Luna C18 column (Phenomenex, Torrance, California) with a run time of 2.2 min. The authors noted that the availability of a sufficient LC/MS/MS throughput is essential. The 384-well technology will not exert a significant impact on the overall throughput unless shorter LC/MS methods such as UPLC, high temperature LC, multiparallel micro-HPLC, and nanoelectrospray infusion are used. For example, a run time of 2.2 min will allow handling of 570 samples in 21 hours, while a 1.5-min run time will facilitate running of 840 samples in the same 21-hour period. With respect to availability of apparatus and disposables, the authors note that SPE using a centrifuge minimizes cross-contamination, but the technique is difficult to automate. On the other hand, one must be careful about cross contamination while using a vacuum. Centrifugation minimizes this contamination. Suitable disposable pipette tips for mixing samples in a deep-well 384-formatted microtiter plate are difficult to locate; only recently was this problem addressed.

The concept of rectangular experimental designs for multiunit platforms (RED-MUPs) as a part of statistical experimental design (also known as design of experiments or DOE) was explored in a recent study[123] aimed at reducing manual preparation and enabling the use of pipetting robots and applied to a reporter gene assay in the 96- and 384-well formats. Further work on this technique continues.

1.2.3.6 Utility of Multisorbent Extraction for SPE High-Throughput Method Development

Since the introduction of the 96-well plate format for SPE, method development for a particular drug or combinations of drugs and/or drug metabolites, impurities, and degradation products from aqueous or biological matrices has normally involved a single sorbent packed in all the wells. A generic method (universal set of extraction conditions, commonly recommended by a manufacturer) usually serves as the starting point.

The complexity of the method in terms of number of steps and solvents needed depends on the sorbent chemistry. The development in a simplified scenario involves running an analyte in several concentrations in multiple replicates and assaying for recovery and performance. This procedure is described in detail for several silica and polymeric sorbents by Wells.[42] However, if a number of sorbents are to be evaluated, the process becomes time-consuming if multiple 96-well plates (each with one sorbent packed in all the wells) must be screened separately. This process may take a week or more and consume an analyst's precious time as well. The most plausible solution is to pack different sorbents in the same well plate and use a universal procedure that applies to all of them. An example of such a multisorbent method development plate is the four-sorbent plate recently introduced by Phenomenex demonstrated[124] to require only 1 to 2 hr to determine optimal sorbent and SPE conditions.

Four polymeric sorbents with different chemistries and interaction mechanisms are packed in a 96-well plate in a configuration wherein three vertical columns are dedicated to each sorbent (total 24 wells; see Figure 1.18). These sorbents consist of the strata-X neutral polar/non-polar balanced functionalized styrene–divinylbenzene polymer, the strong strata-X-C cation exchanger with sulfonic acid moieties located on the phenyl rings of the same base polymeric skeleton, the weak strata-X-CW cation exchanger with a carboxyl-functionalized PSDVB, and a weak strata-X-AW anion exchanger with primary and secondary amine groups on the PSDVB skeleton. The four sorbents cover all possible types of interactions any analyte can exhibit. The strata-X displayed strong hydrophobic and π–π interactions, coupled with moderate hydrogen bonding and weakly acidic properties. The strata-X-C yielded strong cation exchange and hydrophobic interactions, along with weak hydrogen bonding and moderate π–π interactions. At the same time, strata-X-CW showed weak cation exchange and strong hydrogen bonding properties with much lower hydrophobicity; strata-X-AW exhibited strong anion exchange activity along with moderate hydrophobicity and weak hydrogen bonding.

STRATA-X			STRATA-X-C			STRATA-X-CW			STRATA-X-AW		
ULOQ X NN	ULOQ X AB	ULOQ X BA	ULOQ C NN	ULOQ C AB	ULOQ C BA	ULOQ CW NN	ULOQ CW AB	ULOQ CW BA	ULOQ AW NN	ULOQ AW AB	ULOQ AW BA
ULOQ X NN	ULOQ X AB	ULOQ X BA	ULOQ C NN	ULOQ C AB	ULOQ C BA	ULOQ CW NN	ULOQ CW AB	ULOQ CW BA	ULOQ AW NN	ULOQ AW AB	ULOQ AW BA
ULOQ X NN	ULOQ X AB	ULOQ X BA	ULOQ C NN	ULOQ C AB	ULOQ C BA	ULOQ CW NN	ULOQ CW AB	ULOQ CW BA	ULOQ AW NN	ULOQ AW AB	ULOQ AW BA
ULOQ X NN	ULOQ X AB	ULOQ X BA	ULOQ C NN	ULOQ C AB	ULOQ C BA	ULOQ CW NN	ULOQ CW AB	ULOQ CW BA	ULOQ AW NN	ULOQ AW AB	ULOQ AW BA
EREF X NN	EREF X AB	EREF X BA	EREF C NN	EREF C AB	EREF C BA	EREF CW NN	EREF CW AB	EREF CW BA	EREF AW NN	EREF AW AB	EREF AW BA
EREF X NN	EREF X AB	EREF X BA	EREF C NN	EREF C AB	EREF C BA	EREF CW NN	EREF CW AB	EREF CW BA	EREF AW NN	EREF AW AB	EREF AW BA
BLK X NN	BLK X AB	BLK X BA	BLK C NN	BLK C AB	BLK C BA	BLK CW NN	BLK CW AB	BLK CW BA	BLK AW NN	BLK AW AB	BLK AW BA
LLOQ X NN	LLOQ X AB	LLOQ X BA	LLOQ C NN	LLOQ C AB	LLOQ C BA	LLOQ CW NN	LLOQ CW AB	LLOQ CW BA	LLOQ AW NN	LLOQ AW AB	LLOQ AW BA

FIGURE 1.18 Phenomenex four-sorbent SPE method development plate.

The method development process with the multisorbent plate consists of three steps. In step 1, the sorbent chemistry and the pH for loading, washing, and elution are optimized. In step 2, optimization of the percentage organic for wash and elution and the pH of the buffer needed is carried out. Step 3 is validation; the method developed from the results of the previous two steps is tested for linearity, limits of detection, quantitation of recovery, and matrix effects using a stable isotope-labeled analyte as an IS.

Step 1 utilizes three sets of load and elution conditions: neutral (water) load with 100% methanol elution (the NN condition), loading in acidic buffer (ammonium formate, 25mM, pH 2.5, with formic acid) and elution in methanol containing 5% ammonia (the AB condition), and loading in basic buffer (ammonium acetate, pH 5.5, 25mM, pH adjusted with acetic acid), designated the BA condition. The results for neutral, acidic, and basic drugs are shown in Figure 1.19. For carbamazepine, any sorbent and any of the NN, AB, or BA conditions can be used for SPE. For procainamide, cation exchange under the AB condition is best, and for indomethacin, strata-X-AW with AB is preferred.

The step 2 results for procainamide are shown in Figure 1.20. A single sorbent 96-well plate (strata-X-C) was used and 20% increments of 0 to 100% methanol containing 5% formic acid or 5% ammonia were investigated. Procainamide can be washed with 100% methanol under acidic conditions without any breakthrough. Under basic conditions, the drug starts to elute at 40% or higher methanol content. Thus, the former can be used for wash and the latter for elution with the desired percentage of methanol (elution is 100% at 80% methanol content (Figure 1.20)).

With strata-X, procainamide elutes off under any conditions including >25% methanol in the solvent. With strata-X-CW, the drug can be eluted off under both acidic and basic conditions with >40% methanol content in the solvent. As with strata-X, the drug is eluted off under acidic or basic conditions with >50% methanol using strata-X-AW. Overall, strata-X-C is the best option for procainamide from a biological matrix.

1.2.4 RECENT DEVELOPMENTS IN LIQUID–LIQUID EXTRACTION (LLE) FOR CLEAN-UP OF BIOLOGICAL MATRICES: MINIATURIZATION AND HIGH-THROUGHPUT OPTIONS

Traditional LLE utilizes large volumes of solvents that are often hazardous from an environmental perspective and the process is tedious and time consuming. During the past decade, this technique

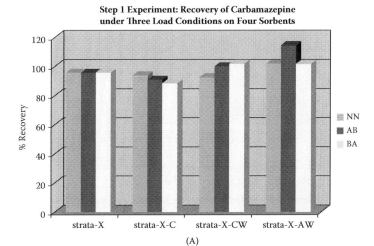

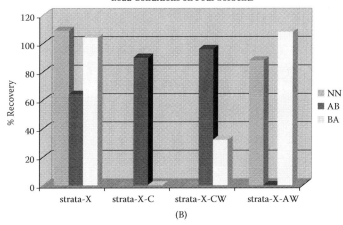

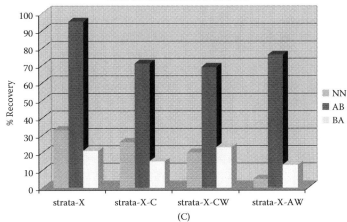

FIGURE 1.19 (A) Load–elution study on moderately hydrophobic, neutral carbamazepine with a four-sorbent method development plate. (B) Load–elution study of procainamide with multisorbent method development plate. (C) Load–elution study of acidic indomethacin with multisorbent method development plate.

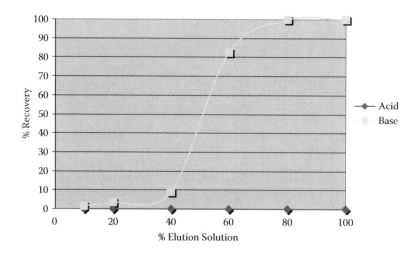

FIGURE 1.20 Wash–elution profile of procainamide on strata-X-C.

has undergone a spectacular transformation in the context of high-throughput screening with the introduction of miniaturized protocols along with revolutionary membrane-based and solid support-based technologies. These advances were reviewed by several authors.[125–135]

The simplest technique is the use of the 96-well collection plate format (analogous to the format used in SPE) in conjunction with a liquid handling robotic system; it follows the same principle as bulk scale LLE. However, immobilization of the aqueous plasma sample on an inert solid support medium packed in a cartridge or in the individual wells of a 96-well plate and percolating a water-immiscible organic solvent to extract the analyte from this medium evoked significant enthusiasm from the pharmaceutical industry.

Several manufacturers introduced products amenable for this solid-supported LLE and for supported liquid extraction (SLE). The most common support material is high-purity diatomaceous earth. Table 1.8 lists some commercial products and their suppliers. The most widely investigated membrane-based format is the supported liquid membrane (SLM) on a polymeric (usually polypropylene) porous hollow fiber. The tubular polypropylene fiber (short length, 5 to 10 cm) is dipped into an organic solvent such as nitrophenyl octylether or 1-octanol so that the liquid diffuses into the pores on the fiber wall. This liquid serves as the extraction solvent when the coated fiber is dipped

TABLE 1.8
Suppliers of Solid-Supported LLE Materials

Manufacturer	Commercialized SLE Product
Varian, Inc., Lake Forest, CA, USA	Chem Elut (96-well plate with 200 mg hydromatrix per well; cartridges of 0.3, 1.0, 3.0, 5.0, 10.0, 20.0, 50.0, 100.0, and 300.0 mL aqueous capacity)
Orochem, Lombard, IL, USA	Aquamatrix (96-well plate with 1- or 2-mL capacity packed with calcined diatomaceous earth)
Biotage, Uppsala, Sweden [International Sorbent Technology, Ltd., Glamorgan, UK]	Isolute SLE+ (96-well plate, 2 mL, packed with modified diatomaceous earth)
Merck, Darmstadt, Germany	Extrelut 1.0, 3.0, and 20.0 mL glass or polyethylene columns

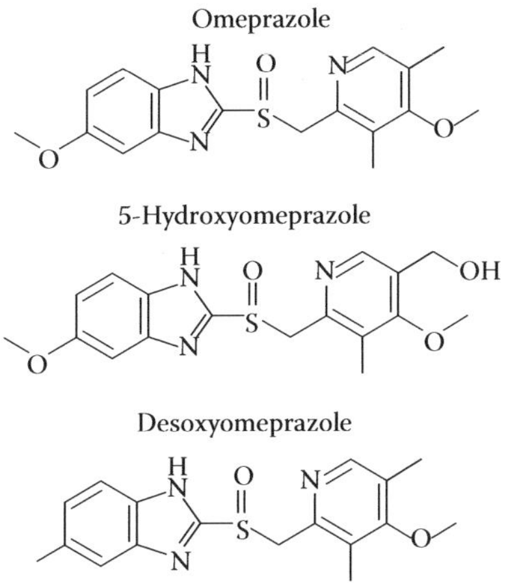

Omeprazole

5-Hydroxyomeprazole

Desoxyomeprazole

FIGURE 1.21 Structures of omeprazole, its 5-hydroxy metabolite, and desoxyomeprazole (internal standard).[136] (Reproduced with permission from Elsevier.)

into an aqueous buffered plasma solution. The analyte is back-extracted from the organic solvent in the fiber pores into an aqueous extractant (acidic solution for bases and basic solution for acids), which is withdrawn for analysis. For neutral analytes, a two-phase extraction is used. The solvent serves as the membrane and as the extractant. The following section illustrates the underlying principles and application of SLE and LPME (liquid phase microextraction with SLM) and includes literature examples for each technique.

1.2.4.1 Automated Liquid–Liquid Extraction without Solid Support

A novel approach for the extraction and LC/MS/MS analysis of omeprazole (used to treat gastroesophageal reflux disease) and its 5-hydroxy metabolite (see Figure 1.21) through automated LLE using the hydrophilic interaction chromatographic mode (HILIC) for HPLC was recently reported by Song and Naidong.[136] Thawed and vortex-mixed plasma sample aliquots were transferred into a 96-deep well collection plate using a Packard Multiprobe II robotic liquid handler. Desoxyomeprazole (internal standard, 100 ng/mL in 1:1 methanol:water, 50 μL) was added to each sample, followed by 10 μL of ammonium hydroxide (2% in water). Ethyl acetate (0.5 mL) was added to each sample and the plate covered with a dimpled sealing mat. The plate was vortex mixed for 10 min, then centrifuged at 3000 rpm and at 4°C for 5 min.

Using the Tomtec Quadra 96 workstation, 0.1 mL of the ethyl acetate layer was transferred to a 96-well collection plate containing 0.4 mL of acetonitrile in each sample well. The solution was mixed 10 times by aspiration and dispersion on the Tomtec. The plate was then covered with a sealing mat and stored at 2 to 8°C until LC/MS/MS analysis. The HILIC-MS/MS system consisted of a Shimadzu 10ADVP HPLC system and Perkin Elmer Sciex API 3000 and 4000 tandem mass spectrometers operating in the positive ESI mode. The analytical column was Betasil silica (5 μm, 50 × 3 mm) and a mobile phase of acetonitrile:water:formic acid with a linear gradient elution from 95:5:0.1 to 73.5:26.5:0.1 was used for 2 min. The flow rate was 1.0 mL/min for the API 3000 and 1.5 mL/min for the API 4000 without any eluent split. The injection volume was 10 μL and a run time of 2.75 min was employed.

The multiple reaction monitoring (MRM) conditions for each analyte were optimized by infusing 0.1 μg/mL of analyte in mobile phase. The Ionspray needle was maintained at 4.0 kV and the turbo gas temperature was 650°C. Nebulizing gas, auxiliary gas, curtain gas, and collision gas flows were set at 35, 35, 40, and 4, respectively. In the MRM mode, collision energies of 17, 16, and 15 eV

TABLE 1.9

Precision and Accuracy of Quality Control Samples for Automated LLE of Omeprazole and 5-OH Omeprazole Metabolite

	Intra-day (n = 6)				Inter-day (n = 18)			
	2.50 ng/ml	7.50 ng/ml	180 ng/ml	1800 ng/ml	10000[a] ng/ml	7.50 ng/ml	180 ng/ml	1800 ng/ml
Omeprazole								
Mean	2.36	7.64	177	1820	10100	7.81	179	1870
R.S.D. (%)	4.0	3.9	1.6	1.8	2.7	4.4	−0.6	+3.9
R.E. (%)	−5.6	+1.9	−1.7	+1.1	+1.0	+4.1	−0.6	+3.9
5-OH omeprazole								
Mean	2.73	7.81	179	1830	10100	7.92	177	1830
R.S.D. (%)	8.1	4.4	0.9	3.3	2.7	4.4	4.5	2.4
R.E. (%)	9.2	+4.1	−0.6	+1.7	+1.0	+5.6	−1.7	+1.7

[a] Samples diluted 10-fold with blank plasma prior to analysis.

Source: Song, Q. and W. Naidong, *J. Chromatogr. B,* 2006, 830, 135. With permission from Elsevier.)

for omeprazole, its 5-OH metabolite, and the IS, respectively, were noted. Transitions monitored were m/z 346 → 198 for omeprazole, 362 → 214 for 5-OH omeprazole, and 330 → 198 for desoxyomeprazole.

The dwell time was 200 msec for the analytes and 100 msec for the IS. At least 500 extracted samples were injected onto each column without any column regeneration. No solvent evaporation and reconstitution steps were involved. Ethyl acetate was preferred over methyl t-butyl ether (MTBE) because MTBE caused pulp-up of the mat. Six blank plasma lots were tested for matrix interference and none was detected in the analyte or IS region. When 100 ng/mL of the analytes were spiked into the blank plasma samples, the relative standard deviations were 1.0 and 1.5% for omeprazole and its metabolite, respectively. Precision and accuracy figures are given in Table 1.9.

A similar automated LLE protocol for a novel ATP-competitive inhibitor for all the vascular endothelial growth factor (VEGF) and platelet-derived growth factor (PDGF) receptor tyrosine kinases (RTKs), named ABT-869 (Abbott Laboratories; see Figure 1.22) was reported during anti-tumor efficacy studies.[137] A fully automated 96-well LLE was developed using a Hamilton liquid handler for ABT-869 and its acid metabolite A-849529 (Figure 1.22).

Hexane:ethyl acetate (1:11) was used as the extraction solvent. The extracted organic layer was transferred automatically into a 96-well injection plate and dried down with nitrogen at room temperature. The residue was reconstituted with 200 μL of 1:1 acetonitrile:water containing 0.1% formic acid. A Symmetry Shield RP8 analytical column (150 × 2.1 mm) was used for LC/MS/MS with an API 3000 mass spectrometer as the detector. The pH of the extraction mixture (diluted plasma solution) was varied by adding 0.2% formic acid, 0.2% acetic acid, or 0.2% acetic acid in ammonium acetate (100mM) or ammonium acetate (100mM) alone. The proportion of ethyl acetate in the organic was varied from 0 to 100% to determine optimal concentration for extracting the drug and its metabolite. The addition of formic or acetic acid in 100mM ammonium acetate buffer significantly improved the extraction recovery of the acid metabolite (see Figure 1.23). The assay protocol developed was applied to clinical samples and the results were satisfactory (see Figure 1.24).

Some additional examples for automated LLE consist of ondansetron in human plasma,[138] muraglitazar in human plasma,[139] boswellic acids in brain tissue and plasma,[140] dextromethorphan, dextrorphan, and guaifenesin in human plasma[141], and dextromethorphan and dextrorphan in human plasma.[142] Details are not discussed due to space considerations.

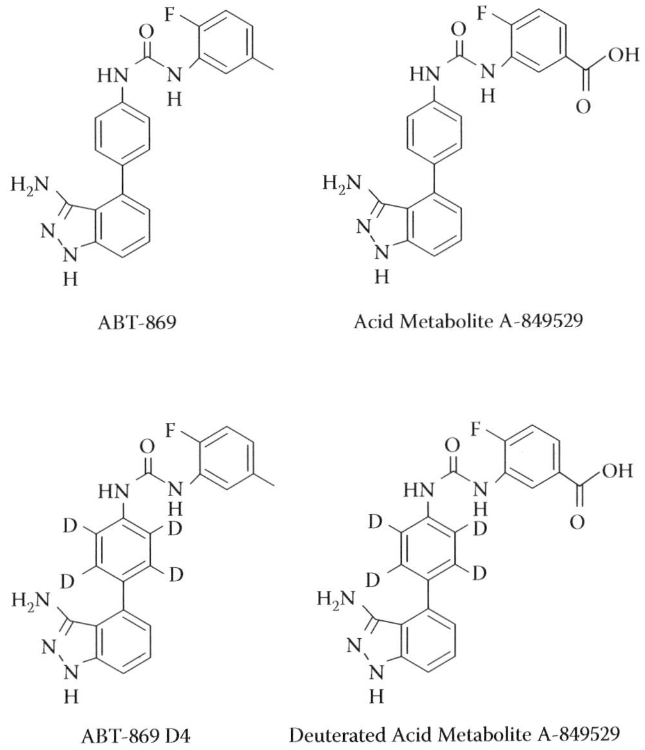

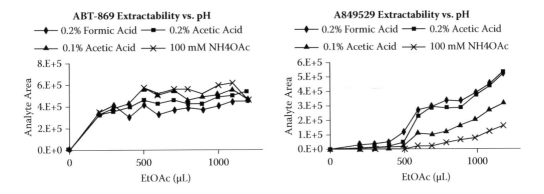

ABT-869

Acid Metabolite A-849529

ABT-869 D4

Deuterated Acid Metabolite A-849529

FIGURE 1.22 Structures of ABT-869, its acid metabolite, and deuterated analogs used as internal standards.[137] (Reproduced with permission from John Wiley & Sons.)

1.2.4.2 Solid-Supported Liquid–Liquid Extraction

Figure 1.25 illustrates the principle underlying LLE in the solid-supported LLE format. In order to facilitate elution with a water-immiscible organic solvent, it is imperative that analytes are in their neutral form during sample load. Thus, for basic analytes, loading should be done in a high pH (9 to 10) buffer and for acidic analytes, a low pH (2 to 3) buffer.

FIGURE 1.23 Dependency of extraction recovery of ABT-869 and acid metabolite AB849529 on buffer pH and proportion of ethyl acetate in organic solvent.[137] (Reproduced with permission from John Wiley & Sons.)

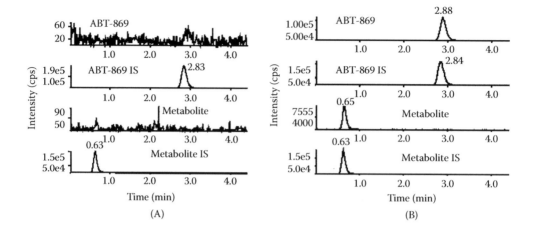

FIGURE 1.24 Ion chromatograms of (A) clinical predose sample and (B) clinical postdose sample.[137] (Reproduced with permission from John Wiley & Sons.)

Wang and coworkers[143] described a rapid and sensitive LC/MS/MS method for the determination of a novel topoisomerase I inhibitor (an indolocarbazole derivative) in human plasma, following SLE on 96-well diatomaceous earth plates. The structures of this inhibitor and the IS used are shown in Figure 1.26. Clinical, QC, and standard plasma samples were thawed at room temperature, vortexed for 30 sec, centrifuged at 3000 g for 10 min; 250 μL aliquots were transferred to

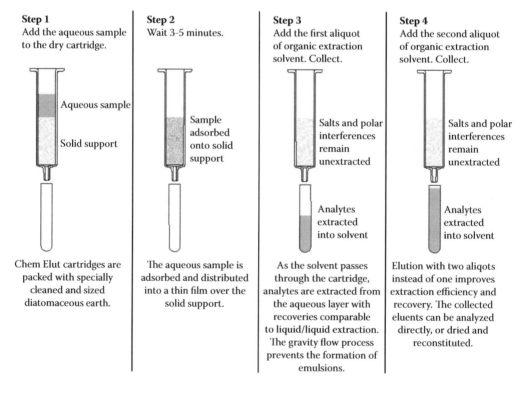

FIGURE 1.25 Principle of operation of solid-supported liquid–liquid extraction from aqueous or plasma samples. (Reproduced with permission from Varian, Inc.)

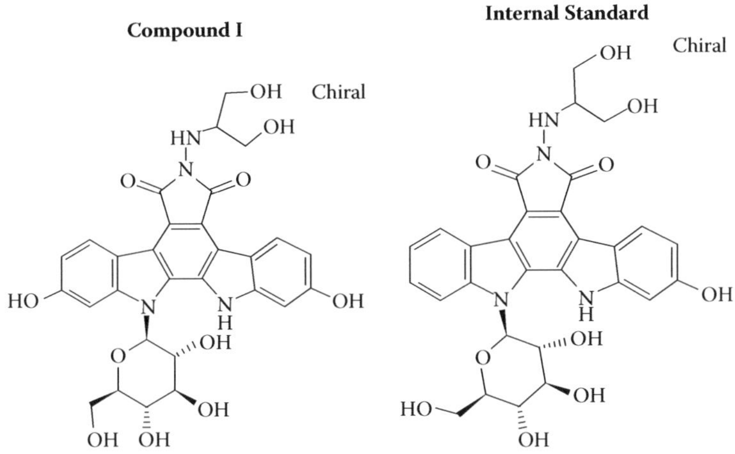

FIGURE 1.26 Structures of indolocarbazole compound I and internal standard.[143] (Reproduced with permission from John Wiley & Sons.)

polypropylene tubes using a Packard MultiProbe II EX robot. The IS solution (25 μL) was added and the mixture pipetted onto a 96-well diatomaceous earth plate using a Matrix Impact 8-channel expandable electronic pipette. The diatomaceous earth was obtained from Varian in bulk and packed into 96-well plates provided by Orochem Technologies.

After 5 min, the analyte and internal standard were eluted from the plate with 2 mL of 9% isopropanol in MTBE under gravity and collected in a 96-well plate. The extracts were evaporated on a SPE dry 96 evaporator (Jones Chromatography) under nitrogen at 50°C and reconstituted in 150 μL 50% methanol:water. A LEAP CTC PAL autosampler was used to inject 10 μL of this reconstituted solution into a Perkin-Elmer Series 200 LC interfaced with an API 3000 mass spectrometer. A Waters XTerra RP18 column (3.5 μm, 50 × 2 mm) was used for the LC part with a mobile phase of 70:30 methanol:6.7mM ammonium hydroxide in water. The negative molecular ion scan and the product ion scan from this parent ion are shown in Figure 1.27 and the precision and accuracy in Table 1.10. Stability and extraction recovery data are shown in Tables 1.11 and 1.12, respectively. Further examples of the solid-supported LLE technique from the literature include toxicity studies on 3-buten-1,2-diol (a major metabolite of 1,3-butadiene used in the synthetic rubber industry) in rat tissue,[144] plasma and urine[145] samples, and geometric isomers of acetyl-11-keto-α (or β)-boswellic acid.[146]

1.2.4.3 Liquid Phase Microextraction (LPME)

In the past decade, several novel solvent-based microextraction techniques have been developed and applied to environmental and biological analysis. Notable approaches are single-drop microextraction,[147] small volume extraction in levitated drops,[148] flow injection extraction,[149,150] and microporous membrane- or supported liquid membrane-based two- or three-phase microextraction.[125,151–153] The two- and three-phase microextraction techniques utilizing supported liquid membranes deposited in the pores of hollow fiber membranes are the most explored for analytes of wide ranging polarities in biomatrices. This discussion will be limited to these protocols.

The principle of a three-phase membrane extraction is illustrated in Figure 1.28. An organic solvent is immobilized in the pores of a porous polymeric support consisting of a flat filter disc or a hollow fiber-shaped material. This supported liquid membrane (SLM) is formed by treating the support material with an organic solvent that diffuses into its pores. The SLM separates an aqueous

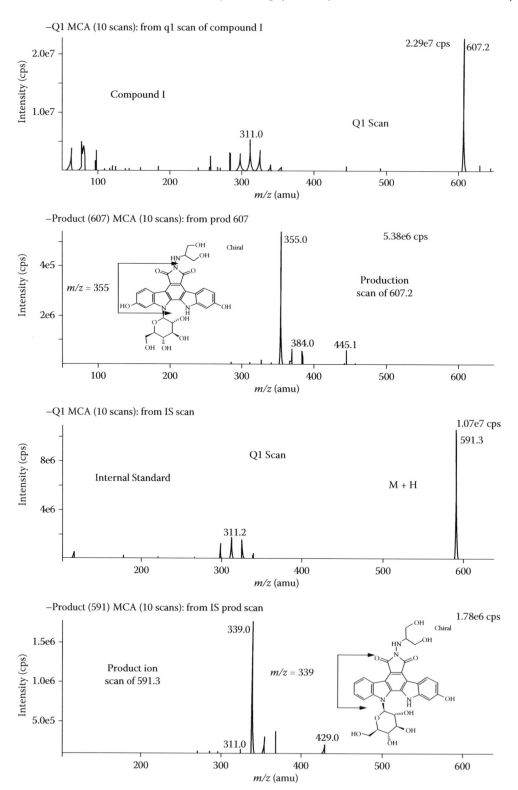

FIGURE 1.27 Negative parent ion and product ion mass spectra of compound I and internal standard.[143] (Reproduced with permission from John Wiley & Sons.)

TABLE 1.10
Precision and Accuracy of Indolocarbazole
Compound I and Internal Standard

Nominal conc. (ng/mL)	Mean conc. (ng/mL)	Precision CV (%)	Accuracy (%)
0.05	0.049	6.2	98.0
0.1	0.11	7.0	107.4
0.5	0.46	5.7	91.1
5.0	5.14	2.2	102.8
25	25.42	7.1	101.7
100	98.07	6.5	98.1
200	201.41	4.3	100.7

Source: Wang, A.Q. et al., *Rapid Commun. Mass Spectrom.,* 2002, 16, 975. With permission from John Wiley & Sons.

TABLE 1.11
Stability Data: Indolocarbazole Compound I

Nominal conc. (ng/mL)	Mean Found Conc.[a] (ng/mL)	Precision CV[b] (%)	Accuracy[c] (%)	Found Conc.[d] after 3 F/T Cycles (ng/mL)	Accuracy (5)
0.2	0.21	3.7	104.3	0.22	108.5
20	20.9	3.7	104.7	20.6	103.2
150	147.3	10	98.2	152.7	101.8

[a] Mean concentrations calculated from weighted (1/x) linear last-squares regression curve after one freeze–thaw.

[b] Percent coefficient of variation (CV) of peak area ratios (n = 5).

[c] Expressed as (mean found concentration/nominal concentration) × 100.

[d] F/T = freeze–thaw; freezing at −20°C (n = 3).

Source: Wang, A.Q. et al., *Rapid Commun. Mass Spectrom.,* 2002, 16, 975. With permission from John Wiley & Sons.

TABLE 1.12
Extraction Recovery of Compound I from Supported
Liquid–Liquid Extraction

Nominal Conc. (ng/mL)	Mean (n = 5) Extracted Peak Area	Mean (n = 4) Spiked Peak Area[a]	Recovery[b] (%)
0.5	4317	7089	60.9
5	47988	72866	65.9
25	218700	350636	62.4
200	1590718	2469997	64.4
2 (IS)	25586 (n = 50)	38286 (n = 40)	66.7

[a] Peak area of standard spiked in extract of plasma double blanks.

[b] Calculated as [(mean extracted peak area)/(mean spiked peak area) × 100] at each concentration.

Source: Wang, A.Q. et al., *Rapid Commun. Mass Spectrom.,* 2002, 16, 975. With permission from John Wiley & Sons.

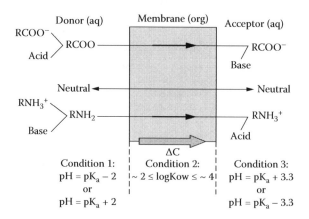

FIGURE 1.28 Principle of extraction of ionizable organic analyte into SLM.[151] (Reproduced with permission from IUPAC.)

donor solution (containing the analyte to be extracted) and an acceptor solution (that may be the same organic liquid as the SLM in the case of a two-phase system or can be aqueous for a three-phase system). The analyte is partitioned into the organic membrane from the donor solution and is then re-extracted into the aqueous acceptor solution inside the hollow fiber (or into the organic solvent for a two-phase system). In order to carry out this kind of double partitioning, an analyte should exist in a nonionic form to achieve its transfer into the organic membrane phase and in an ionic form in the acceptor phase to prevent its back-migration into the organic liquid. These requirements are achieved by pH adjustment of the two aqueous phases. Thus, for extracting a basic analyte, the donor solution pH should be around or above the pKa of the analyte and the acceptor solution should be strongly acidic. The opposite is true for an acidic analyte. The influence of the respective partition coefficients (between the acceptor and donor, $K_{a/d}$; the organic and the donor, $K_{org/d}$; and the acceptor and organic, $K_{a/org}$) in effecting a successful extraction is shown numerically in Table 1.13. It can be easily seen that the larger the value of $K_{a/d}$, the better are the recovery and enrichment figures. [154]

The table also shows that a three-phase LLE (organic extraction followed by back-extraction into aqueous phase) yields lower recoveries and enrichment compared to three-phase LPME, as reflected in peak heights from the two techniques as shown in Figure 1.29. Furthermore, three-phase LLE is sensitive to the magnitude of $K_{a/org}$ and LPME is not.

A simple manual experimental set-up for carrying out three-phase LPME is shown in Figure 1.30. A conventional 2- or 4-mL glass vial fitted with a screw cap containing a silicone rubber septum serves as the extraction apparatus.[155] Two conventional 0.8-mm outer diameter medical syringe needles are inserted through the septum and the two ends of a Q3/2 Accurel KM polypropylene hollow fiber (Membrana, Wuppertal, Germany), inner diameter of 600 μm, length of 8 cm, wall thickness of 200 μm, and pore size of 0.2 μm were inserted into each needle tip. For these dimensions, the internal volume of the acceptor solution inside the fiber is 25 μL and the organic phase volume in the pores of the fiber is 23 μL. For extraction experiments, 1 mL of plasma was basified with 250 μL of 2M sodium hydroxide and diluted with a combined volume of 2.75 mL (water and sample solution together) so that the total volume of the donor solution is 4.0 mL.

The hollow fiber was dipped into dihexyl ether for 5 sec and excess adhering solvent was washed away by ultrasonification in a water bath. Then, 25 μL of 10mM hydrochloric acid (aqueous, acceptor phase) was injected into the lumen of the hollow fiber with a microsyringe. This activated fiber was placed in the vial containing the donor solution and the vial vibrated at 1500 rpm for 45 min. The entire acceptor solution was flushed into a 200-μL micro insert and subjected to capillary electrophoresis or HPLC detection. For 2 mL extractions, 250 μL of plasma sample treated with

TABLE 1.13

Calculated Recovery and Enrichment in Three-Phase LPME and Simple LLE at Different $K_{a/d}$ Values

$K_{a/d}$	$K_{org/d}$	$K_{a/org}$	Three-Phase LPME[a]		Three-Phase LLE[b]	
			Recovery (%)	Enrichment	Recovery (%)	Enrichment
1	1	1	0.6	1.0	1.1	0.3
5	1	5	3.0	4.8	5.1	1.4
	5	1	3.0	4.8	1.7	0.4
10	1	10	5.9	9.4	9.7	2.6
	5	2	5.7	9.1	2.7	0.7
	10	1	5.6	9.0	1.4	0.4
50	1	50	23.7	37.9	34.9	9.3
	5	10	23.4	37.4	6.7	1.8
	10	5	22.9	36.6	6.7	1.8
	50	1	20.0	32.0	1.5	0.4
100	1	100	38.3	61.3	51.7	13.9
	5	20	37.9	60.6	21.7	5.9
	10	10	37.3	59.7	12.6	3.5
	50	2	33.3	53.3	2.9	0.8
	100	1	29.4	47.0	1.5	0.4
500	1	500	75.7	121.1	84.3	22.4
	5	100	75.3	120.5	58.1	15.5
	10	50	74.9	119.8	41.9	11.2
	50	10	71.4	114.2	13.0	3.5
	100	5	67.6	108.2	7.0	1.9
	500	1	47.2	75.5	1.5	0.4
1000	1	1000	86.1	137.8	91.5	24.5
	5	200	85.9	137.4	73.5	19.7
	10	100	85.6	137.0	59.1	15.7
	50	20	83.3	133.3	22.9	6.1
	100	10	80.6	129.0	13.0	3.5
	500	2	64.1	102.6	2.9	0.8
	1000	1	51.0	81.6	1.5	0.4

[a] $V_a = 25$ μl, $V_{org} = 20$ μL, and $V_d = 4$ mL.
[b] $V_a = 150$ μl, $V_{org} = 5$ mL, and $V_d = 2$ mL.

Source: Ho, T.S. et al., *J. Chromatogr. A,* 2002, 963, 3. With permission from Elsevier.

200 μL of 2M sodium hydroxide was used and the total volume made up to 1 mL with water and sample solution. Based on Figures 1.31 and 1.32, the equilibrium time (time for extracting 90% of analyte into the acceptor phase) was attained after 18 to 25 min of extraction for five representative drugs in water matrix; 21 to 40 min were needed for plasma samples.

The set-up was also amenable for protein binding studies, as demonstrated for the five representative drugs that were evaluated with 0, 5, and 50% methanol content in the plasma samples. Maximum recoveries were observed with 50% methanol for promethazine, methadone, and haloperidol, indicating significant protein binding for these drugs. For amphetamine and pethidine, on the other hand, protein binding was significantly lower based on the better recoveries obtained with

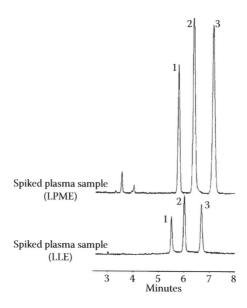

FIGURE 1.29 Three-phase LPME versus three-phase LLE of promethazine.[154] (Reproduced with permission from Elsevier.)

donor sample solutions containing less or no methanol. Figure 1.33 shows the set-up of a flat disc membrane format for a flow-through system.

To enhance automation capacity, a direct transfer of the acceptor phase to a HPLC system can be arranged by setting up a pre-column that allows the injection of as much volume of analyte as possible (Figure 1.33). Pneumatically or electrically actuated valves controlled by a computer provide

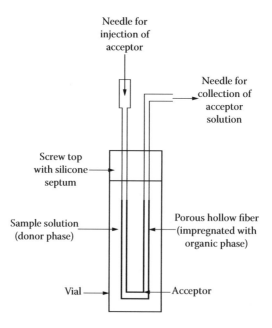

FIGURE 1.30 Simple experimental set-up for three-phase LPME.[155] (Reproduced with permission from the Royal Society of Chemistry.)

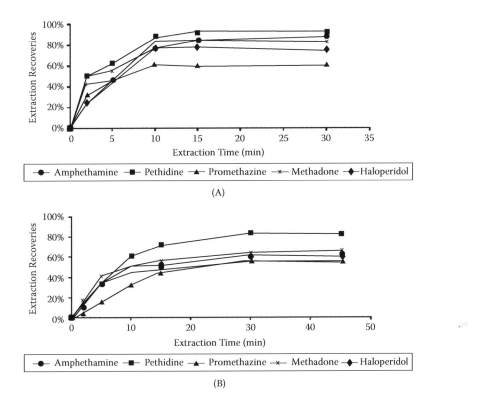

FIGURE 1.31 Extraction–time profiles for five model drugs in 1 mL (A) and 4 mL (B) aqueous solutions.[155] (Reproduced with permission from the Royal Society of Chemistry.)

control of the system. For small sample volumes of 1 mL or less, an analyst can apply membrane extraction equipment based on robotic liquid handlers with syringe pumps, as shown in Figure 1.33. A robotic needle connected to a syringe pump pipettes buffer to adjust the pH of the sample vials, picks up an aliquot, and pumps it through a donor channel with an approximate volume of 10 mL. The entire extract is then transferred from the acceptor channel to an injection loop connected to the HPLC system (Figure 1.34). The entire extract from the 1-mL sample goes to a single chromatographic injection. During the chromatographic separation of one extract, the next sample is extracted. The cycle time of the system is determined by the chromatographic time and the extraction time does not extend total analysis time.[153]

A chiral liquid chromatographic determination of mirtazapine (Figure 1.35) in human plasma through a two-phase LPME sample preparation was reported recently.[156] The manual vial set-up reported (Figure 1.36) is similar to that described above for the three-phase extraction. The organic membrane consisted of toluene coated onto the pores of a 7-cm polypropylene fiber (with the same parameters as mentioned earlier) by dipping the fiber in toluene. The acceptor solution was 22 μL of toluene. The extraction was carried out for 30 min under magnetic stirring at room temperature. The acceptor solution was removed from the fiber with the second syringe, evaporated, and reconstituted with 80 μL of a mobile phase comprised of 98:2 hexane:ethanol containing 0.1% diethylamine; 50 μL of this extract was injected onto a 250 × 4.6 mm Chiralpak AD column. Ultraviolet detection was carried out at 292 nm. The quantitation limit was 6.25 ng/mL. The effect of extraction time and fiber length variation on the two-phase process is shown in Figure 1.37 and the chromatographic analysis in Figure 1.38.

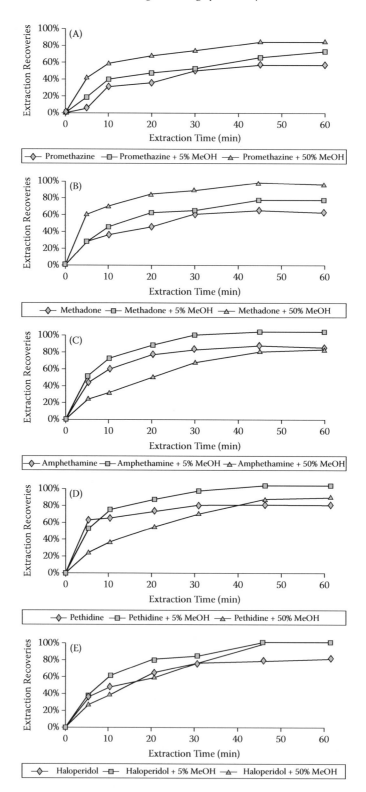

FIGURE 1.32 Extraction–time profiles for promethazine (A), methadone (B), amphetamine (C), pethidine (D), and haloperidol (E) in 1 mL sample volume with 0, 5, and 50% methanol added to plasma sample.[155] (Reproduced with permission from the Royal Society of Chemistry.)

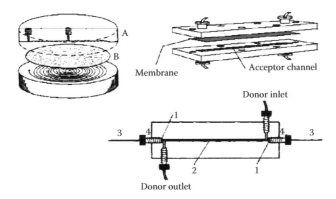

FIGURE 1.33 Top left: membrane unit with 1 mL channel volume (A = inert material; B = membrane); Top right: membrane unit with 10 μL channel volume. Bottom: hollow fiber membrane unit with 1.3 μL acceptor channel (lumen) volume.[151] (Reproduced with permission from IUPAC.)

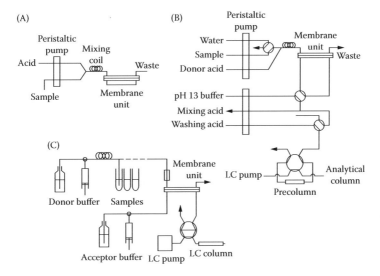

FIGURE 1.34 Apparatus for liquid membrane extraction. (A) Manual off-line instrument based on peristaltic pump. (B) Instrument with online connection to HPLC for environmental studies. (C) Experimental set-up for supported liquid membrane HPLC determination of biomolecules in blood plasma or urine.[153] (Reproduced with permission from the authors.)

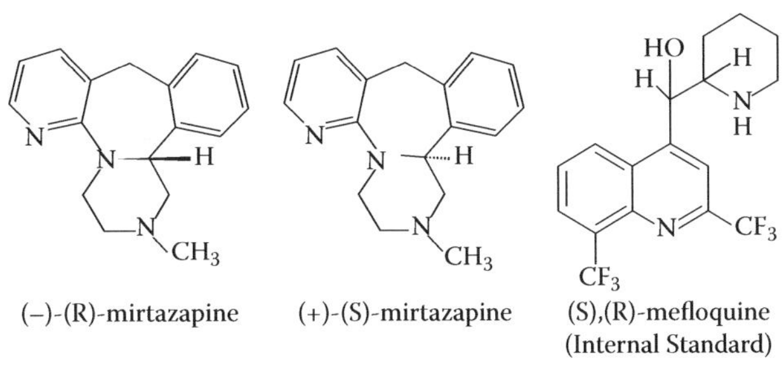

(−)-(R)-mirtazapine   (+)-(S)-mirtazapine   (S),(R)-mefloquine
                                                          (Internal Standard)

FIGURE 1.35 Structures of mirtazapine stereoisomers and internal standard.[156] (Reproduced with permission from Elsevier.)

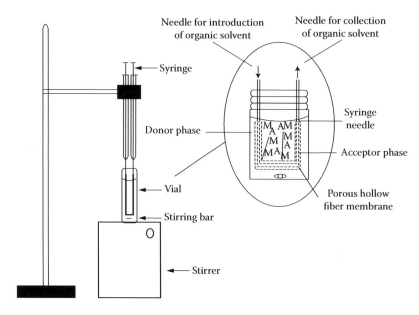

FIGURE 1.36 Experimental set up for two-phase LPME.[156] (Reproduced with permission from Elsevier.)

1.2.5 Protein Precipitation Techniques and Instrumentation for High-Throughput Screening

Protein precipitation is a common protocol for rapid sample clean-up and extraction of pharmaceuticals from plasma samples during drug discovery. It also allows the disruption of drug binding to proteins. Plasmas from various species such as dogs, rats, mice, and humans show variations in their compositions, but often only a single bioanalytical method covers all these species. Protein precipitation denatures the protein and destroys its drug binding ability, depending on the binding mechanism.[157]

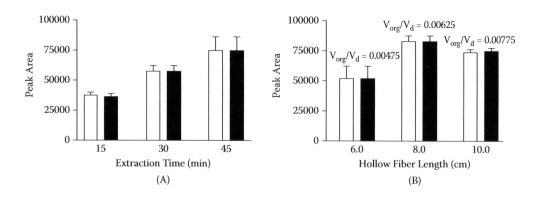

FIGURE 1.37 Effects of extraction time and fiber length on two-phase LPME of mirtazapine stereoisomers.[156] Influence of extraction time (A) and acceptor-to-donor phases volume ratio (B) on the efficiency of LPME. Plots for the (+)-(S)-mirtazapine (white bars) and (–)-(R)-mirtazapine (black bars) enantiomers (response in area counts). Extraction conditions: (A) 1 mL plasma sample; 0.1 mL 10 M NaOH; 3.0 ml deionized water; 7.0 cm fiber length; 22 μL toluene (B) 30 min of extraction; 1 mL plasma sample; 0.1 mL 10 M NaOH; 3.0 mL deionized water; toluene. (Reproduced with permission from Elsevier.)

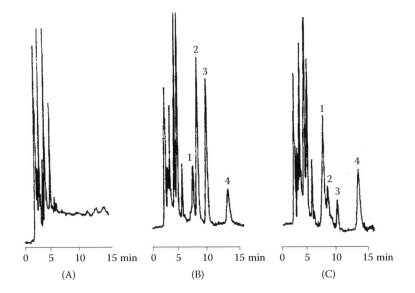

FIGURE 1.38 Chromatograms of LPME-treated drug-free plasma, mirtazapine enantiomers, and mefloquine.[156] Chromatograms refer to drug-free plasma (A); plasma spiked with 62.5 ng mL^{-1} of (+)-(S)-mirtazapine (2) and (–)-(R)-mirtazapine (3) and 500 ng mL^{-1} of (R, S)-mefloquine (1,4) (B); plasma sample from a patient treated with 15 mg/day of *rac*-mirtazapine (C). All samples were pre-treated by LPME. The analysis was performed on a Chiralpak AD column using hexane:ethanol (98:2, v/v) plus 0.1% diethylamine at a flow rate of 1.5 mL min^{-1}, λ = 292 nm. (+)-(S)-mirtazapine (2), (–)-(R)-mirtazapine (3) and (1, 4) internal standard. (Reproduced with permission from Elsevier.)

An organic solvent, acid, salt, and metal have been used for effecting protein precipitation by exerting specific interactive effects on the protein structure. An organic solvent lowers the dielectric constant of the plasma protein solution and also displaces the ordered water molecules around the hydrophobic regions on the protein surface, the former enhancing electrostatic attractions among charged protein molecules and the latter minimizing hydrophobic interactions among the proteins. Thus, the proteins aggregate and precipitate.

Acidic reagents form insoluble salts with the positively charged amino groups of the proteins at pH values below their isoelectric points (pIs). High salt concentrations deplete water molecules from the hydrophobic protein surfaces and allows aggregation through hydrophobic interactions of protein molecules. Binding of metal ions reduces protein solubility by changing its pI. A representative set of protein precipitation efficiency data from Polson et al.[157] is shown in Table 1.14. Variation in ionization effects for different species of plasma using acetonitrile in 50:50 methanol:water is shown in Table 1.15. The study produced interesting observations:

- The most efficient precipitants for protein removal were zinc sulfate, acetonitrile, and trichloroacetic acid (at 2:1 volume of precipitant to plasma, the protein removal values were, respectively, 96, 92, and 91% with <1% RSD for n = 5).
- The chosen precipitants functioned universally for all plasmas.
- Using acidic components exerted significant effects on ionization efficiency.
- Pure mobile phases with organic precipitants produced the largest ionization effects (Table 1.16).

More detailed data on protein precipitation agents and their efficacy has been included in a recent review by Flanagan and coworkers[158] and is presented in Table 1.17.

Protein precipitation by filtration in a 96-well format has been used as a high-throughput, easy-to-automate alternative to the traditional centrifugation-based protocol. However, most filter plates

TABLE 1.14

Comparison of Protein Precipitation Efficiencies of Precipitants upon Treatment with Human Plasma

Precipitant		Percent Protein Precipitation Efficiency[a]						
		Ratio of Precipitant to Plasma						
		0.5:1	1:1	1.5:1	2:1	2.5:1	3:1	4:1
Acids	TCA	91.4	91.8	91.5	91.0	91.2	91.3	91.4
	%RSD ($n = 3$)	4.46	—[b]	3.46	0.20	2.18	5.98	3.96
	m-Phosphoric acid	89.4	90.5	90.3	90.2	90.7	90.5	90.0
	%RSD ($n = 3$)	1.48	4.56	3.52	3.23	12.36	2.35	6.23
Metal ions	Zinc sulfate	89.2	96.8	96.8	99.0	99.0	99.0	>99.9
	%RSD ($n = 3$)	14.73	7.16	14.58	1.70	8.04	3.42	—[c]
Organics	ACN	3.6	88.7	91.6	92.1	93.2	93.5	94.9
	%RSD ($n = 3$)	3.62	2.50	3.63	3.13	5.29	5.91	1.82
	EtOH	0.1	78.2	87.2	88.1	89.8	91.8	92.0
	%RSD ($n = 3$)	2.85	2.43	1.65	9.47	9.56	2.46	1.06
	MeOH	13.4	63.8	88.2	89.7	90.0	91.1	91.5
	%RSD ($n = 3$)	0.95	3.09	3.54	3.50	2.84	5.09	2.46
Salts	Ammonium sulfate	24.8	50.1	64.0	84.2	90.4	90.4	89.0
	%RSD ($n = 3$)	1.80	4.37	3.61	0.53	7.11	3.74	2.45

[a] % Protein precipitation efficiency = [(total plasma protein – protein remaining in supernatant)/total plasma protein] × 100.
[b] One value obtained; samples discarded in error prior to assay.
[c] Concentration of protein in supernatant below quantification limit.

Source: Polson, C. et al., *J. Chromatogr. B,* 2003, 785, 263. With permission from Elsevier.

require a plasma sample to be dispensed before the precipitating solvent is added (the so-called plasma first method). A recent application note from the Argonaut laboratories[159] demonstrates that a solvent-first methodology using its Isolute PPT+ protein precipitation plates avoids the problems of leaking, cloudy filtrates and blocked wells. The Argonaut plates carry functionalized bottom frits that can hold up organic solvents, permitting the dispensing of solvents first. The principle is

TABLE 1.15

Variation of Ionization Effects of Acetonitrile in 50:50 Methanol/Water (2:1)

Plasma Type	Ionization Effect (%)	Duration of Ionization Effect (min)
Human	86.9	0.5 to 7.5
Dog	86.5	0.5 to 6.8
Mouse	92.6	0.5 to 8.6
Rat	92.9	0.5 to 9.9[a]

[a] Ionization suppression evident throughout entire region of analysis. Steady state signal achieved after 12.8 min (data not shown).

Source: Polson, C. et al., *J. Chromatogr. B,* 2003, 785, 263. With permission from Elsevier.

TABLE 1.16
Ionization Effects of Various Protein Precipitant and Mobile Phase Combinations on Human Plasma

Mobile Phase	Acetonitrile Ionization Effect (%)	Duration (min)	Methanol Ionization Effect (%)	Duration (min)	Ethanol Ionization Effect (%)	Duration (min)	Trichloroacetlic Acid Ionization Effect	Duration (min)	m-Phosphoric Acid Ionization Effect (%)	Duration (min)
50:50 methanol:water	76.0	0.4-7.0	85.8	0.4-7.7	86.3	0.4-7.7	-3.0	0.5-1.0	0.3	0.5-1.0
50:50 acetonitrile:water	68.2	0.5-6.4	92.0	0.5-2.6	85.9	0.5-2.6	-9.7	0.5-1.0	-8.1	2.0-5.0
70:30 methanol:water	83.2	0.5-5.2	90.9	0.5-5.2	93.0	0.5-5.2	-2.0	0.5-1.5	-8.3	0.5-1.5
50:50 methanol:0.1% formic acid	-2.9	0.5-2.0	-4.0	0.5-2.5	-2.1	0.5-2.7	13.6	0.5-2	1.0[a]	0.5-1.3
50:50 methanol: 10mM ammonium formate	-194.6[b]	0.5-9.9	52.3	0.5-9.9	-20.2	0.5-9.9	14.7	0.5-9.9	-19.6	0.5-9.9

Analyte retention times of 8.0, 0.7, 0.4, 12.0, and 17.6 min using 50:50 methanol:water, 50:50 methanol:water, 70:30 methanol:water, 50:50 acetonitrile:water, 50:50 methanol:0.1% formic acid, and 50:50 methanol: 10 mM ammonium formate mobile phases, respectively.

[a] Measured for 1.3 min only.
[b] Measured following interface cleaning between blank and human precipitated plasma samples.

Source: Polson, C. et al., . *Chromatogr. B*, 2003, 785, 263. With permission from Elsevier.

TABLE 1.17

Relative Efficacies of Protein Precipitation Methods

Precipitant	Supernatant pH[a]	Plasma Protein Precipitated (%) Volume Precipitant Added/Volume Plasma								
		0.2	0.4	0.6	0.8	1.0	1.5	2.0	3.0	4.0
10% (w/v) Trichloroacetic acid	1.4–2.0	99.7	99.3	99.6	99.5	99.5	99.7	99.8	99.8	99.8
6% (w/v) Perchloric acid	<1.5	35.4	98.3	98.9	99.1	99.1	99.2	99.1	99.1	99.0
Sodium tungstate dihydrate (10% w/v) in 0.3 mol L^{-1} sulfuric acid	2.2–3.9	3.3	35.4	98.6	99.7	99.7	99.9	99.8	99.9	100
5% (w/v) Metaphosphoric acid	1.6–2.7	39.8	95.7	98.1	98.3	98.3	98.5	98.4	98.2	98.1
Copper(II) sulfate pentahydrate (5% w/v) + sodium tungstate dihydrate (6% w/v)	5.7–7.3	36.5	56.1	78.1	87.1	97.5	99.8	99.9	100	100
Zinc sulfate heptahydrate (10% w/v) in 0.5 mol L^{-1} sodium hydroxide	6.5–7.5	41.1	91.5	93.0	92.7	94.2	97.1	99.3	98.8	99.6
Zinc sulfate heptahydrate (5% w/v) in 0.2 mol L^{-1} barium hydroxide	6.6–8.3	45.6	80.7	93.5	89.2	93.3	97.0	99.3	99.6	99.8
Acetonitrile	8.5–9.5	13.4	14.8	45.8	88.1	97.2	99.4	99.7	99.8	99.8
Acetone	9–10	1.5	7.4	33.6	71.0	96.2	99.1	99.4	99.2	99.1
Ethanol	9–10	10.1	11.2	41.7	74.8	91.4	96.3	98.3	99.1	99.3
Methanol	8.5–9.5	17.6	17.4	32.2	49.3	73.4	97.9	98.7	98.9	99.2
Saturated ammonium sulfate	7.0–7.7	21.3	24.0	41.0	47.4	53.4	73.2	98.3	—[b]	—[b]

[a] 0.4 volumes of precipitant and above.

[b] Too cloudy to assay.

Source: Flanagan, R.J. et al., *Biomed. Chromatogr.,* 2006, 20, 530. With permission from John Wiley & Sons.

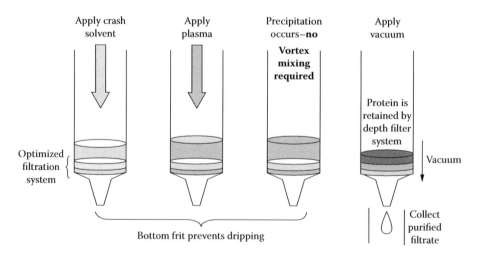

FIGURE 1.39 Solvent-first procedure using ISOLUTE PPT⁺ Protein Precipitation Plates.[159] (Reproduced with permission from Biotage AB.)

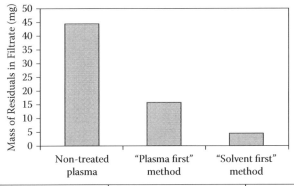

Sample	Mass of residuals in filtrate	% Reduction
Non-treated plasma	44.0 mg	-
'Plasma first' method	15.7 mg	64%
'Solvent first' method	4.40 mg	90%

FIGURE 1.40 Comparison of results from plasma-first and solvent-first protocols.[159] (Reproduced with permission from Biotage AB.)

demonstrated in Figure 1.39 and the results obtained in Figure 1.40. Results similar to the above were reported by Berna and coworkers[160] using Captiva 20-μm polypropylene 96-well plates with top and bottom Duo-seals (see Figure 1.41).

These authors reported greatly improved sample transfer without pipette failure due to plugging caused by thrombin clot formation when a LEAP HTS PAL autosampler was used for liquid transfer automation.

Remedying the problem of clot formation and its detrimental effect on sample transfer has also been addressed in a recent publication from the Johnson & Johnson Pharmaceutical Research and Development Laboratories[161] where workers designed a novel 96-well screen filter plate consisting of 96 stainless steel wire-mesh screen tubes (Figure 1.42).

The advantages of this screen filter, as cited by the authors, consist of its reusability, its standard 96-well format size, and its ability to be used whenever sample transfer or pipetting is needed. After usage, the screen filter can be easily cleaned by rinsing with water and methanol and additionally, by ultrasonication in water or methanol or other appropriate solution. The filter can be inserted into a plasma storage plate before sample transfer by the Tomtec Quadra used by the authors for automation.

FIGURE 1.41 Captiva 20-μm polypropylene 96-well filter plate with top and bottom duo-seals.[160] (Reproduced with permission from the American Chemical Society.)

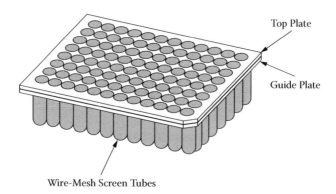

FIGURE 1.42 Example of 96-well screen filter plate design.[161] (Reproduced with permission from the American Chemical Society.)

1.2.5.1 Use of Protein Precipitation in Tandem with Other Sample Preparation Techniques

The principle of using protein precipitation in combination with solid phase extraction is best illustrated through sample pretreatment of a vitamin D metabolite, 25-hydroxyvitamin D, from plasma, as reported recently by Lensmeyer and coworkers.[162] Analysis of vitamin D and its metabolites presents a unique challenge because these highly lipophilic compounds strongly associate with vitamin D binding protein (VDBP) that must be broken for releasing the metabolites for efficient SPE or LLE. At the same time, endogenous lipids readily coextract with the metabolites and produce very dirty extracts that can foul analytical LC/MS instrumentation. For this reason, it is rather difficult to employ LLE.

Furthermore, the compounds are light-sensitive and degrade rapidly. The IS is unstable to heat and temperatures in excess of 35°C must be avoided to prevent evaporation and decomposition of the analytes and the IS. The procedure employed by the authors consists of first precipitating the proteins from serum (blood) samples in disposable glass tubes with acetonitrile (2 mL of precipitant spiked with IS [400 μg/mL in acetonitrile] added to 1.0 mL of serum sample). The mixture is vortex-mixed after 5 min of standing. The resulting mixture is centrifuged and the clean supernatant transferred to a glass vial from which the test solution is transferred to a Gilson ASPEC XL4 autosampler. Four zones (sample, reagent, result, and disposable extraction column [DEC]) are marked in the sample racks. The solvent evaporator (Turbo Vap LV) is set to 35°C, nitrogen flow adjusted to 10 psi, and a typical drying time of 25 min is used. The conditions for automated extraction are shown in Table 1.18. Samples were detected by a UV 3000 integrated system set at 275 nm and a stable bond cyano column (250 × 4.6 mm) was used for HPLC. Representative chromatograms and a comparison of LC/UV and LC/MS are shown in Figures 1.43 and 1.44, respectively.

More information on the comparative evaluation of protein precipitation methods may be obtained from Lei and coworkers.[163] An interesting comparison of protein precipitation (PPT) and solid phase extraction (SPE) methods was presented in a technical library publication from Millipore[164] that describes use of its Multi-SPE-MPC extraction plate and MultiScreen deep well Solvinert filter plate for SPE and PPT, respectively (Figure 1.45). A Biohit Proline multichannel pipette was used to add 400 μL of acetonitrile to each well of the filter plate and then, using the pipette's double aspiration program, 100 μL of spiked serum was aspirated and 100 μL of acetonitrile from the filter plate was aspirated to initiate protein precipitation in the pipette tip. The mixture was deposited back in the filter plate and shaken vigorously for 2 min.

Vacuum filtration at 18 to 20 in. Hg was performed on a MultiScreen HTS vacuum manifold with a deep well collar. For the SPE, the extraction plate was conditioned first with 100 μL of methanol

TABLE 1.18

Conditions for Automated Solid Phase Extraction of Vitamin D Metabolite from Plasma after Protein Precipitation

Program Command	Solvent	Volume (mL)	Solvent Conditions		
			Aspirate (mL/min)	Dispense (mL/min)	Equilibrate (min)
1. Begin loop	NA[a]	NA	NA	NA	NA
2. Rinse needle (inside/outside)	CH_3CN	3.0	20	120	0
3. Condition DEC	CH_3CN	2.0	30	5	0.05
4. Add to DEC	$CH_3CN–H_2O$ (35:65 by volume)	2.0	40	5	0.1
5. Dispense into sample	H_2O	1.0	NA	60	0
6. Load DEC	Diluted serum supernatant	3.5	40	2.5	0.4
7. Rinse needle (inside and outside)	H_2O	4.0	NA	120	0
8. Add to DEC	$CH_3CN–H_2O$ (35:65 by volume)	2.0	40	10	0.1
9. Elute/collect	CH_3CN	2.0	6	3	0.1
10. Rinse needle	Water	4.0	NA	120	0
11. End loop	NA	NA	NA	NA	NA

[a] NA = not applicable.

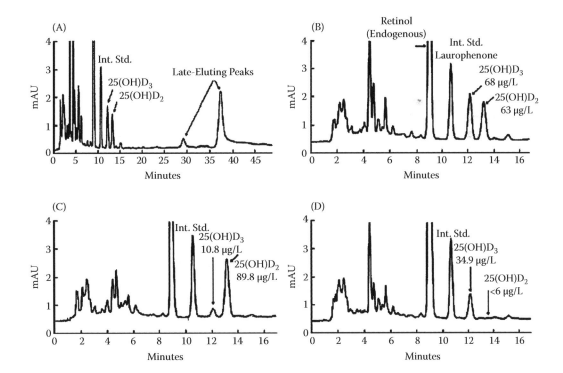

FIGURE 1.43 Representative HPLC chromatograms of vitamin D metabolites.[162] (A) late-eluting peaks; (B) calibrator in extracted serum; (C) sample from patient with low 25(OH)D3 treated with vitamin D2; (D) sample from patient with high concentrations of 25(OH)D3. Int. Std. = internal standard; mAU = milliabsorbance units. (Reproduced with permission from the American Association for Clinical Chemistry.)

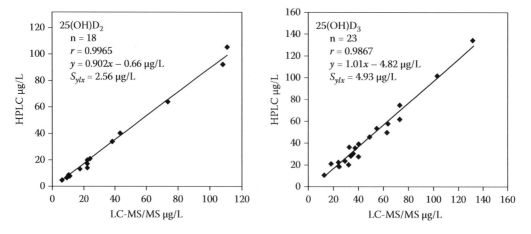

FIGURE 1.44 Comparison of LC/UV method developed with LC/MS/MS results.[162] (Reproduced with permission from the American Association of Clinical Chemistry.)

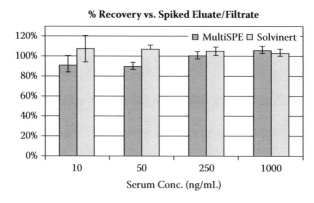

	LOD		LOQ	
	SPE	Protein Crash	SPE	Protein Crash
Starting Serum (ng/mL)	0.8	2	3	7
Analyzed Sample (ng/mL)	0.1	0.1	0.3	0.3
Instrument Control (ng/mL)	0.1	0.1	0.4	0.3

		Serum Conc. (ng/mL)	1	2.5	10	50	250	1000
Accuracy (% nominal)	MultiSPE		<LOQ	<LOQ	89%	95%	101%	106%
	MultiScreen Solvinert		<LOD	<LOQ	104%	115%	105%	103%
Precision (RSD %)	MultiSPE		<LOQ	<LOQ	7%	3%	3%	4%
	MultiScreen Solvinert		<LOD	<LOQ	12%	5%	4%	4%

FIGURE 1.45 Recovery, LOQ, LOD, and precision and accuracy data from Millipore SPE versus PPT comparison experiments.[164] (Reproduced with permission from Millipore.)

for 30 sec and then with 200 μL of water from a Milli-Q Gradient A-10 system. The plate was then subjected to vacuum at 5 to 10 in. Hg. Formic acid (0.001%) in Milli-Q water (200 μL) was added to each well and vacuum filtered to charge the mixed phase cation exchange resin in the well plate. The solution (100 μL) of serum sample was diluted with 900 μL of 0.1% phosphoric acid in each well and mixed by pipetting. After vacuuming at 5 in. Hg, three wash steps were completed with 500 μL of each solution: 0.1% phosphoric acid, 100% methanol, and methanol/ammonia (10%) in water. Elution was performed with methanol containing 2% ammonium hydroxide (100 μL). The elution step was repeated to reach a final elution volume of 200 μL. Analyses were performed with a Sciex API 2000 mass spectrometer coupled to an Agilent 1100 HPLC instrument carrying a Synergi Hydro-RP (4 μm, 50 × 2 mm) with 100% aqueous formic acid (0.1%) for 2 min, then to 20% methanol/80% aqueous formic acid at 3 min, and back to 100% aqueous formic acid at 5 min. The flow rate was 300 μL/min. Comparison of the recoveries, LOQ, LOD, precision, and accuracy of SPE and PPT are shown in Figure 1.45.

The study concluded that: Once wash steps are optimized, samples prepared by solid phase extraction are cleaner than those prepared by protein precipitation. Samples prepared by extraction with a Multi-SPE plate resulted in lower LOQs than samples prepared by solvent precipitation. Drug recoveries were acceptable (>80%) for both the SPE and the solvent precipitation methods. Well-to-well reproducibility of samples was slightly better with extraction with a Multi-SPE plate. Evaporation and reconstitution, while more time-consuming, yield better chromatographic performance, allow analysis of lower concentration samples, and require optimization for good analyte recovery.

1.3   OTHER SAMPLE PREPARATION TECHNOLOGIES: LATEST TRENDS

Among the techniques listed in Section 1.2.1, the two most documented approaches in addition to SPE, LLE, and PPT are solid phase microextraction (SPME) and affinity capture of analytes based on molecularly imprinted polymers (MIPs). Recent developments in these areas are briefly discussed below.

1.3.1   SOLID PHASE MICROEXTRACTION (SPME) AS SAMPLE PREPARATION TECHNIQUE

SPME is a solvent-free extraction technique invented by Pawliszyn and coworkers[165] in 1990. It allows simultaneous extraction and preconcentration of analytes from gas, liquid, or solid samples. It was originally called fiber SPME because it utilized a fused silica fiber coated with an appropriate stationary phase. The device consists of a fiber holder and assembly with a needle that contains the fiber and resembles a modified syringe. The fiber holder includes a plunger, stainless steel barrel, and adjustable depth gauge with a needle. The fused silica fiber is coated with a thin film of a polymeric stationary phase that acts like a sponge, concentrating organic analytes by absorption or adsorption from a sample matrix. The two types of fiber SPME are headspace (HS-SPME) and direct immersion (DI-SPME).

A newer addition is in-tube SPME that makes use of an open capillary device and can be coupled online with GC, HPLC, or LC/MS. All these techniques and their utilization in pharmaceutical and biomedical analysis were recently reviewed by Kataoka.[45] Available liquid stationary fiber coatings for SPME include polydimethylsiloxane (PDMS) and polyacrylate (PA) for extracting nonpolar and polar compounds, respectively. Also in use for semipolar compounds are the co-polymeric PDMS-DVB, Carboxen (CB)-PDMS, Carbowax (CW)-DVB, and Carbowax-templated resin (CW-TPR). A few examples of in-tube SPME extractions from biological matrices are shown in Table 1.19 and drawn from Li and coworkers.[166]

Additional examples may be found in Table 1.20, based on work of Queiraz and Lancas.[167] The effects of fiber chemistry, ionic strength, matrix pH, extraction time, organic additives, temperature, agitation, and derivatization along with the influence of plasma proteins on SPME were reported.[167] Extraction time, pH, salt concentration in sample, and temperature data are presented in Figure 1.46.

TABLE 1.19

In-Tube SPME Applications

Analyte	Sample Matrix	Stationary Phase	Technique	Detection Limit (ng/mL)
Amphetamine	Urine	Omega wax	LC-ESI-MS	038–0.82
Caffeine	Tea	Polypyrrole (PPY)	LC-ESI-MS	0.01
Verapamil	Plasma, urine	Polypyrrole (PPY)	LC-UV	52
			LC-MS	5
Benzodiazepines	Serum	Alkyl-diol	LC-UV	22.29
NSAID	Urine	β-cyclodextrin	LC-UV	18–38
Ranitidine	Urine	Omega wax	LC-ESI-MS	1.4
Ketamine	Urine	Poly(methacrylic acid-ethylene glycol dimethacrylate)	LC-UV	6.4

Source: Li, K.M. et al., *Curr. Pharm. Anal.*, 2006, 2, 95. With permission from Bentham.

Plasma proteins decreased extraction recoveries from SPME by irreversible adsorption onto the fiber. PPT prior to SPME by addition of acid or methanol was used to overcome this problem. SPME sensitivity may also be improved by dilution of plasma samples with buffer or water.

Lord and coworkers[168] illustrated the application of SPME for *in vivo* monitoring of circulating blood concentrations of benzodiazepines in three living male beagles. SPME probes were placed in the cephalic veins of the lower front legs of the animals to accommodate catheters and SPME probes. A polypyrrole-coated stainless steel wire of 0.005 in. thickness was used as the SPME fiber and the coating was achieved by anodic oxidation of pyrrole in lithium perchlorate aqueous medium. *In vitro* extractions of blood samples were performed under equilibrium conditions with static (unstirred) samples and the extraction of benzodiazepines was found to be maximal at 30 min. It was presumed that *in vivo* experiments with flowing intravenous blood should take much less time for extraction.

For optimization, initial *in vitro* experiments were performed on a commercial Carbowax templated resin SPME assembly with samples in different volumes of PBS solution in 96-well microplates. No differences were found between 3.0- and 1.5-mL samples. Concentrations of 10 to 500 ng/mL in PBS, 50 to 5000 ng/mL in dog plasma, and 1 to 1000 ng/mL in whole blood were used for calibration extractions. The LC interface design for probe desorption from the fiber is shown in Figure 1.47.

A six-port valve was used in both manual and semi-automated SPME interfaces and PEEK tubing used to connect the HPLC system to the SPME probe. A Cohesive HTLC 2300 with dual pumps along with a Sciex API 3000 mass spectrometer was used for LC/MS/MS and a Symmetry Shield RP-18 (5 μ, 50 × 2.1 mm) for HPLC. A quaternary pump with flow switching was used for desorption chamber flushing along with MS make-up flow and a binary pump for LC/MS/MS. Acetonitrile/0.1% acetic acid in water (90:10, solvent B) and 10:90 acetonitrile/0.1% aqueous acetic acid (solvent A) were used, with 10% B for 0.5 min ramped to 90% B in 2 min and held at this concentration for 1.5 min before returning to 10% B for 1 min at a flow rate of 0.5 mL/min.

For plasma and blood experiments, LC effluent was directed to waste for the first 1 min. Conventional blood analysis by drawing 1 mL samples from the saphenous catheter was used to validate SPME results. These samples were subjected to PPT with acetonitrile and the supernatant from centrifugation was analyzed. The SPME probes were also evaluated for pharmacokinetic analysis of diazepam and its metabolites, oxazepam and nordiazepam. Good correlation was obtained for conventional blood drawn from saphenous and cephalic sites of the animals, as shown in Figure 1.48. Although the analytical parameters for the automated study need improvement, the authors cite the study as a first demonstration of SPME technology for *in vivo* analysis.

TABLE 1.20
Applications of SPME to Plasma Samples

Analyte	Extraction Mode Fiber Coating (thickness, mm)	Analytical System (LOQ or LOD)	Remarks	References
Valproic acid	Direct immersion PDMS (100)	GC–FID (LOD:1 mg/mL)	Equilibrium dialysis followed SPME	Krogh et al., 1995 (7)
Aniline, phenols, nitrobenzenes	Direct immersion PA (85)	GC–MS	Protein binding study, determination of free concentrations	Vaes et al., 1996 (8)
Antidepressants	Direct immersion PDMS (100)	GC–NPD, GC-MS (LOQ:90–200 ng/mL)	Theoretical model for influence of proteins	Ulrich and Martens, 1997 (9)
Diazepam	Direct immersion PA (85) PDMS (7, 100)	GC–FID (LOQ:0.25 nmol/mL)	1-Octanol-modified PA fiber, pretreated plasma (TCA)	Krogh et al., 1997 (10)
Benzodiazepines	Direct immersion PA (85)	GC–FID (LOQ: 0.01–0.48 mmol/mL)	1-Octanol-modified PA fiber, pretreated plasma (TCA)	Reubsa et al., 1998 (11)
Clozapine	Direct immersion PDMS (100)	GC–NPD (LOD: 30 ng/mL)	Influence of proteins and triglycerides	Ulrich et al., 1999 (12)
Lidocaine and 3 metabolites	Direct immersion CW-DVB (65) PA (85) PDMS (100)	GC–NPD (LOQ: 8–21 ng/mL)	Effect of different fiber coating	Abdel-Rehim et al., 2000 (13)
Lidocaine	Direct immersion PDMS (100)	GC–FID (LOD: 5 ng/mL)	Analysis of free, protein-bound, and total amount of lidocaine in human plasma	Koster et al., 2000 (14)
Anesthetics	Direct immersion CW-DVB (65) PA (85) PDMS (100)	GC–NPD (LOQ:0.5 mmol/mL)	Study of protein-binding ultra filtrate plasma	Abdel-Rehim et al., 2000 (15)
GABA	Derivatization headspace	GC-PICI-MS (LOQ: 1 mg/mL)	Conversion of gamma-hydroxybutyric to gamma-butyrolactone	Frison et al., 2000 (16)
Methadone and main metabolite	Direct immersion PDMS (100)	GC–MS (LOD: 40 ng/mL)	Application to methadone -treated patients	Bermejo et al., 2000 (17)
Levomepromazine	Direct immersion PDMS (100)	GC-NPD (LOQ: 5 ng/mL)	Application to therapeutic drug monitoring	Kruggel and Ulrich, 2000 (18)
Midazolam	PA (85)	GC–MS (SIM) (LOD: 1.0 ng/mL)	Application to therapeutic drug monitoring	Frison et al., 2001 (19)

(Continued)

TABLE 1.20 (CONTINUED)
Applications of SPME to Plasma Samples

Analyte	Extraction Mode Fiber Coating (thickness, mm)	Analytical System (LOQ or LOD)	Remarks	References
Anticonvulsants	Direct immersion CW-TPR (50)	LC–UV (LOQ:0.05–1.0 mg/mL)	Off-line desorption	Queiroz et al., 2002 (20)
Anticonvulsants	Direct immersion CW-DVB (65)	GC–TSD (LOQ:0.05–0.2 mg/mL)	Application to therapeutic monitoring	Queiroz et al., 2002 (21)
Thymol	Headspace PDMS-DVB (65)	GC–FDI (LOQ 8.1 ng/mL)	Enzymatic cleavage of thymol sulfate	Kohlert et al., 2002 (22)
Sulfentanil	Direct immersion PDMS-DVB (65)	GC–MS (LOQ: 6.0 ng/mL)	Influence of pH and ionic strength	Paradis et al., 2002 (23)
Amitraz	Direct immersion PDMS (100)	GC-TSD (LOQ: 20 ng/mL)	Application to toxicity studies in dogs	Queiroz et al., 2003 (24)
Busulphan	Direct immersion CW-DVB (65)	GC–MS (LOQ: 20 ng/mL)	In-vial derivatization	Abdel-Rehim et al., 2003 (25)

PDMS = polydimethylsiloxane. PA = polyacrylate. CW = Carbowax. DVB = divinylbenzene. FID = flame ionization detection. NPD = nitrogen-phosphorus detection. TSD = thermionic-specific detection. LOQ = limit of quantitation. LOD = limit of detection. TCA = trichloroacetic acid. PICI-MS = positive ion chemical mass spectrometry. SIM = selected ion monitoring.

Source: Queiroz, M.E.C. and F.M. Lancas, *LC-GC North America,* 2004, 22, 970. With permission from the authors.

1.3.2 SAMPLE CLEAN-UP THROUGH AFFINITY PURIFICATION EMPLOYING MOLECULARLY IMPRINTED POLYMERS

This technique adopts the molecular recognition principles of natural receptors such as antibodies and enzymes. A molecularly imprinted polymer (MIP) carrying specific recognition sites on its backbone through the use of templates or imprint molecules is utilized to trap an analyte of interest. Molecular imprinting is achieved through co-polymerization of functional and cross linking monomers in the presence of a target molecule that acts as a molecular template.

The functional monomers initially form a complex with the imprint molecule. After polymerization, these functional groups are held in position by the highly cross linked polymeric structure. Subsequent removal of the imprint molecule yields binding sites that are complementary in size and shape to the desired target or analyte (see Figure 1.49). Thus, a molecular memory introduced into the polymer can be used to selectively rebind the target molecule. The interactions of the target molecule with the functionalized monomers may lead to a reversible covalent bond or may be held by the monomers through polar interactions such as hydrogen or ionic bonds or hydrophobic interactions (van der Waals forces).

A recent review by Pichon and Haupt[169] summarizes the progress in the area of utilization of MIPs for sample preparation purposes and cites several examples of solid phase extraction (MISPE) from biological matrices. The requirements and applications of MIPs are reviewed in a recent book[170] and other literature.[171–176]

This section will illustrate the MIP technique for sample preparation by presenting examples of diazepam and its metabolites in hair samples.[177] An anti-diazepam molecularly imprinted polymer

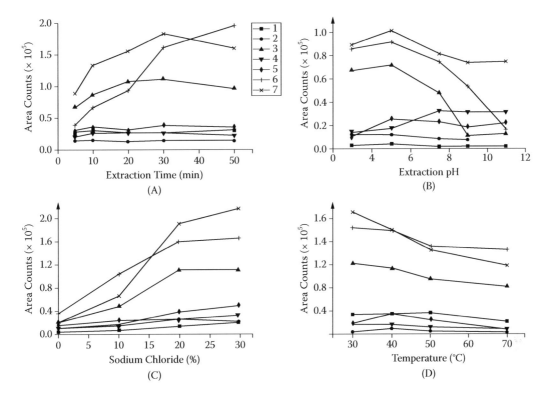

FIGURE 1.46 Effects of major experimental variables: (A) extraction time, (B) pH, (C) sodium chloride concentration, and (D) temperature on the efficiency of direct SPME of anticonvulsants in plasma sample.[167] (Reproduced with permission from the authors.)

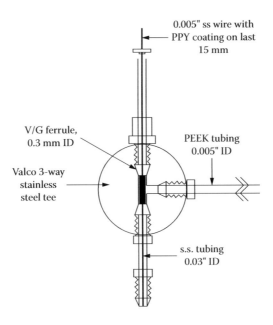

FIGURE 1.47 Interface used for SPME probe desorption and transfer of analytes to LC/MS/MS for quantification.[168] (Reproduced with permission from the American Chemical Society and the authors.)

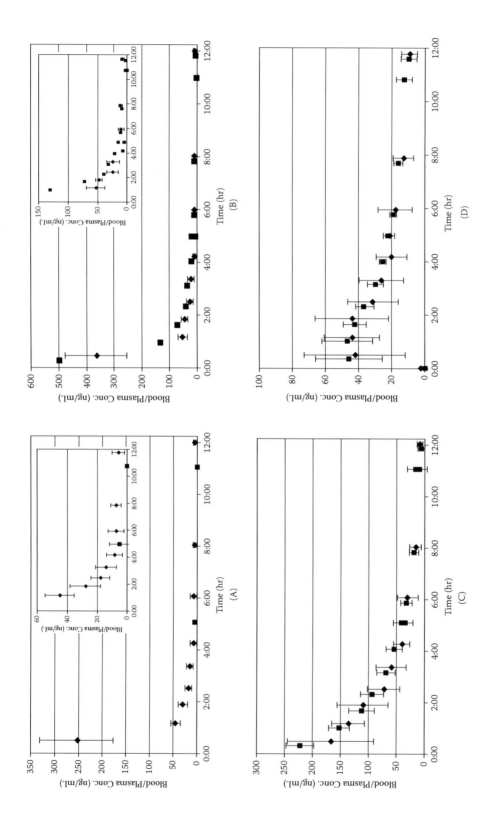

FIGURE 1.48 (A) Averaged diazepam profiles from pharmacokinetic studies. (B) Diazepam profile for one study with one dog, both catheters cephalic. (C) Averaged nordiazepam profiles from pharmacokinetic studies. (D) Averaged oxazepam profiles from pharmacokinetic studies.[168] (Reproduced with permission from the American Chemical Society and the authors.)

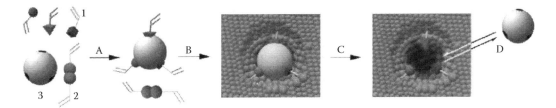

FIGURE 1.49 Principle of molecular imprinting.[169] 1 = functional monomers; 2 = cross-linking monomer; 3 = molecule whose imprint is desired (molecular template). In (A), 1 and 2 form a complex with 3 and hold it in position; in (B), polymerization involving 1 & 2 occurs and the template (imprint molecule) is held in the polymeric structure; in (C) and (D) the imprint molecule is removed leaving a cavity complementary to its size and shape into which a target analyte of similar dimensions can fit. (Reproduced with permission from Taylor & Francis.)

(MIP) was synthesized by dissolving diazepam, inhibitor-free methacrylic acid, and ethylene glycol dimethacrylate in ethanol-free chloroform and adding the initiator 2,2′-azobisisobutyronitrile, sparging the solution with oxygen-free nitrogen while cooling in an ice bath. The container (tube) was sealed and thermostatted at 4°C to facilitate template–monomer complex formation and then irradiated with an ultraviolet lamp for 24 hr. The polymer monolith obtained was transferred to a water bath set at 60°C for 24 hr to complete the cure of the polymer.

A nonimprinted polymer (NIP) was prepared in the same manner but without the diazepam template molecule. The polymers were crushed, mechanically ground, and wet-sieved to generate an MIP of 25- to 38-μm particle size. The template was extracted by extensive washing with a mixture of methanol and acetic acid (9:1) for 24 hr. The polymer particles were dried *in vacuo* at 60°C prior to use. Cartridges (1 mL, polypropylene) were packed with these MIP and NIP polymeric materials separately. Any bleeding of the template was checked by washing with 1 mL acetonitrile containing 5 to 25% acetic acid, starting with the lowest percentage. LC/MS/MS analysis showed that the last traces of diazepam template were removed in the washing step using 10% acetic acid. In all subsequent experiments, washing was done with 15% acetic acid containing acetonitrile to ensure complete removal of the template.

The solvent used for binding diazepam to the templated MIP plays an important role in the recognition step and hence three different solvents were investigated in this study.[177] The results are presented in Table 1.21. Toluene showed the best binding for diazepam on the MIP.

A number of benzodiazepine derivatives were evaluated to probe the selective binding of diazepam to the templated MIP. Samples of 20 and 50 ng of each drug were prepared in 1 mL of toluene.

TABLE 1.21
Role of Solvent in Binding Diazepam to Templated MIP

Solvent	Dielectric Constant, ε	MIP	NIP
		Percent Bound	
Toluene	2.4	94.0	22.0
Chloroform	4.8	72.8	29.5
Dichloromethane	9.1	70.6	48.0

Source: Ariffin, M.A. et al., *Anal. Chem.*, 2007, 79, 256. With permission from the American Chemical Society and the authors.

TABLE 1.22

Recoveries of Benzodiazepines Bound to MIP, Limit of Detection (LOD) and Limit of Quantitation (LOQ)

Analyte	Mean Percent Recovery (n = 5; 50 ng Added)	
	MIP	NIP
7-Aminoflunitrazepam	91.9 (1.5)	71.0 (7.5)
Oxazepam	73.4 (8.7)	41.4 (34.3)
Lorazepam	97.2 (17.0)	48.1 (22.0)
Chlordiazepoxide	61.6 (13.5)	32.0 (86.8)
Temazepam	89.6 (13.4)	54.3 (52.9)
Flunitrazepam	39.0 (5.1)	2.3 (67.8)
Nordiazepam	102.9 (9.8)	65.0 (14.3)
Nitrazepam	92.3 (5.4)	60.6 (33.9)
Dazepam	93.0 (1.5)	16.3 (17.1)
Analyte	**LOD (ng/30 mg)**	**LOQ (ng/30 mg)**
7-Aminoflunitrazepam	0.03	0.06
Oxazepam	0.13	0.21
Lorazepam	0.66	1.11
Chlordiazepoxide	0.33	0.57
Temazepam	0.39	0.63
Flunitrazepam	0.78	1.32
Nordiazepam	0.21	0.33
Nitrazepam	0.06	0.11
Diazepam	0.09	0.14

Source: Ariffin, M.A. et al., *Anal. Chem.*, 2007, 79, 256. With permission from the American Chemical Society and the authors.

The cartridge was preconditioned with 0.5 mL toluene and each of the above benzodiazepine solutions passed through it. Analytes retained on the MIP were eluted with 0.5 mL of 15% acetic acid in acetonitrile. Internal standard (corresponding deuterated benzodiazepine) was added and subjected to LC/MS/MS. The results obtained for recovery, limit of detection (LOD), and quantitation (LOQ) are shown in Table 1.22. The binding capacity of diazepam to the templated MIP was found to be 110 ng/mg polymer. The same results were obtained for postmortem hair samples.

Diazepam and its nordiazepam, oxazepam, and temazepam metabolites are well retained by the MIP, while they are much less retained on NIP, also exhibiting large RSD. Other benzodiazepines of similar structures (Figure 1.50) were well retained on the MIP, showing that this template can be used for the general class of benzodiazepines. Two benzodiazepines studied, chlordiazepoxide and flunitrazepam, were poorly retained, indicating poor fit of these structures into the templated MIP.

1.4 PURIFICATION OF SYNTHETIC COMBINATORIAL LIBRARIES

Yan et al.[178] stated in a 2004 article:

> One of the driving forces to apply combinatorial chemistry in drug discovery is to accelerate lead discovery and preclinical research in order to find the next drug. It is important that these combinatorial library compounds are as pure as possible when performing lead discovery screening. At this stage,

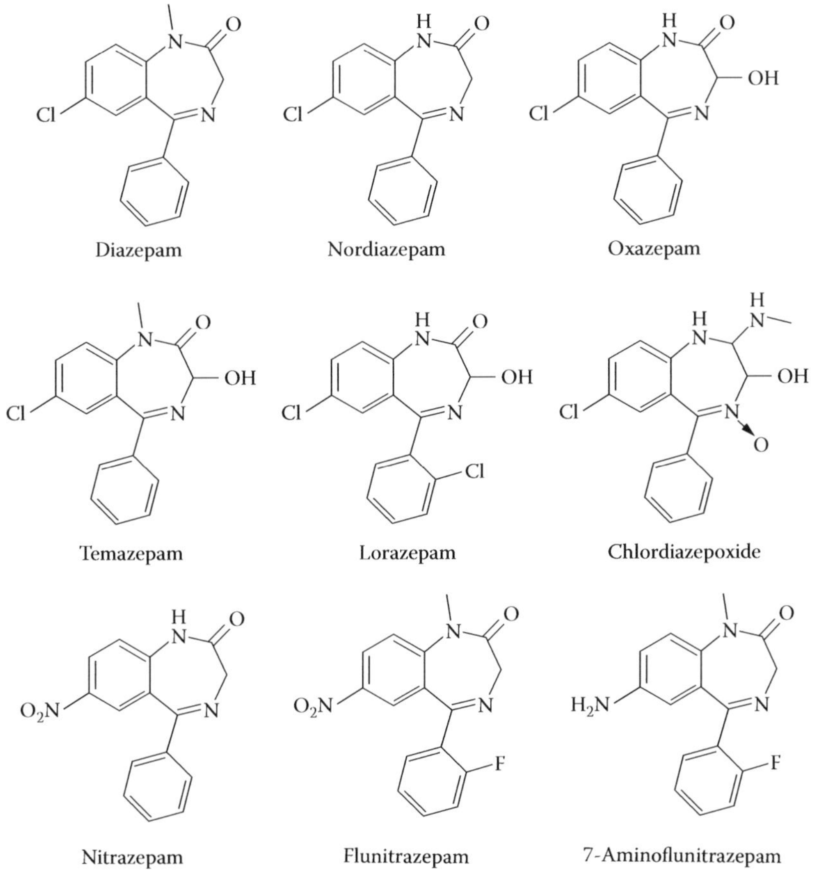

FIGURE 1.50 Structures of benzodiazepines.[177] (Reproduced with permission from the American Chemical Society and the authors.)

any impurities in samples may lead to false positive results. Even with the rapid advances in solid phase and solution phase synthesis methods and intensive reaction optimization, excess reagents, starting materials, synthetic intermediates and byproducts are often found along with the desired product. Furthermore, strong solvents for swelling the resin bead used for solid-phase synthesis or scavenging treatment in solution-phase reactions can often bring in additional impurities extracted from resins and plastic plates. The requirement for the *absolute purity* of combinatorial library compounds demands the development of high-throughput purification methods at a scale that matches combinatorial or parallel synthesis. An HTP method for purifying combinatorial libraries must possess three qualities: high throughput, full automation and low cost.

1.4.1 HPLC-Based High-Throughput Separation and Purification of Combinatorial Libraries

The two major approaches for HPLC purification are fast gradient separation and parallel purification. Yan et al.[178] utilized the former (Figure 1.51). The purification lab received a 96-well plate containing synthesized products at 0.1 to 0.2 mmol/well. A Hydra 96-probe liquid handler prepared QC plates for all samples that were analyzed with a MUX-LCT eight-channel parallel LCMS instrument at a throughput of 2000 samples/day. Only samples with purities above 10% were purified on a

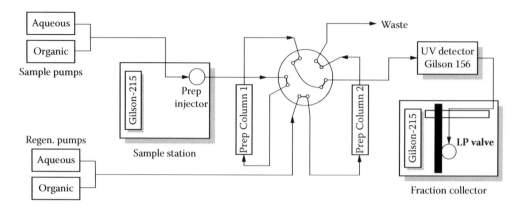

FIGURE 1.51 Preparative HPLC system.[178] (Reproduced with permission from the American Chemical Society.)

Gilson dual column preparative HPLC system. The retention time of each compound from the analytical LC/MS was used to calculate the specific gradient segment targeted to elute the compound about 2.3 min. A 1 min window was set around 2.3 min for fraction collection. The analytical peak heights were used to calculate the thresholds for collection triggers on the basis of UV or ELSD. All test tubes for fraction collection were preweighed automatically on Bohdan weighing stations. Purification of 96 samples took about 7 hr.

Post-purification QC was carried out again on another MUX-LCT eight-channel parallel LC/UV/MS system, after QC plates were made from collected tubes using a Tecan liquid handling system. The final purities and identities of all fractions were determined at a throughput of 2000 samples/day. Test tubes containing purified compounds were dried in a lyophilizer or centrifugal vacuum evaporation system and final tube weights were measured automatically on a Bohdan weighing station. On the basis of the final identity, purity and weight measurements of the purified compounds, equimolar solutions of purified compounds were made in the final plates. The whole high-throughput purification (HTP) process is summarized in Figure 1.52.

An accelerated retention window (ARW) starting from a certain proportion of organic ($\phi A\%$ acetonitrile) as the start of the gradient mobile phase composition was used to calculate preparative retention time on the basis of analytical retention time. The same procedure was used to adjust the retention times of all compounds to facilitate collection at the same predetermined retention time. This allowed set-up of a device for collecting any HPLC peaks that surpassed a certain threshold defined by UV or ELSD.

The MUX-LCT eight-channel parallel LC/UV/MS system consists of an autosampler with eight injection probes, two pumps for generating a binary gradient, eight UV detectors and an eight-way MUX with a TOF mass spectrometer. The LC/UV instrumentation consists of a Gilson pump system, autosampler, and eight UV detectors. The solvent delivered at 16 mL/min was split into eight LC columns (4.6×50 mm) and eight samples were injected simultaneously into eight columns separated by the same gradient and detected by individual UV detectors at 214 nm. Two mobile phases (A: 99% water, 1% acetonitrile, 0.05% TFA; B: 99% acetonitrile, 0.05% TFA) with a gradient time of 3 min (10 to 100% B) and a flow of 50 μL/min from each column post-UV was introduced into an eight-channel multiplexed electrospray ion source (MUX), while the remaining flow was directed to waste.

The mass spectrometer was a Micromass LCT orthogonal acceleration time-of-flight instrument equipped with an eight-channel MUX. The interface consisted of eight electrospray probes and a sampling aperture positioned co-axially to the sampling cone. Each of the probes within the MUX source was indexed using an optical position sensor and selected using a programmable stepper

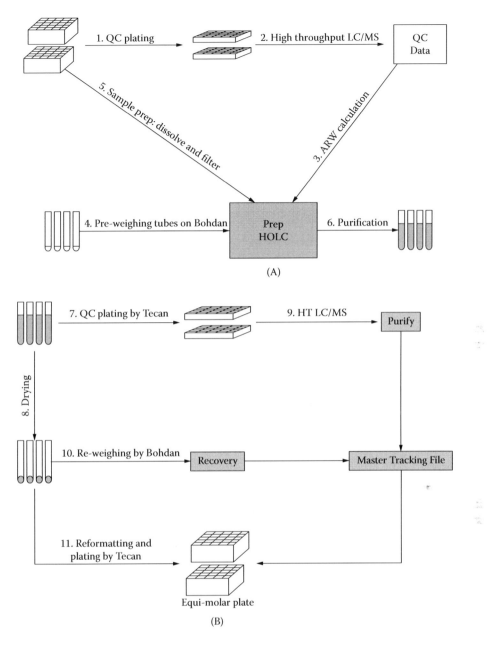

FIGURE 1.52 (A) Initial phase of HTP process. (B) Final phase of HTP process.[178] (Reproduced with permission from the American Chemical Society.)

motor controlled by MasLynx software. The position of the sampling aperture was controlled by the stepper motor that allowed ions from only one probe at a time to enter the sampling cone of the spectrometer. This arrangement made it possible to acquire discrete data files of electrospray ion current sampled from each channel. The eight-channel MUX-LCT worked like eight individual ESI-MS systems.

Preparative reversed-phase HPLC purification was carried out using Gilson liquid handlers and HPLC equipment controlled by Unipoint Version 3.2 software. Initial and final HPLC gradient conditions and the threshold setting for the fraction collector were set according to analytical LCMS data for each compound. Four pumps were used to control mobile phase flow through two 21.2×50 mm Hydro-RP columns (Phenomenex, Torrance, CA). Liquid streams were mixed in a Gilson 811C dynamic mixer and pressure spikes were moderated using Gilson 806 manometer modules. Sample injection and fraction collection were automated using two separate 215 liquid handlers. HPLC purification and post-purification processing were facilitated by software that tracked desired fractions and coordinated the consolidation, post-purification analysis, drying, and final plating. Additional examples of automated HPLC purification can be found in References 179 through 182.

1.4.2 SCAVENGER-BASED PURIFICATION OF COMBINATORIAL LIBRARIES GENERATED BY SOLUTION PHASE SYNTHESIS

Combinatorial libraries can be synthesized by solid phase organic synthesis (SPOS) or solution phase reaction protocols. The advantages of SPOS include ease of purification (through simple washing and/or filtration) and the ability to use excess reagents to push the reaction to completion. However, SPOS also has certain disadvantages, such as additional steps to attach and then remove the linker from the solid support at the end, difficulty in carrying out multistep syntheses due to non-optimization of the chemistries employed, linker compatibility, points of substrate attachment of the linker to the solid support, and achieving efficient and complete reaction conversions.[183–185]

The major advantages of solution phase synthesis are the availability of a vast selection of well documented reactions and the lack of extra steps needed for SPOS. On the other hand, the major drawbacks are need for purification of the desired product from impurities, excess reagents, and by-products and its isolation. A key approach to alleviate these issues is the use of solid-supported reagents and scavengers that incorporate the advantages of SPOS, especially use of excess reagents and easy work-up. Sorbents and formats designed for solid phase extraction (SPE) have been found useful for purifications of crude products from parallel solution synthesis. On the other hand, a variety of specialty polymeric materials with specific functionalities, targeted for removal of specific classes of organic molecules, are commercially available. Several reviews and publications detail the preparation and applications of scavengers.[186–196] Table 1.23 lists examples of scavenging resins and their application areas. Examples of purification of solution phase products using scavengers and regular solid phase extraction sorbents are presented below.

Amino groups of amino acids and amino alcohols are typically protected as their tertiary butoxycarbonyl (tboc) derivatives. After reaction involving the other functionality, the tboc group, is cleaved with an excess of trifluoroacetic acid (Figure 1.53), the liberated free amine compound must be separated from this strongly acidic reagent. A strong cation exchange SPE sorbent can be employed to hold the amine on its surface while the trifluoroacetic acid is washed off. The amino compound can be subsequently recovered from the cation exchanger by eluting with methanol containing 5% ammonium hydroxide and evaporating off the eluting solvent (Figure 1.54).

Another example of the use of an SPE sorbent is for the purification of a quaternary amine from its mixture with primary, secondary, and tertiary amines employing strata-X-CW, a weak cation exchange polymeric sorbent with carboxylic functionalities. The principle of this clean-up consists of adsorbing all the four amines on the sorbent surface and after washing off all other impurities with methanol:water (1:1), selectively eluting the primary, secondary, and tertiary amines with methanol containing 5% ammonium hydroxide—a compound basic enough to disrupt the ion exchange interactions of these amines (that are neutralized) with the carboxylic acid moiety of the sorbent. The quaternary amine is retained on the sorbent under these conditions. Elution with a strong organic such as formic acid in methanol that protonates the carboxylic moieties and prevents ionization

TABLE 1.23
Scavenger Resins

PS-Benzaldehyde	Nucleophile	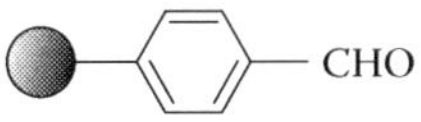	Scavenges nucleophiles, Scavenger including primary amines, hydrazines, reducing agents
PS-DEAM	Metal	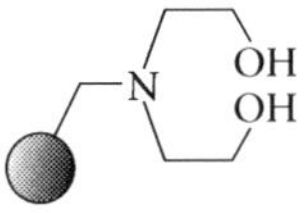	Scavenger for scavenger titanium (IV) chloride, titanium (IV) isoproposide, boronic acids
PS-Isocyanate	Nucleophile scavenger	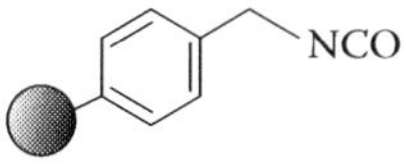	Scavenging nucleophiles, including amines and alkoxides
MP-Isocyanate	Nucleophile scavenger	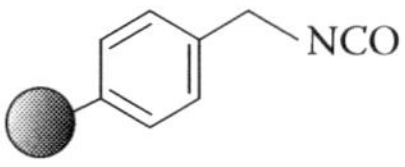	Scavenging nucleophiles, including amines and alkoxides
PS-NH2	Electrophile scavenger	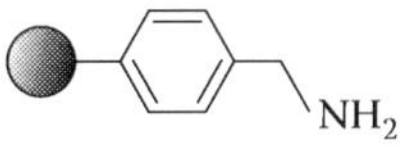	Scavenging acid chlorides, sulfonyl chlorides, isocyanates, and other electrophiles
PS-Thiophenol	Electrophile scavenger	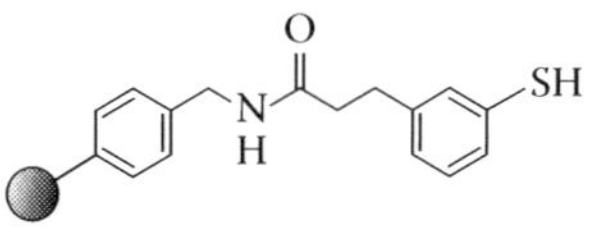	Scavenging alkylating agents
MP-TMT	Palladium scavenger	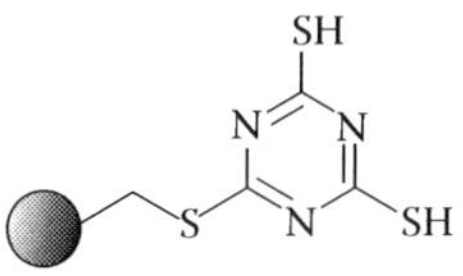	Scavenger for palladium
PS-Trisamine	Electrophile scavenger	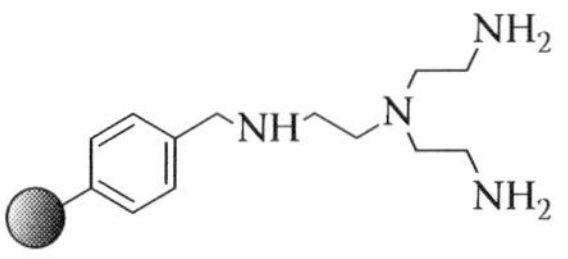	Scavenging acid chlorides, sulfonyl chlorides, isocyanates and other electrophiles

Source: Biotage AB. With permission.

allows recovery of the quaternary amine. Figure 1.55 depicts the structures of the amines used. Figure 1.56 contains chromatographic data, and Figure 1.57 illustrates the principle of scavenging.

Figure 1.58 shows scavenging of a catalyst via an organic synthetic reaction. The reductive amination of a ketone is catalyzed by titanium tetra-isopropoxide added in a molar excess. A scavenger,

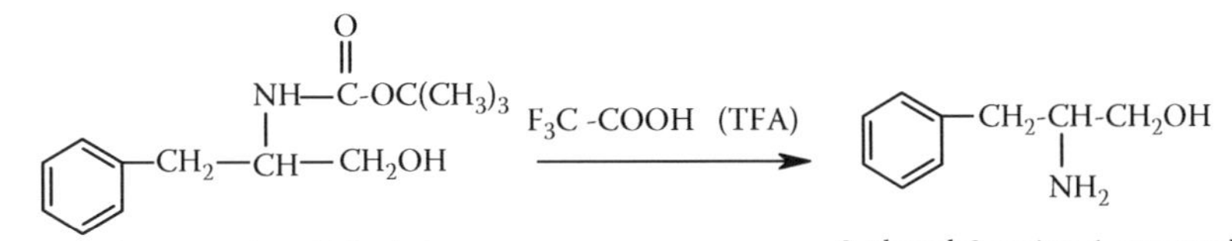

Tertiarybutoxycarbonyl (tboc) derivative
of 3-phenyl-2-amino-1-propanol

3-phenyl-2-amino-1-propanol

FIGURE 1.53 Cleavage of t-BOC derivative with trifluoroacetic acid.

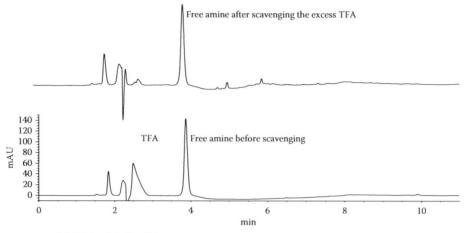

LC-UV Analysis Conditions:
Column: Gemini C18, 5u, 150 × 4.6 (150 × 2.0 for MS)
Mobile Phase: 20 mM KH$_2$PO$_4$, pH = 2.5 (0.1% formic acid/water, for LC-MS)/acetonitrile = 90/10
  (hold for 1 min) to 40/60 in 5 mins, hold for 5 mins.
Flow rate: 1.0 ml/min (0.2 mL/min for MS), UV detection: 210 nm; Injection vol: 50 uL (1.0 uL for MS) for
  reconstituted (from acidic synthesis) solution.

Analyte	Analyte conc. (ug/mL)	Log P	%Recovery (MeOH wash)	% Recovery	% RSD (n = 4)
2-amino-3-phenyl-1-propanol (m/z = 152)	2.00.	0.37	None	79%	9.6%

FIGURE 1.54 HPLC chromatogram of 2-amino-3-phenyl-1-propanol generated by cleavage of its t-BOC derivative with trifluoroacetic acid and clean-up on strata-X-C.

MP borohydride catches one equivalent of the titanium catalyst, while the polystyrene-bound diethanolamine resin (PS-DEAM) can scavenge the remaining titanium catalyst. The borohydride reagent also assists in the reductive amination reaction. Final purification of the crude amine product is achieved with a polystyrene-bound toluene sulfonic acid resin scavenger that holds the amine through an ion exchange reaction, while impurities are washed off. The pure amine can be recovered with methanol containing 2M ammonium hydroxide.

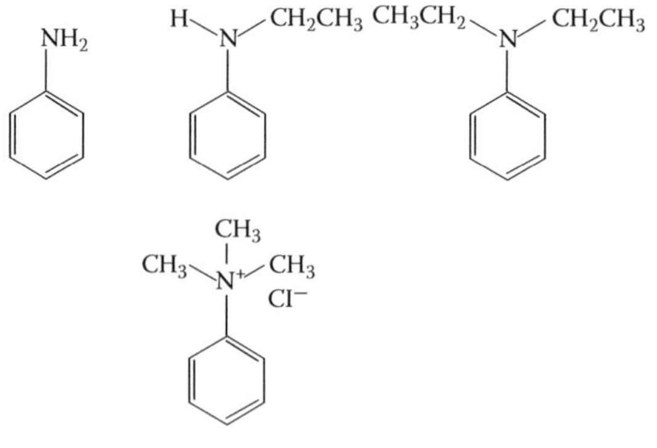

FIGURE 1.55 Structures of primary, secondary, tertiary, and quaternary amines separated on strata-X-CW.

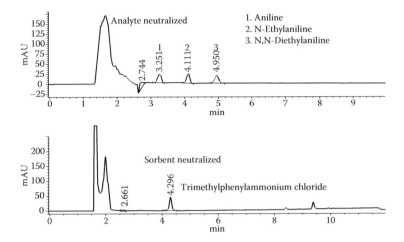

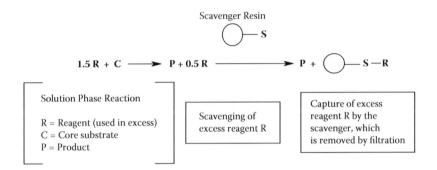

FIGURE 1.56 HPLC chromatograms of basified methanol (top) and acidified methanol (bottom) elution fractions from strata-X-CW.

FIGURE 1.57 Scavenging of excess reagent with scavenger resin. (Reproduced with permission from Biotage AB.)

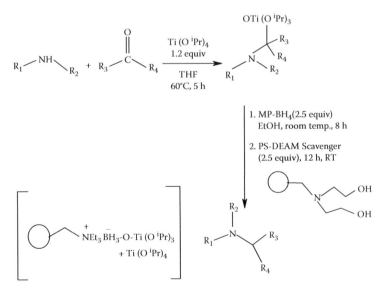

FIGURE 1.58 Scavenging of excess catalyst from crude synthetic reaction product. (Reproduced with permission from Biotage AB.)

1.5   CONCLUDING REMARKS

This chapter presented several sample preparation and purification techniques that may be used for manual or automated high-throughput applications. Each technique has its own unique features, advantages, and disadvantages. Overall, solid-phase extraction (SPE) appears to be the most popular sample preparation technique owing to its universality for all analytes ranging from the most polar to the most hydrophobic, the selectivity it offers, the wide choice of chemistries, and automatability for high-throughput applications.

In real life, sample preparation techniques are utilized in tandem with HPLC/MS/MS. Due to the high selectivity of LC/MS/MS in the SRM mode, small quantities of impurities or interferences in the purified fractions from sample preparation protocols dealing with biological matrices can be tolerated. This is especially true for protein precipitation, which is widely used because of simplicity and rapidity, when high-throughput turn-around is required. High-throughput HPLC is discussed elsewhere in this book.

1.6   ADDITIONAL READING

Recent articles about new sorbents and applications of high-throughput sample preparation methods in conjunction with HPLC/MS/MS analysis of analytes from biological matrices are included at the end of the reference list below. Spatial considerations prevented detailed discussions in this chapter. The articles appear as References 197 through 212 on the list below.

REFERENCES

1. G. Jung (Ed.), *Combinatorial Chemistry: Synthesis, Analysis, Screening*, Wiley-VCH, Weinheim, 2001.
2. A.W. Czarnik, B. Yan, and Y. Yan, *Optimization of Solid-Phase Combinatorial Synthesis*, Marcel Dekker, New York, 2002.
3. S.V. Ley et al., *J. Chem. Soc. Perkin Trans. 1*, 3815.
4. D.G. Hall, S. Manku, and F. Wang, *J. Comb. Chem.*, 2001, 3, 125.
5. L. English (Ed.), *Combinatorial Library: Methods and Protocols*, Humana Press, Totowa, NJ, 2002.
6. G. Fassina and S. Miertus, *Combinatorial Chemistry and Technologies: Methods and Applications*, CRC Press, Boca Raton, FL, 2005.
7. P. Seneci, *Solid-Phase Synthesis and Combinatorial Technologies*, Wiley-Interscience, New York, 2000.
8. P.H. Seeberger, *Solid Support Oligosaccharide Synthesis and Combinatorial Carbohydrate Libraries*, Wiley, New York, 2004.
9. W. Bannwarth and B. Hinzen, *Combinatorial Chemistry: From Theory to Application*, Wiley-VCH, New York, 2006.
10. F.Z. Dorwald, *Organic Synthesis on Solid Phase*, Wiley-VCH, New York, 2002.
11. K.C. Nicolaou, R. Hanko, and W. Hartwig, *Handbook of Combinatorial Chemistry, Volumes 1 and 2*, Wiley, New York, 2002.
12. M.C. Pirrung, *Molecular Diversity and Combinatorial Chemistry: Principles and Applications*, Elsevier, Oxford, 2004.
13. Literature highlights in combinatorial science, *QSAR Comb. Sci.*, 2005, 25, 803.
14. H. Fenniri, *Combinatorial Chemistry: A Practical Approach*, Oxford University Press, Oxford, 2000.
15. A. Beck-Sickinger and P. Weber, *Combinatorial Strategies in Biology and Chemistry*, Wiley, Chichester, 2002.
16. P. Wipf, *Handbook of Reagents for Organic Synthesis: Reagents for High Throughput Solid-Phase and Solution-Phase Organic Synthesis*, Wiley, Chichester, 2005.
17. H. An and P.D. Cook, *Chem. Rev.*, 2000, 100, 3311.
18. M. Chorghade (Ed.), *Drug Discovery and Development, Vol. 1*, Wiley, Hoboken, 2006.
19. C. Gunaratna, *Curr. Separations*, 2001, 19, 87.
20. C. Gunaratna, *Curr. Separations*, 2000, 19, 17.

21. M.D. Garrett et al., *Progr. Cell Cycle Res.*, 2003, 5, 145.
22. J.S. Lazo and J. Wipf, *J. Pharmacol. Exp. Therap.*, 2000, 293, 705.
23. B. Testa et al. (Eds.), *Pharmacokinetic Optimization in Drug Research: Biological, Physicochemical, and Computational Strategies*, Wiley-VCH, Weinheim, 2002.
24. J.P. Devlin and D.P. Devlin, *High Throughput Screening: The Discovery of Bioactive Substances*, Marcel Dekker, New York, 1997.
25. W. Purcell, *Approaches to High-Throughput Toxicity Screening*, Taylor & Francis, London, 2000.
26. W.P. Janzen (Ed.), *High Throughput Screening: Methods and Protocols*, Humana, Totowa, NJ, 2002.
27. H. Lee, *J. Liquid Chromatogr. Rel. Technol.*, 2005, 28, 1161.
28. Y. Hsieh and W.A. Korfmacher, *Curr. Drug Metabol.*, 2006, 7, 479.
29. S. Zhou et al., *Curr. Pharm. Anal.*, 2005, 1, 3.
30. A. Triolo et al., *J. Mass Spectrom.*, 2001, 36, 1249.
31. J.L. Herman, *Int. J. Mass Spectrom.*, 2004, 238, 107.
32. K. Whalen, J. Gobey, and J. Janiszewski, *Rapid Commun. Mass Spectrom.*, 2006, 20, 1497.
33. B. Yan (Ed.), *Analysis and Purification Methods in Combinatorial Chemistry*, Wiley Interscience, Hoboken, 2004.
34. G. Eldridge et al. *Anal. Chem.*, 2002, 74, 3963.
35. A. Ganesan, *Pure Appl. Chem.*, 2001, 73, 1033.
36. K.T. Kate and S.A. Laird (Eds.), *The Commercial Use of Biodiversity: Access to Genetic Resources and Benefit Sharing*, Earthscan, Kew, UK.
37. P.M. Abreu and P.S. Branco, *J. Braz. Chem. Soc.*, 2003, 14, 675.
38. R.E. Dolle et al., *J. Comb. Chem.*, 2006, 8, 597.
39. I. Paterson and E.A. Anderson, *Science*, 2005, 310, 451.
40. J. Ortholand and A. Ganesan, *Curr. Opin. Chem. Biol.*, 2004, 8, 271.
41. S. Mitra (Ed.), *Sample Preparation Techniques in Analytical Chemistry*, Wiley-Interscience, Hoboken, 2003.
42. D.A. Wells, *Progress in Pharmaceutical and Biomedical Analysis, Vol. 5,* Elsevier, Amsterdam, 2003.
43. J. Rosenfeld (Ed.), *Sample Preparation for Hyphenated Analytical Techniques*, Blackwell, Oxford, 2004.
44. J. Pawliszyn (Ed.), *Sampling and Sample Preparation in Field and Laboratory: Fundamentals and New Directions in Sample Preparation*, Elsevier, Amsterdam, 2002.
45. H. Kataoka, *Curr. Pharm. Anal.*, 2005, 1, 65.
46. M. Gilar, E.S.P. Bouvier, and B.J. Compton, *J. Chromatogr. A*, 2001, 909, 111.
47. E.M. Thurman and M.S. Mills, *Solid-Phase Extraction: Principles and Practice*, Wiley, New York, 1998.
48. J.S. Fritz, *Analytical Solid-Phase Extraction*, Wiley-VCH, New York, 1999.
49. C.F. Poole, *Trends Anal. Chem.*, 2003, 22, 362.
50. M. Hennion, *J. Chromatogr. A*, 1999, 856, 3.
51. C. Misl'anova and M. Hutta, *J. Chromatogr. B*, 2003, 797, 91.
52. A. Jiye et al., *Anal. Chem.*, 2005, 77, 8086.
53. T. Zhang et al., AAPS Annual Meeting and Exposition, Nashville, TN, November 2005.
54. D.F. Wendelborn, K. Seibert, and L.J. Roberts II, *Proc. Natl. Acad. Sci. USA*, 1988, 85, 304.
55. D.J. Kennaway and A. Voultsios, *J. Clin. Endocrinol. Metabol.* 1998, 83, 1013.
56. M. Ruggiero et al., *Pathophysiol. Haemost. Thromb.*, 2002, 32, 44.
57. T.A. Annesley, *Clin. Chem.*, 2003, 49, 1041.
58. C.R. Mallet, Z. Lu, and J.R. Mazzeo, *Rapid Commun. Mass Spectrom.*, 2004, 18, 49.
59. B.K. Matuszewski, M.L. Constanzer, and C.M. Chavez-Eng, *Anal. Chem.*, 2003, 75, 3019.
60. M. Jemal, *Biomed. Chromatogr.*, 2000, 14, 422.
61. J. Schuhmacher et al., *Rapid Commun. Mass Spectrom.*, 2003, 17, 1950.
62. Y. Hsieh et al., *Rapid Commun. Mass Spectrom.*, 2001, 15, 2481.
63. R. King et al., *J. Am. Soc. Mass Spectrom.*, 2000, 11, 942.
64. C. Muller et al., *J. Chromatogr. B*, 2002, 773, 47.
65. R. Bonfiglio et al., *Rapid Commun. Mass Spectrom.*, 1999, 13, 1175.
66. F.A. Kuhlmann et al., *J. Am. Soc. Mass Spectrom.*, 1995, 6, 1221.
67. J. Eshraghi and S.K. Chowdhury, *Anal. Chem.*, 1993, 65, 3528.
68. P.K. Bennett, M. Meng, and V. Capka, 17th Int. Mass Spectrometry Conf., Prague, August 2006.
69. V. Capca and S.J. Carter, ASMS Conf., Seattle, May 2006.

70. X. Xu, J. Lan, and W.A. Korfmacher, *Anal. Chem.*, 2005, 77, 389A.
71. C.K. Hull et al., *J. Chromatogr. B*, 2002, 772, 219.
72. K. Suenami et al., *Anal. Bioanal. Chem.*, 2006, 384, 1501.
73. I. Fu, E.J. Woolf, and B.K. Matuszewski, *J. Pharm. Biomed. Anal.*, 2004, 35, 837.
74. T. Unno, Y.M. Sagesaka, and T. Kakuda, *J. Agric. Food Chem.*, 2005, 53, 9885.
75. N.V.S. Ramakrishna et al., *Rapid Commun. Mass Spectrom.*, 2005, 19, 1970.
76. Q. Song et al., *J. Chromatogr. B*, 2005, 814, 105.
77. J. Chapdelaine et al., Poster 65, ASMS Conf., Seattle, 2006.
78. T. Perez-Ruiz et al., *J. Pharm. Biomed. Anal.*, 2006, 42, 100.
79. D.G. Watson et al., *J. Pharm. Biomed. Anal.*, 2004, 35, 87.
80. G.L. Lensmeyer et al., *Clin. Chem.*, 2006, 52, 1120.
81. Varian Consumable Products Application Note A02102, August 2004.
82. R.J. Motyka, P. Gerhards, and S. Sadjadi, Varian Application Note 57, February 2004.
83. T. Mroczek, K. Glowniak, and J. Kowalska, *J. Chromatogr. A*, 2006, 1107, 9.
84. M. Wood et al., *Forensic Sci. Int.*, 2005, 150, 227.
85. F. Saint-Marcoux, G. Lachatre, and P. Marquet, *J. Am. Soc. Mass Spectrom.*, 2003, 14, 14.
86. N. Fontanals et al., *J. Separation Sci.*, 2006, 29, 1622.
87. Y. Iwasaki et al., *J. Chromatogr. B*, 2006, April 15.
88. Y. Cheng, LCGC North America, June 2004.
89. A.R. McKinney et al., *J. Chromatogr. B*, 2004, 811, 75.
90. S.M.R. Wille et al., *J. Chromatogr. A*, 2005, 1098, 19.
91. S. Huq and K. Kallury, *LC-GC Appl. Notebook*, September 2006, 42.
92. E. Bermudo et al., *J. Chromatogr. A*, 2006, 1120, 199.
93. S. Huq et al., *LC-GC Appl. Notebook*, September 2005, 64.
94. S. Huq et al., *LC-GC Appl. Notebook*, February 2005.
95. K. Kallury and C. Sanchez, *LC-GC Europe*, March 2, 2006.
96. N. Fontanals, R.M. Marce, and F. Borrull, *Trends Anal. Chem.*, 2005, 24, 394.
97. S. Huq et al., *J. Chromatogr. A*, 2005, 1073, 355.
98. S. Huq, M. Garriques, and K.M.R. Kallury, *J. Chromatogr. A*, 2006, 1135, 12.
99. S. Huq, A. Dixon, and K.M.R. Kallury, Poster, Eastern Analytical Symposium, Somerset, NJ, November 2005.
100. K.M.R. Kallury et al., Poster, Pittcon, Orlando, FL, 2006.
101. T. Soriano et al., *J. Anal. Toxicol.*, 2001, 25, 137.
102. F. Saint-Marcoux, G. Lachatre, and P. Marquet, *J. Amer. Soc. Mass Spectrom.*, 2003, 14, 14.
103. J. Yawney et al., *J. Anal. Toxicol.*, 2002, 26, 325.
104. M. Wood et al., *Forensic Sci. Int.*, 2005, 150, 227.
105. N. Fontanals et al., *J. Separation Sci.*, 2006, 29, 1622.
106. P.I. Dobrev and M. Kaminek, *J. Chromatogr. A*, 2002, 950, 21-29.
107. A. Nordstrom et al., *Anal. Chem.*, 2004, 76, 2869.
108. D. Richard et al., *Anal. Chem.*, 2005, 77, 1354.
109. M. Josefsson and A. Sabanovic, *J. Chromatogr. A*, 2006, 1120, 1.
110. S.M.R. Wille et al., *J. Chromatogr. A*, 2005, 1098, 19.
111. J.X. Shen et al., *J. Pharm. Biomed. Anal.*, 2005, 37, 359.
112. J.X. Shen et al., *J. Pharm. Biomed. Anal.*, 2006, 40, 689.
113. Q. Song et al., *J. Chromatogr. B*, 2005, 814, 105.
114. L.C. Romanoff et al., *J. Chromatogr. B*, 2006, 835, 47.
115. J.X. Shen, C.I. Tama, and R.N. Hayes, *J. Chromatogr. B*, 2006, 843, 275.
116. C. Hasegawa et al., *Rapid Commun. Mass Spectrom.*, 2006, 20, 537.
117. X. Zang et al., *Rapid Commun. Mass Spectrom.*, 2005, 19, 3259.
118. Y. Alnouti et al., *Anal. Chem.*, 2006, 78, 1331.
119. T.K. Nevanen et al., *Anal. Chem.*, 2005, 77, 3038.
120. G. Rule, M. Chappe, and J. Henion, *Anal. Chem.*, 2001, 73, 439.
121. R.A. Biddlecombe, C. Benevides, and S. Pleasance, *Rapid Commun. Mass Spectrom.*, 2001, 15, 33.
122. M.S. Chang, E.J. Kim, and T.A. El-Shourbagy, *Rapid Commun. Mass Spectrom.*, 2006, 20, 2190.
123. I. Olsson et al., *Chemometr. Intell. Lab. Syst.*, 2006, 83, 66.
124. K.M. Kallury et al., Oral and poster presentations, Pittcon, Chicago, February 2007.
125. R.E. Majors, *LCGC Europe*, 2006, 19, 1.

126. S. Pedersen-Bjergaard et al., *J. Separation Sci.*, 2005, 28, 1195.
127. T.S. Ho et al., *J. Chromatogr. A*, 2005, 1072, 29.
128. O.B. Jonsson, E. Dyremark, and U.L. Nilsson, *J. Chromatogr. B*, 2001, 755, 157.
129. T.S. Ho et al., *J. Chromatogr. A*, 2002, 963, 3.
130. C. Deng et al., *J. Chromatogr. A*, 2006, 1131, 45.
131. G. Ouyang, W. Zhao, and J. Pawliszyn, *Anal. Chem.*, 2005, 77, 8122.
132. H.G. Ugland, M. Krogh, and L. Reubsaet, *J. Chromatogr. B*, 2003, 798, 127.
133. L. Chimuka, E. Cukrowska, and J.A. Jonsson, *Pure Appl. Chem.*, 2004, 76, 707.
134. K. Schugerl, *Adv. Biochem. Eng. Biotechnol.*, 2005, 92, 1.
135. B.E. Smink et al., *J. Chromatogr. B*, 2004, 811, 13.
136. Q. Song and W. Naidong, *J. Chromatogr. B*, 2006, 830, 135.
137. R.C. Rodila et al., *Rapid Commun. Mass Spectrom.*, 2006, 20, 3067.
138. Y. Dotsikas et al., *J. Chromatogr. B*, 2006, 836, 79.
139. Y.J. Xue, J. Liu, and S. Unger, *J. Pharm. Biomed. Anal.*, 2006, 41, 979.
140. K. Reising et al., *Anal. Chem.*, 2005, 77, 6640.
141. T.H. Eichhold et al., *J. Pharm. Biomed. Anal.*, 2007, 43, 586.
142. R.D. Bolden et al., *J. Chromatogr. B*, 2002, 772, 1.
143. A.Q. Wang et al., *Rapid Commun. Mass Spectrom.*, 2002, 16, 975.
144. C.L. Sprague et al., *Toxicol. Sci.*, 2004, 80, 3.
145. R.A. Kemper, A.A. Elfarra, and S.R. Myers, *Drug Metabol. Dis.*, 1998, 26, 914.
146. B. Buchele et al., *J. Chromatogr. B*, 2005, 829, 144.
147. M. Ma et al., *J. Pharm. Biomed. Anal.*, 2006, 40, 128.
148. R. Tuckermann, *J. Acoustic Soc. America*, 2004, 116, 2597.
149. T. Blanco et al., 1998, 123, 191.
150. B. Karlberg, *Fresenius J. Anal. Chem.*, 1988, 329, 660.
151. L. Chimuka, E. Cukrowska, and J.A. Jansson, *Pure Appl. Chem.*, 2004, 76, 707.
152. N. Jakubowska et al., *Crit. Rev. Anal. Chem.*, 2005, 35, 217.
153. J.A. Jonsson and L. Mathiasson, *LC-GC* Europe, 2003, 2.
154. T.S. Ho, S. Pedersen-Bjergaard, and K.E. Rasmussen, *J. Chromatogr. A*, 2002, 963, 3.
155. T.S. Ho, S. Pedersen-Bjergaard, and K.E. Rasmussen, *Analyst*, 2002, 127, 608.
156. F.J. Malagueno de Santana, A.R. Moraes de Oliveira, and P.S. Bonato, *Anal. Chim. Acta*, 2005, 549, 96.
157. C. Polson et al., *J. Chromatogr. B*, 2003, 785, 263.
158. R.J. Flanagan et al., *Biomed. Chromatogr.*, 2006, 20, 530.
159. Argonaut Technical Note 130 (from Website), 2004.
160. M. Berna et al., *Anal. Chem.*, 2002, 74, 1197.
161. S.X. Peng et al., *Anal. Chem.*, 2006, 78, 343.
162. G.L. Lensmeyer et al., *Clin. Chem.*, 2006, 52, 1120.
163. J. Lei, H. Lin, and M. Fountoulakis, *J. Chromatogr. A*, 2004, 1023, 317.
164. V. Joshi et al. *Millipore Technical Library Bulletin* (from Website).
165. C.L. Arthur and J. Pawliszyn, *Anal. Chem.*, 1990, 62, 2145.
166. K.M. Li, L.P. Rivory, and S.J. Clarke, *Curr. Pharm. Anal.*, 2006, 2, 95.
167. M.E.C. Queiroz and F.M. Lancas, *LC-GC North America*, 2004, 22, 970.
168. H.L. Lord et al., *Anal. Chem.*, 2003, 75, 5103.
169. V.Pichon and K. Haulpt, *J. Liq. Chromatogr. Rel. Technol.*, 2006, 29, 989.
170. L.I. Andersson, in *Solid Phase Extraction on Molecularly Imprinted Polymers: Requirements, Achievements and Future Work*, S. Piletsky and A. Turner (Eds.), Landes Biosciences, New York, 2006.
171. E. Caro et al., *Trends Anal. Chem.*, 2006, 25, 143.
172. J. Haginaka, *Anal. Bioanal. Chem.*, 2004, 379, 332.
173. K. Ensing and R.E. Majors, *LC-GC Europe*, January 2002, 1.
174. K. Moller, Ph.D. thesis, 2006, University of Stockholm, Sweden.
175. Y. Shi et al., *Pharm. Biomed. Anal.*, 2006, 42, 549.
176. F. Chapius et al., J. *Chromatogr. A*, 2006, 1135, 127.
177. M.M. Ariffin et al., *Anal. Chem.*, 2007, 79, 256.
178. B. Yan et al., *J. Comb. Chem.*, 2004, 6, 255.
179. J.J. Isbell et al., *J. Comb. Chem.*, 2005, 7, 210.
180. M. Schaffrath et al., *J. Comb. Chem.*, 2005, 7, 546.

181. T. Karancsi et al., *J. Comb. Chem.*, 2005, 7, 58.
182. P.A. Searle, K.A. Glass, and J.E. Hochlowski, *J. Comb. Chem.*, 2004, 6, 175.
183. P. Wipf and C.M. Coleman, *Drug Dis. World*, Winter 2003, 62.
184. B.C. Bookser and S. Zhu, *J. Comb. Chem.*, 2001, 3, 205.
185. H. Han et al., *Proc. Natl. Acad. Sci. USA*, 1995, 92, 6419.
186. S.V. Ley et al., *J. Chem. Soc. Perkin Trans.*, 1, 2000, 3815.
187. P.J. Edwards, *J. Comb. Chem. High-Through. Screen.*, 2003, 6, 11.
188. J.C. Hodges, *Synletters*, 1999, 1, 152.
189. J.C. Hodges, L.S. Harikrishnan, and S. Ault-Justus, *J. Comb. Chem.* 2000, 2, 80.
190. B. Boughtflower et al., *J. Comb. Chem.*, 2006, 8, 441.
191. A.H. Harned et al., *Aldrichimia Acta,* 2005, 38, 3.
192. L. Williams and S.M. Neset, Fourth Int. Conf. on Synthetic Organic Chemistry, September 1–30, 2000 www.mdpi.org/ecsoc-4.htm.
193. W. Zhang, D.P. Curran, and C.H. Chen, *Tetrahedron*, 2002, 58, 3871.
194. *Fluorous Technologies Bulletin* on Fluorous Scavenging, www.fluorous.com/html.
195. H.N. Weller, *Mol. Diversity*, 1998, 4, 47.
196. A.G.M. Barrett, M.L. Smith, and F.J. Zecri, *Chem. Commun.*, 1998, 2317.
197. R. Majors, *LC-GC North America*, 2007, 25, 16.
198. K. Georgi and K. Boos, *LC-GC Europe*, 2004, 17, 21.
199. M.J. Berna, B.L. Ackermann, and A.T. Murphy, *Anal. Chim. Acta*, 2004, 509, 1.
200. M.I. Churchwell et al., *J. Chromatogr. B*, 2005, 825, 134.
201. S. Notari et al., *J. Chromatogr. B*, 2006, 831, 258.
202. J.L. Little, M.F. Wempe, and C.M. Buchanan, *J. Chromatogr. B*, 2006, 833, 219.
203. J.M. Castro-Perez, *Drug Dis. Today*, 2007, 12, 249.
204. C. Pan et al., *J. Pharm. Biomed. Anal.*, 2006, 40, 581.
205. L.D. Penn et al., *J. Pharm. Biomed. Anal.*, 2001, 25, 569.
206. J. Stevens et al., *J. Chromatogr. A*, 2007, 1142, 32.
207. S. Dai et al., *Rapid Commun. Mass Spectrom.*, 2005, 19, 1273.
208. A. Kaufmann et al., *Anal. Chim. Acta*, 2007, 586, 13.
209. S. Mathur et al., *J. Biomol. Screen.*, 2003, 8, 136.
210. S. Fox et al., *J. Biomol. Screen.*, 2006, 11, 864.
211. C. Guintu et al., *J. Biomol. Screen.*, 11, 933.
212. J.M. Lazar, J. Grym, and F. Foret, *Mass Spectrom. Rev.*, 2006, 25, 573.

2 Online Sample Extraction Coupled with Multiplexing Strategy to Improve Throughput

Katty X. Wan

CONTENTS

ABSTRACT

The coupling of high performance liquid chromatography (HPLC) with tandem mass spectrometry (MS/MS) has revolutionized the process of quantitative bioanalysis. The continually shortening timelines in drug discovery and development demand high-throughput methodologies to quantify drugs, metabolites, and endogenous biomolecules in complex biological matrices. This chapter reviews the most recent advances in mass spectrometry, liquid chromatography, and sample preparation techniques aimed at achieving high throughput and discusses online solid phase extraction and multiplexed front end HPLC for quantitative bioanalysis in detail.

Two case studies (regulated and conducted under good laboratory practices [GLP]) will be presented. Each was conducted with a flexible LC/MS front end designed to operate two HPLCs in parallel by staggering injections and MS acquisition times. The system was configured to carry out two types of tasks with great flexibility. It can perform regular paralleled LC/MS analysis with guard column regeneration and also operate in online SPE mode, alternating sample extraction and LC/MS analysis between the two paralleled systems. Switching between the operational modes requires minimum re-plumbing and loading of corresponding macros. The system has been proven to be flexible and reliable and provides high-throughput platforms for all our current bioanalytical needs.

2.1 INTRODUCTION

Quantitative bioanalysis in the pharmaceutical industry has been revolutionized by the use of liquid chromatography coupled with tandem mass spectrometry (LC/MS/MS). The inherent high specificity and sensitivity of LC/MS/MS established this technique as the primary analytical tool for determining the concentrations of drugs, metabolites, and endogenous biomolecules in biological matrices (blood, plasma, serum, urine, etc.). Because quantitative bioanalysis plays a key role in drug discovery and development, significant investments have been made to minimize bioanalysis times. Over the past decade, sample throughput has become the focus of quantitative bioanalysis in order to meet ever-shortening timelines.

Numerous efforts have focused on increasing sample throughput for the purposes of drug discovery and drug development. Higher throughput during drug discovery has allowed development of generic, quantitative assays for large libraries of compounds and provides vital pharmacokinetic information at early stages of the drug candidate screening process. The method development phase of drug development has been greatly shortened because of the superb sensitivity and specificity of LC/MS/MS. Faster run cycle times have not only increased sample throughput, but also provided rapid turnarounds for sample analysis. This is particularly important for clinical studies requiring real-time analytical results. One major difference of bioanalysis for drug discovery and for development is the regulation governing the respective processes. Drug development work is normally regulated under GLPs. The data generated are included in submissions to regulatory agencies and the assays must be validated according to international guidelines.

This chapter will review recent advances in mass spectrometry, liquid chromatography, and sample preparation techniques that aim at achieving high throughput. In particular, online solid phase extraction and multiplexed HPLC front ends for quantitative bioanalysis will be discussed in detail.

2.1.1 MASS SPECTROMETRY

Recent innovations in ionization techniques have allowed the development of ambient mass spectrometry. Mass spectra can be determined for samples in their native environment without sample preparation. Although the ambient mass spectrometry technique is still in its infancy, its potential for serving as a tool of choice for high-throughput bioanalysis is very encouraging.

A new family of ionization techniques allows ions to be created under ambient conditions and then collected and analyzed by MS. They can be divided into two major classes: desorption electrospray ionization (DESI) and direct analysis in real time (DART).

In DESI, a fine spray of charged droplets hits the surface of interest, from which it picks up analytes, ionizes them, and then delivers them into a mass spectrometer. DESI is a novel ionization method that allows use of MS to acquire spectra of condensed-phase samples under ambient conditions. It can be used on solid samples including complex biological materials such as tissue slides. It can also be applied to liquids, frozen solutions, and adsorbed gases. DESI has high sensitivity, responds almost instantaneously, and is applicable to both small organic molecules and large biomolecules. DESI allows organic molecules present at sample surfaces to be analyzed by MS without requiring sample preparation in most cases. Ionization is achieved through the impacts of charged microdroplets, gas-phase solvent ions, and ionic clusters produced by an electrospray emitter onto condensed-phase samples. The charged droplets dissolve sample molecules as they splash off the surface, and the secondary charged droplets carrying sample molecules produce gaseous ions by a well known ESI mechanism. Released ions are then vacuumed through the MS interface for detection. Detailed applications using DESI are discussed in recent reviews.[1,2]

In DART, a stream of gas composed of excited-state atoms and ions is directed to the surface of interest, from which it desorbs low molecular weight molecules. Rapid and noncontact analysis of solids, liquids, and gases at ambient pressure and ground potential can be achieved. The most commonly used gases such as nitrogen and helium have high ionization potentials. When subjected to high electrical potential, they form a plasma of electronic excited-state species (helium) or vibronic excited-state species (nitrogen). The DART gas flow can be aimed directly to the MS orifice or it may be reflected off a sample surface and into MS. Several ionization mechanisms have been proposed including surface Penning in which ionization of the sample occurs by energy transfer from an excited atom or molecule to the sample. Direct ionization was achieved without sample preparation. Details of applications using DART can be found in a recent review.[3] Unlike DESI, DART exposes a sample to a stream of excited gas and does not require electrosprayed liquid solvent. Comparisons of DESI and DART were made using common drugs and samples of biological origin.[4] Although the full potential of ambient mass spectrometry in high-throughput bioanalysis remains to be determined, a wide variety of applications and automations have already been explored.

Another MS-based approach used in high-throughput bioanalysis utilizes a mass spectrometer equipped with several API spray probes. Each of the analytical columns in parallel is connected to a separate spray probe and each spray is sampled in rapid successions for data acquisition by the MS. A separate data file for each spray is recorded. Several samples can be analyzed simultaneously on parallel columns[5,6] in the course of a single chromatographic run.

2.1.2 LIQUID CHROMATOGRAPHY

Because mass spectrometric detection time consumes such a small fraction of the overall LC/MS/MS cycle time (chromatographic run time plus autosampler injection time), the focus on increasing sample throughput has been given to chromatographic techniques. This section reviews the strategies for developing fast HPLC methods and utilizing parallel HPLCs.

2.1.2.1 Fast HPLC

The use of high flow and fast gradient HPLC has gained a lot of popularity because of the ability to reduce LC/MS/MS cycle times during bioanalysis. In the case of fast gradient HPLC, peak shapes were improved and method development times were minimized, especially when multiple analytes with diverse functionalities had to be separated. Flows as high as 1.5 to 2 mL/min were achieved on a 2.1 × 30 mm Xterra C18 column.[7] Details are discussed in a recent review.[8]

Another approach to increase HPLC speed is the use of higher temperatures. The viscosity of a typical mobile phase used in reversed-phase separation decreases as the column temperature is increased. This allows an HPLC system to operate at a higher flow rate without suffering too much from increased back pressure. Zirconia-based packing materials provide excellent physical and chemical stability. They have been used successfully for high-throughput bioanalysis at elevated temperatures.[9]

Ultra-performance HPLC (UPLC) utilizes sub-2-μm porous particles inside packed microbore columns up to 150 mm long. Significant improvements in terms of resolution, analysis time, and detection sensitivity have been reported. A side-by-side comparison of HPLC and UPLC was made to determine concentrations of alprazolam in rat plasma.[10] UPLC provided a four-fold reduction in terms of LC/MS/MS cycle time that translated into higher sample throughput. Another important benefit of UPLC is production of narrow peaks that can provide better resolution from endogenous compounds in a matrix. Less ion suppression and better sensitivity can be achieved with the same extracts.

The utility of packed column supercritical and enhanced fluidity liquid chromatography (pcSFC) for high-throughput applications has increased during the past few years.[11] In contrast to traditional reversed-phase liquid chromatography, the addition of a volatile component such as CO_2 to the mobile phase produces a lower mobile phase viscosity and high diffusivity. This allows the use of high flow rates that translate into faster analysis times. In addition, the resulting mobile phase is considerably more volatile than the aqueous-based mobile phase typically used with LC/MS/MS, allowing the entire effluent to be directed into the MS interface. An analysis speed of approximately 10 min/96-well plate was achieved for determination of dextromethorphan in human plasma.

Another important approach for achieving fast separation via high-throughput bioanalysis is the utilization of a monolithic stationary phase. Monolithic columns are characterized as polymeric interconnected skeletons with pores. The macropores (approximately 2 μm in diameter, also called through pores) of a monolithic column provide channels for high permeability that allow higher flow rates than conventional particle-based columns. The mesopores (approximately 13 nm in diameter) provide an extended surface area comparable to traditional columns packed with 3-μm particles. Monolithic HPLC columns are silica- or organic polymer-based and both can operate at high flow rates up to 10 mL/min. Many applications originally developed with packed columns can also utilize monolithic columns while reducing analysis times by a factor of 5 to 10. Detailed applications can be found in a recent review.[12]

2.1.2.2   Parallel HPLC

In many cases, analyte separations may only occur within a very small fraction of total MS acquisition time. The paralleled (also known as multiplexed) LC/MS/MS systems take advantage of the elution window by allowing numerous separations to occur simultaneously while staggering detection of the analytes. The goal is to maximize the MS time spent on detecting analytes by reducing the time wasted on collecting the baselines between samples. Parallel HPLC can double or quadruple sample throughput without compromising separation. Successful implementations of both commercial (Aria LX4)[13] and custom-built[5] parallel HPLC systems have been reported. We developed a staggered parallel HPLC system that provides great flexibility. Section 2.2 of this chapter details case studies using this system for high-throughput bioanalysis in a GLP environment.

2.1.2.3   No HPLC

Tandem mass spectrometric methods have demonstrated superb specificity because of their ability to isolate analytes selectively in the presence of endogenous interferences. Attempts to further increase sample throughput led to the idea of using LC/MS/MS without the LC. Traditional chromatographic separations were replaced with flow injection analysis (FIA) or nanoelectrospray infusion techniques. The MS-based columnless methods attracted a lot of attention because of their inherent fast cycle times and no need for LC method development.

Compared with LC/MS/MS methods, nanoelectrospray/MS/MS methods offer additional benefits such as no sample carry-over and low sample and solvent consumptions. The major concerns surrounding columnless analysis of biological samples are matrix ion suppression and direct interference from endogenous components or metabolites of the dosed compound. Therefore, an extensive sample clean-up process must be in place to ensure the accuracy and precision of the assay. A nine-fold gain in terms of sample throughput was achieved with a nanoelectrospray/MS/MS method that produced accuracy, precision, and detection limits comparable to those of a traditional LC/MS/MS method.[14]

2.1.3 SAMPLE PREPARATION

Automation using a robotic liquid handling system eliminated most of the tedious steps encountered with traditional manual extraction procedures. Automated 96-well SPE and LLE techniques using robotic liquid handlers have been successfully implemented to support high-throughput bioanalysis.[5]

Direct injection of pretreated biological samples (also called online sample cleanup) greatly simplified sample preparation for LC/MS/MS analysis. The normal process involves sample aliquot steps, internal standard addition, and centrifugation. Compared to traditional off-line LLE and SPE sample preparation procedures, online methods are easier and faster. Two types of online SPE columns are commercially available. One is the restricted access media (RAM) column. The other is the turbulent flow chromatography (TFC) column.

A RAM column functions through a size exclusion mechanism. Large biomolecules such as proteins are restricted from the adsorptive surfaces inside silica particles. Small analyte molecules are able to penetrate into the inner surfaces of the particles. As a result, protein molecules pass through the column rapidly and analytes of interest are retained on the adsorptive sites. Depending on the application, the analyte molecules are directed to MS for detection or transferred onto an analytical column for separation prior to MS detection. Detailed applications are discussed in a recent review.[8]

TFC is a high flow chromatographic technique that takes advantage of unique flow dynamics occurring when relative high flow rates (2 to 4 mL/min) are applied to columns of small internal diameters (1 mm or 0.18 μm) and packed with large particles (20 to 60 μm). When the mobile phase flows through a TurboFlow column, high linear velocities 100 times greater than those typically seen in HPLC columns are created. The large interstitial spaces between column particles and the high linear mobile phase velocity create turbulence within the column. Because small molecular weight molecules diffuse faster than large molecular weight molecules, the small compounds diffuse into the particle pores. The turbulent flow of the mobile phase quickly flushes the large molecular weight compounds such as proteins through the column to waste before they have an opportunity to diffuse into the particle pores. Of the sample molecules that enter the pores, those that have an affinity to the chemistry inside the pores bind to the internal surfaces of the column particles. The small molecules that have lower binding affinities quickly diffuse from the pores and are flushed to waste. TFC columns with a variety of chemistries to accommodate different analyte types are available. A mobile phase change elutes the small molecules bound by the TFC column to the mass spectrometer or to a second analytical column for further separation.

Direct injections using RAM or TFC have simplified sample preparation and increased throughput. Matrix ion suppression was greatly reduced or eliminated in several cases compared with traditional off-line sample cleanup procedures such as PPT, SPE, and LLE. Method development time was minimized with generic methods[15] that suit most applications. Detailed applications can be found in a recent review.[8]

2.2 CASE STUDIES

We developed a staggered parallel HPLC system with a CTC HTS PAL autosampler equipped with trio valves. The system consists of four (six if gradient is needed) independent HPLC pumps. Parallel analysis is achieved by an offset dual-stream system with a time delay that allows efficient staggering of MS acquisition times.

The system is configured to carry out two types of tasks with great flexibility. It can perform regular paralleled LC/MS analysis with guard column regeneration. In this mode, the guard column can be switched off-line for regeneration after the sample is eluted from it. Guard column lifetime can be extended; the risk of over-pressure due to buildups can be minimized; and the late elutors that interfere with sample analysis can be eliminated. The system can also be operated in online SPE mode in which samples are extracted in line with LC/MS analysis. The two streams alternate between online SPE extraction and LC/MS analysis, making efficient use of MS acquisition time.

Timing and triggering of injections, analytical pumps, loading and washing pumps, and data collection are controlled by Cycle Composer software. Switching between the two operational modes requires minimum re-plumbing and the loading of corresponding macros in Cycle Composer. The system can also run in single-stream mode for ease of method development. Two case studies conducted with this flexible LC/MS/MS system are discussed.

2.2.1 CASE I: ANALYSIS OF PARICALCITOL USING DUAL HPLC FOR CLINICAL STUDY M04-693

2.2.1.1 Introduction

Zemplar® (paricalcitol) injection is a synthetically manufactured selective vitamin D receptor activator (SVDRA) indicated for the prevention and treatment of secondary hyperparathyroidism associated with chronic kidney disease (CKD) stage 5. The U.S. Food & Drug Administration (FDA) approved a capsule form of Zemplar for development to satisfy a need for an oral formulation. The objective of study M04-693 was to assess the bioequivalencies of several dosage strengths of paricalcitol capsules under fasting conditions.

2.2.1.2 Analytical Method

Instrumentation — A parallel LC/MS/MS system was operated under dual HPLC with a guard column regeneration mode. Figure 2.1 is a flow diagram for the dual stream. Figure 2.2 shows the staggered timing scheme. Because of the interference from endogenous components, the LC run time had to be relatively long (approximately 11 min). The guard wash was also essential for eliminating late elutors. Taking advantage of the paralleled injections, a throughput of approximately 6 min/sample was achieved. Figure 2.3 shows the autosampler setup in action.

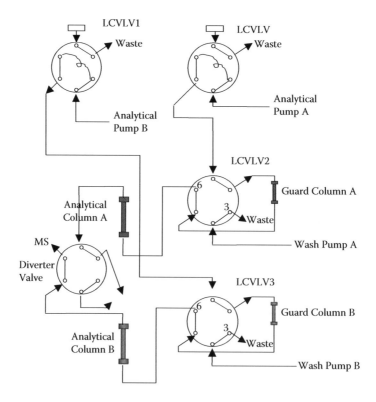

FIGURE 2.1 Flow diagram for dual-stream HPLC with guard column regeneration.

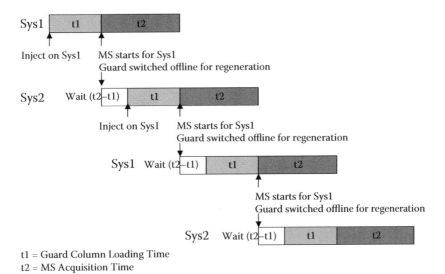

t1 = Guard Column Loading Time
t2 = MS Acquisition Time

FIGURE 2.2 Timing scheme for dual-stream HPLC with guard column regeneration.

HPLC condition — A reversed-phase HPLC column (Synergi Max RP C12, 4 µm, 80 Å, 2.0 × 150 mm) from Phenomenex was used in conjunction with its matching guard column (Synergi Max RP Security Guard C12, 2.0 ×4 mm). An 0.3 mL/min isocratic flow of 40/40/20 v/v/v methanol/acetonitrile/0.5mM ammonium acetate was used as the mobile phase. Approximately 1 min after each injection, the guard column was switched off-line for regeneration and 1 mL/min of 40/40/20 v/v/v THF/methanol/water was used to wash the column. The mobile phase was used to re-equilibrate the guard column prior to the next injection.

MS condition — An API 4000 equipped with a Turbo Ionspray from Applied Biosystems was used as the mass detector and $[M + NH_4]^+$ was chosen as the precursor ion for multiple reaction monitoring (MRM) due to the lack of protonated molecular ions. A transition of m/z 434.4 → 273.2 was chosen for paricalcitol and m/z 450.5 → 379.2 was selected for the structure analog internal standard.

Sample preparation — Analytes of interest were extracted from human plasma using the LLE technique. The following steps were followed:

1. Accurately pipette 600 µL plasma into clean borosilicate glass 12 × 75 mm extraction tubes.
2. Supplement the plasma with 100 µL internal standard solution (approximately 1000 ng/mL in 50/50 v/v methanol/water) and mix well.
3. Add extraction solvent (1000 µL of 50/50 v/v hexane/ethyl acetate) to each tube.
4. Cap the individual tubes; mix on a reciprocating shaker for a few minutes.
5. Separate the phases by centrifugation.
6. Remove the caps; transfer most of the upper organic layer to a clean 96-well polypropylene injection plate.
7. Evaporate the organic solvent in the injection plate under a stream of heated nitrogen in a 96-well format evaporator.
8. Reconstitute the extracted residues with 120 µL of 70/30 v/v methanol/1mM ammonium acetate (pH adjusted to 5.5).
9. Vortex the injection plate to thoroughly dissolve the extracted samples.
10. Centrifuge the plate and inject 50 µL for LC/MS/MS analysis.

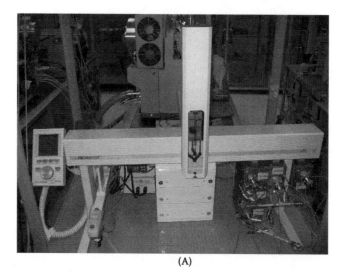

(A)

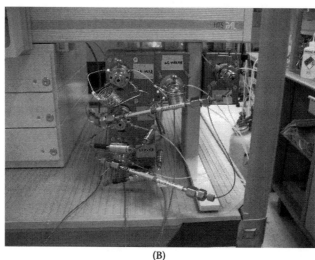

(B)

FIGURE 2.3 Autosampler set-up for performing dual-stream HPLC with guard column regeneration: (A) overall view, (B) zoomed view of the valving configuration.

2.2.1.3   Validation Results

Accuracy, precision, and linearity of standards — The linear dynamic range was established as 10.22 pg/mL to 2.037 ng/mL with coefficient of determination (r^2) below 0.996190 when using $1/x^2$ weighing for three consecutive accuracy and precision runs. The lower limit of quantitation (LLOQ) was accurate (inter-run mean bias = 1.1%) and precise (inter-run mean CV = 14.1%). Three levels of QCs were prepared. The inter-run mean bias varied from –4.3 to 1.0% and the inter-run mean CV varied from 4.6 to 6.6% for all QC levels.

   Selectivity — Assay selectivity was extensively tested during method development due to endogenous interferences. During validation, 12 individual lots of human plasma were screened. Additionally, three pre-dose (0 hour) samples from a previous study in which no results were reportable due to the lack of chromatographic selectivity were tested. No substantial peaks were observed at the MRM retention time for paricalcitol in any individual plasma lots tested. Figure 2.4 is a representative chromatogram. Interferences from metabolites, drug-related, and drug-induced

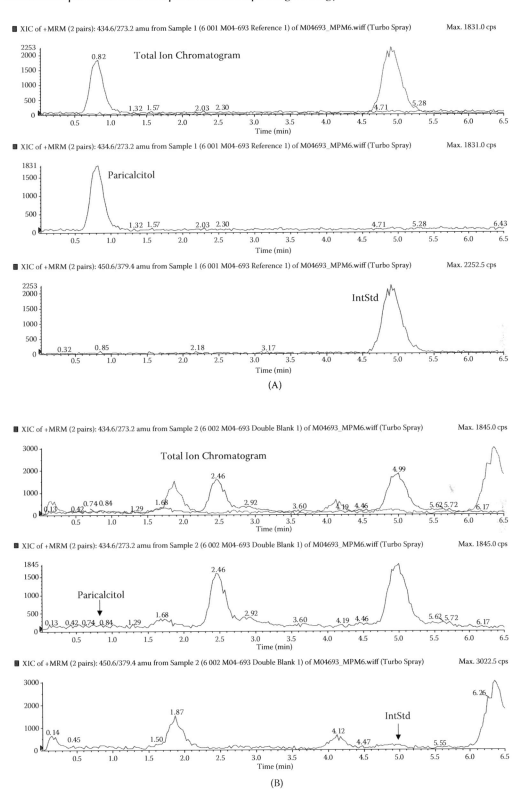

FIGURE 2.4 Representative chromatograms of reference and blank: (A) reference solution, (B) double blank sample.

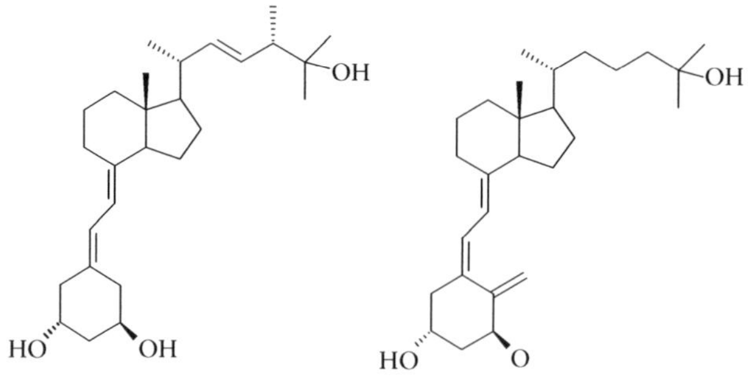

FIGURE 2.5 Chemical structures of paricalcitol and calcitriol.

materials were also tested. Endogenous calcitriol (see Figure 2.5 for chemical structure) and (R) 24-OH paricalcitol and (S) 24-OH paricalcitol metabolites were evaluated by injecting neat solutions. No peaks were observed at the MRM retention time for paricalcitol.

Matrix effect — To demonstrate that the assay performance was independent from the sample matrix, QC samples were prepared using two different lots of matrix. The QC samples were evaluated using the same calibration curve. With regard to analytical recovery, no significant difference was observed for the QCs prepared in two lots of plasma.

Recovery — Overall procedural recovery was evaluated. The results from spiked plasma QC (evaluation) samples that went through the analytical procedure were compared to the results from neat spiking (control) solution samples. The neat spiking solutions used to prepare the plasma evaluation samples were evaporated and reconstituted at the same volumes as the extracted samples. The analyte was tested at three concentration levels and the internal standard was tested at one. Mean recovery for the analyte was approximately 122.9%; the level was 55.2% for the internal standard.

Stability — Samples remain stable for at least 468 days when frozen at –20°C. They are stable for at least five simulated freeze-and-thaw cycles and approximately 28 hr at room temperature. The analyte is viable for at least 6 days in a reconstitution solution stored in the autosampler (temperature set point at 10°C). A dried-down batch (sample process stopped at dry-down step) was stable at least 5 days in a refrigerator (temperature varied from 4 to 8°C). A stock solution of paricalcitol is stable for at least 11 months. Stock solution of internal standard is stable about 4.5 months under refrigeration.

2.2.1.4 Assay Performance

Sample throughput — Sample analysis (including analytical repeats and reassays for pharmacokinetic purposes) was performed between 25 August 2004 and 26 October 2004. Analytical results from 6875 samples were finalized during this period.

Standards performance — The assay performed well. Mean bias at LLOQ was –0.2% and for other standards varied from –1.2 to 1.0%. Coefficient of determination (r^2) was greater than 0.989372 for all batches run.

QC performance — The mean bias for three levels of QCs varied from –0.7 to 0.8%. Representative diagrams of QC performance are shown in Figure 2.6.

2.2.2 Case II: Analysis of Fenofibric Acid Using Dual Online SPE for Clinical Study M06-830

2.2.2.1 Introduction

A variety of clinical studies have demonstrated that elevated levels of total cholesterol (total-C), low-density lipoprotein cholesterol (LDL-C), and apolipoprotein B (apo B) are associated with

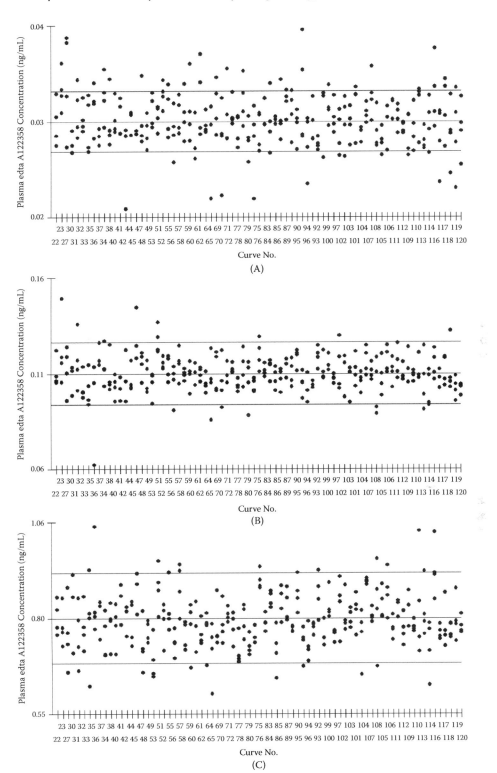

FIGURE 2.6 Plot of QC performance for Study M04-693: (A) lowQC (0.03 ng/mL), (B) mid QC (0.11 ng/mL), (C) high QC (0.8 ng/mL).

human atherosclerosis. Epidemiologic investigations established that cardiovascular morbidity and mortality vary directly with the levels of total-C, LDL-C, and triglycerides and inversely with the level of HDL-C.

Several clinical studies revealed that administration of fenofibrate produces reductions in total-C, LDL-C, apo B, total triglycerides, and triglyceride-rich (very low density) lipoprotein (VLDL) in treated patients. In addition, treatment with fenofibrate results in increases in HDL-C, apo AI, and apo AII. However, since fenofibrate is rapidly converted to fenofibric acid during absorption and fenofibric acid, but not fenofibrate, is found circulating in plasma, the effects of fenofibric acid have been extensively evaluated in these studies.

The objective of study M06-830 is to evaluate the bioavailability of fenofibric acid from the fenofibric acid choline salt formulation manufactured at full production scale at Abbott Laboratories' Puerto Rico facility relative to the bioavailability of (1) the fenofibric acid choline salt formulation used in Phase 3 trials and manufactured at the Abbott Park facility, and (2) the 200 mg micronized fenofibrate capsule.

2.2.2.2 Analytical Method

Instrumentation — The parallel LC/MS/MS system for this application was operated under the dual online SPE mode. Figure 2.7 is a flow diagram for the dual stream and Figure 2.8 depicts the staggered timing scheme. Figure 2.9 shows the autosampler setup in action.

HPLC condition — A Waters reversed-phase HPLC column (Symmetry Shield RP C18, 5 µm, 2.1 × 50 mm) was used in conjunction with a Regis SPS guard column (ODS, 5 µm, 100 Å,

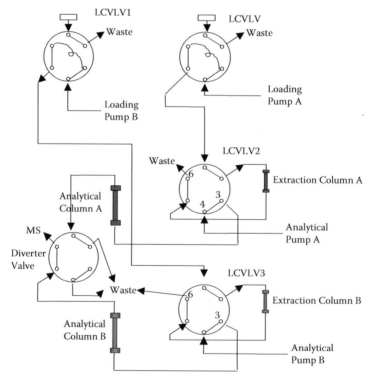

Analytical Pump Goes in 3 = Forward Elution
Analytical Pump Goes in 4 = Backward Elution

FIGURE 2.7 Flow diagram for dual-stream online SPE.

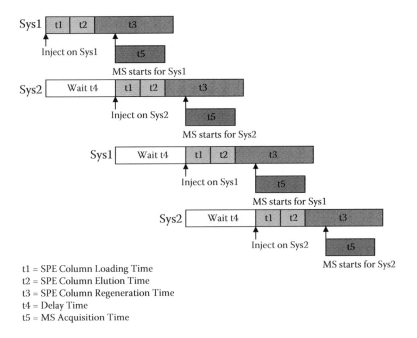

t1 = SPE Column Loading Time
t2 = SPE Column Elution Time
t3 = SPE Column Regeneration Time
t4 = Delay Time
t5 = MS Acquisition Time

FIGURE 2.8 Timing scheme for dual-stream online SPE.

10×3 mm). An 0.3 mL/min isocratic flow of 1/2 v/v acetonitrile/buffer (4.7mM ammonium and 6 mM acetate in H_2O) was used as the mobile phase. Cohesive Technologies' Turbo HTLC (C18, 1.0×50 mm) column was selected to perform online solid phase extraction of fenofibric acid from human plasma.

Online SPE — One HPLC system was used to perform both SPE and HPLC separation. A column switching valve was used to direct the flow from the extraction column to waste (off-line) or the analytical column (online). A second column switching valve directed flow from the analytical column to the mass spectrometer or to waste.

For one stream, the loading solution (2mM acetic acid) initially flowed at 4.0 mL/min through the extraction column to waste for approximately 1 min. The column switching valve then switched so that the mobile phase eluted analytes from the extraction column onto the analytical column. During the elution step, the loading pump transmitted the wash solution directly to waste. After elution, the column switching valve returned to its original position to wash (weak base solution followed by 100% MeOH) and re-equilibrate (2mM acetic acid) the extraction column with a 4.0 mL/min flow rate. At the mean time, the analytical pump continued to elute analytes from the analytical column. With approximately 2.0 min remaining in the run, the mass spectrometer began to acquire data and the second column switching valve directed the analytical column flow from waste into the mass spectrometer. At the end of acquisition, the system was ready for the next injection.

During method development, different elution modes (forward and backward) were investigated. During backward elution (Figure 2.7), peak shapes were generally sharper and the carryover caused by the residue analyte trapped in the extraction column was smaller. However, the guard column tended to over-pressure or started channeling by showing split peaks after approximately 200 injections. It was speculated that matrix residue in the extraction column was transferred more easily to the analytical column during backward elution and the decision was made to adopt forward elution for this assay.

MS condition — An API 3000 equipped with a Turbo Ionspray from Applied Biosystems was used as the mass detector. $[M - H]^-$ was chosen as the precursor ion for multiple reaction monitoring

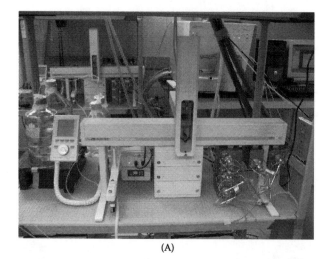

(A)

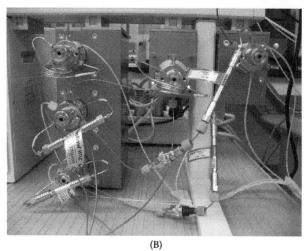

(B)

FIGURE 2.9 Autosampler set-up for performing dual-stream online SPE: (A) overall view, (B) zoomed view of the valving configuration.

(MRM). A transition of m/z 317.0 → 231.0 was chosen for fenofibric acid and m/z 267.0 → 195.0 was selected for the Pestanal structure analog internal standard.

Sample preparation — Analytes of interest were extracted from human plasma using the online solid phase extraction technique. The steps required are noted below:

1. Add 150 μL of internal standard (1 μg/mL Pestanal in ACN) solution to each appropriate well of a 96-well plate. Manually add 150 μL of ACN to the double blank.
2. Add 50 μL of standards, QCs, blank plasma, and unknowns (if applicable) to appropriate wells of a clean 96-well plate.
3. Add 50 μL of diluent (2mM acetic acid) to appropriate wells of a 96-well plate.
4. Cover the plate and vortex on low/medium speed for approximately 2 min.
5. Centrifuge for approximately 10 min at approximately 3400 rpm.
6. Inject 30 μL, extract, and analyze by SPE/HPLC/MS/MS.

2.2.2.3 Validation Results

Accuracy, precision, and linearity of standards — The linear dynamic range was established as 0.017 to 5.472 µg/mL with a coefficient of determination (r^2) below 0.993860 when using $1/x^2$ weighing for three consecutive accuracy and precision runs. The lower limit of quantitation (LLOQ) was accurate (inter-run mean bias = –0.6%) and precise (inter-run mean CV = 5.9%). Four levels of QCs were prepared. The inter-run mean bias varied from –3.5 to 9.9% and the inter-run mean CVs were less than 5.7% for all QC levels.

Selectivity — To demonstrate selectivity, six lots of matrix with and without IS were screened for interference from endogenous matrix components. No interference was observed for fenofibric acid. Figure 2.10 is a representative chromatogram.

Matrix effect — To demonstrate that the assay performance was independent from the sample matrix, low QC samples were prepared using six different lots of matrices. The QC samples were evaluated using the same calibration curve. With regard to analytical recovery, no significant differences were observed for the QCs prepared in six lots of plasma.

Recovery — Recovery control (RC) solutions were prepared in 10/90 v/v ACN/water. Recovery evaluation (RE) samples were prepared in human plasma. Aliquot of RC solutions into assay plates followed sample preparation procedure steps 1 and 2. Instead of adding 50 µL of diluent, wells containing RC solutions were dried down under a steady stream of room temperature N_2. The dried wells were then reconstituted with 250 µL of diluent. Reconstituted RC solutions were directly injected onto an HPLC analytical column, bypassing the extraction column. RE samples were aliquoted into an assay plate following normal sample preparation. RE samples were analyzed using the full extraction procedure (with extraction column). The analyte was tested at three concentration levels and the internal standard was tested at one. Mean extraction recovery for fenofibric acid varied from 93.2 to 111.1%, and mean extraction recovery for the Pestanal internal standard was 105.2%.

Stability — Samples remained stable for at least 220 days when frozen at –20°C. They were stable for 7 simulated freeze-and-thaw cycles and approximately 44 hr at room temperature. The analyte was viable at least 2 days in the autosampler (temperature set point at 10°C). The batch was stable for 5 days in a refrigerator (temperature varied from 4 to 8°C). A stock solution of fenofibric acid is stable at least 50 days; the stock solution for internal standard is stable at least 10 days under refrigeration.

System carryover — Because a lot more valving and plumbing are involved with online SPE compared to regular LC/MS/MS analysis, wash solvents and wash sequences were extensively screened during method development to minimize system carryover. Table 2.1 shows the chronology of a validated wash program. The autosampler flush was also optimized to minimize carryover. Two organic washes (with MeOH) and two aqueous (water) washes were used on the injection ports. One aqueous wash of the injection port was put in place prior to sample aspiration to minimize the risk of sample precipitation in the injection port. System carryover was evaluated during validation by comparing the peak area of the single blank injected immediately after the ULOQ to that of the LLOQ. The peak area of the single blank was less than 30% of the area of the LLOQ. The system carryover was deemed acceptable.

2.2.2.4 Assay Performance

Sample throughput — Sample analysis (including analytical repeats and 5% repeats of samples required to determine method reproducibility) was performed between 18 October 2006 and 16 November 2006. Analytical results from 3293 samples were finalized.

Standard performance — The assay performed well during study M06-830. Mean bias at LLOQ was –1.9%, and mean bias for other standards varied from –1.9 to 1.3%. Coefficient of determination (r^2) exceeded 0.995917 for all batches run.

QC performance — The mean bias for four levels of QCs varied from –0.7 to 1.0%. The mean bias for dilution QCs was 2.9%. Figure 2.11 plots QC performance.

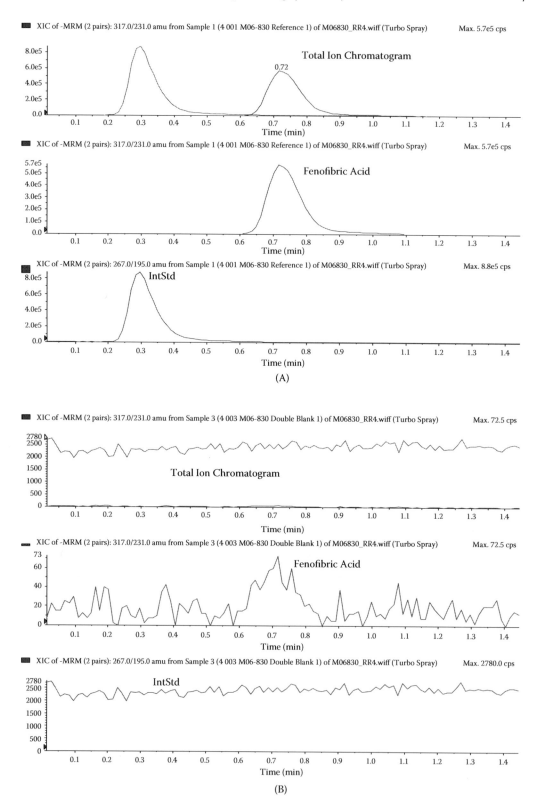

FIGURE 2.10 Representative chromatograms of reference and blank: (A) reference solution, (B) double blank sample.

TABLE 2.1
Example Wash Program

Time	Parameter	Setting	Description
0.00	Flow	4.0 mL/min	Ramp flow up to 4.0 mL/min
0.05	%B, %C, %D	Off, Off, 0%	Load extraction column with 2mM acetic acid
1.10	%B, %C, %D	Off, Off, 0%	
1.20	%B, %C, %D	100%, Off, Off	Wash extraction column with weak base solution
2.10	%B, %C, %D	100%, Off, Off	
2.20	%B, %C, %D	Off, Off, 100%	Wash extraction column with MeOH
3.10	%B, %C, %D	Off, Off, 100%	
3.20	%B, %C, %D	Off, Off, 0%	Re-equilibrate extraction column with loading solution
4.20	%B, %C, %D	Off, Off, 0%	

A = Loading solution, 2mM acetic acid

B = Weak base solution, 7.7 g NH_4OAc + 5 L H_2O + 1 mL NH_4OH

C = NA

D = MeOH

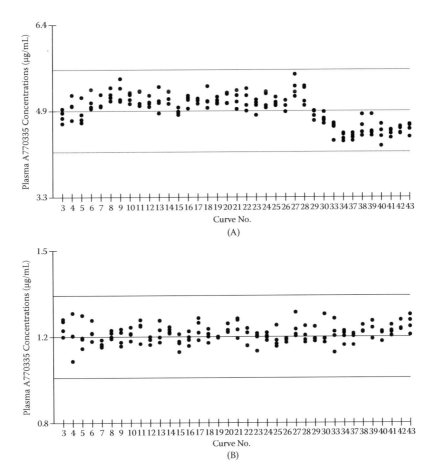

FIGURE 2.11 Plot of QC performance for Study M06-830: (A) QC1 (4.869 μg/mL), (B) QC2 (1.169 μg/mL), (C) QC3 (0.234 μg/mL), (D) QC4 (0.047 μg/mL), (E) Dil QC (8.115 μg/mL).

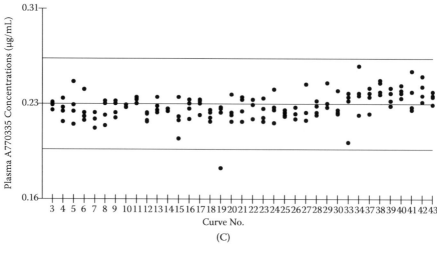

(C)

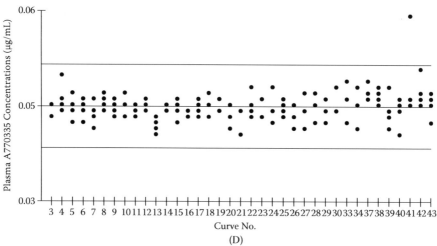

(D)

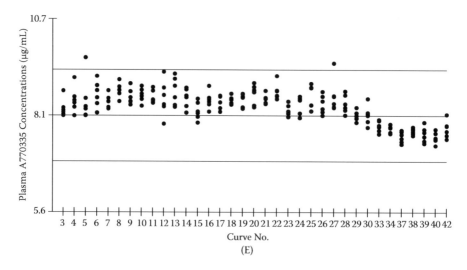

(E)

FIGURE 2.11 (Continued).

TABLE 2.2
Reassay Statistics for Study M06-830

Subject	Percent Bias Range	Mean Percent Bias	Overall Mean Percent Bias
101	–6.3 to 16.7	6.4	2.0
123	–11.1 to 5.8	–3.0	
145	–8.4 to 8.6	0.1	
166	–4.2 to 18.7	4.4	

Percent bias = (second analysis value – first analysis value)/first analysis value × 100.

Method reproducibility — Individual incurred samples from four subjects (approximately 5% of all samples) were re-assayed individually to evaluate reproducibility. The four samples set for reanalysis and evenly spaced throughout the study were designated 101, 123, 145, and 166. The values generated from the reassays were used only to assess reproducibility and were not used in pharmacokinetic calculations. Table 2.2 summarizes the method reproducibility results. The analytical method used in study M06-830 was accurate, precise, and reproducible.

REFERENCES

1. Takats, Z., Wiseman, J.M., and Cooks, R.G., *Journal of Mass Spectrometry*, 2005, 40, 1261.
2. Cooks, R.G. et al., *Science*, 2006, 311, 1566.
3. Cody, R.B., Laramee, J.A., and Durst, H.D. *Analytical Chemistry*, 2005, 77, 2297.
4. Williams, J.P. et al., *Rapid Communications in Mass Spectrometry,* 2006, 20, 1447.
5. Jemal, M., *Biomedical Chromatography*, 2000, 14, 422.
6. Yang, L. et al., *Analytical Chemistry*, 2001, 73, 1740.
7. Cheng, Y.F., Lu, Z., and Neue, U. *Rapid Communications in Mass Spectrometry*, 2001, 15, 141.
8. Berna, M.J., Ackermann, B.L., and Murphy, A.T., *Analytica Chimica Acta*, 2004, 509, 1.
9. Hsieh, S. and Selinger, K. *Journal of Chromatography, B*, 2002, 772, 347.
10. Mazzeo, J.R. et al., *Analytical Chemistry*, 2005, 77, 460A.
11. Hoke, S.H. et al., *Analytical Chemistry*, 2001, 73, 3083.
12. Cabrera, K., *Journal of Separation Science*, 2004, 27, 843.
13. King, R.C. et al., *Rapid Communications in Mass Spectrometry*, 2002, 16, 43.
14. Chen, J. et al., *Journal of Chromatography B*, 2004, 809, 205.
15. Herman, J.L., *Rapid Communications in Mass Spectrometry*, 2002, 16, 421.

3 Optimizing LC/MS Equipment to Increase Throughput in Pharmaceutical Analysis

Michael G. Frank and Douglas E. McIntyre

CONTENTS

3.1 INTRODUCTION

It is a well accepted fact that the pharmaceutical industry faces a productivity problem in delivering new efficient drugs.[1,2] Although spending for research and development has increased continuously, the numbers of new drug submissions continue to decrease. Optimizing the overall process is a major initiative for all large pharmaceutical companies—both in a global sense and in relation to the details.

Participants in the drug discovery arena responded to the hype surrounding the brute force approach known as combinatorial synthesis by enlarging the sizes of compound libraries for high-throughput screening aimed at finding promising new drug motifs. This led to compound repositories containing more than 10^6 compounds, each. Enormous endeavors were undertaken to industrialize chemical synthesis to generate these numbers of compounds. Parallel pipetting robots and fully automated sample workup and preparation systems found theirs way into chemical synthesis labs.

In recent years, a new trend toward smaller compounds—drug fragments—is taking shape,[3] but all these undertakings still require analyses of the compounds produced. Furthermore, at each downstream step, we see analyses of small chemical compounds starting with *in vitro* assays, *in vivo* assays during drug optimization, and finally in clinical trials. Each analysis serves as a filtering step

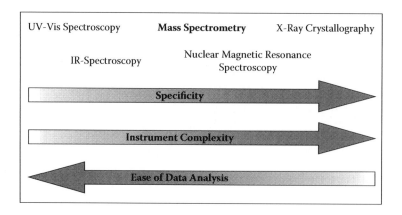

FIGURE 3.1 Comparison of atmospheric pressure ionization mass spectrometry and other widely used analytical methods for analysis of small organic molecules.

on the road to finding a suitable drug candidate. Of all known analytical methods, the most commonly performed are liquid chromatography/mass spectrometry (LC/MS) analyses, usually reverse phase liquid chromatography coupled with atmospheric pressure ionization mass spectrometry. Why is this so? Because LC/MS represents a good balance of achievable data quality and required effort (Figure 3.1).

If we look to light absorption-based methods like ultraviolet or visible light absorption spectroscopy (UV/Vis), the specificity of MS is much greater. MS can determine for a given compound the nominal mass of the molecular ion. Certain types of mass spectrometers even reveal its accurate mass, thus narrowing the number of possible total formulas significantly. In cases where tandem mass spectrometry (MS/MS) is used, unique fragmentation patterns may be obtained. The specificity is comparable to that of nuclear magnetic resonance spectroscopy (NMR) but data evaluation is usually significantly easier. Specificity is greater only with x-ray crystallography that delivers exact molecular structures but requires must more analytical effort in instrumentation and sample preparation.

The introduction of atmospheric pressure ionization (API) interfaces such as the electrospray interface (ESI) or atmospheric pressure chemical ionization interface (APCI) in combination with liquid chromatography as reliable and reproducible means of separating mixtures into individual compounds exploded the usage of MS. In contrast, combining NMR analysis with liquid chromatography for separation may be troublesome because solvent suppression is required and sensitivity may become an issue. Another big advantage of LC/MS is the ability to quantify very precise amounts of a specific compound. Despite the advantages noted, LC/MS never reached the throughput achievements of light absorption and fluorescence spectroscopy.

Certain plate readers require less than a minute to analyze a complete 384-well microtiter plate through the use of fast moving optical mechanics for scanning microtiter plates and relatively simple detection electronics. Some readers even image a complete microtiter plate in one operation. Not even MALDI/MS—probably the fastest MS technique available now—can reach this throughput. A speed of approximately 20 to 30 min for a 384-well microtiter plate is a reasonable rate for optimized high-throughput conditions.[4] The fastest flow injection analysis (FIA) MS systems without chromatographic separation and parallel sampling achieve cycle times in the range of seconds per sample. This translates to a throughput of a 384-well microtiter plate in about 1 hr.[5] Finally, LC/MS systems allowing chromatographic separations of samples operate at rates of a minute to several hours—the latter is certainly not a high-throughput application.

LC/MS covers a broad application area in the pharmaceutical development field. Table 3.1. provides a brief overview of such application areas. It should be emphasized that a negative

TABLE 3.1

Typical Applications of LC/MS in Pharmaceutical Industry

Development Stage	Task	Focus	HT	Matrix Complexity
Drug Discovery	Chemical library synthesis	Purity and identity determination	Yes	Low
	Natural product identification	Identification	Yes	Medium to high
	Medicinal chemistry drug optimization	Purity and identity determination	No	Low
	ADME properties	Quantitation	Yes	Medium
	Metabolite identification	Identification	No	Medium to high
Drug Development	Metabolite identification	Identification and quantitation	No	Medium to high
	Impurity and degradant identification	Identification and quantitation	No	Low
	Pharmacokinetic profiling	Quantitation	Yes	Medium
Clinical Trials	Human PK and metabolite identification	Identification and quantitation	Yes	Medium to high
Manufacturing	Impurity and degradant identification	Identification and quantitation	No	Low

HT = high throughput.

high-throughput indication does not necessarily mean that high speed analysis is not favored. For example, a medicinal chemist would be very happy to receive analytical results of a synthesis optimization promptly in order to proceed with his work. In this case, turnaround time is more important than numbers of samples run.

The complexity of a sample matrix increases if the sample is derived from a biological system. Note that the number of analyses per compound dramatically increases as the complexity of determining compound properties increases, while the number of individual compounds decreases by filtering out unsuitable structures. At the end of a clinical trial phase, perhaps 100,000 LC/MS analyses may have been performed for one compound (Figure 3.2).

These issues have implications for the use of LC/MS. During drug discovery and the early stages of drug development, generic LC/MS methods for separation and identification are used along with continuously changing methods for quantitative applications with single ion or multi-reaction monitoring. Later in the process, highly specialized and fully validated methods are used thousands of times for only a few compounds.

The issue is how LC/MS analysis—by nature a serial procedure—can meet the demands of parallelized sample generation and preparation. This can be achieved by speeding up the serial LC/MS, parallelizing parts of the analysis, or both. Several aspects of optimization should be considered:

- How good is the data quality; what are the consequences of incorrect answers?
- What overall speed/throughput covering the total workflow can be achieved?

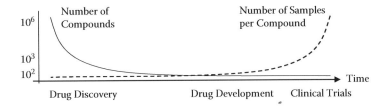

FIGURE 3.2 Development of compounds and samples per compound.

- How complex are the hardware and software systems?
- Are the hardware and software system components robust?
- What is the cost of the solution including training, maintenance, and other issues?
- Is the final methodology easy to use, to operate, and to maintain?

Finally, optimization means dealing with time and other improvements spanning the overall process. Optimizing the speed of the analysis is obvious, but optimizing resolution can improve the process as well (as we will see later). An economic optimization of individual analyses will result in time improvements throughout the process because it will liberate resources for other tasks.

3.2 OPTIMIZING SERIAL LC/MS OPERATIONS

When we talk about optimization of serial LC/MS operations, we consider the genuinely serial sequences of actions necessary to perform such analyses including equilibration of columns, sample aspiration, sample injection, isocratic or solvent gradient sample separation, detection, and column washout.

3.2.1 PEAK CAPACITY

Obviously, the most time consuming subtask is the separation although subsequent subtasks may compete with separation in that regard. What determines the chromatographic separation time? A useful concept is the use of chromatographic peak capacity n_c to determine required separation times. Peak capacity is a measure of how many peaks can be separated during a chromatographic run time. Equation 3.1 defines peak capacity for a gradient separation. Equation 3.2 calculates peak capacity under isocratic conditions. Peak capacity depends on the available time for separation and the widths of detected peaks.

$$n_c = 1 + \frac{t_g}{w_b} ; \qquad t_g = \text{gradient time}, \; w_b = \text{peak width} \tag{3.1}$$

$$n_c = 1 + \frac{t_r - t_A}{w_b} ; \qquad t_r = \text{retention time of the last peak}, \; t_A = \text{column dead time} \tag{3.2}$$

The peak capacity approach is very similar to the use of the column efficiency to determine the expected quality of a chromatographic separation, but it conveys the advantage of being easily combinable with the peak capacities of orthogonal methods, for example mass spectrometry, to determine total peak capacity ($n_{c,total} = n_{c,LC} \cdot n_{c,MS} \cdot n_{c,x} \cdot \ldots$). We can define the total peak capacity for LC/MS analysis as the product of individual peak capacities of all compound separating means. This covers both the separation in the time domain by chromatography and also separation in the mass domain by the mass spectrometer. Because peak capacity in the mass domain is usually tightly bound to the type of spectrometer and a significant increase usually requires a change in MS instrumentation, this section focuses only on optimization in the chromatographic domain.

A reduced peak capacity in one domain may be counterbalanced by an increased peak capacity in another domain. If we know the average peak width of a chromatographic separation and the gradient duration, we can calculate the maximum number of peaks that can be separated. (*Note:* peak capacity does not mean that this number of compounds in a sample will be separated; they may still co-elute). That means we can operate between two limits: (1) a peak capacity of zero representing a flow injection analysis and (2) a minimal required peak capacity that defines the peak capacity to separate all compounds in a given mixture. Unfortunately, especially in the early stages of drug discovery when new compounds must be handled, this number is usually unknown. Because the main reason for this process is to filter out many compounds, no major effort such as LC method optimization will be undertaken to determine the required peak capacity for each separation problem. Researchers in this environment usually select an arbitrary number for the available peak capacity of their chromatographic systems—usually determined by other factors like maximum acceptable run time. This means that a specific column (often one preferred for historical reasons) is used along with a generic gradient covering a wide range, for example going from 5 to 95% organic in a given gradient time or (even worse) with a set of standardized isocratic conditions. The resulting peak capacity is then accepted as is, but slight changes can make the analysis much more efficient, reduce potential for co-eluting peaks, and improve data quality.

Some LC/MS users adhere to isocratic separation because of the myths around gradient elution (it is complex to develop and transfer between instruments and laboratories, it is inherently slower than isocratic methods because of re-equilibration, and other reasons summarized by Carr and Schelling[6]). A researcher may have a very good reason to use an isocratic method, for example, for a well defined mixture containing only a few compounds. The isocratic method would certainly not be useful in an open access LC/MS system processing varying samples from injection to injection.

Carr and Schelling[6] state in detail how the reservations many chromatographers still have toward gradient elution no longer hold true: "…gradient elution provides an overall faster analysis, narrower peaks and similar resolution of the critical pair compared to isocratic elution without loss in repeatability of retention time, peak area, peak height, or linearity of the calibration curve."

If a chromatographer decides to use gradient separation, optimization can achieve better peak capacity. Neue[7,8] clearly summarized all factors related to optimization of peak capacity. Neue takes the known relations of peak width to column efficiency and capacity factors, the relation of the capacity factor in gradient separation to gradient time, flow rate, and gradient steepness, and finally the dependence of column efficiency ($N = L/H$ where $N =$ efficiency, $L =$ column length, and $H =$ theoretical plate height) as described in the well known van Deemter curve to reach Equation 3.3.[8]

$$n_c = 1 + \frac{1}{4} \cdot \sqrt{\frac{L}{a \cdot d_p + \frac{b \cdot D_M \cdot t_0}{L} + \frac{d_p^2}{c \cdot D_M} \cdot \frac{L}{t_0}}} \cdot \frac{B \cdot \Delta c}{B \cdot \Delta c \cdot \frac{t_0}{t_g} + 1} \tag{3.3}$$

Where a, b, and $c =$ van Deemter coefficients, $d_p =$ particle size of column, $L =$ column length, $D_M =$ diffusion coefficients of analytes, $t_0 =$ column dead time (depends on flow rate F), $t_g =$ gradient time (determines analysis time via $t_A = t_g + t_0$), $\Delta c =$ difference in concentrations of the organic modifier at the end and the beginning of the gradient (a continuous linear gradient is assumed), and $B =$ slope of the linear relationship between the logarithm of the retention factor and the solvent composition.

This complex-looking equation now contains all the variables we need to optimize the chromatographic separation by improving peak capacity without changing the run time (improving resolution), by reducing the run time and maintaining peak capacity (improving speed), or a combination of both approaches.

3.2.2 Optimizing Speed of Chromatographic Separation

One problem is how to optimize throughput (analysis time) without losing peak capacity. Different approaches have been suggested and led to different developments by instrument and column manufacturers. This section will concentrate on the usage of totally porous particle columns for chromatographic separation only. Alternatives are monolithic columns[9] and shell packing materials such as Halo or Poroshell.[10–13]

A van Deemter plot for a given particle size d_p and diffusion coefficient D_M shows the relation of the theoretical peak height H to the linear velocity u that can be expressed as column length L

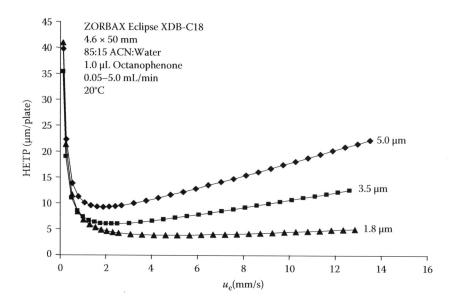

FIGURE 3.3 Van Deemter plot of columns with different particle sizes. (Courtesy of Agilent Technologies, Inc.)

divided by column dead time t_0 ($u = L/t_0$). We see that plate height decreases with smaller particles that exhibit lower H_{min} levels at a higher optimal linear velocity and that the slope of the *van Deemter* curve is flatter than for larger particle sizes (Figure 3.3). Simply stated, by using smaller particles to pack a chromatographic column we improve efficiency.

We can reduce column length to maintain constant efficiency and save analysis time because compounds will elute earlier. For example, if we consider a 4.6 × 50 mm column with 5 μm particles and change to modern sub-2-micron (e.g., 1.8 μm) particles, we should achieve the same efficiency with a column only 18 mm long because $N \sim L/d_p$. With the same flow rate, the compounds should elute 2.8 times earlier (5 μm/1.8 μm).

This simplified view neglects the fact that smaller particles reach their maximum plate counts at higher linear velocities than larger particles. The full scope of optimizing chromatography can be deduced from Equation 3.3. In Figure 3.4, the peak capacities of a 4.6 × 50 mm column with a conventional 5 μm particle size stationary phase are calculated for different flow rates and gradient times. With the conventional column, a rather long gradient time of ~60 min is required to achieve the maximum peak capacity of ~190 (consistent with the intuitive maxim that a longer gradient time will always result in a better separation). The flow rate is ~0.6 mL/min.

But what happens after a switch to a stationary phase with 1.8 μm particle sizes, if we first consider the same column length as a conventional column and then an 18 mm column length? Generally, with 1.8 μm particles, the achievable peak capacity is much higher, as expected. Calculating the peak capacity for a 50 mm length sub-2-micron column shows for the maximal peak capacity of the conventional column ($n_{c,max} \approx 190$) a curve with a minimal gradient time of only ~5 min at a flow rate of ~2.5 mL/min. This means in theory a 12-fold gradient time reduction and about a 3-fold reduction in solvent consumption (conventional = 36 mL, sub-2-micron = 12.5 mL). For a 18 mm length column with the same length-to-particle-size ratio as a conventional column, we find the same peak capacity at that flow rate (0.6 mL/min) with a gradient time of ~13 min. This is 4.6 times faster than conventional techniques and saves even more solvent: only 7.8 mL would be consumed.

At a first glance, a very promising procedure for speeding up LC/MS analysis without compromising peak capacity of the separation would involve (1) reduction of particle size, (2) reduction of

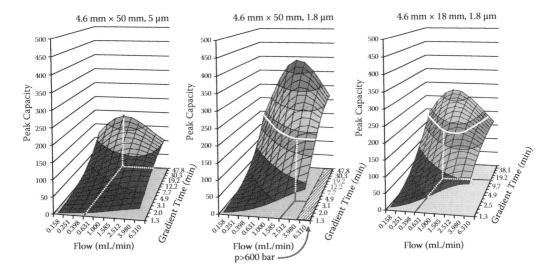

FIGURE 3.4 Calculated peak capacities dependent on flow rate and gradient time. Left: conventional column using 5-μm particles. Middle: same column dimension with sub-2-micron particles (1.8 μm). Right: sub-2-micron particles in column with same L/d_p ratio as conventional column on left. Parameters for typical applications have been estimated. Note: logarithmic scale of flow rate and time axis.

column length, and (3) increasing the flow rate. Figure 3.4 reveals one problem: the pressure drop of sub-2-micron columns is much higher than with larger particles. The pressure drop of a column follows an inverse proportionality to the square of the particle size:

$$\Delta p = \frac{u \cdot \eta \cdot L \cdot \Phi}{d_p^2} \tag{3.4}$$

where u = linear velocity, η = viscosity, L = column length, Φ = flow resistance factor, and d_p = particle size.

Several commercial HPLC manufacturers addressed this issue by increasing the operation ranges of their equipment either by increasing the maximum allowed backpressure and/or increasing the

TABLE 3.2

Commercially Available LC Instrumentation Optimized for Sub-2-Micron Particle Column Use

Manufacturer	Brand	Year Introduced	Maximum Pressure (bar/psi)	Maximum Temperature (°C)	Maximum Flow Rate (mL/min)
Agilent	1200 Series Rapid Resolution LC	2006	600/8700	100	5
Dionex	Ultimat 3000	2006	500/7250	85	10
Jasco	X-LC	2005	1000/15000	65	2
Thermo-Fisher	Accela	2006	1000/15000	95	1
Waters	Acquity UPLC	2004	1000/15000	90	1[a]

[a] 2 mL/min below 600 bar

temperature range of the column oven to reduce the viscosity of the solvent. Table 3.2 provides an overview of HPLC instruments optimized for sub-2-micron particle sizes (specifications as of the time of writing). For an overview of the Agilent 1200 Series Rapid Resolution LC system, the Thermo-Fisher Accela system, and the Waters Acquity system, see Cunliffe et al.[14] Researchers often refuse to increase temperatures in LC separations, even though this produces several beneficial effects such as reducing backpressure, producing narrower peaks (better signal-to-noise ratio), requiring less organic modifier to achieve the same retention ("green chemistry"), and reducing the retention times (faster analysis). This is due to concerns about possible decomposition of the column and/or the analyte. If columns are used under the specified conditions, especially at low pH, their stability is usually sufficient to allow continuous operation.[15] On the other hand, it is difficult to generalize about analyte stability. Note that conditions inside a chromatographic column are different from those when high temperatures are applied to a sample under "normal" conditions, that is, in a solution or a solid state. Decomposition often is related to oxidative processes or dehydration. During LC analysis, the analytes are maintained in an aqueous solution under strictly anaerobic conditions and the exposure time of a sample to elevated temperatures is decreased by the resulting higher speed of analysis. See Smith[16] for a recent review on using only water as mobile phase at very high temperatures.

Both LC modules and columns must withstand higher generated backpressures. Table 3.3 presents an overview of commercially available columns for use with particle sizes below 2.0 μm. Wu[17] cites additional manufacturers and the numbers increase continuously. According to Kofmann et al.,[18]

TABLE 3.3
Commercially Available Columns Packed with Sub-2-Micron Particles

Manufacturer	Particle Size (μm)	Maximum Pressure (bar/psi)	Maximum Temperature (°C)	Available Phases	Available Dimensions
Agilent Zorbax RRHT	1.8	600/8700	90 (Stable Bond) 60 (all others)	Eclipse Plus C8, C18, PAH; StableBond C8, C18, CN, Phenyl, AQ; Extend C18, Eclipse XDB C8, C18; Rx-Sil (HILIC)	1.0, 2.1, 3.0, 4.6 mm ID; 30, 50, 100, 150 mm length
Grace Alltech Vydac	1.5	N/A	N/A	Platinum C18, C18-EPS; HP HILIC, HiLoad, C18-AQ; ProZap C18	N/A
Macherey-Nagel Nucleodur	1.8	N/A	N/A	Gravity C8, C18; ISIS C18, Pyramid C18, Sphinx RP	2, 3, 4, 4.6 mm ID; 50 mm length
Restek Pinnacle	1.9	N/A	80	C18, PFP-Propyl, Biphenyl, Aq-C18, Silica	2.1 mm ID; 30, 50, 100 mm length
Thermo-Fisher Hypersil	1.9	1000/15000	N/A	C18, PFP, AQ	1.0, 2.1, 3.0 mm ID; 20, 30, 50, 100 mm length
Waters Acquity	1.7 (BEH) 1.8 (HSS-T3)	1000/15000	N/A	BEH C8, C18, Shield RP-18, Phenyl, HILIC; Atlantis T3	1.0, 2.1 mm ID; 30, 50, 100, 150 mm length

Note: not all combinations of stationary phase, column internal diameter, and column length are available. For analytical comparisons, see References 36 and 37. Names of columns and/or stationary phases may be trademark protected. N/A = not available. ID = inner diameter.

sub-2-micron columns do not achieve their theoretically expected efficiencies even under optimized instrument conditions. This may be due to failure to fully optimize packing conditions since this technology is relatively new and column manufacturers are still learning. Nevertheless, sub-2-micron columns are still far better than 3.5 or 5 μm particle columns of the same length.

Based on Table 3.2, we can deduce from the available flow rates that some manufacturers clearly optimized their systems for 2.1 mm inner diameter (ID) columns, namely the Jasco X-LC, the Thermo Accela, and the Waters Acquity. By sacrificing flexibility in column ID, these systems have been completely optimized to these 2.1 mm columns. Achieving that will be explained below. The instruments of Agilent and Dionex mentioned in Table 3.2 involve a more flexible approach. The higher flow rate limits of their systems allow the use of columns with larger Ids. Users benefit from the better efficiency obtained with 4.6 or 3.0 mm ID columns instead of 2.1 mm ID columns of similar length; and their systems are fully compatible with existing conventional methods.

3.2.3 Extra-Column Band Broadening

While the foregoing information may appear straightforward, note that Equations 3.3 and 3.4 are valid only for the column bed and do not apply to the LC instrument. The LC hardware around the column may be responsible for sometimes dramatic reductions of calculated column efficiency or peak capacity by extra-column dispersion (*Note:* dispersion occurs inside a column as well and is covered by the van Deemter relation). Dispersion is the sample band spreading or dilution that occurs in connecting tubing, injection valves, flow cells, and in column hardware such as frits and end fittings. It begins with the injector and ends at the last detector in the system. Dispersion reduces measurable column efficiency and degrades resolution.

Obviously, extra-column dispersion exerts its most adverse effect on peaks with very small volumes. Very low volume peaks appear under the following conditions:

- Small column volume (internal diameter and length)
- Low retention factor k' ($k*$ in gradient separations)
- Use of small particles (highest N)
- Low mobile phase viscosity (highest N)
- Optimized linear velocity (highest N)

Several of these points are met by applying the optimization steps discussed above, e.g., using smaller particles, shortening column lengths, and reducing solvent viscosity to reduce backpressure. When we consider a virtual column—a packed bed in a purely theoretical sense— we commonly accept that reducing particle size proportional to column length results in columns with at least the same theoretical efficiency. This is true, but only in the theoretical world.

After a column is installed in a HPLC instrument, and depending on the amount of dispersion caused by the many connecting capillaries and fittings, a reduction in achieved efficiency will generally occur after column volume and/or particle size are reduced. The greater efficiency of a column can be realized only if system dispersion does not substantially degrade column performance. The small particle size columns used in low volume configurations are most difficult to use and require the greatest attention with respect to plumbing of the LC system. Estimating the possible efficiency of a column non-empirically is easily accomplished via a simple equation. Column length divided by particle size multiplied by the reduced plate height (a constant of typically 2 to 2.3) results in a value for efficiency [$N_{theor.} = L/(h \cdot d_p)$]. This does not consider viscosity, particle size distributions, imperfections in packed bed density, or other inherent design factors, but provides a benchmark of expectation for column dimension and particle size. In practical use, the empirical result will always be lower. Equation 3.5 can be used to estimate the peak volume of a given column and a compound with a given capacity factor. Table 3.4 shows resulting values for the peak volumes for different

TABLE 3.4
Calculated Peak Volumes, Theoretical Column Efficiencies without Dispersion, and Column Efficiencies with Assumed Dispersion of 20 μL for Different Dimensions and Particle Sizes

Column Dimension and Particle Size	Peak Volume without Dispersion (μL)	Theoretical Efficiency	Efficiency with 20 μL Dispersion
4.6 × 150 mm, 5 μm	229	13000	11000 (85%)
4.6 × 50 mm, 1.8 μm	79	12100	7700 (64%)
3.0 × 50 mm, 1.8 μm	34	12100	4800 (40%)
2.1 × 50 mm, 1.8 μm	17	12100	2500 (20%)

$k' = 2.5$. Reduced plate height = 2.3.

column dimensions along with their theoretical (isocratic) efficiencies and effective efficiencies by system dispersion as calculated using Equation 3.6.

$$V_{peak} = \frac{Vm \cdot (k' + 1)}{\sqrt{\frac{N}{25} + \frac{\sigma}{1000}}} \quad (3.5)$$

$$N_{eff} = 25 * \left(\frac{Vm \cdot (k' + 1)}{V_{peak}} \right)^2 \quad (3.6)$$

where V_m = column void volume, k' = capacity factor, N = theoretic efficiency, N_{eff}, and σ = dispersion. Simply by replacing the conventional column in a conventional HPLC system with a new sub-2-micron column would produce disappointing results. The shorter and narrower the column, the worse the outcome as shown in Figure 3.5 with calculated values for several different column dimensions

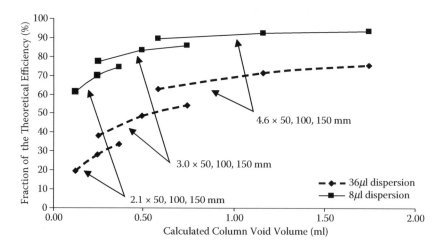

FIGURE 3.5 Calculated efficiency yields (fraction of efficiency with and without dispersion) for different column dimensions, always assuming sub-2-micron particles. Solid lines: assumed 36-μL dispersion. Broken lines: 8-μL dispersion. (Courtesy of Michael Woodman, Agilent Technologies, Inc.)

using sub-2-micron particles with a poorly designed system having 36 μL dispersion (a decent design for a conventional system) and an optimized system of 8 μL dispersion. This means additional care must be taken in setting up an LC/MS system if we wish to obtain results close to the theoretical predictions. Even more effort is required when small ID columns are used. Extra-column dispersion must be kept at a minimum. Unfortunately, a typical LC/MS system has many sources of dispersion:

Component	Dispersion Source
Sampler	Injection volume variation
	Sample aspirating needle and loading/transfer port
	Contact between sampler switching valve and sample
General set-up	Interconnecting tubing (ID, length, internal surface)
	Connectors (unions, tees, bulkhead fittings)
	Switching valves for automated sample treatment
	or alternating column regeneration
	Pre-column filter
	Column frits (inlet and outlet)
Detection	Inlet heat exchangers
	Flow cell volume and geometry
	MS ion sources
	Sprayers (e.g., in evaporative light scattering detectors)
	Data filtering effects in high speed applications

Much system dispersion originates from poor choices of tubing and connections. Tubing may be longer than necessary and have too large an internal diameter (typically 0.17 mm ID, color coded green). It is advisable to shorten tubing length to the minimum needed to connect the components and if possible narrower tubing (0.12 mm ID, color coded red) should be used. Precut and polished tubing should be chosen because self-cutting will usually result in jagged and non-planar ends, creating additional dead volume. Utmost care should be taken when attaching capillaries. Absolutely no gaps must be inside fittings and each capillary must strictly abut against its counterpart.

Because a flow cell acts as very good mixer, care also should be taken for proper selection. Achieving minimum dispersion may not be possible due to analytical issues. For example, if very high sensitivity is required, it may be necessary to use a larger than optimal flow cell with a long path length and higher internal volume. When setting up a system to fit sub-2-micron columns, every part should be carefully considered to determine any contributions to extra column dispersion. Figure 3.6 shows the steps required to optimize a *conventional* LC system with a large dispersion to obtain improved results for accommodating sub-2-micron columns. On such systems, 3.0 or 4.6 mm ID columns would produce results approaching the theoretical maximum (Figure 3.5) even though the full power of sub-2-micron columns would probably not be fully exploited due to system backpressure limitations.

It is usually easy to rearrange the modules of modular LC systems to achieve a short distance from the column to the detectors. For example, a DAD/MS system might be arranged so that the DAD detector sits right next to the MS interface and the column oven right next to the DAD. Figure 3.7 shows how even a very complex system may be optimized for short flow paths. In this case, a diode array detector, evaporative light scattering detector, chemical luminescence nitrogen detector, and single quadrupole MS are linked. As another example, Waters designed its Acquity LC system to include a swing-out column compartment. This allows placement of the column outlet right next to the MS interface, minimizing the amount of connection tubing required. This system also includes length-optimized and very narrow capillaries to minimize dispersion. Agilent's 1200 Series Rapid Resolution LC system includes specially designed low volume heat exchangers in its

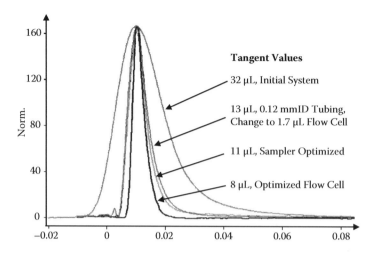

FIGURE 3.6 Decreasing peak dispersion of a conventional Agilent 1100 Series HPLC instrument by stepwise optimization of components. System dispersion determined by injecting low volumes of acetone without a column. (Courtesy of Michael Woodman, Agilent Technologies, Inc.)

column ovens to minimize tubing lengths and offers specially manufactured capillary sets to accommodate every possible module arrangement with the shortest possible capillary lengths. Agilent also abandoned the use of all so-called zero dead volume unions in its Rapid Resolution LC system and replaced them with special female–male connectors in which the capillaries abut each other.

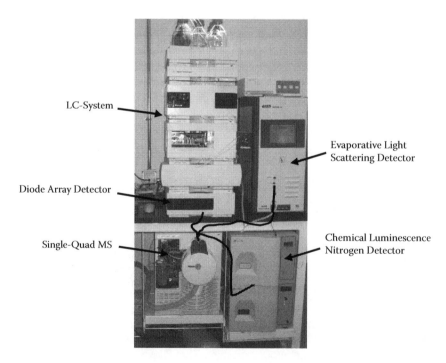

FIGURE 3.7 Optimizing arrangement of LC modules for very complex systems. Four-detector (DAD, ELSD, CLND, and SQ-MS) LCMS system. The capillary connections from the diode array detector have been highlighted for better visibility. (Courtesy of Kenneth Lewis, OpAns Plc.)

A big impact on extra-column band broadening can originate from the injection volume. In gradient separations, a focusing of the compounds at the column head will occur for highly retained compounds and the injection volume is not that important for band spreading. For weakly retained compounds and isocratic separations, the injected volume of sample will exert significant influence on extra-column dispersion. The smaller the column ID and length, the lower the highest acceptable injection volume. For sub-2 micron particle columns of 50 to 100 mm in length with internal diameters of 2.1 to 3.0 mm, the injection volumes should be in the range of 1 to 5 μL.[17]

Following these principles to minimize extra-column band broadening and using a suitable detector will produce good results by utilizing sub-2-micron columns with a 400 bar/6000 psi HPLC system. It is possible to upgrade an existing system by (1) using a sub-2-micron column and low dispersion optimization, (2) upgrading the detector, and finally (3) upgrading the pump and autosampler to higher pressure ranges to achieve higher gradient speeds.

3.2.4 SYSTEM BACKPRESSURE

As with the peak capacity calculation (Equation 3.3), the backpressure calculation in Equation 3.4 also applies only to the column. The flow through LC instrument components will generate additional backpressure by friction. The interconnecting capillaries also add significant backpressure to a system. The backpressure of a capillary under the flow rates used in HPLC is inversely proportional to the fourth power of the capillary diameter ($\Delta p \sim 1/d_{cap}^{4}$). Thus, as column diameter is narrowed, narrower capillary diameters should be chosen to minimize dispersion, but system-generated backpressure grows over-proportionally compared to the column-generated backpressure. In some commercially available LC systems optimized for the use of 2.1 mm ID sub-2-micron columns, the low dispersion values are achieved by the use of very narrow capillaries that unfortunately generate significant backpressure.

By replacing conventional 3.5 or 5 μm columns with sub-2-micron columns, gradient time can be reduced dramatically. The flow rate must be increased for optimal conditions as well but solvent consumption will be less than the amount used by the original method. To use the full power of these columns, an LC instrument must be thoroughly optimized toward lowest extra-column dispersion. The smaller the column (small ID and short length), the more sensitive the performance is to dispersion. With smaller internal diameter columns, the injection volumes and internal diameters of the capillaries should be reduced.

3.3   OPTIMIZING DETECTORS

Our optimized LC/MS system includes detection devices. The most common types are ultraviolet (UV) detectors and mass spectrometers, but additional devices such as evaporative light scattering detectors (ELSDs), charged aerosol detectors (CADs), chemical luminescence nitrogen detectors (CLNDs), fluorescence detectors, radioactivity detectors, and others are available. Especially in early stages of drug discovery, complex instruments including several different types of orthogonal detectors (see Figure 3.7) may be set up as generically as possible and aim to detect everything.

Separation is only as good as the detection of the separated compounds. It is nowadays possible to achieve better separation of compounds than many detectors employed can measure. This means that a close examination of the detection device is required to truly optimize LC/MS. To check the utility of a detector for inclusion in a fast LC system, the following points may help:

- Does the detector induce significant band broadening by its design?
- What are the flow rate limits of the detector?
- What is the highest data acquisition rate?
- What is the signal-to-noise ratio under fast LC conditions?
- What is the data quality under fast LC conditions?
- Is the detector robust enough for the increased samples load?

3.3.1 CONNECTING DETECTORS

When a fast LC system is connected to a detector, care must be taken to ensure that the detector is well suited for the expected flow ranges and peak widths. Most manufacturers, especially those offering dedicated systems for sub-2-micron particle columns, offer efficient UV detectors. Flow rate is usually not an issue for UV and other flow-through cell-based detection systems. However, flow rate can become limiting for dead-end detectors that alter the column effluent, mainly by eliminating mobile phases such as ELSD, CAD, CLND, and mass spectrometers.

Determine that a detector does not cause significant band spreading by its design. For example an ELSD detector may have a relatively large spray chamber. Check on its influence on peak widths. Some manufacturers offer only one spray chamber for all applications; others present choices for low to high-flow applications. The same principle applies to CLND detectors. Most MS interfaces available today and especially the most commonly used ESI sources are specified for a flow range of 1 mL/min or less. Those that offer higher flow rates include Agilent's Multimode Source (simultaneous ESI and APCI ion generation up to 2 mL/min), Applied Biosystems'/Sciex Turbo-V ion source with the TurboIonSpray (up to 3 mL/min), and the Thermo Ion Max Source for APCI (2 mL/min reduced to 1 mL/min for ESI). The dedicated source of the Thermo Surveyor MSQplus MS can handle 2 mL/min in ESI as well as can Waters' IonSABRE APCI source.

Staying within ion source specifications by direct connection without splitting the column effluent is possible only with short 2.1 and 3.0 mm ID columns or long 4.6 mm ID columns if sub-2-micron columns are used and their advantage of reducing gradient time is to be exploited. In fact, 3.0 mm ID columns represent good compromises for achieving unsplit MS connections while avoiding the problems encountered with narrower diameters (particularly sensitivity to extra-column volume). Splitting should be always avoided because of the extra-column volume introduced by an active or passive splitter, the peak shape disturbance that often occurs, and possible blocking issues with a sprayer probe—in case an alternative flow path is available, precipitants at the sprayer needle will not be washed away. Furthermore, when mass-sensitive probes like the APCI are used for quantitative measurement, the split ratio must be precisely controlled.

3.3.2 DATA ACQUISITION CONSIDERATIONS

If the LC part is optimized to deliver peaks in a shorter time or more peaks in the same time when compared to a conventional method, we must consider the system's ability to handle data. Because the speed optimization described above will produce much narrower peaks, widths below 1 sec can be achieved easily. However, the **data acquisition rate** and data filtering steps must be considered.

To identify a compound, five data points per peak may be sufficient. Quantitation may require at least 10 data points across a peak. Many of today's laboratories still house standard detectors (UV, ELSD, fluorescence, etc.) with maximum data acquisition rates at or below 20 Hz. Many conventional LC/MS methods acquire data at rates of 5 Hz or less. As shown in Figure 3.8, this is not sufficient for modern speed optimized chromatography. Obviously, selecting the wrong data acquisition rate will nullify all attempts to optimize chromatography.

Single quadrupole mass spectrometers in scanning mode achieve scan rates around 5000 to 10000 m/z/sec. A scan range of 1000 m/z results in a data acquisition rate of 5 to 10 Hz (actually below 5 to 10 Hz because interscan delays must be considered). Probably the fastest single quad mass spectrometers at the time of writing are the Agilent 6140 MSD (10000 m/z/sec), the Thermo Fisher Surveyor MSQPlus (12000 m/z/sec), and Waters' Acquity SQD (10000 m/z/sec). Even with these fast scanning single quad MS systems, data acquisition rates may be too low to properly capture narrow peaks.

Careful setting of the scan range is recommended. The narrower the scan range, the higher the number of data points per peak. This is not necessarily a linear relation; it can depend on the MS. If a large mass range and fast data acquisition rate are required, a time-of-flight (ToF) MS would be

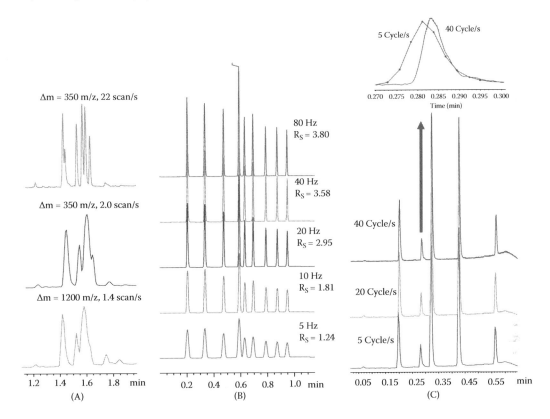

FIGURE 3.8 Comparison of different data acquisition rates under fast LC conditions for (A) single-quad MS varying from 1.4 to 22 scans/sec; (B) diode-array detector varying from 5 to 80 Hz; and (C) time-of-flight MS, varying from 5 to 40 cycles/sec. (Courtesy of Agilent Technologies, Inc.)

the best choice. ToF mass spectrometers do not scan but acquire per transient the entire ion mass range almost instantaneously. An Agilent publication[19] discusses the principles of operation. These devices are the fastest acquiring mass spectrometers with acquisition rates of up to 40 Hz at broad mass ranges exceeding 1000 m/z.

Even if high data rates are supported by instruments, **data quality** must be carefully evaluated. For example, based on physics principles, ToF mass spectrometers lose sensitivity at very high data acquisition rates because fewer transients can be accumulated per data point. Orbital trapping instruments are not so sensitive but they lose one of their great advantages, mass resolution, because the shorter orbiting times mean fewer harmonic oscillations can take place. Resolution can drop from 100,000 at a 1.9-sec scan cycle time down to 7500 at scan cycle times of 0.25 to 0.3 sec.[20] ToF instruments do not show this dependency of resolution and data acquisition rate. Despite the fact that they cannot reach the extreme resolutions as effectively as orbital trapping instruments, they will outperform them at fast acquisition rates. Resolving powers of 14,000 for mass values ~200 m/z are achievable even at a 20 Hz acquisition rate. Typically, no loss in mass accuracy (a key value) is observed. For triple quad and other MS instruments with collision cells, the cells must be cleared quickly enough to prevent cross-talk at high data acquisition rates. Fast scanning single quad mass spectrometers should be able to deliver comparable data quality at high scan speeds, for example, correct isotope ratios.

Another data acquisition consideration is **data file size**. A high speed LC/MS data file can easily reach dimensions of 20 MB/min if maximal information is required and the detectors are set to broadest scan ranges and highest sampling rates without data reduction. LC/MS systems capable of acquiring MS/MS or even MS^n will produce even larger data files. This will stress an acquisition computer especially if other tasks are performed in parallel. Depending on the instrument connection, an installed virus scanner might examine each incoming data package from the detector. Network traffic will increase if data files are stored centrally on a server.

Some optimization may involve storing UV spectra only in regions where a peak is detected or centroiding of mass peaks, preferably done on a firmware level inside a mass spectrometer as an option to high data volume profile spectra. Look for optimization possibilities in your acquisition software or check whether such options are supported if you decide to buy a new instrument. Always consider what data acquisition rate is really required. For example, it makes no sense to use an 80 Hz data acquisition rate with a 150 mm length column even if it is packed with sub-2-micron particles because peak widths will never drop below several seconds even with a perfectly optimized system and at flow and pressure capabilities no commercially available instrument could achieve. Also, attention should be paid to parameters like scan ranges (wavelength range or mass range) and depths of MS^n levels for data-dependent MS/MS acquisitions.

A detector plays an important part in achieving overall performance when optimizing for very short run times with reasonable peak capacities. The detector should match the optimized LC conditions in terms of flow rate and dispersion. The electronics must accurately capture the peak form produced by the column; this is most important for quantitative analyses. The data quality may be limited by the laws of physics at high speed. The amount of data produced over time can become an issue.

3.4 CYCLE TIME OPTIMIZATION

We have discussed individual analyses and the demands to achieve optimization of instrumentation. However, an analytical laboratory must deal with series of samples and we must consider another factor if we want to optimize complete workflows: cycle time optimization. Cycle time is defined as the time from finishing the analysis of one sample to the time the next sample is finished. This can be easily determined on Microsoft Windows®-based operating systems by examining the data file creation time stamps of two consecutive samples. A better way is calculating the average of a reasonable number of samples.

Cycle time consists of several individual components. One is the separation time of a sample. Another component is instrument overhead time that may be subdivided into conditioning, sample preparation, and post-separation phases. The final component is system overhead time that covers delays caused outside the LC modules (Figure 3.9). These times do not necessarily have to follow the fixed order shown in the figure. In particular, the position of the instrument conditioning may vary and the tasks do not have to be arranged linearly. Cycle times in early chromatographic systems

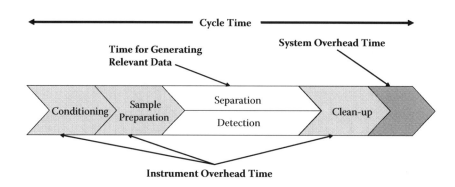

FIGURE 3.9 Breaking down cycle time in serial LCMS analysis.

were strictly serial processes. Refinements allow parallelization of many subtasks as will be discussed in the next section. This section focuses on optimizing the individual parts.

Conditioning relates mainly to the equilibration of the column but also includes balancing of detectors (usually very fast) and resetting temperatures if temperature gradients are used. A classic rule is that columns should be conditioned with at least 10 to 15 column volumes. For a 2.1 × 50 mm column, this means at least 1.0 mL (approximately 100 μL column void volume). Assuming a reasonably fast analysis with a flow rate of 0.8 mL/min we would have to wait 1.25 min just for the conditioning. For some fast LC applications the wait would be longer than the gradient time.

Full column conditioning may be very important if a method transfer is planned (for example, a transfer from method development to a quality assurance or quality control laboratory). For analytical laboratories using generic methods to analyze many samples per day, fewer column volumes may be used and still yield excellent run-to-run repeatability and retention time precision. A single column volume may be sufficient to establish a highly reproducible state of equilibration on a column, as examined in detail by Carr et al.[21] Series of analyses with methods using few column volumes for equilibration should always start with two or three blank runs to ensure that the column is in a steady-state before the first sample run starts. The equilibration state can be effectively monitored on the pressure read-back of the instrument as illustrated in Figure 3.10. Using only two column volumes for a 2.1 × 50 mm column would require only 200 μL for equilibration and only 0.25 min equilibration time at 0.8 mL/min instead of 1.25 min—a significant time saving.

Although detector balancing takes only a few seconds on typical instruments, the need to do so should be assessed. The time and numbers of samples lost over time to balance a detector may be considerable. Perhaps only a detector balance between sample plates will do. This can be accomplished by so-called pre-run macros or scripts that can be executed between individual sample lists by many acquisition software packages.

The sample preparation step in Figure 3.9 in the simplest case would include only the aspiration, but often steps like dilution, pre-concentration, and other treatments will be performed before injection of the sample onto the separation column. This is easily performed on modern instruments that allow injector programming and the use of additional valves. This will be discussed further in the next section about parallelizing steps. If parallelizing is not possible and a purely serial analysis is required, the required treatment should be abandoned or performed off-line.

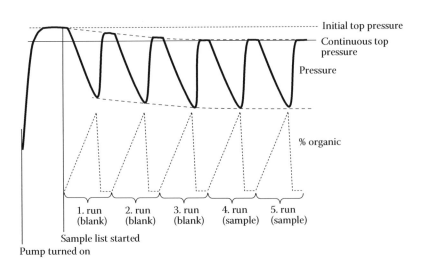

FIGURE 3.10 Development of column backpressure when a series of incompletely equilibrating methods are applied (illustration). After an initial phase of blank runs, the pressure curve shows a continuous pattern (illustration).

The clean-up step mainly consists of washing the column and the process is sample-dependent. If samples are known to contain highly retained impurities, a thorough wash-out is required, usually by applying several column volumes of a very high organic content mobile phase (usually 95 to 98% organic). If the samples are known to be relatively pure (for example, samples coming from preparative purification for a final quality check), a thorough column cleaning only after a certain number of samples may be sufficient. Again, this can be accomplished by special wash methods or macros within a sample list.

The final issue is system overhead time. This is a complex matter that depends on many factors and often requires improvement. System overhead time derives mainly from communication of the acquisition software with the instrument modules (for example, downloading methods to the individual modules), electronic handshaking among modules, data processing and file handling (writing data to a disk drive, locally or via intranet). A key point for optimizing this contribution to the cycle time is not to perform data processing directly after data acquisition. At a minimum, a chromatographic data system should be capable of processing data off-line.

Good chromatographic data systems allow users to perform data processing online but asynchronously. This means that data acquisition for the next sample on the list starts directly and the processing of the prior sample starts as soon as computer resources are available.

Another feature of the control software should be the ability *not* to download the method at the beginning of every run unless the method changes. Each time the control software downloads method parameters, typically 2 to 5 sec are consumed in a situation where every second counts. The more complex the instrument is (second pump, switching valves, additional detectors), the longer the initialization phase will be before a run starts.

Two other critical issues related to system overhead time are the amount of data that must be saved and the way the system stores data on the disk drive. Table 3.5 shows the cycle times measured for an ultra-high-throughput DAD/ToF-LC/MS system using the same chromatographic method for individual tests and acquiring different data amounts during individual runs. The run time of the high-throughput method was 39 sec plus a 2-sec pre-run balance of the diode array detector, theoretically resulting in a shortest cycle time of 41 sec. The DAD was used in spectral acquisition mode (with bandwidths of 1 and 2 nm and two different wavelength ranges) and in signal acquisition mode acquiring one or two wavelengths. All runs were done at a data acquisition rate of 80 Hz. The ToF

TABLE 3.5

Comparison of Cycle Times and Achievable Daily Sample Throughput of DAD/ToF/LC/MS System at Different Detector Settings

	DAD	ToF				
Type	Wavelength (Bandwidth)	Centroid	Profile	Data Rate (Hz)	Cycle Time (sec)	Throughput (samples/day)
Spectral	190 to 900 (1)		x	20	62	1394
Spectral	190 to 900 (1)	x		20	62	1394
Spectral	190 to 400 (2)		x	20	59	1464
Spectral	190 to 400 (2)		x	40	59	1464
Spectral	190 to 400 (2)		x	30	58	1490
Signal	210/254		x	20	50	1728
Signal	210	x		30	49	1763

Method: 5 to 90% B in 0.5 min, water/ACN, flow 1.8 mL/min, 80°C, alternating column regeneration with $2 \times 2.1 \times 50$ mm, Zorbax SB C18, 1.8 μm, stop time = 0.65 min = 39 sec, pre-run balance of DAD, 80 Hz data acquisition rate on DAD, mass range 100 to 1000 amu.

MS operated at 20 to 40 Hz data acquisition rates in centroid or profile mode and at a mass range of 900 amu. The resulting cycle times varied from 49 to 62.[22] The longest system overhead time was 21 sec and the shortest was 8 sec—a difference of 260 percent! Translated into throughput, the difference equals almost 400 samples per day!

Interestingly, little difference occurs if a switch is made between profile or centroid data acquisition of a ToF instrument (rows 1 and 2) even though the data file size in profile mode is much larger. As soon as DAD acquisition parameters are changed to produce less data by switching from full spectral acquisition with a narrow bandwidth to lower ranges and finally to single wavelength data acquisition, a significant drop in overhead time occurs. The reason is that with the here used system, ToF data are continuously written to the disk drive during acquisition, and DAD data are buffered to be added to the file at the end of the run. This is done during system overhead time. The larger the DAD data, the longer the system overhead time. Such system details may exert big impacts on throughput if very high sample numbers per day must be processed.

The throughput of fast LC analyses may be significantly improved if cycle times are optimized. The cycle time involves different instrument-dependent and -independent parts. Optimizing the equilibration time and data processing may produce the greatest influence but small additions (seconds) to the cycle time should not be overlooked because optimization of cycle time can increase throughput significantly.

3.5   PARALLELIZING LC/MS WORKFLOW STEPS

After discussing many points on optimizing serial LC/MS analysis, we will now consider the possibilities of parallelizing steps. Parallelization makes the biggest impact on sample throughput by dramatically reducing cycle time, but almost any kind of parallelization comes with a corresponding disadvantage. This section is about balancing the advantages and disadvantages of parallelization. Figure 3.11 shows a review of Figure 3.9 with tasks parallel to the separation and multiplied tasks to increase throughput. Already in Figure 3.9, the detection is parallel to the separation in what has become the standard method of performing LC/MS today. This was not the case in the past when mixtures were separated and fractions were analyzed off-line.

It is common in high-throughput analyses to perform sample preparation (Figure 3.11-1), which in its simplest form is aspiration of a sample by an autosampler parallel to analysis of the prior sample. Most commercial autosamplers support so-called overlapped injection or inject-ahead functions and can significantly reduce instrument overhead time. Some autosamplers can handle more operations. For example, a sample dilution can certainly be performed off-line by a pipetting robot, but it may also be done online using injector programming, utilizing solvent troughs or large volume vials to aspirate solvent, ejecting the solvent into sample vials, mixing by aspirating and ejecting, and finally injecting the resulting diluted sample into the injector valve.

In our laboratories, a cycle time of 90 sec can be achieved with a dilution factor of 1:25 for a given sample concentration, allowing the purity and identity control of two and a half 384-well microtiter plates per day. The online dilution eliminated an external step in the workflow and reduced the risks of decomposition of samples in the solvent mixture (weakly acidic aqueous solvent) required for analysis. Mao et al.[23] described an example in which parallel sample preparation reduced steps in the workflow. They described a 2-min cycle time for the analysis of nefazodone and its metabolites for pharmacokinetic studies. The cycle time included complete solid phase extraction of neat samples, chromatographic separation, and LC/MS/MS analysis. The method was fully validated and proved rugged for high-throughput analysis of more than 5000 human plasma samples. Many papers published about this topic describe different methods of sample preparation. Hyötyläinen[24] has written a recent review.

Even for analyses that do not require high throughputs, online sample preparation will reduce the risk of mixing up samples (which is a throughput reduction in a much wider sense).

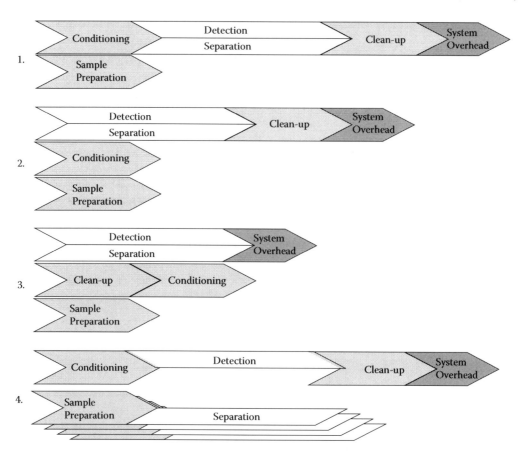

FIGURE 3.11 Different cycle time optimization possibilities achieved by parallelizing individual steps of LCMS-analysis.

Typical instrument set-ups for online sample preparation, for example, solid phase extraction and sample pre-concentration, require the control of a six-port valve with an additional special column and a second pump. Ideally, this system will be fully controlled by the data acquisition software (Figure 3.12 (A)).

A technically simpler approach to increase throughput is the use of a ten-port two-position valve, an identical second column, and a second pump for alternating column regeneration (ACR; see Figure 3.11-2 and 3.11-3). The idea is to parallelize column conditioning and possibly column clean-up and it is supported by certain instrument manufacturers including Agilent, Dionex, and Shimadzu (Figure 3.12). With short equilibration times involving only one or two column volumes (see above), parallel equilibration will save even more time. The minimal requirement is an additional isocratic pump. If a gradient pump is used, it is also possible to perform column wash-out in parallel. In the complete procedure, analysis is performed on one column while the second column is rinsed in parallel, first with high organic content and then under gradient starting conditions. At the end of the gradient on the first column the ten-port valve switches and the second column is used to separate the components of the next sample. When combined with parallel sampling, this technique allows dramatic reductions in cycle time. However one drawback in addition to operating a more complex system must be accepted. The retention time precision is usually reduced because two different columns are used alternately during analyses.

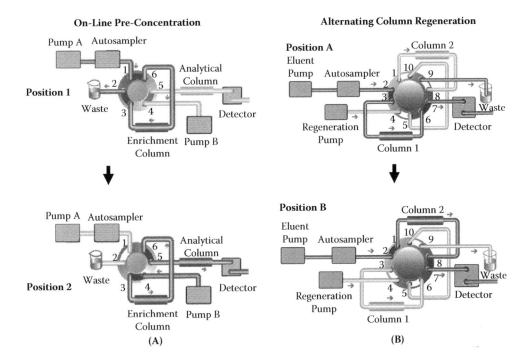

FIGURE 3.12 Examples of valve solutions for optimizing cycle time by (A) online pre-concentration of samples using a six-port, two-position valve and (B) alternating column regeneration using a ten-port, two-position valve. (Courtesy of Agilent Technologies, Inc.)

With alternating column regeneration, the cycle time can be reduced to 50 to 60% of single-column fast LC operations. Obviously, the longer the gradient time is, the smaller the time saving will be. The extent of time saving will ultimately be determined by the sample preparation step and/or the time required for clean-up and conditioning. Commercially available serial autosamplers still require 10 to 30 sec for aspiration, injection, and cleaning of the autosampler. Shorter gradients would not achieve shorter cycle times. Care must be taken with optimizations involving an autosampler. Carry-over may become an issue and should be monitored.

The highest throughput can be achieved when the whole separation step is parallelized (Figure 3.11-4). Commercially available dedicated parallel HPLC systems are the Eksigent Express 800, the Sepiatec Sepmatic, and the Nanostream Veloce 24; see Welch et al.[25] for a comparison. These systems use several standard or capillary columns (typically 8 but up to 24) and several individual pumps or flow splitting after a single pump (typically a binary one). Autosamplers used for parallel sampling are typically the 8-channel Gilson 215, CTC Analytics' Dual-HTSPAL, or special devices. Unfortunately, very few truly parallel detectors are available. Waters offers 4- or 8-channel fully parallel UV detector. Sedere offers a 4 channel ELSD detector. The dedicated parallel HPLC systems mentioned above use parallel UV or diode array detection.

On the LC/MS side, the offerings are limited. Other than certain prototype instrumentation[26] only Waters offered with its MUX technology a type of parallel mass spectrometer ion interface. The Waters MUX technology does not truly operate in parallel, but multiplexes among combinations of four or eight liquid streams. Combined with Waters' LockSpray technology even an additional fifth or ninth channel to introduce a reference mass was used.

The MUX interface uses a dedicated ESI sprayer for each liquid stream. The sprayers are arranged perpendicular to the MS orifice. A sampling rotor within the source cuts parts of the spray out and allows entrance into the MS for analysis. The operation may be considered parallel for the chromatographic time scale if the peaks do not become too narrow. Since one detector must split the available duty time to several LC streams, data quality will be reduced accordingly when compared to single stream LC/MS.

One disadvantage of all types of ultra-high-throughput instruments with multiple separation channels is their linearly increasing complexity. A user may encounter problems like blocking channels or unevenly distributed solvent flows, resulting in varying retention times of a given compound per LC stream (where only one pump is used with a flow splitter). Other types of channel dropouts may occur and the problem of performing re-analysis of missing samples will have to be solved. Perhaps the most important issue for ultra-high-throughput systems is how to analyze a backlog of samples after a system breaks down. Using more systems that are less complex means the breakdown of one instrument just limits a lab in its throughput but does not stop its operation. The most recent fast LC systems can almost the achieve same throughput as parallel detecting systems if comparable data quality is assumed.

Most manufacturers of MSs offer another kind of parallelization of the workflow: instead of repeated analyses of samples with different detection techniques, the analyses are combined into one run. Fast positive and negative switching is now a common capability for many MS devices and different types of ionization techniques can be applied to reduce the number of repeats if a sample does not ionize with a given technique. Agilent for example offers a simultaneously operating ESI–APCI ion source that contains a zone in which ESI-type ions are generated; in a spatially separated area, molecules not ionized by ESI may be ionized by APCI. Waters took a similar approach with its ESCi source that switches electronically between ESI and APCI. Applied Bioscience/Sciex has a DuoSpray source that switches between ESI and APCI ionization mechanically. Thermo and Waters claim to achieve parallel ionization by APCI and atmospheric pressure photo-ionization (APPI) by combining their APCI sources with Syagen PhotoMate APPI technology to increase the chances of ionizing all compounds in a single analytical run.

For non-simultaneously operating multi-sources it is necessary to check whether the data quality achieved is sufficient, especially for fast LC systems with very narrow peaks. A switching source splits the available acquisition time for the offered ionization types and therefore reduces the true acquisition rate of the mass spectrometer. However, simultaneously operating sources do not explicitly show which compound ionized with what technique. If this information is really necessary, two separate runs must be performed.

Finally, parallelization can be achieved before LC/MS analysis when multiple samples are pooled. This can be done directly before the analysis or even earlier in the workflow. Dunn-Meynell[27] et al. determined pharmacokinetic properties by dosing rats simultaneously with six new chemical entities (NCEs). Their LC/MS/MS analysis was done with a fast LC (85-sec cycle time) even though columns with large particles were used. The data quality and reliability were discussed in the literature and concerns were expressed about drug–drug interactions[28] and ion suppression effects for pooled and co-eluting compounds. (*Note:* Korfmacher et al. described the workflow as a cassette-accelerated rapid rat screen (CARRS). See Korfmacher, W.A. et al., *Rapid Commun. Mass Spectr.*, 15, 2001, 335.)

3.6 OPTIMIZING THROUGHPUT OF HIGH RESOLUTION LC/MS

The previous chapters have dealt mainly with LC/MS analysis involving short run times, many samples, and relatively small numbers of compounds in samples. What about samples containing very complex compound mixtures, for example, natural products, samples from biomarker discovery, protein digests, and QA/QC method development or metabolite identification samples requiring detection of every component? Such workflows often require several analysis steps with different columns and different mobile phases and pH values to increase the separation probability by changing the selectivities of individual runs.

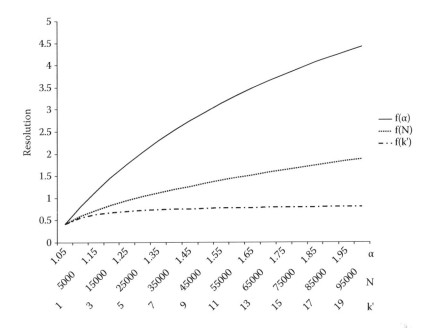

FIGURE 3.13 Dependence on the resolution of two adjacent peaks from the separation selectivity, column efficiency, and capacity factors of peaks. Curves were calculated by keeping values of two parameters constant at the starting value and varying the third parameter.

As can be seen from Equation 3.7 and in Figure 3.13 describing the relation of the resolution of two neighboring peaks to selectivity α, column efficiency N, and capacity factor k', modifying the selectivity exerts the biggest impact on resolution. Modifying k' helps only for very low values of k', but the efficiency N greatly impacts resolution even though its influence grows only by its square root. In contrast to selectivity, the influence on resolution of changing the efficiency of the column is absolutely predictable.

$$R_s = \frac{\alpha - 1}{\alpha} \cdot \frac{\sqrt{N}}{4} \cdot \frac{k_2'}{k_2' + 1} \tag{3.7}$$

where α = selectivity, N = efficiency, and k_2' = capacity factor of second peak.

The greater the efficiency, the better the separation probability. With very high efficiencies we might reduce or even avoid the time-consuming changes of selectivity—time-consuming because additional runs with other columns and/or other solvents must be done (even if automated and/or performed in parallel). We have learned above that sub-2-micron particles increase the efficiency of a column of the same length when compared to a conventional column with larger particles. Can we save overall analysis time by avoiding selectivity changes by implementing sub-2-micron columns? Yes, but only to a certain degree if we take experimental conditions, mainly the available backpressure of the LC equipment and the acceptable run time, into account.

Using very small particles in long columns will indeed improve the efficiency ($N \sim 1/d_p$) but it will also over-proportionally generate higher backpressure ($\Delta p \sim 1/d_p^2$). As a result, lower flow rates must be used to meet the limits of the LC equipment and efficiency will be lost as soon as we reach the left side of the minimum of the van Deemter curve in the diffusion-controlled region. On the other hand, larger particles can be packed in extremely long columns to compensate for their lower resolving power—at the cost of long run times. This can be seen in kinetic Poppe plots that show time per

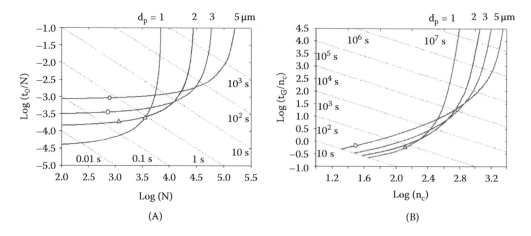

FIGURE 3.14 (A) Isocratic Poppe plot for packed bed columns with different particle sizes. $\Delta P = 400$ bar; $T = 40°C$; $\varphi = 500$; $\eta = 0.69$ cP; $Dm = 1 \times 10^{-5}$ cm²/sec. Coefficients in reduced van Deemter equation were measured on a 2.1×50 mm 3.5 μm Zorbax SB-C18 column using heptanophenone in 40% acetonitrile v/v at 40°C ($k' = 20$): $A = 1.04$; $B = 15.98$; $C = 0.033$. Open triangles represent points where column length is 10 mm. Open circles represent points where flow rate is at 5.0 mL/min. Dotted lines represent constant column dead times. (B) Gradient conditional peak capacity Poppe plot for mixture of 11 peptides on packed bed columns with different particle sizes. Same conditions as in (a) except that the diffusion coefficients of peptides were estimated using the Wilke-Chang equation. Dotted lines represent constant gradient times. (Reprinted from Wang, X. et al., *J. Chromatogr. A,* 1125, 2006, 177. With permission from Elsevier.)

plate—that is actually a measure of the speed of an analysis—versus the required efficiency. Carr et al.[29] modified these plots for gradient elution and to a more intuitive representation in which gradient time per peak capacity (again, a measure for speed) is plotted against peak capacity (Figure 3.14).

It can be clearly seen in the original Poppe plot and in the modified version for gradient separations (right) that for low resolutions (low efficiencies and peak capacities), small particles have a clear advantage in time but high resolutions (more than 100,000 plates, peak capacities above 1000) require extremely long analysis times—up to 10^2 days! The solution is to use very long columns with larger particles. This theory was proven by Sandra et al. who coupled up to eight columns of 2.1×250 mm and 4.6×250 mm packed with 5-μm particles together, achieving an overall column length of 2 m. This approach enabled them to achieve 200,000 plates and peak capacities of 1000 to separate tryptic digests[30] for analyzing PCB mixtures and drugs for impurities.[31]

Although the tryptic digestion analysis took 500 min, the overall time saving becomes obvious upon reviewing the chromatograms. It is not difficult to imagine how many conventional separations under different method conditions would be required to separate the same number of peaks. These analyses were performed with commercial HPLC equipment. Even higher peak capacities were reported in the literature. They involved the use of experimental instruments to deliver extremely high pressures, for example, using capillary columns packed with 1-μm particles and operated at pressures of 6800 bars/100000 psi.[32]

In addition to increasing the separation power in the chromatographic time domain, processing can be expanded into the mass domain by using better resolving MS equipment. The highest resolving powers can now be achieved by the *Fourier* transform ion cyclotron resonance mass spectrometer (FT/ICR/MS) that reaches total peak capacities exceeding 10^7 when combined with LC separation because of its very high resolving power in the range of 10^6 full width at half mass (FWHM). The resolution of a mass spectrometer is mass-dependent. Care must be taken when comparing instrument specifications. Mass resolution can also depend on data acquisition rate, which is not very important for high resolution LC applications but may become important for fast LC separations.

Orbital trapping mass spectrometers achieve resolutions of up to 10^5 and would be the next choice after ToF mass spectrometers if resolving powers above 10^4 are required. In addition to mass resolution, the selectivity of an MS can be critical to distinguish between co-eluting and not mass-resolved compounds. For example, typical triple-quad mass spectrometers usually cannot achieve better than unit–mass resolution. However, special operation modes like neutral loss scans and precursor ion scans can filter out compounds of interest even if neither LC separations nor MS scans would be sufficient to resolve these compounds (note that this is a filtering step).

Increasing the dimensions in LC/MS separations is another option to reduce time in the workflow by enhanced separation power, but it also increases the complexity of the instrumentation. The use of comprehensive two-dimensional LC (2DLC)—using orthogonal separation methods like normal- and reversed-phase LC together—allows extremes in chromatographic separation power to be achieved. Ideally, the column used for the second LC dimension is operated under fast LC conditions to attain good peak capacity in a short time. At the time of writing, only a few instruments and appropriate software for 2DLC data acquisition and data processing were commercially available.

One problem with 2DLC is the need to match the two orthogonal separation mechanisms. To date, the technique has not found a broad distribution but is a promising technology and "opens new perspectives for the separation of complex mixtures."[33] 2DLC combined with MS and LC combined with ion mobility spectrometry mass spectrometry (IMS/MS) are examples of expanding separations into three dimensions (two in the time domain, one in the mass domain). Ion mobility spectrometry allows separation of ions based on their drift times in a gas-filled electrostatic potential.[34] The time scale is much shorter (microsecond to millisecond range) than the chromatographic time scale and fits ideally as an additional separation dimension. Because of its time range, it is usually equipped with fast analyzing ToF MS. Besides offering additional separation power IMS/MS also offers additional selectivity like separation based on molecule conformation. However, the disadvantages of IMS are poor sensitivity and a somewhat limited peak capacity.[35]

3.7   SUMMARY

Modern technologies provide many techniques for expanding the throughput of an analytical laboratory. The task that needs to be accomplished and the possible drawbacks should be carefully considered. Optimized LC equipment can utilize columns packed with much smaller stationary phase particles to achieve significant reductions in gradient time while still achieving the same or even better peak capacities than conventional methods.

Such columns are excellent "filters" and require more sample preparation to ensure the removal of all solids. To benefit from the full power of LC optimization, the detectors must be optimized as well. Data rates and duty times must match the narrower peaks in very fast (and well resolving) separations. Careful consideration and optimization of all instrument components and software can produce significant cycle time improvements of fast LC separations and further increase throughput. An important aspect of cycle time improvement is parallelization of components of individual analyses.

LC/MS analyses requiring high resolving power to separate all compounds present in a sample may be optimized as well to increase throughput. Optimizing in the LC dimension utilizes smaller particles as well; more radical approaches may involve a change in workflow toward extremely high column efficiencies and peak capacities in contrast to the present common work flow of many individual runs with modified selectivities.

In general, LC peak capacity can be reduced if it is counter-balanced by an increased MS peak capacity in the orthogonal mass domain. Increasing the dimensions in LC/MS separations is another option to increase overall peak capacity in a given time, but comes at a cost of increased complexity of instruments and data evaluation. In the mass domain, highly resolving mass spectrometers round out the choices. Nevertheless, every throughput optimization step should be viewed in relation to data quality, instrument complexity, robustness, ease of use (including training and maintenance), and price—and always related to the complete workflow that may start or end outside the analytical laboratory.

REFERENCES

1. Bormann, S., *Chem. Eng. News*, 84, 2006, 56.
2. U.S. Department of Health and Human Services, *White Paper: Stagnation, Challenge and Opportunity on the Critical Path to New Medical Products*, U.S. Government Printing Office, Washington, D.C., 2004.
3. Hadjuk, P.J. and Greer, J., *Nat. Rev. Drug Dis.*, 6, 2007, 211.
4. Van den Boom, D. et al., *Int. J. Mass Spectrom.*, 238, 2004, 173.
5. Kassel, D.B., *Chem. Rev.*, 101, 2001, 255.
6. Schelling, A.P. and Carr, P.W., *J. Chromatogr. A*, 1109, 2006, 253.
7. Neue, U.D., *J. Chromatogr. A*, 1079, 2005, 153.
8. Neue, U.D., Cheng, Y.F., and Lu Z., in *HPLC Made to Measure: A Practical Handbook for Optimization*, Kromidas, S., (Ed.), Wiley-VCH, New York, 2006, 59.
9. El Deeb, S., Preu, L., and Wätzig, H., *J. Separation Sci.*, 30, 2007, 1993.
10. Gritti, F. et al., *J. Chromatogr. A*, 1157, 2007, 289.
11. Cavazzini, A. et al., *Anal. Chem.*, 79, 2007, 5972.
12. Guillarme, D. et al., *J. Chromatogr. A*, 1149, 2007, 20.
13. Xu, R.N. et al., *J. Pharm. Biomed. Anal.*, 44, 2007, 342.
14. Cunliffe, J.M. et al., *J. Separation Sci.*, 30, 2007, 1214.
15. Liu, Y. et al., *Anal. Chim. Acta*, 554, 2005, 144.
16. Smith, R.M., *J. Chromatogr. A*, 1184, 2007, 441.
17. Wu, N. and Clausen, A.M., *J. Separation Sci.*, 30, 2007, 1167.
18. Kofman, J. et al., *Am. Pharm. Rev.*, 9, 2006, 88.
19. Agilent Technologies, Inc., Application Compendium, 2005, Publication 5989-2549EN, *Time-of-Flight Solutions in Pharmaceutical Development: The Power of Accurate Mass*.
20. Makarov, A. et al., *Anal. Chem.*, 78, 2006, 2113.
21. Schellinger, A.P., Stoll, D.R., and Carr, P.W., *J. Chromatogr. A*, 1064, 2005, 141.
22. Frank, M.G. et al., PittCon 2006, CO-1152.
23. Mao, Y. et al., *J. Pharm. Biomed. Anal.*, 43, 2007, 1808.
24. Hyötyläinen, T., *J. Chromatogr. A*, 1153, 2006, 14.
25. Welch, C.J. et al., *J. Liquid Chromatogr. Rel. Technol.*, 29, 2006, 2185.
26. Schneider, B.B., Douglas, D.J., and Chen, D.D.Y., *Rapid Commun. Mass Spectrom.*, 16, 2002, 1982.
27. Dunn-Meynell, K.W., Wainhaus, S., and Korfmacher, W.A., *Rapid Commun. Mass Spectrom.*, 19, 2005, 2905.
28. White, R.E. and Manitpisitkul, P., *Drug Metabol. Distrib.*, 29, 2001, 9570.
29. Wang, X., Stoll, D.W., Carr. P.W., and Schoenmakers, P.J., *J. Chromatogr. A*, 1125, 2006, 177.
30. Sandra, P. and Vanhoenacker, G., *J. Separation Sci.*, 30, 2007, 241.
31. Lestremau, F. et al., *J. Chromatogr. A*, 1109, 2006, 191.
32. Patel, K.D. et al., *Anal. Chem.*, 76, 2004, 5777–5786
33. François. I., de Villiers, A., and Sandra, P., *J. Separation Sci.*, 29, 2006, 492.
34. Guevremont, R., *J. Chromatogr. A*, 1058, 2004, 3.
35. McLean, J.A. et al., *Int. J. Mass Spectrom.*, 240, 2005, 301.
36. Peterson, P. and Euerby, M.R., *J. Separation Sci.,* 30, 2007, 2012.
37. Nguyen, D.T. et al., *J. Chromatogr. A*, 1128, 2006, 105.

4 Throughput Improvement of Bioanalytical LC/MS/MS by Sharing Detector between HPLC Systems

Min Shuan Chang and Tawakol El-Shourbagy

CONTENTS

4.1   INTRODUCTION

Like any businesses, bioanalytical laboratories perform operations that transform starting materials (samples and supplies) into products of higher value (quality reports continuing accurate sample concentration data). To maximize productivity and stay ahead of competition, bioanalytical scientists continuously invent, reinvent, and implement processes and techniques that generate more accurate and better quality reports with fewer resources (labor, time, capital, energy, and consumable goods). These continuous optimizations of laboratory operations drove the bioanalytical laboratories to begin

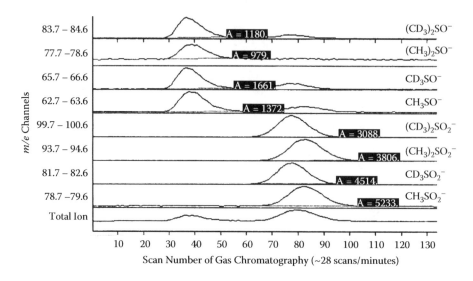

FIGURE 4.1 Gas Chromatogram of a 3 μL injection of a solution containing 10 μg each of DMSO, DMSO2, and their deuterated analogues. The chromatogram was obtained using a 2 mm $\times$ 3 ft glass column packed with Porapak R at 210°C. Helium was used as the carrier gas.

using high performance liquid chromatography (HPLC) in the 1970s. They automated sample preparation in the 1980s. In the 1990s and early 2000s, advances included replacing absorbance detectors with mass spectrometers (MSs), using LIMS systems for data management, preparing samples in SBS 96-well formats, and multiplexing LC/MS systems. This chapter describes the approaches to increase MS throughput and provides fundamental information about the processes.

4.1.1 DEVELOPMENT OF MASS SPECTROMETRY AS LIQUID CHROMATOGRAPHIC DETECTOR

The mass spectrometer has been successfully used as a detection device for gas chromatography (GC) for a long time. It is suitable to quantify analytes in complex matrices because the mass-to-charge ratio (m/e)-based MS technique provides sufficient selectivity in the presence of potentially interfering background compounds. The ability of MS to distinguish an analyte from an isotopically labeled analog has also been recognized. Isotope analogs have been used to compensate for variations in sample preparation and ensure specificity (Figure 4.1).[1] Because most early MS devices provided only limited vacuum pumping capability, most GC/MS interfaces reduced the amounts of carrier gases entering the MS. For traditional packed GC columns, the task was accomplished via differential pumping (jet separator) or a membrane separator.

The lack of sufficient vacuuming capability made interfacing LC with MS an even greater challenge. Instead of 3 to 5 mL/min of helium from a typical GC, 2 mL/min of mobile phase from a LC is equivalent to 1.1 L (CH_3CN) to 2.5 L (water) vapor at atmospheric pressure. The flight path of the ions in a MS requires a high vacuum to avoid collision. Therefore, early LC/MS interfaces such as the moving belt interface which removes the HPLC solvent outside the vacuum chamber,[2] and direct capillary interface limited the liquid volume and removed the bulk of the solvent with cryogenic "fingers" around ion sources.[3] These approaches were successfully applied but were not user friendly.

After the development of larger and more efficient vacuum pumps, more user-friendly LC/MS interfaces of thermospray,[4,5] and atmosphere pressure ionization,[6,7] LC/MS earned its place in bioanalytical laboratories. The resulting device was a powerful instrument that required significantly more capital investment than HPLC/UV, GC, or GC/MS.

Mass spectrometry detection gained the acceptance of bioanalytical scientists primarily based on its higher selectivity compared to detection that relies on UV/visible absorbance. Absorption spectra of aqueous solutions usually appear as broad absorbance bands. The selectivity provided by UV/visible absorbance for a colorless analyte is usually very low. To detect a colorless analyte, a wavelength setting below 210 nm is usually used. UV absorbance in this region is not specific because most compounds containing hetero-atoms and multiple bonds absorb UV below 200 to 210 nm.

To achieve adequate specificity, a highly selective and efficient chromatographic separation is required to ensure the observed peak is the intended analyte. On the other hand, MS detection is very selective. MS detection is based on *m/e* and can achieve a unit mass resolution from several hundred in the MS mode to tens of thousands in the MS/MS mode if a tandem mass spectrometer is used. Therefore, the mass chromatogram of a sample extract usually contains only the intended analytes and a simple chromatographic separation provides sufficient selectivity. In addition, LC/MS with its structural related selectivity gives scientists confidence that the chromatographic peak detected is from the intended analyte. The absence of background interference also allows a shorter separation and sometimes a simpler sample preparation method. Figure 4.2 presents typical chromatograms of plasma extracts for an Abbott compound from HPLC/UV and LC/MS assays.

Lengthy chromatographic separations are required for some assays such as assays for compounds that generate steroidal isomeric metabolites, assays that require gradient elution, online solid phase extraction, and assays for analytes that have different chemical and physical properties. Even for these difficult assays, however, the chromatographic peaks from the analytes occupy only a small fraction of total run time. Most mass chromatograms require a solvent front and time for the autosampler to prepare for the next injection (Figure 4.3). These slots of time before and after the chromatographic peak is eluted can be utilized by sharing a detector with other HPLC systems or, in the absence of ion suppression, the next sample.

4.2 APPROACHES FOR IMPROVING THROUGHPUT OF MASS SPECTROMETER

A mass spectrometer is usually the most costly detector in a bioanalytical laboratory. In the 1990s, most bioanalytical laboratories did not have a sufficient number of MSs to meet the needs of every project and scientist. Therefore, the optimum use of this type of equipment was a goal for many bioanalytical laboratories and bioanalytical scientists adopted many approaches intended to improve MS productivity. Two are described below.

4.2.1 DEVELOPMENT OF FAST CHROMATOGRAPHY

Figure 4.4 depicts a LC/MS assay utilizing fast chromatography.[8] The throughput of the assay, which used a 4 × 20 mm C4 column, was limited only by the cycle time of the autosampler. Heart-cut column switching was not required because of the little background interference above *m/e* 300 in a thermospray mass chromatogram.

Unlike thermospray, most of the common ionization techniques for LC/MS do not proceed in a vacuum (where ion densities are low). Compared to thermospray, the ionization efficiencies of the modern techniques are greater and more molecules are ionized. The omnipotent ability to create ions increased the chances of unseen interferences from the background peaks, commonly known as effect of matrix or ionization suppression. Shortening the length of LC run time for modern LC/MS is limited by the need to keep the analyte peaks away from potential interferences, seen and unseen and the cycle time of the autosampler. Developing a short chromatograph is the first step for a high-throughput LC/MS or LC/MS/MS assay. Timeshare MS is a scheme to obtain greater efficiency beyond that obtained from optimized methods.

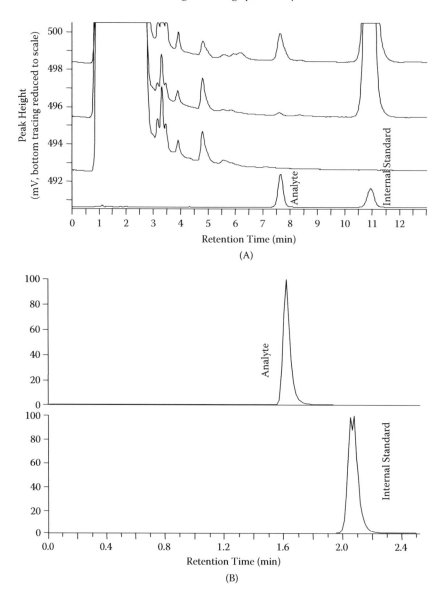

FIGURE 4.2 (A) Chromatographic profile from extracts for an Abbott compound. The detection was UV at 205 nm. The identity of tracings (from bottom) are a reference standard, a blank plasma extract, a low limit of quantitation sample, and a dosed sample (12 hour) from a study subject. A run time of 13 minutes and a liquid-liquid extraction with back extraction is required for a rugged assay. (B) LC/MS tracing of a dosed sample (not the same sample). A 2.5 minute run time is sufficient for the assay.

4.2.2 DETECTOR TIMESHARING

The first reported case of timesharing for a mass spectrometer[9] involved the design of an Ionspray® interface with multiple sprayers to support the analysis of effluents from multiple columns. This approach led to the development of a multiplexed electrospray interface (MUX)[10] using an LC/MS interface and multiple (identical) sprayers linked to a HPLC system and a spinning screen to allow the output of only a single sprayer to enter the MS (Figure 4.5). The injections of the HPLC systems

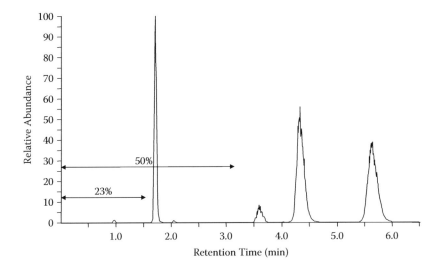

FIGURE 4.3 Total LC/MS ion chromatogram of an Abbott compound, the analog internal standard, its metabolites and impurities. Depending on the need to assay the polar metabolite, 23 to 50% of the mass chromatogram will not show useful information (arrows).

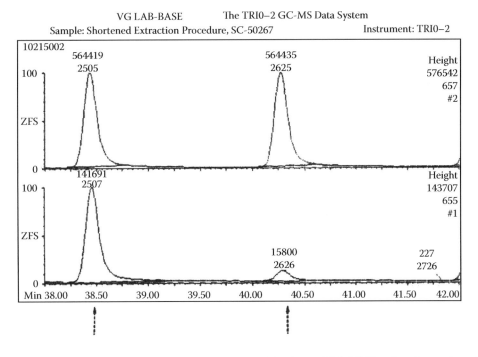

FIGURE 4.4 Arrow indicates chromatograms from three injections of SC-50267 and an internal standard [D2]SC-50267. The last injection appears near the end of the display. A VG Trio-2 mass spectrometer equipped with a thermospray LC interface was used. Samples were injected every 1.85 min—the cycle time of the autosampler (Waters WISP). (*Source:* Chang, M. and G. Schoenhard, presentation at Pittsburgh Conference and Exposition, 1993. With permission).

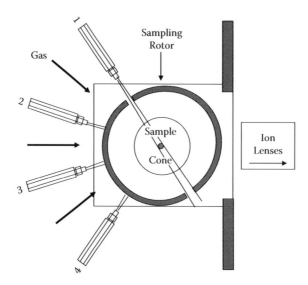

FIGURE 4.5 Graphic presentation of MUX (not to scale). (*Source:* Sage, A.B. et al., *Curr. Trends Develop. Drug Dis.*, 2000, 15, S20. With permission.)

were synchronized. LC/MS with MUX allowed the collection of data from each HPLC system in sequence into individual designated files and folders.

Using this technology, two or more complete mass chromatograms could be generated at the same time although the dwell time for each data point and the sampling rate (number of data points in time) are reduced as shown in Table 4.1. MUX and other multiple sprayer approaches require new interfaces and MS software. The multiple sprayer approach is only available in limited models of LC/MS instruments.

TABLE 4.1
Effects of Monitoring More Ions and Channels on Data Points and S/N

	Set Dwell Time to 200 msec # of Data Points across the Peak			Monitor 16 Data Points per Peak Dwell Time (msec) for Each Channel			Relative S/N		
	2 Analytes	**3 Analytes**	**4 Analytes**	**2 Analytes**	**3 Analytes**	**4 Analytes**	**2 Analytes**	**3 Analytes**	**4 Analytes**
1 Channel	73	48	36	933	620	464	100	82	71
2 Channels	32	22	17	416	260	182	67	53	44
3 Channels	21	15	11	261	157	105	53	41	33
4 Channels	16	11	8	183	105	66	44	34	27
5 Channels	13	9	6	137	74	43	38	28	21
6 Channels	10	7	5	105	53	27	34	24	17
7 Channels	9	6	4	83	39	16	30	20	13
8 Channels	8	5	4	67	28	8	27	17	9

Estimated results for 30-sec wide peak with interion time of 5 msec and interchannel time of 50 msec. S/N calculated using assumption that it is proportional to square root of dwell times.

Most other techniques to enhance MS throughput such as cassette dosing[11] and pooling of samples (sample extracts)[12] do not require the modification of the spectrometer or specific spectrometry software. These indirect approaches timeshare the MS by requiring it to monitor extra channels for all the analytes. The data parcel size is measured in milliseconds and similar to MUX except that the inter-channel time is shorter. Unlike MUX that stores chromatograms from each sprayer in different files, all data from cassette dosing are kept in the same file. Therefore, the cassette dosing and the sample mixing approaches are only applicable for analytes that do not interfere with each other. Cassette dosing and sample pooling techniques are frequently used to improve bioanalytical throughput in drug discovery and early development.

Another approach for increasing the throughput from an MS that does not require purchasing a new MS or a new interface is LC_nMS or staggered parallel chromatography. Unlike cassette dosing and sample pooling, the approach does not require the mixing of samples. Therefore, each sample maintains the integrity and data from each LC is stored in different files and folders. The data parcel size is in minutes and the entire usable region of the LC/MS chromatogram is generally collected in a single continuous file.

4.3 IDENTIFYING APPLICABLE ASSAYS

Certain assays may benefit from staggered parallel chromatography, for example, when (1) the same assay must be performed for a large number of samples in a short time and (2) if the analytes of interest that elute in a narrow window account for only a fraction of a chromatogram.

The LC/MS throughput enhancement approach developed in our laboratory in 1997 and 1998 used multiple HPLC systems, each of which had dedicated HPLC pumps, autosamplers, and columns.

The effluent from each column was linked to a column switching valve as presented in Figure 4.6. The valve was used to select effluent from multiple HPLC systems that fed into a single sprayer. The injection of samples was staggered to allow sequential monitoring of the data-rich window of the LC/MS chromatogram of each system. The data window was determined from a chromatogram of the reference mixture and peak drifting information obtained previously or experimentally. Section 4.6.1 discusses determination of the column switching window. Retention time stabilization techniques such as guard column regeneration are very useful in maintaining the switching window without impacting the overall scheme. Guard column regeneration is described in Section 4.7. The following sections describe staggered parallel chromatography in detail.

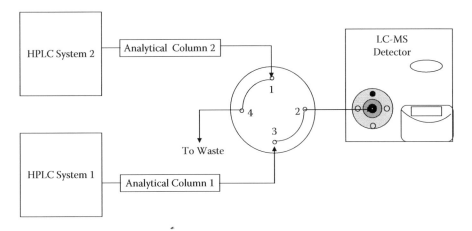

FIGURE 4.6 Quantification of effluent from System 1. Effluent from System 2 is diverted to waste.

4.4 THEORY

Figure 4.7 presents a chromatogram that hastens implementation of the LC_nMS program. The LC separation took about 13.5 min to separate metabolites that were structural isomers. The implementation of LC_2MS made it possible to use a single MS to support a large clinical study.

The general LC/MS chromatogram contains four regions: (1) a data region that includes sufficient baseline for appropriate integration, (2) a pre-data region from sample injection (S_{inject}) to the start of data (D_{start}), (3) a post-data region from the end of the data region (D_{end}) to the end of the HPLC run ($HPLC_{end}$), and (4) a MS data process region (M) that allows the MS to finalize the current data file and prepare for the next file. Figure 4.8 shows the LC_2MS approach including the controlling events. It is possible to arrange the packing differently (Figure 4.8a) to gain more efficiency. However, the approach of interlacing chromatographic peaks requires absolutely stable retention times, is difficult to program, and is not widely applicable. The sequence of events for best data packing depends on a combination of factors. Figure 4.9 illustrates the variations that lead to the following formulas.

For MS linked to a single LC, Equation 4.1 represents the cycle-time of HPLC:

$$Cycle\ time_{LC\text{-}MS} = HPLC_{end} + M \qquad (4.1)$$

and the MS run length is

$$MS_{runtime} = cycle\ time_{LC\text{-}MS} \qquad (4.2)$$

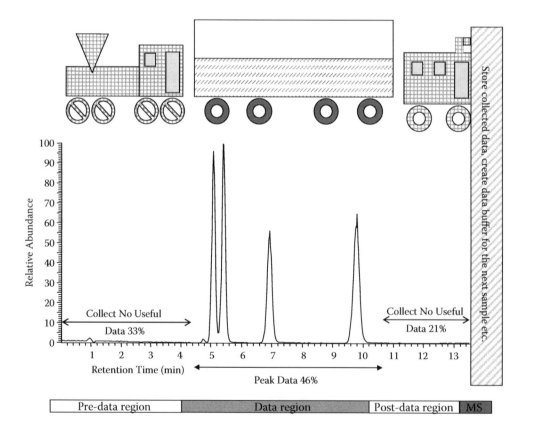

FIGURE 4.7 Typical mass chromatogram.

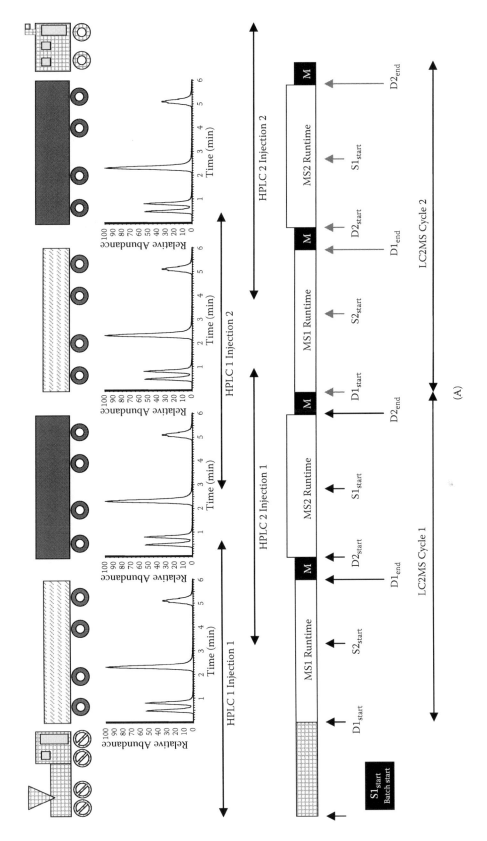

FIGURE 4.8 (A) Staggering two HPLCs to utilize MS more efficiently. Gray lines indicate repetition of first cycle. The batch start time (0) is a single event.

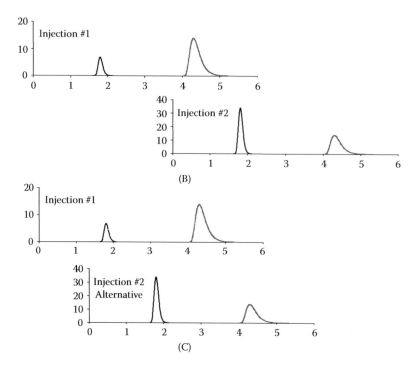

FIGURE 4.8 (B) Injections from HPLC system 1 and system 2 are packed according to the LC_2 MS approach. (C) Injections from HPLC system 1 and system 2 are packed in an interlaced fashion. Although interlaced packing of chromatographic peaks is more efficient, the packing pattern is difficult to apply.

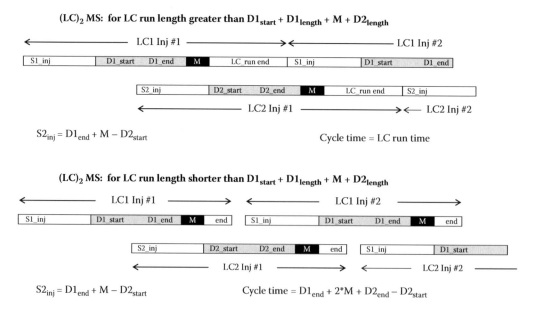

FIGURE 4.9 Graphic depiction of timing schedule.

To save time, a bioanalytical scientist may end the MS data collection early such that M is included in the post-data region. If this approach is used, Equation 4.1 is replaced by Equations 4.1A and 4.2A:

$$\text{Cycle time}_{\text{LC-MS}} = \text{HPLC}_{\text{end}} \tag{4.1A}$$

$$\text{MS}_{\text{runtime}} = \text{cycle time}_{\text{LC-MS}} - \text{M} \tag{4.2A}$$

For an MS linked to two LC systems to perform staggered parallel chromatography, the MS data processing region is at the end of data region for LC Systems 1 and 2. The events are presented in Figure 4.9 and represented by the following equations:

$$\text{Injection time}_{\text{LC\_system\_1}} = 0 \tag{4.3}$$

$$\text{Cycle time}_{\text{LC2MS}} = \text{maximal of } [(\text{data region}_{\text{LC\_system\_1}} + \text{data region}_{\text{LC\_system\_2}} + 2*\text{M}) +$$
$$\text{HPLC}_{\text{end\_LC\_system\_1}} + \text{HPLC}_{\text{end\_LC\_system\_2}}] \tag{4.4}$$

$$\text{MS starting time}_{\text{LC\_system\_1}} = \text{D}_{\text{start\_LC\_system\_1}} \tag{4.5}$$

$$\text{Injection time}_{\text{LC\_system\_2}} = \text{cycle time}_{\text{LC2MS}} - \text{cycle-time}_{\text{LCMS\_LC\_system\_2}} \tag{4.6}$$

$$\text{MS start time}_{\text{LC\_system\_2}} = \text{injection time}_{\text{LC\_system\_2}} + \text{D}_{\text{start\_LC\_system\_2}} \tag{4.7}$$

Since the eluent selection valve must direct the data-rich effluent from the LC system to the MS and divert the effluent from the other LC system to waste, the timing of the valve control must follow Equations 4.8 through 4.10.

$$\text{Initial valve direction: MS receives effluent from system 2} \tag{4.8}$$

$$(\text{D}_{\text{start\_LC\_system\_1}} - \text{M}) < \text{switching time to receive effluent from system 1} < (\text{D}_{\text{start\_LC\_system\_1}}) \tag{4.9}$$

$$(\text{MS start time}_{\text{LC\_system\_2}} - \text{M}) < \text{switching time to receive effluent from system 2}$$
$$< (\text{MS start time}_{\text{LC\_system\_2}}) \tag{4.10}$$

Note that the parameters for LC Systems 1 and 2 do not have to be equal and that LC System 1 may perform a different assay from System 2 (column, mobile phase, assay run time, and data window may be different). The batches performed on both systems may also contain different numbers of samples.

4.5 BASIC CONSTRUCTION

4.5.1 Fluid Diagram

Figure 4.10 presents fluid path diagrams for LC_nMS. Figure 4.6 illustrates the LC_2MS system concept and Figure 4.11 depicts the actual fluid path for LC_2MS. Based on Figure 4.6, the simplest LC_nMS system consists of two independent HPLC systems and a detector. Each HPLC system contains a pump, an autosampler, and a column. However, instead of installing individual MS devices for each HPLC system, the two HPLC systems share the same MS. The effluents from the HPLC systems are linked to a three-way column switching valve that may stand alone or be the diverting valve attached to the MS. The valve is used to collect the effluent from one of the HPLC systems for detection. The fluidic system is similar to other HPLC column switching operations.

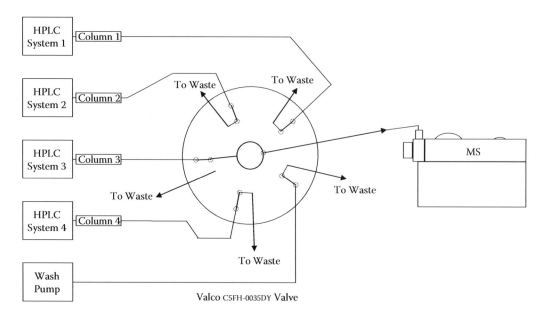

FIGURE 4.10 Switching valve for four HPLC systems.

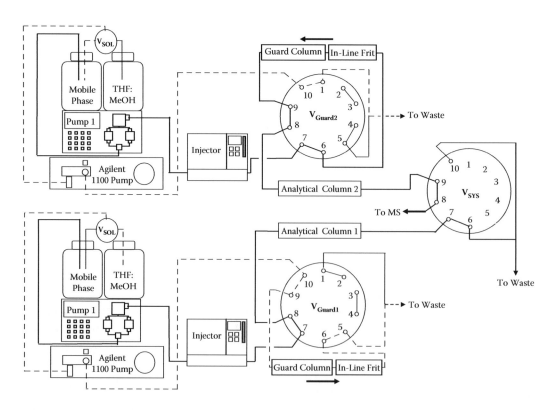

FIGURE 4.11 Timeshare of MS without MS modification. Flow diagram also shows guard column regeneration set-up. (*Source:* Chang, M. et al., presentation at Pittsburgh Conference and Exposition, 2000. With permission.)

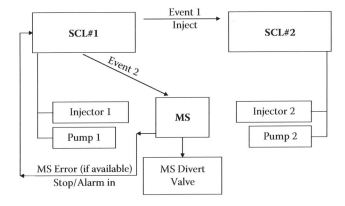

FIGURE 4.12 Control diagram for Scheme A1 showing use of MS bypass valve.

4.5.2 ELECTRONIC DIAGRAM

For an LC_2MS system to function, the components must be synchronized and started according to the parameters obtained from Equations 4.1 through 4.10. It is also critical to ensure that the HPLC performance does not drift outside the set window. Many schemes for synchronizing LC_nMS components can be used. The scheme for LC_2MS serves as an example.

4.5.2.1 Schemes for System Control

4.5.2.1.1 Use of Individual Autosamplers to Trigger MS
The time events are synchronized by one autosampler, and the diverting valve of the MS is used to select column effluent to monitor. Figure 4.12 shows an electronic diagram for this scheme.

Hardware requirements — The system controller responsible for synchronizing the events is defined as LC System 1. It requires at least two time event outputs to trigger the injection of LC System 2 and start MS data collection. If MS fails, the injection of LC System 1 should be inhibited. Autosampler with "ready-in," "alarm-in," and "stop" inputs indicate capability to be stopped remotely. The autosampler of LC System 2 must be able to prepare a sample before the run from LC System 1 is finished and hold the sample in the injector loop until an injection signal is received. A manual injection input devices indicates that the autosampler can perform the required function.

Programming — The LC systems are programmed normally except that LC System 1 run time should be the run time of LC_2MS as computed by Equation 4.4. The time events for LC System 2 injections and triggers to start MS data collection for both systems should also be programmed in LC System 1. The MS sequence should collect the data from each batch in different folders and use appropriate acquisition MS methods (run time = data region$_{LC\_system\_1}$ or data region$_{LC\_system\_2}$). The methods should also position the diverting valve on the MS appropriately. The run time of LC System 2 can be the HPLC run time of System 2 or an arbitrary entry (< HPLC run time of System 2 but after the content of the injection loop is loaded to the analytical column). LC System 2 should be programmed to prepare the samples prior to the end of the run and the remote injection mode should be selected.

Behavior — The LC_2MS set up presented in Figure 4.12 will continue to trigger the MS to collect the data files designated for LC System 2 even if System 2 fails. However, the files after the point of failure contain only baseline information and are easily distinguished. The data files for LC System 1 are not impacted by the failure of System 2. If stopping the injection for both systems is the preferred action for the failure of LC System 2, System 2 must be able to communicate the

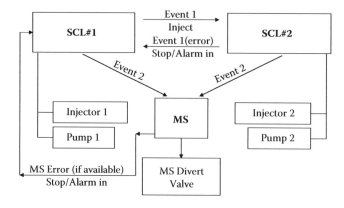

FIGURE 4.13 Control diagram for Scheme A2 showing use of MS bypass valve.

error state to System 1 and stop the run as presented in Figure 4.13. If LC System 1 fails, System 2 will not be triggered. Thus, if the batch sizes for both systems are not equal, we recommend that the larger batch be placed on LC System 1. The user may choose to collect extra empty files for LC System 2 for ease of programming.

Alternative set-up — A dedicated column switching (column selecting) valve may be used in place of the MS diverting valve or for MS systems lacking built-in diverting valves. The column switching valve is controlled by an additional event or events. Thus, LC System 1 must have at least three time event outputs instead two. If guard column regeneration is desired, additional time event output is needed. Figure 4.14 shows that both LC systems may be operated as a single entity without rewiring or re-plumbing. However, LC System 1 must be interrupted to prevent collection of data in an incorrect folder if LC System 2 fails.

4.5.2.1.2   Use of Separate Timing Device to Synchronize MS and HPLC Systems
Figure 4.15 presents a diagram for this scheme that was developed based on a lack of sufficient time event output from older equipment, leading to loss of data.

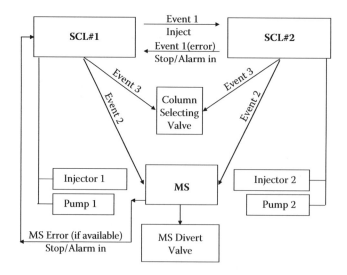

FIGURE 4.14 Control diagram for Scheme A3 showing valve not linked to MS.

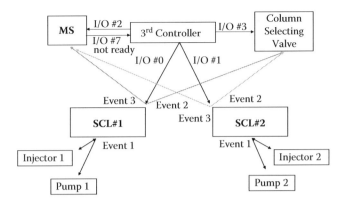

FIGURE 4.15 Control diagram showing use of third controller.

Hardware requirements — This approach relocates the synchronization of components to a third controller, initially a HPLC system controller and later a National Instrument DAQ card inserted in a PC or Macintosh computer to control the MS. A simple program written in Labview® was used to schedule the output events. The third controller must have at least three I/O channels if the MS is equipped with a diverting valve, and at least four I/O channels if an external column switching valve is used. Additional I/O channels may be used as back-up channels and for monitoring LC_2MS components. The output of the DAQ card is TTL high or low. A device may be required to convert the TTL signal to contact closure if the LC_2MS components require such closures.

Programming — The third controller is programmed to trigger both LC systems, the start of MS data acquisition, and the direction of the column selecting valve.

Behavior — Behavior issues follow the previous scheme except that the failure of LC System 1 will no longer automatically interrupt the batch. The user may decide to allow the batch to continue for LC System 2.

Function verification of user program written in Labview® — We designed a simple device that uses a chart recorder, a battery, and four resistors to verify that the program performs as intended. Each arm of the programming tree was tested using different parameter entries. Figure 4.16 presents the raw data from one of the experiments.

4.5.2.1.3   Building Intelligence in Third Controller

This function is an extension of the Labview® program. The equations used to calculate controlling parameters may be used within the Labview® environment. Instrument monitoring, positioning of the system for data window determination, appropriate handling of user decisions such as pausing a batch, efficient handling of different batch sizes, handling component failures, and running log generation may also be pre-programmed. Figure 4.17 depicts a control screen.

Behavior — This scheme is similar to the previous one except that all control parameters are calculated within the program.

4.5.2.1.4   Building Program Shell to Control (Download Operating Parameters to) MS and HPLCs and Monitor Component Performance

This level of control requires a controlling script from the instrument vendor and is for more serious programmers. Direct HPLC control by the MS and multiple staggered LC capabilities provided by several autosampler and instrument vendors fulfill the function but may limit the hardware used or require purchase of software/hardware.

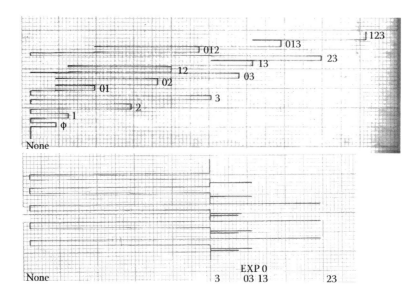

FIGURE 4.16 Function verification of programmable interfaces. The top chart recorder tracing presents resistor value equivalent to contact closure setting. The bottom tracing shows results of a verification experiment and indicates that the program performs as designed. Entry with five runs for LC System 1 and three runs for LC System 2. Event 0 (momentary) triggers the injection of System 1. Event 1 (momentary) triggers the start of System 2. Event 2 (momentary) triggers MS acquisition. Event 3 (on/off) sets column-selecting valve.

4.5.2.1.5   Building Toaster System or Black Box

This hypothetical system requires minimal user inputs to determine a data window automatically from reference injections or prior history of the method and coordinate the LC and MS components as well as the LIMS system. This scheme is achievable with a small set of components but requires an industry-wide standard for component flexibility.

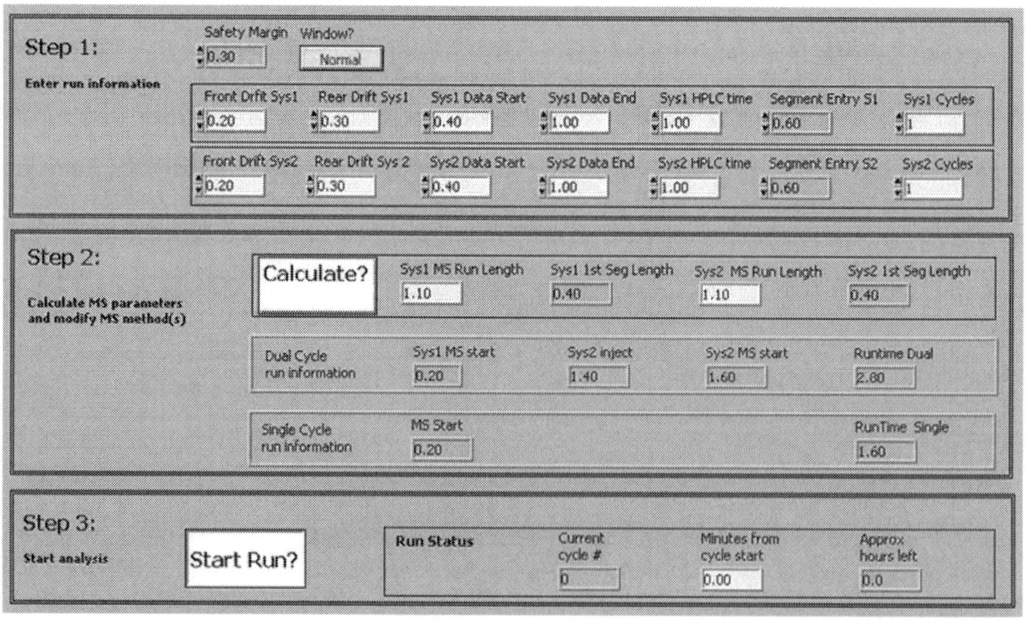

FIGURE 4.17 Control screen of a program written in Labview®.

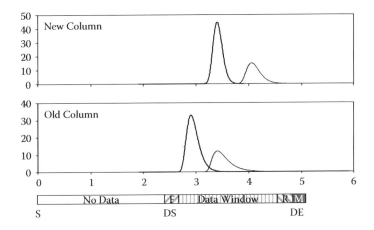

FIGURE 4.18 Determination of data window. Final data window should span (include) data windows of new and old columns. It should also include a region on each end of the data window to establish a chromatographic baseline. The illustrations were created with Microsoft Excel®.

4.6   OPERATION

4.6.1   Window Determination

The most important task for successful implementation of LC_nMS is defining the correct data window. It was recognized in the early phases of development that a narrow data window setting risked chopping the peaks and invalidating the results. Conversely, an excessively wide window setting resulted in a loss of efficiency. To set a data window appropriately, a bioanalytical scientist must determine windows using a new column and an "aged" column. The data window is then set to encamp the two data windows plus margins for setting baseline and random drift.

Figure 4.18 presents a hypothetical chromatogram to illustrate the importance of window setting. The figure also illustrates the importance of maintaining good chromatography. Placement of a separate in-line frit before guard columns, frequently changing guard columns, and use of column switching to regenerate a guard column between runs are approaches that maintain chromatographic performance. They are discussed in Section 4.7.

4.6.2   Arranging Mass Spectrometer Sequence

An LC_nMS batch consists of several runs of samples. Each run is prepared separately and injected on a separate HPLC system. Each run contains a set of calibration standards, at least two sets of quality control samples and unknowns. Most LIMS systems are designed to generate sequence files for single runs. Because an LC_nMS approach injects alternatively from multiple sequences, the direct output or exported files from the LIMS system must be interleafed to create sequence files for the LC_nMS batch. Sequence files ensure that the MS data from different HPLCs are stored in appropriate files or folders. Using a spreadsheet or validated program as an intermediate to interleave individual run sequences to form the final sequence is recommended. LC_nMS supports HPLC systems that run different assays as long as the source and the interface temperature remain the same. We used this approach to assay a compound and metabolite in the same LC_2MS batch using HPLC systems running two different LC methods.

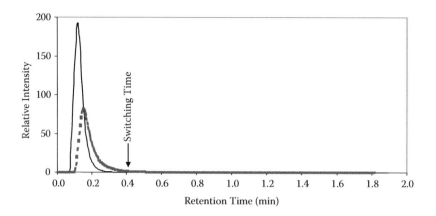

FIGURE 4.19 Setting the switching time using a chromatogram of analytes with only a guard column. Analytes eluted before switching time is loaded to the analytical column. Compounds eluted after switching time are diverted to waste.

4.7   GUARD COLUMN REGENERATION

Guard column regeneration was used to maintain the performance of analytical column.[13] The technique was developed independently in our laboratory to maintain backpressure and the performance of an HPLC assay using online SPE.[14] We applied the technique to LC/MS assays when SPE or PPT was used for sample preparation.[15] When the approach was applied to an LC/MS method, it provided the added benefit of eliminating ion suppression from late-eluting background peaks. Figure 4.11 is the fluid diagram.

The early elution compounds off the guard column are loaded onto the analytical column. If the retention mechanisms of the guard column and the analytical column are the same, the compounds eluted before the analyte from the guard column will elute before the analyte from the analytical column and does not affect quantitation. The compounds that elute significantly later than the last analyte are not loaded to the analytical column and are washed off the guard column to waste in the reverse direction. Therefore, the analytical column is not exposed to phospholipids, lipids, lipoproteins, and other late-eluting compounds that cause ionic suppression and deterioration of column performance. The switching time of guard column regeneration is determined by using analyte injections with a guard column only. Figure 4.19 presents an example for setting switching parameters.

4.8   SELECTION OF HIGH-THROUGHPUT SCHEMES FOR LC/MS

Bioanalytical scientists responsible for method development should attempt to develop an efficient and rugged method with short run times. In addition to higher throughput and less mobile phase consumption, a method with a shorter run time usually yields greater response and better sensitivity. Use of staggered parallel chromatography should be considered if sample numbers are large and peaks of interest account for only a small portion of the entire chromatogram. We suggest that laboratories setting up staggered parallel chromatography for the first time consult this chapter and select an appropriate route. The Figure 4.20 flow diagram may serve as a guide.

4.8.1   SYSTEM VARIATIONS

This section discusses the development of custom-built systems of varying complexities. If a scientists decides to set up an in-house staggered parallel chromatography system, it is beneficial to

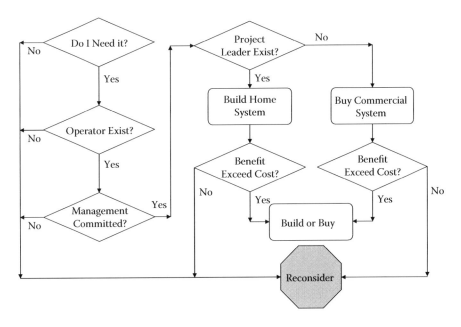

FIGURE 4.20 Process selection flow chart.

determine the complexity required. Figure 4.21 compares costs and returns for typical LC_2MS and LC_nMS systems. The figures indicate that the additional benefit of adding an LC unit decreases as the number of LC units increases. For example, if used at capacity, the addition of the first LC unit to a LC/MS unit decreases the assay cost attributed to the LC/MS by 40%. The additions of second and third LC units decrease the costs by 22 and 14%, respectively. If the LC_nMS system was used as LC_1MS part of the time, the benefit of the LC_nMS decreases rapidly. For the LC_2MS set-up, the benefit from an additional LC unit reaches 20% if the unit operates as LC_1MS 50% of the time.

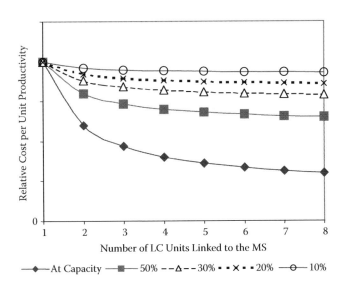

FIGURE 4.21 Each line represents a decrease of cost attributed to purchasing, depreciating, and operating an LC_nMS system at the utilization rate. The system is assumed to operate as an LC_1MS system at other times.

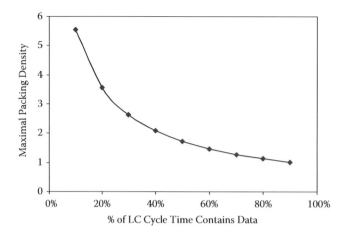

FIGURE 4.22 Maximal packing of chromatographic data.

Figure 4.22 presents the maximum number of LCs that may be used as a function of the width of data window and might be helpful for determining the number of LC units needed.

4.9 USER FRIENDLINESS

It was noted during the early stages of implementation that the LC_2MS system did not always operate as LC_2MS. The unit was used in a staggered parallel chromatography mode operated only 20 to 30% of the time. Although operating at that capacity was still profitable (Figure 4.21), using components of the LC_nMS to meet the specific requirements of other assays in between LC_nMS operations can be costly. It was often necessary to alter plumbing, electronic connections, and function verification prior to use in LC_2MS mode after use for other assays. An approach to reducing the need to re-plumb is to design different levels of simpler instruments within a complicated system. An example is to add a few manual switches or controller programs to allow LC_2MS to operate as LC/MS with or without guard column regeneration.

4.10 IMPLEMENTATION OF LC_nMS SYSTEM

It is sometimes necessary to describe the best scenario to initiate a process change. However, improvements in capability or capacity (that may be predicted accurately) should be combined with the predicted utilization rate of a multiplex system to provide realistic expectations to management. A project may be judged a failure if the benefit produced falls short of the expected improvement even if the process produced a net improvement over an old process. This is frequently the "missing link" of an improvement initiative. This topic was addressed recently by Bremer.[16] A user must evaluate the potential risks and benefits of implementing an LC_nMS scheme in his or her own circumstances. Setting realistic goals, obtaining management understanding and support, and providing sufficient training for analysts to ensure the project is beneficial are the main steps in implementing a successful system.

4.11 COMPARISON OF TIME SHARING MODES

MUX, cassette dosing, and sample pooling (parallel chromatography) approaches time-share MS by reducing dwell windows or reducing the number of data points across a chromatographic peak. These approaches in which the data parcels in milliseconds can monitor multiple channels rapidly

TABLE 4.2
Mass Spectrometer Time Sharing Modes

MS Time Sharing Method	Laboratory and Instrument Activity	Data Packet Size	Limitations
Cassette analysis (performed by user)	Mix samples, injected to same LC-MS or infused to MS	0.05 to 2 sec (dwell time of ion)	Analytes must not have same m/e (MS, or MSn); Analytes must be suitable for same HPLC method; Analyte must be ionizable in similar sprayer conditions; Ion suppression between co-eluting compounds; Lower sensitivity or precision due to shorter dwell time or data points across chromatographic peak
Mix effluents of HPLCs (performed by user)	Mix effluents of several HPLC columns before introducing to MS; HPLCs operated in parallel; may or may not be simultaneous	0.05 to 2 sec (dwell time of ion)	Ion suppression between co-eluting components; Analyte must be ionizable in similar sprayer conditions; Lower sensitivity or precision due to shorter dwell time or data points across chromatographic peak; HPLCs and MS must be synchronized
Multiple sprayer ion source (performed by instrument vendor or advanced user)	Purchase or build LC-MS interface containing two or more sprayers in same interface housing; each sprayer receives effluent from HPLC column; HPLCs operated in parallel and simultaneously	0.05 to 2 sec (dwell time of ion)	Ion suppression between co-eluting compounds; Lower sensitivity or precision due to shorter dwell time or data points across mass chromatographic peak; HPLCs and MS must be synchronized
Multiple sprayer ion source with rotatory ion beam chopper (MUX) (performed by instrument maker)	Place two to four sprayers in same interface housing; each sprayer receives effluent of HPLC; spinning chopper protects orifice of MS to allow ions only from one sprayer to enter at a time; special MS program separates data from sprayers; HPLCs operated in parallel and simultaneously	0.05 to 2 sec (dwell time of ion); relatively longer interchannel time is required for rotation	Closest approach for parallel chromatography; Lower sensitivity or precision due to shorter dwell time or data points across mass chromatographic peak; HPLCs and MS must be synchronized
Staggered parallel chromatography (performed by user or commercial supplier)	Effluents from two to four HPLCs fed to multiposition valve; sprayer of MS receives effluent from only one HPLC at a time; HPLCs operated in parallel and offset from each other	1 to 10 min (data containing portion of mass chromatogram)	Mass chromatogram retention time offset from real retention time; Throughput improvement less than parallel and simultaneous approaches; Analytes must be ionizable in similar sprayer conditions; HPLCs and MS must be synchronized; Drifting chromatographic peaks will lead to loss of data
Offline-surface mediated approach	Collect effluent from multiple LCs on surface; place surface in MS interface; use fast, surface-mediated ionization methods such as MALDI, DESI, etc. to sample effluents	Milliseconds dependent on repetition sampling required to achieve acceptable quantitation	Potential for low cost and very high throughput; Technology for quantitative purposes in development; Sensitivity, precision, and accuracy to be proven; Can tolerate different HPLC methods if multiple HPLC pumps are used

for injections performed simultaneously. Entire chromatograms including portions that do not contain usable information are monitored. The retention time presented on mass chromatograms is correct for these approaches.

LC_nMS approaches (staggered parallel chromatography) parse the work by assigning the MS to an HPLC that is expected to generate useful data (at or near the retention time windows of the analytes), resulting in minute-sized data packages. LC_nMS approaches do not monitor portions of chromatograms that do not contain analyte information. The retention times displayed are offset (delayed) by the values from the sample injection until the start of data collection. Solvent front and effluents after the analyte of interest may be diverted to waste and the MS data in the regions are not collected. Both parallel and staggered parallel approaches are reliable and are used to determine analytes in biological fluids. Unlike MUX in which several HPLCs may be operated simultaneously, the staggered parallel approach relies on a predetermined data window and synchronization of the HPLC, valve, and MS. Table 4.2 presents the different modes of time sharing and the characteristics of each scheme.

4.12   CONCLUSIONS

Staggered parallel chromatography is an efficient and capable tool for linking multiple HPLC systems to a serial detection device, the mass spectrometer. The LC_2MS version of the LC_nMS approach has been in continuous use since 1998 in our laboratory and has supported hundreds of studies. The LC_nMS scheme is now available from many commercial venders. However, for users who want to utilize existing equipment and not purchase additional equipment, it is beneficial to configure an ad hoc system from standard HPLC equipment as presented in this chapter.

REFERENCES

1. Chang, M.S., paper presented at Ninth Annual FACSS Meeting, Philadelphia, 1982.
2. Eckers, C. et al., *Biomed. Mass Spectr.*, 9 (1982) 162.
3. Henion, J.D. and Maylin, G.A., *Biomed. Mass Spectr.*, 7 (1980) 115.
4. Blakley, C.R., McAdams, M.J., and Vestal, M.L., *J. Chromatogr.* 158 (1978) 264; *Anal. Chem.* 52 (1980) 136; *J. Am. Chem. Soc.* 102 (1980) 5931.
5. Blakley, C.R. and Vestal, M.L., *Anal. Chem.* 55 (1983) 750.
6. Dole, M. et al., *J. Chem. Phys.*, 49 (1968) 2240.
7. Henion, J.D., Thomson, B.A., and Dawson, P. *Anal. Chem.*, 54 (1982) 451.
8. Chang, M. and Schoenhard, G., paper presented at Pittsburgh Conference and Exposition, March 1992.
9. Zeng, L. and Kassel, D.B., *Anal. Chem.*, 70 (1998) 4380.
10. Yang, L. et al., *Anal. Chem.* 73 (2001) 1740.
11. McLoughlin, D.A., Olah, T.V., and Gilbert, J.D., *J. Pharm. Biomed. Anal.* 15 (1997) 1893.
12. Olah, T.V., McLoughlin, D.A., and Gilbert, J.D., *Rapid Commun. Mass Spectrom.* 11 (1997) 11.
13. Steinborner, S., *J. Anal Chem.*, 71 (1999) 2340.
14. Chang, M. et al., paper presented at AAPS annual meeting, Long Beach, CA, 1994.
15. Chang, M.S., Kim, E.J., and El-Shourbagy, T.A., *Rapid Commun. Mass Spectrom.*, 21 (2007) 64.
16. Bremer, M., *Profit Bus. Technol.*, 11 (2006) 37.

5 High-Throughput Strategies for Metabolite Identification in Drug Discovery

Patrick J. Rudewicz, Qin Yue, and Young Shin

CONTENTS

5.1 INTRODUCTION

The drug discovery and development processes are time consuming and costly endeavors. It has been reported that on average it takes 10 to 15 years and costs more than $800 million to bring a molecule from discovery to market.[1,2] Compounds fail for various reasons. One that accounts for a reported 40% of failures in clinical trials is poor pharmacokinetics.[3] In an effort to improve the number of compounds that exhibit optimal absorption, distribution, metabolism, elimination (ADME), and pharmacokinetic (PK) properties and reach development, drug metabolism and pharmacokinetic scientists continually implement new technologies and compound screening approaches.

In a typical small molecule drug discovery paradigm, thousands of compounds are screened for activity using high-throughput techniques to identify chemical hits. When hits are found, lead series of compounds that are identified undergo hit-to-lead screening to determine selectivity, *in vitro* efficacy, physicochemical characteristics, and ADME properties. As part of ADME screening, *in vitro* assays are performed to determine stability in the presence of common metabolizing enzymes like cytochrome P450, uridine 5′-diphosphate glucuronosyltransferase, and sulfotansferase. Metabolic screens involve incubations with hepatic microsomal, S9, or hepatocytes and are highly automated using 96-, 384-, or 1536-well plates. It becomes important to identify the structures of metabolites formed in these assays so that the sites of metabolism ("metabolic soft spots") on candidate molecules can be identified. Once identified, such structures can be modified by a medicinal chemist to produce molecules with more desirable metabolic properties.[4]

In vivo PK studies are performed in the course of drug discovery to assess parameters such as clearance and bioavailability. *In vivo* metabolites of drug candidates are identified for a number of reasons including identification of metabolic labile sites on the molecules and determination of *in vivo–in vitro* correlations. Another reason for performing *in vivo* metabolite identification studies is to determine whether any abundant circulating metabolites must be synthesized and subsequently quantitated during toxicology and clinical studies. Any information about the biotransformation of molecules formed *in vitro* or *in vivo* will aid in the elucidation of metabolite structures in radiolabeled mass balance studies and help select the proper toxicology species for drug development submissions. Screening for potential reactive metabolite formation is conducted during the discovery stage.[5–8] Hence, metabolite identification plays a large role in the overall drug discovery process.

In terms of throughput, metabolite identification techniques have not kept pace with other ADME screening procedures. In DMPK, liquid chromatography/tandem mass spectrometry (LC/MS/MS) is used for the quantitation of drug candidates from *in vitro* and *in vivo* studies. These quantitative experiments have reached a level that could be described as high throughput in that sample processing is automated and the analysis is performed with triple quadrupole mass spectrometers in the multiple reaction monitoring (MRM) mode with fast run times (shorter than 5 min). Identification of the biotransformation pathways of drug candidates usually involves a separate group of experiments that require longer LC gradients and run times. Recent developments in MS instrumentation coupled with software improvements paved the way for dramatic increases in the speed by which metabolite identification can be accomplished in complex biological matrices.

It is not the goal of this chapter to provide complete coverage of all the aspects of metabolite identification using LC/MS/MS in drug discovery and development. Rather, we wish to summarize the instrumentation and software advancements that led to and enabled a high-throughput approach to metabolite identification in support of drug discovery. The interested reader is referred to several excellent literature sources for more comprehensive coverages of metabolite identification in drug discovery and development.[9–12]

5.2 TRADITIONAL APPROACH TO METABOLITE IDENTIFICATION

The metabolism of xenobiotic compounds is generally divided into two broad categories: Phase I and Phase II. Phase I pathways of metabolism (or biotransformation) include oxidation, reduction, hydration, isomerization, and other miscellaneous reactions. Phase II reactions, also known as conjugation reactions, include glucuronidation, sulfation, methylation, acetylation, amino acid conjugation, glutathione conjugation, and others. Phase I biotransformations often serve to introduce a polar functional group into a molecule that may act as a site for Phase II conjugation with a suitable endogenous moiety, for example, glucuronic acid. In the 1960s and early 1970s, structural elucidation of metabolites involved extensive sample clean-up and metabolite isolation followed by analysis using physicochemical techniques such as MS and nuclear magnetic resonance (NMR). The introduction of the triple quadrupole mass spectrometer by Yost and Enke in 1978 for mixture analysis[13] paved the way for metabolite identification to be performed in a relatively rapid manner in drug metabolism laboratories.

Perchalski, Yost, and Wilder first described the use of triple stage quadrupole mass spectrometers for primary drug metabolite analysis by first obtaining a product ion scan of the parent drug molecule and subsequently using precursor or parent ion scans to search for metabolites in biological matrices containing common substructural features.[14] The technique was limited to the analysis of primary drug metabolites since chemical ionization with direct probe insertion and gas chromatography/mass spectometry (GC/MS) sample introduction were employed. Rudewicz and Straub reported the use of LC/MS/MS on a triple quadrupole for drug conjugate analysis.[15] They described the use of a neutral loss scan of 176 amu for glucuronide conjugates and a neutral loss of 80 amu for the detection of aryl sulfate esters in biological matrices. The added separation power of LC with the specificity of the triple quadrupole scanning functions greatly facilitated the identification of drug metabolites including thermally labile drug conjugates such as sulfates, glucuronides, and glutathione conjugates.[16,17]

The introduction of pneumatically assisted electrospray (ionspray) and the combination of this technology with triple quadrupole MS greatly increased the applicability of this methodology for the detection and identification of thermally labile drug conjugates.[18] Although drug conjugates could be detected with thermospray ionization, the sensitivity was poor: thermally labile conjugates degraded during the ionization process. For example, with ionspray, the sensitivity for glucuronide and sulfate conjugates of the androgen receptor antagonist, zanoterone, was approximately two orders of magnitude greater than when thermospray was used.[19] The combination of electrospray ionization with triple quadrupole mass spectrometry has been used successfully since the early 1990s for metabolite identification in DMPK. Since then, significant developments in both hardware and software have increased the speed of metabolite identification and the quantities of information acquired in LC/MS/MS experiments.

5.3 LC/MS INSTRUMENTATION DEVELOPMENTS: ION TRAP TECHNOLOGIES

A major advancement that exerted a large impact on metabolite identification is the commercial development of ion trap mass spectrometers. A triple quadrupole MS selects ions by changing the ratio of RF/DC voltages applied to quadrupole rods. Only ions of a particular m/z are stable at any one time and able to traverse the lengths of the quadrupoles. All other ions with unstable trajectories are lost and therefore wasted. A three-dimensional (3-D) ion trap contains two end cap electrodes and a ring electrode and can serve as an ion storage device as well as a scanning MS. Consequently, the sensitivity of an ion trap in full scan and product ion mode is better than that obtained via a triple quadrupole mass spectrometer. A 3-D ion trap provides sensitive full scan spectra as well as MS/MS and MS^n fragmentation. There is, however, a low mass cut-off of 30% for obtaining product ion spectra. Another limitation of ion traps is space charging effects created when the ion density becomes too large inside the trap. Three-dimensional ion traps are especially susceptible to space charging because of their geometry. Although space charging can be mitigated by software developments like Automatic Gain Control, ion traps are not used routinely for robust quantitative analysis in the pharmaceutical industry. In addition, ion traps do not scan in the neutral loss or precursor ion mode in a conventional fashion. For a thorough description of 3-D ion trap mass spectrometers and their scanning modes, the reader is referred to a review article by March.[20]

The development of a two-dimensional (2-D) linear ion trap based upon a triple quadrupole ion path rail as first describe by Hagger[21] significantly impacted metabolite identification capabilities.[22] This instrument has a QqQ rail design; the third quadrupole works as either a normal quadrupole mass filter or a 2-D ion trap capable of axial ion extraction (Figure 5.1). This type of linear ion trap design, commercially known as the QTRAP, is capable of performing scan functions normally associated with a triple quadrupole MS such as precursor or constant neutral loss scans. In addition, sensitive enhanced product ion (EPI) spectra may be acquired using the ion trapping capability of Q3, without the usual 30% low mass cut-off of a 3-D ion trap. Sequential MS^3 data can also be obtained with a QTRAP. For metabolite identification purposes, the QTRAP combines attractive features of both triple quadrupole and ion trap MS. Since the QTRAP can also serve as a normal triple quadrupole, it can be used for routine quantitation and consequently offers DMPK scientists more flexibility.

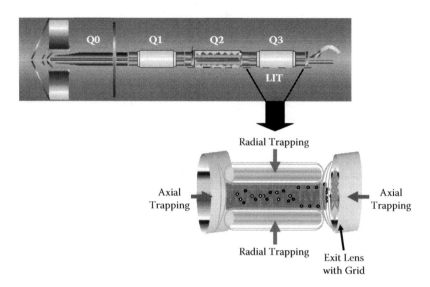

FIGURE 5.1 Ion path for 4000 QTRAP, showing linear ion trap design in Q3. (Courtesy of Applied Biosystems, Foster City, California.)

5.4 LC/MS INSTRUMENTATION DEVELOPMENTS: HIGH RESOLUTION

During the 1960s and 1970s, high resolution sector mass spectrometers were used to obtain accurate data for metabolite identification purposes. Often, the metabolites of interest had to be isolated and purified with the aid of a radiolabel. Due to the nature of the electron and chemical ionization techniques employed, this approach was mostly limited to the analysis of primary drug metabolites. For the analysis of thermally labile drug conjugates, enzymatic or chemical hydrolysis followed by the analysis of the primary metabolite or parent drug was performed. Derivatization for certain types of drug conjugates was also done to enhance volatility. With the introduction of fast atom bombardment (FAB) in the 1980s, it became possible to directly obtain accurate mass measurements for glucuronide and sulfate drug conjugates using magnetic sector MS. Routine LC/MS high resolution experiments using sector instruments were somewhat difficult to carry out because of the high voltage potential of the ion source and the susceptibility of the mass spectrometer to contamination.

With the advent of hybrid orthogonal quadrupole time-of-flight instruments (qToF), more drug metabolism laboratories could perform high resolution experiments using electrospray LC/MS and MS/MS.[23] The reflectron ToF mass analyzer allowed routine mass resolution of 10,000 full width at half maximum (FWHM) peak height to be achieved in the full scan and product ion mode. Accurate mass data expedites assignment of correct elemental compositions to molecular ions and product ions of metabolites. Hence metabolite structures can be assigned with greater confidence and in many instances differentiation of isobaric ions becomes possible. Indeed, using a qToF MS at a resolution of 10,000 full with half maximum (FWHM) means accurate masses and elemental compositions can be assigned within 5 parts per million.

A recent development in high resolution mass spectrometry that has the potential to make a large impact on metabolite identification is the introduction of the LTQ-Orbitrap mass spectrometer.[24,25] As the name implies, this MS is actually a combination of two mass spectrometers, an LTQ, and a new type of mass analyzer design (the Orbitrap component; see Figure 5.2). An Orbitrap mass analyzer consists of two electrodes, an inner spindle electrode and an outer, coaxial, barrel electrode. Ions are trapped and oscillate with characteristic frequencies using electrostatic fields within the Orbitrap. The radial frequency is used to measure mass-to-charge ratio using a fast Fourier transform. The LTQ-Orbitrap possesses all the properties of a linear ion trap with high sensitivity MS and

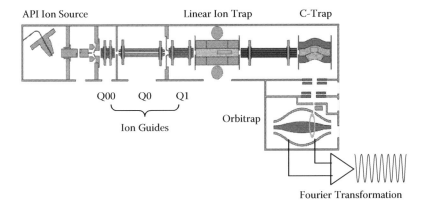

API Ion Source   Linear Ion Trap   C-Trap

Q00   Q0   Q1

Ion Guides   Orbitrap

Fourier Transformation

FIGURE 5.2 Ion path for LTQ-Orbitrap. (Courtesy of Thermo Fisher Scientific, Waltham, Massachusetts.)

MS^n capabilities. However, ions may then subsequently be detected at unit resolution using an electron multiplier or, alternatively, focused in a C-Trap (Figure 5.2) and then transferred and detected at high resolution using the Orbitrap. In our experience with the LTQ-Orbitrap, ions may be measured with a resolution of approximately 60,000 with online LC/MS in the full scan mode.

5.5 SOFTWARE IMPROVEMENTS FOR METABOLITE IDENTIFICATION

To meet the demands of high-throughput metabolite identification in drug discovery, not only have new liquid chromatography and mass analyzers been introduced but also software programs for the automation of metabolite detection, acquisition of MS/MS spectra, and interpretation of MS/MS data have greatly improved. More than a decade ago, Cole et al. developed a rapid automated biotransformation identification (RABID) procedure that represented a breakthrough in the development of automated LC/MS/MS methodology for the identification of drug metabolites.[26] RABID detects any component whose molecular ion's mass-to-charge ratio differs from that of the parent drug by a common biotransformation modification and records the molecular ions of these putative metabolites with their retention times. This information is then utilized in a second analysis that constitutes a retention time-dependent automated LC/MS/MS procedure that allows LC/MS/MS data on predicted metabolites of a drug in a complex mixture to be obtained in two chromatographic analyses.

In a subsequent development, a novel, real-time peak detection algorithm known as intelligent automated LC/MS/MS (INTAMS) was developed for the analysis of samples generated by *in vitro* systems.[27] INTAMS also requires two separate chromatographic runs for each sample. It allows the user to detect the two most abundant ions of all components and also any predetermined metabolite precursor ions. In the second chromatographic run, INTAMS conducts automatic product ion scanning of molecular ions of metabolites detected in the first full scan analysis.

As the demand for metabolite screening in drug discovery has increased, MS vendors have developed metabolite identification software packages such as MetaboLynx (Waters Corporation), Xcalibur/Metabolite ID (Thermo Finnigan), and Analyst/Metabolite ID (Applied Biosystems) to facilitate metabolite detection and LC/MS/MS acquisition automation. These software packages contain lists of editable common biotransformation reactions for the rapid detection of drug metabolites through comparison of reconstructed ion chromatograms of test and control samples. Samples are screened for expected metabolites according to predicted gains and losses in molecular masses relative to the molecular mass of the parent drug. After metabolites are detected, the software automatically constructs retention time-based product scan methods with which one can acquire MS/MS data with additional injections. Due to scan speed limitations, usually only three analytes with a triple quadrupole or eight analytes with ToF MS can be analyzed simultaneously in each period.

Data-dependent acquisition ability has been developed and incorporated into most software packages [MetaboLynx, Xcalibur, and Analyst Information Dependent Acquisition (IDA)]. In data-dependent acquisition mode, a mass spectrometer decides "on the fly" whether to collect MS/MS or MS^n data, remain in full scan MS mode, or conduct other survey scans based upon user-defined criteria. Product ion spectra of potential metabolites can be automatically acquired in a single LC/MS run. However, false positives may be generated due to highly intense matrix ion signals that may inadvertently trigger MS/MS or MS^n scan functions.

As an alternative, a targeted analysis based on a list of the m/z values for potential metabolites (list-dependent data acquisition) can provide product ion spectra for many metabolites within a single LC run. The list of potential metabolites can be predicted based on experience or by utilizing cheminformatic-based data mining and substructural similarity search software packages, such as MDL Metabolite, Accelrys Metabolism, MetabolExpert, METEOR, and META. The integration of knowledge-based predictions of drug metabolites with data-dependent LC/MS/MS allows one to rapidly identify major metabolic pathways of a given drug.[28,29]

More advanced automated software algorithms, such as Waters' MetaboLynx™, Applied Biosystems' Lightsight,™ and Thermo's MetWorks™ have recently become available to detect biotransformations for metabolites. Essentially, these software algorithms are designed to compare and contrast each sample with a control sample. Nevertheless, searching for unexpected metabolites can be performed in the absence of a suitable control using multi-dimensional data searches. For example, MetWorks contains options such as component search, isotope pattern, chromatogram search, and biotransformation search. Both Lightsight™ and MetWorks have integrated mass fragment functions capable of predicting CID fragmentation patterns for organic compounds. Lightsight incorporates MS manager developed by Advanced Chemistry Development (ACD) Laboratories. MetWorks uses Mass Frontier developed by Thermo to accomplish this task. Ideally, these software packages may eventually provide the opportunity to automate the entire identification process from metabolite detection, to MS/MS or MS^n acquisition, spectrum interpretation, and report generation.

Also in the area of software development, a complementary data processing method called mass defect filtering (MDF) introduced by Zhang et al. takes advantage of the high resolution accurate mass measurements of drug metabolites.[30] MDF is based on the premise that under high resolution/accurate mass conditions, most predicted Phase I and II metabolites with molecular weights in the 200- to 1000-Da range have mass defects below 50 mDa when they are compared to the value of the parent compound. (One notable exception is the formation of a glutathione adduct that results in a mass defect of +68 mDa.) MDF is useful in simplifying ion chromatograms, thereby making it easier to locate metabolites. Several MDF applications for high-throughput metabolite identification have been published.[31–33]

5.6 RECENT DEVELOPMENTS IN DRUG DISCOVERY METABOLITE IDENTIFICATION

One obvious starting point for enhancing throughput for metabolite identification is to shorten the chromatographic run times. This must be accomplished, however, without a loss of chromatographic integrity that is so often a crucial aspect of identifying metabolites in biomatrices. Good chromatographic separation is often needed to reduce matrix ion suppression effects[34] and obtain adequate separation of metabolites, especially isobaric or isomeric metabolites.[35] As an example, Dear et al. reported using monolithic columns for the identification of the metabolites of desdebrisoquine.[36] Monolithic columns contain silica rods instead of silica particles and allow much higher flow rates with lower HPLC backpressures relative to normal 2.1- or 4.6-mm internal diameter HPLC columns. The authors reported the separation of seven hydroxylated metabolites with a run time of 1 min. In another study, fast gradient elution using short (2 cm) HPLC columns with a triple quadrupole or qToF mass spectrometer was employed to increase the throughput of *in vitro* drug discovery metabolite identification.[37]

Castro-Perez et al. described the use of ultra-performance liquid chromatography (UPLC) coupled to a quadrupole ToF MS.[38] UPLC columns are packed with sub-2 μm particles, allowing greater separation efficiency at higher HPLC flow rates. The authors compared results obtained for the separation and detection of N- and O-dealkylated metabolites of dextromethorphan and the hydroxylated metabolites of prochlorpemazine using HPLC/MS and UPLC/MS. With UPLC, better chromatographic separation was achieved with faster run times and better sensitivity. Several manufacturers now sell versions of UPLCs, including Thermo (Accela), Waters (Acquity), and Agilent (1200 Rapid Resolution).

Castro-Perez and co-workers also described the use of a five-channel MUX interface to a qToF mass spectrometer for parallel LC/MS high resolution analysis for *in vitro* and *in vivo* metabolite identification.[39] They reported a four-fold throughput increase. However, inter-channel cross-talk and a three-fold reduction in sensitivity relative to a single sprayer system were comparable to results reported by Yang et al. in quantitative bioanalytical applications.[40] In general, implementing this technology in a metabolite identification laboratory for routine support of drug discovery is challenging. Nagele and Fandino reported the use of a conventional single electrospray qToF mass spectrometer combined with column switching using a two-position ten-port valve for the simultaneous determination of the metabolic stability and metabolite identification for buspirone.[41] The metabolic stability determinations were performed with a short (2.1 × 50 mm, 1.8 μm particle size) C18 column. Metabolite identification experiments were performed with a longer (2.1 × 150 mm, 1.8 μm particle size) C18 column with a 17-min gradient elution. Herman reported the use of turbulent flow chromatography for the identification of *in vivo* metabolites in rat plasma and bile.[42] The samples were diluted and the metabolites of interest were focused on an analytical column using a Cohesive system.

5.7 SIMULTANEOUS QUANTITATION OF PARENT DRUG AND METABOLITE IDENTIFICATION

The introduction of new MS instrumentation led to a paradigm shift allowing the quantitation of new chemical entities and the simultaneous structural characterizations of their metabolites. These operations can often be performed in the same timeframe as a high-throughput quantitative LC/MS/MS experiment. The generated data are then used to characterize the biotransformations of new chemical entities, enhancing the overall speed of the drug discovery process within DMPK. Cai et al. carried out this approach using LC/MS/MS with a 3-D ion trap mass spectrometer for the *in vitro* analysis of dog microsomal incubations of α-1a receptor antagonists.[43] The samples were pooled after incubation into four cassette groups. A computer program was used to generate the pooled groups to avoid isobaric metabolite interference. The metabolic stability for each compound was determined by comparing the peak intensity of the molecular ion for each compound at the 0- and 60-min incubation time points. Since full scan data were acquired, ion chromatograms for putative metabolites for each compound were reconstructed from the 60-min incubation run. After metabolites were detected from the full scan data, MS/MS and MS[3] analyses were used to confirm their identities in a subsequent run. A similar approach was used for quantitation of drugs and metabolite identification in pooled plasma samples from *in vivo* discovery pharmacokinetic studies.[44] Kantharaj et al. also used a 3-D ion trap for the simultaneous determination of the metabolic stability of drugs in microsomal incubations and metabolite identification.[45] Zhang et al. used ToF MS to perform rapid quantification and simultaneous metabolite biotransformation elucidation for compounds dosed to rats via cassette.[46] The accurate mass capability of the ToF analyzer was utilized to identify metabolites. Quantitation of a parent drug with a ToF instrument is possible but suffers from a low dynamic range when compared to a triple quadrupole MS.

Newer hybrid instrumentation including the MDS Sciex 4000 QTRAP and the Thermo Electron LTQ-Orbitrap allows quantitation of drug candidates with simultaneous structural characterization of their metabolites within the timeframe of a single high-throughput quantitative LC/MS/MS experiment. For example, with the 4000 QTRAP, during the quantitative analysis of a drug candidate, MRM transitions derived from the parent compound are automatically compiled and scanned from mass shifts of common biotransformations. When putative metabolites are found, the linear ion trap is used to obtain enhanced sensitivity product ion spectra during the same chromatographic run. These data are then used to characterize the biotransformations of new chemical entities. This methodology using an API 4000 QTRAP for drug discovery PK studies[47] and an LTQ-Orbitrap for *in vitro* samples[48] has been presented and several other studies have also been published.[49–51] The next section describes an application of this protocol in a drug discovery pharmacokinetic study in rats.

5.8 QUANTITATION OF DRUG CANDIDATE WITH SIMULTANEOUS METABOLITE IDENTIFICATION: APPLICATION

The hybrid triple quadrupole linear ion trap, 4000 QTRAP, may be used for the simultaneous quantitation of a parent compound and characterization of metabolites by combining MRM-triggered information-dependent acquisitions (IDAs) using ion trap product ion scans in an iterative looped experiment. As a first step, the parent compound to be administered *in vivo* is infused into the 4000 QTRAP and a product ion spectrum is obtained. Using fragment ions that represent major substructures of the compound and common neutral loss masses for secondary drug conjugates, a preformed list of potentially formed MRM transitions is designed. If any transition is detected above a pre-defined peak intensity threshold, enhanced product ion (EPI) spectra are obtained using the linear ion trap on the two most abundant ions in each scan.

In a typical protocol, the sequence of experiments would be as follows: (1) MRM for the parent compound; (2) a search for hits using the MRMs included in the biotransformation table; and (3) performance of EPI scans on the two most abundant ions found in the MRM scans from the biotransformation tables. An exclusion list is incorporated for those transitions that one wants to eliminate from enhanced product scan data acquisition.

Experimental — Compound A was administered orally by gavage to rats at a dose of 5 mg/kg. Samples were collected at 0, 2, 5, 15, 30, 60, 120, 240, and 480 min. Aliquots of 25 μL of plasma were processed using protein precipitation with 100 μL of acetonitrile. A Phenomenex Gemini, 2×50 mm, C18 column was used along with a 5-min gradient. The mobile phases were 0.1% formic acid in water as A and 0.1% formic acid in acetonitrile as B. The flow rate was 0.2 mL/min. The mass spectrometer was an Applied Biosystems, 4000 QTRAP hybrid MS with a Turbo ionspray source. The acquisition methods were MRM/IDA/EPI and MRM quantitation. The MRM channels were based upon a biotransformation table constructed from the product ion spectrum of the parent drug and common as well as predicted Phase I and Phase II metabolic pathways. The IDA threshold was set at 500 cps, above which EPI spectra of the two most intense ions were collected. For quantitation of the parent compound A, a weighted $1/x$ (x = concentration) quadratic regression was used to construct a calibration curve containing ten standards ranging from 5 to 10,000 nM. Structural information was obtained using EPI scans of metabolites with ion counts greater than 500 cps in MRM survey scans.

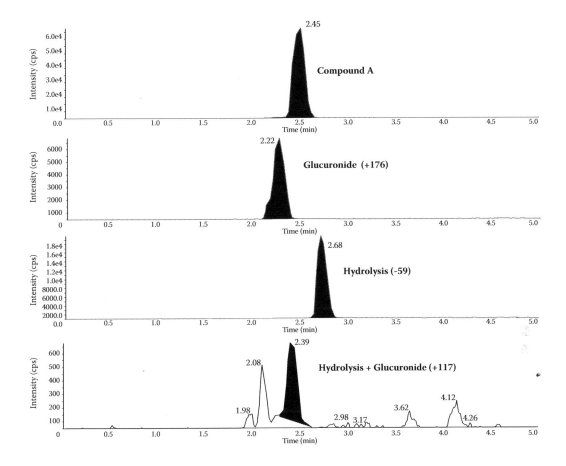

FIGURE 5.3 MRM chromatograms of compound A and its metabolites.

Results and discussion — The top panel of Figure 5.3 is the ion chromatogram corresponding to the MRM transition for the parent compound. The individual ion chromatograms for the MRM channels of three major metabolites detected with this method are also shown. They correspond to a glucuronide of the parent compound, a hydrolysis product, and the glucuronide of the hydrolysis product. Structural data for these three metabolites were also acquired using MRM/IDA/EPI. These enhanced product ion spectra acquired using the linear ion trap mode were similar to those obtained with a triple quadrupole, with better sensitivity. The EPI spectra for the parent compound and three metabolites are shown in Figure 5.4.

The calibration curve for the parent compound obtained using normal MRM only is shown in Figure 5.5; it is very similar to that obtained using MRM/IDA/EPI (Figure 5.6). Moreover, the quantitative results for the parent compound from this pharmacokinetic study are very similar to those obtained using the normal MRM-only measurement. In Figure 5.7, the plasma level concentrations using both approaches are compared by plotting MRM (x axis) versus MRM/IDA/EPI (y axis). The line has a slope of 0.99 and an r^2 of 0.98 indicating that the two methods produced very similar results. This is further demonstrated in Figure 5.8 plotting the pharmacokinetic profiles of the parent compound using both the MRM and the MRM/IDA/EPI measurements. They are in agreement and are acceptable for drug discovery pharmacokinetic support. The MRM/IDA experiment allows identification of biotransformation pathways and plotting of the time profiles of their abundances

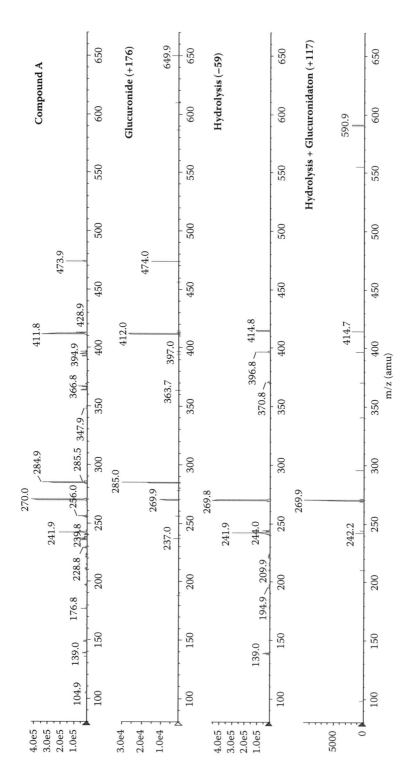

FIGURE 5.4 Product ion spectra of compound A and its metabolites acquired via MRM IDA.

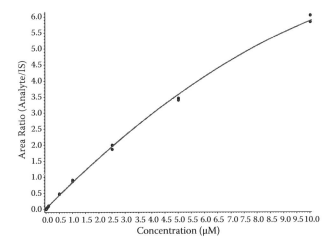

FIGURE 5.5 Standard curve of compound A in rat plasma generated by MRM.

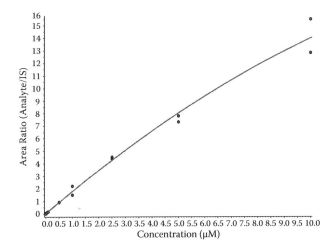

FIGURE 5.6 Standard curve of compound A in rat plasma generated by MRM-IDA.

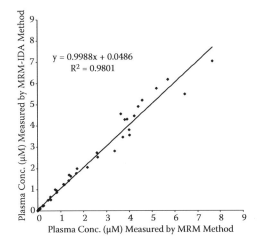

FIGURE 5.7 Correlation of concentrations of compound A in rat plasma generated by MRM and MRM-IDA.

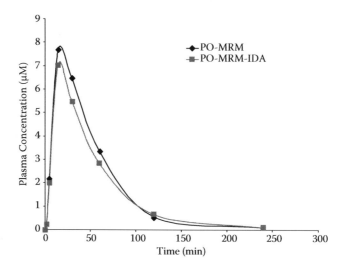

FIGURE 5.8 Comparison of PK profiles using MRM and MRM-IDA after PO administration in rats.

(Figure 5.9). The responses for the three metabolites are plotted as a ratio of the metabolite to the internal standard used to quantitate the parent compound as shown in the figure. This is a relative response, not an absolute quantitative measurement. Nevertheless, for a drug discovery application, it is useful to gain as much information as possible about major pathways of biotransformation. For high clearance compounds, metabolic soft spots may quickly be elucidated and addressed by blocking or modifying sites of metabolism. Moreover, the propensity to form reactive or potentially toxic metabolites may also be discerned. Hence, as demonstrated by this example, it is possible to both quantitate a parent compound and also identify and characterize metabolites within a single experiment.

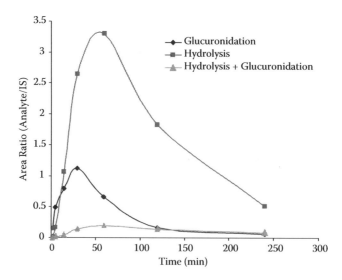

FIGURE 5.9 Time course profile of metabolites of compound A after PO administration in rats.

5.9 CONCLUSIONS

Since the original outline of the general methodology for metabolite identification in the 1980s, the application of LC/MS/MS for metabolite identification has followed the paths of both software and LC/MS hardware developments. Along the way, the speed, throughput, and information content for drug discovery support have markedly improved. Nevertheless, some of the same difficulties and limitations from the 1980s still exist.

To obtain absolute quantitative metabolite levels, one needs either a radiolabeled parent compound or a synthetic metabolite standard—often not available in early drug discovery. Hence, the metabolite abundances are estimates based upon their relative electrospray ionization efficiencies that may vary greatly with structure. Ion suppression may also greatly reduce or totally mask the ion intensity for a given metabolite. Another limitation of these techniques is the possibility of missing a major biotransformation pathway. The biotransformation tables used to search for metabolites are based either upon (1) the substructures of the parent compound and theoretical changes that can occur to these substructures or (2) the difference in mass that would result from a conjugation reaction. Unusual changes like a ring cleavage may be difficult to predict and consequently missed entirely.

As is often the case, further collaboration between LC/MS/MS instrumentation vendors and DMPK scientists will undoubtedly help to reduce these disadvantages in the future. For example, nanoelectrospray ionization has been shown to reduce ion suppression and also yield more uniform responses with compounds that have widely varied structures.[52,53] In addition, software that more accurately predicts putative metabolites and works in conjunction with extensive libraries of metabolic pathways for existing compounds will help reduce uncertainties in search algorithms. Technology advancements in mass analyzer design and the introduction of hybrid instruments with high resolution capabilities along with further software development will pave the way for a greater realization of high-throughput metabolite identification of new chemical entities in support of drug discovery.

REFERENCES

1. DiMasi, J., Hansen, R.W., and Grabowski, H.G. *J. Health Econ.*, 2003, 22, 151.
2. Rowlins, M.D. *Nat. Rev. Drug Discov.*, 2004, 3, 360.
3. Thomson, T.N. in *Using Mass Spectrometry for Drug Metabolism Studies*, Korfmacher, W.A., Ed., 2005, Boca Raton, FL: CRC Press, p. 35.
4. Diana, G.D. et al. *J. Med. Chem.*, 1995, 38, 1355.
5. Baillie, T.A. et al. *Toxicol. Appl. Pharmacol.*, 2002, 182, 188.
6. Ma, S. and Subramanian, R. *J. Mass Spectrom.*, 2006, 41, 1121.
7. Argoti, D. et al. *Chem. Res. Toxicol.*, 2005, 18, 1537.
8. Yan, Z. et al. *Rapid Commun. Mass Spectrom.*, 2005, 19, 3322.
9. Ma, S., Chowdhury, S.K., and Alton, H.B. *Curr. Drug Metab.*, 2006, 7, 503.
10. Prakash, C., Shaffer, C.L., and Nedderman, A. *Mass Spectrom. Rev.,* 2007, 26, 340.
11. Chen, Y., Monshouwer, M., and Fitch, W.L. *Pharm. Res.,* 2007, 24, 248.
12 Chowdhury, S.E., Ed. *Progress in Pharmaceutical and Biomedical Analysis Volume 6: Identification and Quantification of Drugs, Metabolites and Metabolizing Enzymes by LC-MS,* 2005, New York: Elsevier.
13. Yost, R.A. and Enke, C.G. *J. Am. Chem. Soc.*, 1978, 100, 2274.
14. Perchalski, R.J., Yost, R.A., and Wilder, B.J. *Anal. Chem.*, 1982, 54, 1466.
15. Rudewicz, P.J. and Straub, K.M. *Anal. Chem.*, 1986, 58, 2928–2934.
16. Rudewicz, P.J., Garvie, C., and Straub, K.M., *Xenobiotica*, 1987, 17, 413–422.
17. Rudewicz, P.J. and Straub, K.M. In *Drug Metabolism from Molecules to Man*, Benford, D. et al., Eds., 1987, New York: Taylor & Francis.
18. Bruins, A.P., Covey, T.R., and Henion, J.D. *Anal. Chem.*, 1987, 59, 2642.

19. Stack, R.F. and Rudewicz, P.J. *J.Mass Spectrom.*, 1995, 30, 857.
20. March, R.E., *J. Mass Spectrom.*, 1997, 32, 351.
21. Hagger, J.W., *Rapid Commun. Mass Spectrom.*, 2002, 16, 512.
22. Hopfgartner, G., Husser, C., and Zell, M., *J. Mass Spectrom.*, 2003, 38, 138.
23. Chernushevich, I.G., Loboda, A.V., and Thomson, B.A., *J. Mass Spectrom.*, 2001, 36, 849.
24. Hu, Q. et al. *J. Mass Spectrom.*, 2005, 40, 430.
25. Lim, H.K. et al. *Rapid Commun. Mass Spectrom.*, 2007, 21, 1821.
26. Cole, M.J. et al. *Proceedings of 43rd ASMS Annual Conference on Mass Spectrometry and Allied Topics*, Atlanta, May 1995, p. 569.
27. Yu, X., Cui, D., and Davis, M.R. *J. Am. Soc. Mass Spectrom.*, 1999, 10, 175.
28. Reza-Anari, M. and Baillie, T.A. *Drug Discov. Today*, 2005, 10, 711.
29. Reza-Anari, M. et al. *Anal. Chem.*, 2004, 76, 823.
30. Zhang, H., Zhang, D., and Ray, K. *J. Mass Spectrom.*, 2003, 38, 1110.
31. Bateman, K.P. et al. *Rapid Commun. Mass Spectrom.*, 2007, 21, 1485.
32. Castro-Perez, J. et al. *Rapid Commun. Mass Spectrom.*, 2005, 19, 798.
33. Zhu, M. et al. *Drug Metabol. Dis.*, 2006, 34, 1722.
34. Buhrman, D.L., Price, P.I., and Rudewicz, P.J., *J. Am. Soc. Mass Spectrom.*, 1996, 7, 1099.
35. Castro-Perez, J. *Drug Discov. Today*, 2007, 12, 249.
36. Dear, G.J., Plumb, R.S., and Mallet, D.N. *Rapid Commun. Mass Spectrom.*, 2001, 15, 152.
37. Hop, C.E., Tiller, P.R., and Romanyshyn, L. *Rapid Commun. Mass Spectrom.*, 2002, 16, 211.
38. Castro-Perez, J. et al. *Rapid Commun. Mass Spectrom.*, 2005, 19, 843.
39. Leclercq, L. et al. *Rapid Commun. Mass Spectrom.*, 2005, 19, 1611.
40. Yang, L. et al. *Anal. Chem.*, 2001, 73, 1740.
41. Nagele, E. and Fandino, A.S. *J. Chromatogr. A*, 2007, 1156, 156.
42. Herman, J.L. *Rapid Commun. Mass Spectrom.*, 2005, 19, 696.
43. Cai, Z., Sinhababu, A.K., and Harrelson, S. *Rapid Commun. Mass Spectrom.*, 2000, 14, 1637.
44. Cai, Z. et al. *Rapid Commun. Mass Spectrom.*, 2001, 15, 546.
45. Kantharaj, E. et al. *Rapid Commun. Mass Spectrom.*, 2003, 17, 2661.
46. Zhang, N. et al. *Anal. Chem.* 2000, 72, 800.
47. Shin, Y., Ubhayakar, S., and Rudewicz, P.J. paper presented at 13th ISSX Meeting, Maui, October 2005; *Drug Metab. Rev.* 2005, 37, Abstr. 176.
48. Du, A. et al. paper presented at 13th ISSX Meeting, Maui, October 2005; *Drug Metab. Rev.* 2005, 37, Abstr. 521.
49. Shou, W.Z. et al. *J. Mass Spectrom.*, 2005, 40, 1347.
50. Li, A.C. et al. *Rapid Commun. Mass Spectrom.*, 2005, 19, 1943.
51. Li, A.C., Gohdes, M.A., and Shou, W.Z. *Rapid Commun. Mass Spectrom.*, 2007, 21, 1421.
52. Chen, J. et al. *J. Chromatogr. B*, 2004, 809, 205.
53. Hop, C. E., Chen, Y., and Yu, L.J., *Rapid Commun. Mass Spectrom.*, 2005, 19, 3139.

6 Utilizing Microparallel Liquid Chromatography for High-Throughput Analyses in the Pharmaceutical Industry

Sergio A. Guazzotti

CONTENTS

6.1 INTRODUCTION

High-throughput techniques for qualitative and quantitative analysis are currently in great demand due to requirements for increased productivity in the pharmaceutical industry. Researchers employ these techniques to eliminate bottlenecks in drug discovery, subsequently enabling them to make informed decisions about drug candidates earlier in the discovery process. Early drug discovery has notoriously been a source of bottlenecks and frustration for the industry. Analytical chemistry, which constitutes a significant component of the drug discovery process, is an area where progress was limited by available techniques and no significant improvements occurred for decades.

The advent of combinatorial chemistry produced compound libraries that significantly increased in size and diversity. Consequently, the expansion and diversification of compound libraries generated greater demand for technology that increased analytical throughput.[1] In addition, early determination of solubility, purity, log P, and other physiochemical properties limited downstream attrition of compounds due to poor ADME properties,[2] thus leading analytical chemists to evaluate technologies that provide higher quality data along with increased throughput.

Assay development and high-throughput screening (HTS) are also critical processes utilized early in the discovery phase. These processes have suffered from data uncertainty resulting from traditional methods of developing biochemical assays. The ensuing high frequency of false positives generated from HTS exposed the importance of gathering information about every compound in a library to improve confidence in screening results. Advances and developments in early discovery, in particular those involving high-throughput techniques, aid in the optimization of the processes described above.

Additional technologies for high-throughput screening and combinatorial chemistry have failed to deliver on the promise to increase productivity and left many skeptical about their future uses.[3] Drug discovery companies that employ HTS have come to realize that these technologies created bottlenecks in compound management and lead optimization. Questions about the time involved in developing an assay appropriate for a high-throughput screen and the reliability of information obtained from screening campaigns led to a trend to explore separation-based assays. Therefore, high-throughput separation techniques are emerging as new ways to reliably and reproducibly identify active compounds.

One technique often referred to as the workhorse of analytical chemistry and providing many advantages for assay development is high performance liquid chromatography (HPLC).[4] Improvements made to conventional HPLC systems allow greater accuracy and precision. To increase throughput for many analytical applications, conventional HPLC systems can be employed in a serial fashion with shorter columns and faster cycle times. However, these systems suffer from their intrinsic low-throughput natures when used in serial mode in addition to resulting in applications that are time-consuming. Further, conventional HPLC instrumentation presents a significant bottleneck in profiling large sets of samples and analyzing data. In recent years, ultra-throughput LC systems have significantly reduced analysis time. These changes have only been incremental to a linear process for sample analysis that eventually faces a finite limit.

Recent technological developments in automation combined with advances in microfluidics led to the development of a novel microparallel liquid chromatography (μPLC) system (Nanostream, Inc.) that enables high-throughput chemical analysis and a separation-based approach to biochemical assays. The advantage of microfluidic technology is that it enables complex chemical and biological reactions to be executed and analyzed with microliters (and submicroliters) of samples. This technology is especially applicable to drug discovery, where the manipulation of complex and expensive biological fluids is required. Furthermore, pharmaceutical companies desire to conserve expensive samples and seek technologies that allow minimum use of valuable resources. With microfluidic technology at its core, μPLC miniaturizes HPLC analysis by enabling 24 simultaneous separations and real-time UV and fluorescence detection. In this way, and operating in a truly parallel fashion, μPLC systems allow for the analysis of more samples in less time.

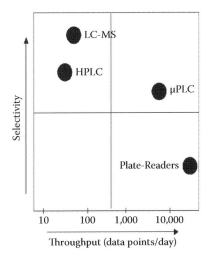

FIGURE 6.1 Tradeoff between throughput and selectivity in evaluating analytical technologies.

This microparallel analysis provides the ability to analyze a large number of compounds, increase the number of replicates or conditions used in a study, and reduce solvent consumption and mixed waste generation. Nanostream, Inc. has developed instrumentation for μPLC with different capabilities suited for diverse applications (e.g., the Nanostream CL system for high-throughput LC, the Nanostream LD for assay detection, and the Nanostream CX for sample preparation).

When evaluating available techniques, scientists have had to determine trade-offs between data selectivity and analytical throughput as exemplified in Figure 6.1. When these parameters (selectivity versus throughput) are considered, it is clear that HPLC and LC/MS deliver highly selective information, but yield lower throughput when used in a serial format. Plate readers, the traditional tools of biochemistry assays, produce uncertain results because they are not sensitive to compound purity and identity. However, this uncertainty in data quality is a trade-off for throughput. To maximize speed, biochemists design homogeneous assay formats, but this approach increases the number of potential false positive results. μPLC affords researchers increased sample analysis capacity without compromising data quality.

Chemical analysis applications such as characterization of ADMET, physiochemical property profiling, compound library purity, and routine analyte quantitation are ideal uses for μPLC. For example, Wielgos and Havel[4] validated the use of μPLC for routine measurement of lactic acid and demonstrated an increase in throughput without a data quality compromise. μPLC offers the advantages of a separation-based approach while enabling high-throughput chemical analysis and reducing sample consumption, solvent usage, and waste generation. Typically, analytical chemists in both compound library management and in discovery analytical laboratories have utilized μPLC with UV absorbance detection and, in cases where greater sensitivity was desired, fluorescence detection.

For biochemical assays, μPLC allows direct quantification of substrates and products using a much-valued separation-based approach that allows development and optimization of challenging enzymatic assays faster and with fewer false positives. The separation-based approach employed by μPLC dramatically reduces assay development time from months to a few days. Since substrate and enzymatic products are separated prior to detection, μPLC enables development of difficult assays, such as analyzing enzymes with low kinetic activities and enzymes that cannot be analyzed on existing platforms.

Separations allow quantitation of each component, yielding greater degrees of certainty in results. This approach is less prone to interference because the signals due to substrates and products are followed and other components present in the samples can be separated from those of interest (i.e., substrates and products). This produces more reliable data. In addition, the separation-based approach offers greater flexibility in attaching a tag anywhere on a substrate, which in turn, speeds development time.

μPLC also yields advantages for secondary screening applications because it allows the identification of false positives by repeating analysis of hits from primary screens. For example, if an inhibitor appears fluorescent, the assay results could potentially be distorted. By using μPLC for a secondary screen, the technology detects the inhibitor because it is quantitated separately. In therapeutic areas where high-throughput processes are typically unnecessary, biologists can employ μPLC to develop assays for hard-to-target substrates. As an example of this type of application, the NIH Chemical Genomics Center reported data from secondary screening results for active verification employing Nanostream's μPLC system coupled with HTS,[5] and determined that the combination of concentration–response screening and chromatographic follow-up via μPLC dramatically reduced the number of false positives. Jezequel-Sur et al.[6] described the results of successful evaluations of a Nanostream μPLC system as a platform for secondary screening on kinase drug targets. As with the results presented by Yasgar et al., μPLC in this case permitted the identification of false positive hits previously obtained by FP and TR-FRET.

In addition, μPLC with time-triggered fraction collection is used for chromatographic sample preparation in bioanalytical applications, i.e., drug metabolism and pharmacokinetic profiling. For the clean-up of complex samples, μPLC allows LC fractionation prior to MS analysis. Sample preparation is followed by analysis using MS/MS. The increased analytical throughput offered by μPLC allows users to significantly increase productivity of their existing MS/MS equipment. The use of the μPLC could potentially yield up to a ten-fold increase in MS/MS productivity compared to online LC/MS/MS. For example, Mehl et al.[7] demonstrated the rapid sample clean-up of complex biological mixtures employing μPLC with fraction collection capabilities coupled to MS. When compared to other methods such as SPE, LLE, and PPT, only the μPLC method did not produce a level of ion suppression, therefore providing a "more effective sample clean-up method." Lloyd et al.[8] demonstrated that this approach improved MS efficiency and allowed the delivery of samples on MS time scales.

The next section describes the utilization of μPLC for different applications of interest in the pharmaceutical industry. The part discusses the instrumentation employed for these applications, followed by the results of detailed characterization studies. The next part focuses on particular applications, highlighting results from the high-throughput characterization of ADMET and physicochemical properties (e.g., solubility, purity, log P, drug release, etc.), separation-based assays (assay development and optimization, real-time enzyme kinetics, evaluation of substrate specificity, etc.), and sample preparation (e.g., high-throughput clean-up of complex samples prior to MS (FIA) analysis).

6.2   INSTRUMENTATION OVERVIEW

6.2.1   GENERAL OVERVIEW

The μPLC system described in this chapter is equipped with 24 parallel columns for liquid chromatography, each with its own sample introduction port and exit port for connection to detectors of choice (UV absorbance and/ or fluorescence). Flow from a binary solvent delivery system is divided evenly across 24 channels and results in 1/24 of the programmed pump flow rate through each column (i.e., total flow of 300 μL/ min will produce a flow of 12.5 μL/ min in each column). Samples are introduced to the columns by a multichannel autosampler configured to sample from either 96- or 384-well SBS standard plates. Figure 6.2 depicts a general view of the system.

The truly parallel approach of this technology permits many different samples to be separated simultaneously using a minimum number of common system components such as pumps and pulse dampers. To increase the automation of the system, a plate loader and exchanger that accommodate

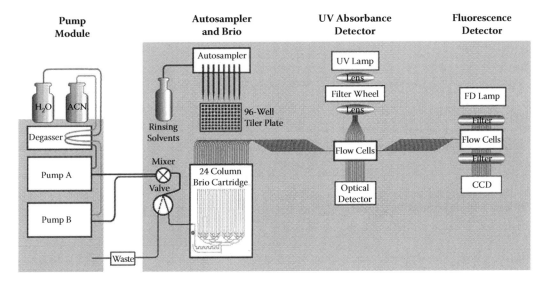

FIGURE 6.2 μPLC system.

SBS standard microplates may be integrated as shown in Figure 6.3.

6.2.2 SOLVENT DELIVERY

Mobile phases employed for the separations are housed in a cartridge and delivered to the LC columns through a set of binary HPLC pumps (Shimadzu Corporation), as shown in Figure 6.2. The pumps provide a flow rate accuracy of $\pm2\%$ or 2 μL (whichever is greater) in constant flow pumping mode, with a flow rate precision of $\pm0.3\%$. A degasser (two channels; internal volume of 195 μL/channel) is also housed in the pump module employed to minimize the occurrence of air bubbles.

Mobile phase programming (for isocratic or gradient modes) is accomplished through the software that provides full control of the modules and components of the μPLC system. The total flow for the runs (that may be variable) is set via the system software. The total flow from the binary pumps is split evenly into 24 streams; this delivers 1/24 of

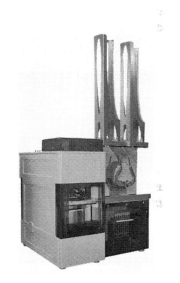

FIGURE 6.3 μPLC system with automatic plate exchanging mechanism.

the total programmed flow to each column within the cartridge (see Section 6.2.3 below). If the total flow (combined flow from both pumps) is set at 240 μL/ min for a particular run, the actual flow received in each column will be 10 μL/min. As shown by the preceding example, the per-column flow rates employed in this system are significantly smaller than those commonly employed in traditional HPLC and LC/MS instrumentation. This difference in flow rate results in a significant saving of mobile phases needed for the analysis.

6.2.3 BRIO CARTRIDGES

At the heart of the system is a cartridge that houses 24 liquid chromatography columns. This cartridge enables multiple samples to be analyzed in parallel. The 24 incorporated columns may be packed with many different standard stationary phase materials. Each parallel column has its own sample introduction port and exit port for connection to detectors of choice. Mixing and distribution of the mobile phase to each of the 24 columns is precisely controlled in each cartridge. The cartridges are made of polymeric materials and subjected to very rigorous and controlled manufacturing processes in which many different layers of material are bonded to generate the desired solvent and packing channels, as shown in Figure 6.4.

This construction method yields chemically resistant cartridges with high bond strengths. These properties allow the cartridges to withstand column packing processes and subsequent operation after insertion into the μPLC system. Figure 6.5 is a composite diagram of a cartridge. As shown, different channels are created in a single cartridge to house the packing material to be employed in the separations and transfer the mobile phase to the LC columns. The packing inlet present in the cartridge (Figure 6.5) is employed to distribute the packing material into the 24 columns during a slurry pack-

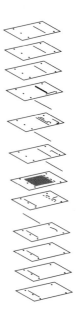

FIGURE 6.4 Different layers of polymeric material employed in the manufacturing of Brio cartridges employed in μPLC systems.

ing process. After the device is packed, all the packing structures are filled with packing material with the exception of the channel leading from the waste frit to the waste outlet. The samples to be analyzed are placed onto the columns (by an autosampler; see Section 6.2.4) at the injector frits and transported downward to the outlet frits.

The frits in the cartridges are intended to retain stationary phase material in the separation channels (columns) while permitting the passage of the mobile phase during separations. Each frit is constructed from a permeable polypropylene membrane with an average pore size smaller than the

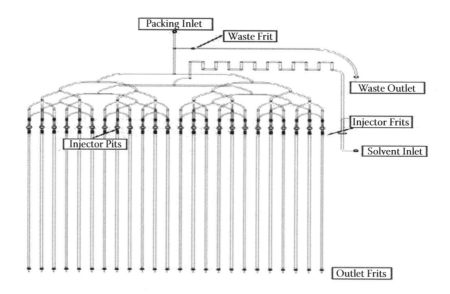

FIGURE 6.5 Packing and solvent channels present in Brio cartridges employed in μPLC systems.

FIGURE 6.6 Brio cartridges with different designs employed in μPLC systems.

average particle size of the particulate to be packed within the cartridge, to ensure that the packing material is successfully retained within the cartridge. When separation takes place, the mobile phase is transferred from the pump module into the solvent inlet present in the cartridge (Figure 6.5). The mobile phase pushes through the splitter network, through the first set of frits, through the injector pits, and then through the injection frits and onto the columns.

The columns present in the cartridges have specific dimensions, depending on the type of design employed. For example, in some of the most common cartridges, the columns are 80 mm long, with rectangular cross sections of 1 mm width and 0.2 mm height, resulting in a equivalent circular cross section of ca. 0.5 mm. Other designs accommodate columns with equivalent circular cross sections of 1.0 mm. These cartridges allow for housing of more packing material on a per-column basis, leading to increased practical resolution during applications.

Another type of cartridge design employs shorter 30-mm columns with equivalent cross sections of 0.5 mm, ideal for faster separations. The volume of sample that can be injected into these cartridges also varies with design and depends on the application sought. Some designs allow a maximum injection volume of 1.0 μL; other designs increase this range to accommodate up to 5.0 μL of sample. Figure 6.6 shows the most common Brio cartridges. The cartridges are packed with many different stationary phases, such as C18, modified C18 for hydrophilic compounds, C8, C4, phenyl, etc. The ultimate choice of cartridge design and packing material depends on the desired performance for specific experimental protocols and throughput requirements.

6.2.4 Autosampler

After the cartridge is inserted into the system, samples are introduced to the columns housed in a cartridge by a multichannel autosampler configured to sample from 96- or 384-well SBS standard plates. When 96-well plates are employed, the autosampler is configured with six needles. For 384-well plates, the autosampler is designed for eight needles. The autosampler is coupled to an 8-channel (or 6-channel for sampling from 96-well plates) syringe pump module.

Figure 6.7 depicts an autosampler employed in a μPLC system. Figure 6.8 details the autosampler component. Samples are transferred from the desired well in the microtiter plate into the columns of the Brio cartridge. If a 384-well plate is employed, the autosampler will carry out 3 sets of 8 injections into the columns, for a total of 24 columns. The solvent (mobile phase) does not circulate in the cartridge but is diverted into a backpressure regulator located in the waste line (Figure 6.2). This process of injection is known as stop-flow injection. After all samples are placed into the injection pits of the 24 columns in the cartridge (Figure 6.5), a clamp containing a seal

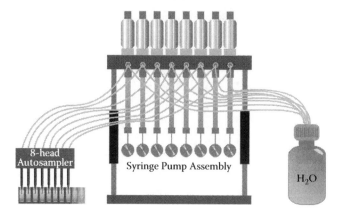

FIGURE 6.7 Autosampler employed in μPLC system.

gasket is displaced down to seal the cartridge, after which the mobile phase starts to circulate through the cartridge, evenly split through the solvent tree within the cartridge, as represented in Figure 6.5. After the mobile phase reaches the injection pits where the samples are, it pushes the samples into the columns where separation takes place.

The automation process is controlled by a user-friendly software application, with an intuitive graphical advanced user mode for tuning and set-up. The depth of the autosampler needles used for sample transfer into the analytical columns can be carefully controlled to sample the solution at the desired height from a microtiter plate (e.g., to sample supernatant solution without disturbing precipitate if present). As with any liquid handler, the autosampler allows control of desired top-off volumes, volumes employed for rinsing (aspiration and dispensing), number of iterations employed during washing, and other factors. The system is designed with a total of three rinsing stations that can accommodate different compositions of washing solutions required to minimize sample carry-over. The selection of washing solution depends on the composition of the samples used.

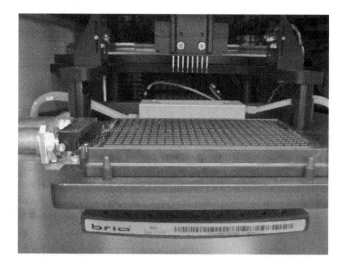

FIGURE 6.8 Autosampler employed in μPLC system. Note eight-head autosampler above the rinsing station.

FIGURE 6.9 Individual flow cell employed for UV absorbance detection in μPLC systems. The shaded channels represent fluid (sample + mobile phase) pathways; the clear channels represent the light pathways. The flow cell counts with a U-shape design since liquid is transported in a U-shape within the flow cell.

6.2.5 DETECTION: ULTRAVIOLET ABSORBANCE AND FLUORESCENCE

While separation takes place in the cartridge housed in the instrument, samples elute from each column through 24 individual exit ports as shown in Figure 6.5. The ports are connected to a bank of 24 flow cells employed for UV absorbance detection. Figure 6.9 depicts one of the individual cells employed. Figure 6.10 shows an overall view of the UV absorbance detection mechanism.

The light source employed for UV absorbance detection is a deuterium lamp that provides a usable range of 200 to 350 nm. Filters are housed in a wheel that allows automatic filter selection (controlled by the system software) for single wavelength determinations. A total of five filters can be accommodated in the filter wheel; the most common filters for several applications are 214, 254, and 280 nm. As shown in Figure 6.10, light from a deuterium lamp passes through the filter of choice and is coupled through optical fibers to the individual flow cells. Light from the optical fibers reaches the sample present in the U-shaped flow cell and the amount of light that passes through after absorption in the sample (also transferred by individual optical fibers) is determined by photomultiplier tubes.

Some μPLC systems are equipped with UV absorbance detection, and other systems allow for both UV absorbance and fluorescence detection. Fluorescence detection increases the sensitivity and selectivity of certain applications and is the method of choice in many separation-based assays. The liquid (mobile phase + sample) leaving the individual flow cells designated for UV detection is transferred through capillaries to a bank of 24 flow cells designated for fluorescence detection.

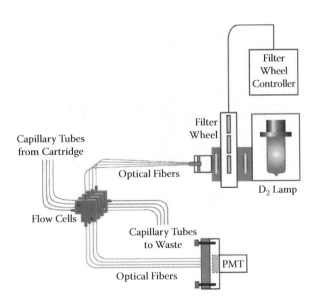

FIGURE 6.10 Ultraviolet absorbance detection.

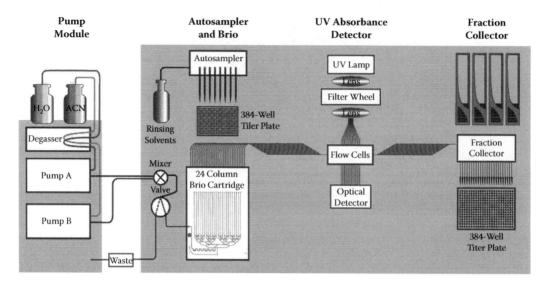

FIGURE 6.11 μPLC system with time trigger fraction collection capabilities.

The light source is a 75 W xenon arc lamp. Excitation and emission filters are chosen according to the particular application and based on the fluorescent properties of the analytes of interest. For example, for tetramethylrhodamine (TAMRA) labeled compounds, it is common to use an excitation filter of 525 nm (40 nm bandwidth) and an excitation filter of 585 nm (40 nm bandwidth). Light from the xenon arc lamp is transferred to the flow cells employing a liquid light guide. Light emitted by the fluorescent species after relaxation to the ground state passes through the emission filter prior to reaching a back-illuminated CCD used for detection.

6.2.6 TIME TRIGGER FRACTION COLLECTION

For certain bioanalytical applications such as those described in the previous section, it is desirable to collect analytes of interest while separation in the μPLC system takes place (e.g., for follow-up by MS). To collect the analytes, the μPLC system may be equipped with a time-triggered fraction collection mechanism. This system allows the selection of time intervals in which samples will be diverted (after passing through the detectors) to the selected wells of microtiter plates. Figure 6.11 shows an overall view of a system with time-triggered fraction collection capabilities. Figure 6.12 depicts the system along with a detailed view of the collection sampler employed.

As shown, the system incorporates an integrated plate changer that accommodates plates for analysis as well as plates for collecting fractions of interest. Plates can be of different formats for sampling and collection, for example, a 384-well plate could be used for samples and a 96-well plate for collection (of course, the same plate type may be used for both sampling and collection). The system also incorporates a dedicated rinse station at the fraction collection end. The number of fractions and the time intervals for collection are defined by the user and automatically controlled by the software. In this way, analytes can be isolated and collected using μPLC. Collected fractions can, for example, be injected onto MS instrumentation with minimal cycle time by employing a flow injection analysis approach.

6.2.7 SOFTWARE

The two main software applications are used with a μPLC system. One controls the instrument and acquisition of data and the other processes the acquired data (chromatograms). The system software

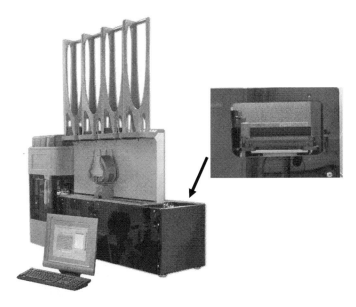

FIGURE 6.12 μPLC system with time trigger fraction collection capabilities.

enables fully integrated control of sample entry, sample sequences, autosampler parameters, chromatography parameters, data acquisition, and chromatogram visualization. The software used for data analysis offers immediate peak integration results with the possibility of applying batch analysis to rapidly process large amounts of data and report results.

Since the system described in this chapter operates in a truly parallel fashion, with every run (a run is a composite run employing a cartridge and 24 columns), 24 separations occur simultaneously. Twenty-four chromatograms are generated in each run when UV absorbance detection is employed and a total of 48 chromatograms when both UV absorbance and fluorescence detections are employed (24 chromatograms with UV signals and 24 chromatograms with fluorescence signals). Since the number of chromatograms generated per unit of time increases significantly with μPLC (in comparison with traditional HPLC in which only a single chromatogram is generated per run), it is critical for the analysis software to provide a user-friendly and effective way to carry out data analysis and reduction for each application.

Users can also select assay-specific advanced analysis software modules, depending on their needs. The analysis software modules support a range of applications including compound library purity assessment, log P, log D, solubility and permeability assessments, and ratiometric assay analyses. For example, the solubility module automatically extracts peak areas from both standards and unknowns, calculates standard curves, and outputs the final solubility result along with statistics. The ratiometric assay analysis software module enables researchers to process large amounts of data and rapidly determine percent conversion and/or percent inhibition for compounds using separation-based assay data from the μPLC system, as exemplified in Figure 6.13. The software was developed in an open architecture format. Thus, if desired, exporting raw data from the system into any existing software platforms may be accomplished in a seamless manner.

6.3 PERFORMANCE CHARACTERIZATION

As with any analytical instrumentation, processes that involve both operation qualification and performance qualification are implemented during manufacture of the instrumentation for μPLC described in this chapter to ensure that the instrument will perform according to specifications.

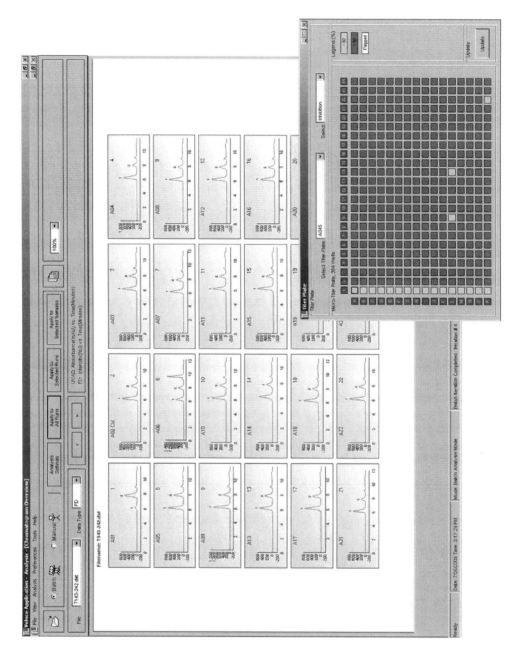

FIGURE 6.13 Example of selected graphical output for ratiometric assay analysis.

During operation qualification, all components of the instrument are tested individually along with integral parts of the overall instrumentation. In this section, results of some of the most common performance parameters are presented with a brief description of the methods used for evaluation.

6.3.1 Flow Reproducibility and Delay Volume

In order to evaluate pump flow rate reproducibility and pulsation, one method is commonly used to assess gradient formation capability. A certain amount of an analyte with adequate molar absorptivity at the wavelength employed for detection is introduced into one of the mobile phases employed to create the gradient. In the case described, 5% acetone was introduced into the mobile phase, distributed to the system by pump B. No UV-absorbing analyte was introduced into mobile phase A. The fractional flow rate of pump B relative to the total flow rate of the system (mandated by the sum of the flow rates of pumps A and B) was increased in individual steps to account for 0, 3, 6, 12.5, 25, 50, and 100% fractional rates. The total flow for the system was maintained at 300 μL/ min (for 24 columns), resulting in a per column flow rate of 12.5 μL/min/column.

Figure 6.14 shows the results obtained from flow rate reproducibility studies. Results obtained for a randomly selected column over a total of three consecutive runs are presented. The inset highlights results obtained for pump B during these evaluations. As shown, the pump pulsation observed was consistently lower than 10 nL/min for each column.

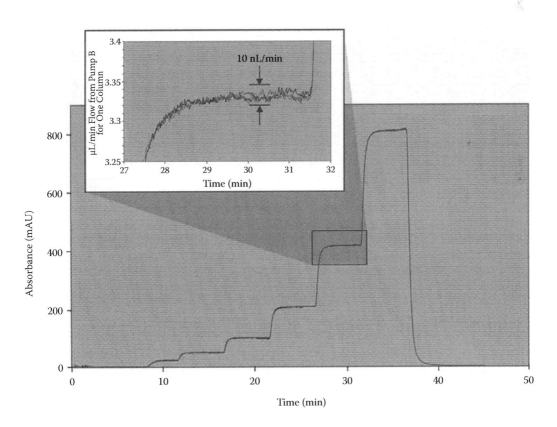

FIGURE 6.14 Results of flow reproducibility evaluations showing the overlay of three consecutive runs on a single column (chosen randomly). The flow rate employed during these evaluations was 12.5 μL/min/column. Inset: magnified view of an overlay of three consecutive runs for a single detector at 400 mAU. Absorbance values were converted to flow rate values to highlight the results obtained.

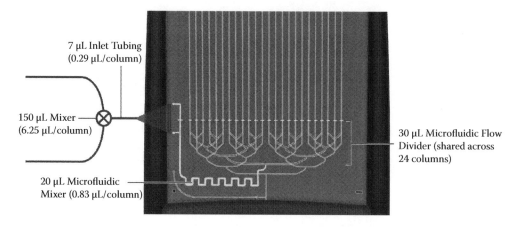

FIGURE 6.15 Evaluation of delay volume in a μPLC system equipped with a cartridge with a per-column internal cross section of 0.5 mm.

The delay volume, defined as the volume between the mixer and the top of the column, may be evaluated from the design of the cartridge and the components present in the instrumentation that deliver the solvent (mobile phase) to the cartridge (Figure 6.15). As shown in the figure, several components contribute to the total delay volume present in the μPLC system equipped with a cartridge, i.e., inlet tubing, mixer, microfluidic mixer, and microfluidic flow divider. As with any LC instrumentation, it is desirable to keep the delay volume to a minimum. Multiple columns share portions of the total delay volume of the system. When the evaluation is conducted on a per-column basis, the resulting delay volume is 5.0 μL/column.

6.3.2 Retention Time and Peak Area Reproducibility

The ability of a μPLC system to produce the "same" values of retention time and peak areas for analytes of interest is determined by evaluating the precision obtained under standardized conditions and analytical methods. The precision (reproducibility) values obtained are functions of the autosampler, cartridge, and detectors employed. Due to the parallel design of the μPLC system described in this chapter, reproducibility evaluations of retention time and peak area involved comparisons of results obtained for these parameters for consecutive runs performed in the same column and across different columns.

A total of 24 composite runs (each consisting of runs across the 24 columns, i.e., 24 equivalent HPLC runs) were performed. The analytes used were methyl paraben (0.10 mg/mL), propyl paraben (0.10 mg/mL), and rhodamine 110 chloride (100 nM). To conduct the evaluations, consecutive 1 μL injections of a test mixture containing the analytes described above were executed utilizing a cartridge packed with a C18 stationary phase and per-column dimensions of 0.5 mm circular cross section and 80 mm length. Signals for methyl paraben and propyl paraben were monitored by UV detection at 254 nm while the signal for rhodamine 110 chloride was monitored via fluorescence detection with an excitation filter of 482 nm (35 nm bandwidth) and emission filter of 535 nm (40 nm bandwidth). A gradient method (Figure 6.16) was used for these evaluations. Compositions of mobile phases A and B were 5:95 $H_2O:CH_3CN$ with 0.1 HCOOH, and CH_3CN with 0.085% HCOOH, respectively, with a total flow rate of 300 μl/ min (corresponding to 12.5 μL/min for each of 24 columns).

Figure 6.17 is an example overlay of the chromatograms obtained for 24 consecutive runs within a single column (chosen at random). The average retention time reproducibility and peak area reproducibility values obtained for all 24 columns for evaluating these parameters within a single column are presented in Table 6.1 (run to run within column). Reproducibility values are expressed as percentages relative standard deviations [%RSD = (standard deviation/average) × 100].

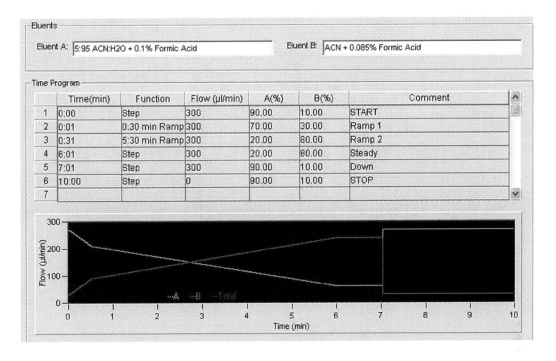

FIGURE 6.16 Gradient method employed for evaluation of retention time and peak area reproducibility. Solvent A: 5:95 H$_2$O:CH$_3$CN with 0.1 HCOOH. Solvent B: CH$_3$CN with 0.085% HCOOH.

Figure 6.18 presents example chromatograms obtained during the evaluation of retention time and peak area reproducibility across the 24 columns (within a single composite run). Table 6.1 lists the average retention time reproducibility and peak area reproducibility values obtained across 24 columns for a total of 24 composite runs (column to column within run). Reproducibility values are expressed as percentage relative standard deviation (%RSD).

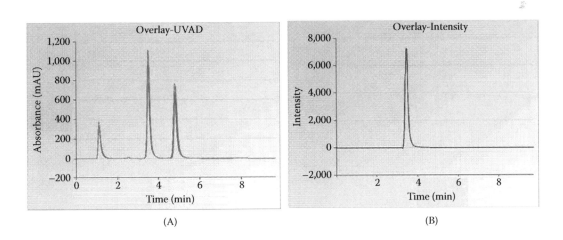

FIGURE 6.17 Chromatogram overlay for 24 consecutive runs performed on a single column. (A) results of overlay for the chromatograms obtained with UV absorbance detection. Peaks are identified as (with increasing retention time) uracil (dead volume marker), methyl paraben, and propyl paraben. (B) results of overlay for chromatograms obtained from fluorescence detection (peak identified as rhodamine 110 chloride).

TABLE 6.1
Results of μPLC Evaluations of Retention Time and Peak Area Reproducibility

	%RSD Peak Area	%RSD Retention Time
Methyl Paraben		
All runs, all columns	2.4	0.6
Column to column within run	2.3	0.6
Run to run within column	1.4	0.3
Propyl Paraben		
All runs, all columns	2.4	0.4
Column to column within run	2.4	0.4
Run to run within column	1.4	0.2
Rhodamine 110 Chloride		
All runs, all columns	1.5	0.9
Column to column within run	1.5	0.9
Run to run within column	1.1	0.2

Total of $24 \times 24 = 576$ runs were performed.

An additional parameter may be considered when evaluating performance of instrumentation regarding retention time and peak area reproducibility, i.e., overall reproducibility obtained for all columns during all composite runs (all runs, all columns). A total of $24 \times 24 = 576$ runs were performed after 24 consecutive composite runs were carried out (24 columns used in each run). Results are presented graphically in Figure 6.19 and numerically in Table 6.1.

As demonstrated by the results summarized in the table, retention time reproducibility values below 1.0% were consistently obtained during all evaluations. Peak area reproducibility values

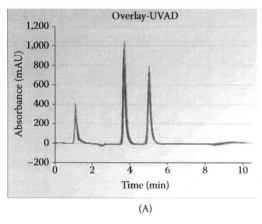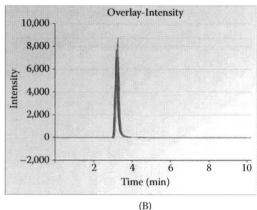

(A)   (B)

FIGURE 6.18 Overlay of 24 chromatograms obtained simultaneously during a single run employing μPLC. (A) results of overlay for chromatograms obtained with UV absorbance detection. Peaks are identified as (with increasing retention time) uracil (dead volume marker), methyl paraben, and propyl paraben. (B) results of overlay for chromatograms obtained with fluorescence detection (peak identified as rhodamine 110 chloride).

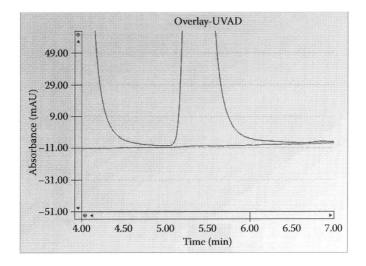

FIGURE 6.19 Magnified view of overlay of two chromatograms obtained during consecutive injections of a propyl paraben solution followed by injection of $CH_3CN:H_2O$ (50:50) solution.

below 2.5% were obtained in each case. As expected, the peak area reproducibility values improved when results from consecutive runs were evaluated on a per-column basis, yielding results consistently below 1.5%. This is very evident if it is perceived that results obtained for peak areas for consecutive runs on a column are compared for this evaluation (similar to the way evaluations are carried out with traditional HPLC). In the other evaluations, all 24 columns are considered (similar to comparing results obtained with 24 HPLC instruments). Of course, relative peak area values can also be evaluated (i.e., evaluating the area ratios of an analyte of interest and of a compound employed as an internal standard), and corresponding reproducibility parameters may be calculated. When such a calculation is executed for the peak area ratio between methyl paraben and propyl paraben, the peak area reproducibility values are consistently less significant, as expected (e.g., 0.4% for run-to-run within-column evaluations).

Also, evaluations of peak area reproducibility have been conducted by users of the instrumentation in the pharmaceutical industry for other analytes and methods employed in assays. Wielgos and Havel[4] reported the results of their evaluations for the determination of lactic acid (utilizing hydroxyisobutyric acid as an internal standard) employing the Nanostream μPLC system, with average peak area %RSD for two cartridges tested (for a total of $2 \times 120 = 240$ equivalent HPLC runs) of 1.55% (run-to-run within-column reproducibility). Similarly Liu et al.[9] evaluated peak area reproducibility for the evaluation of drug release profiles in OROS® tablets employing μPLC, reporting consistent RSD% values below 3%.

6.3.3 CARRYOVER

As with any analytical instrumentation that incorporates an autosampler, it is essential to evaluate the percent carryover obtained for a particular analyte under particular rinsing conditions. For these evaluations, a cartridge packed with a C18 stationary phase (80×0.5 mm/column) was employed. Gradient and detection conditions were the same as those described for the evaluation of retention time and peak area reproducibility (see Section 6.3.2).

The autosampler was set to rinse the needles with a mixture 50:50 $CH_3CN:H_2O$ for three consecutive times between injections, with an aspiration volume of washing solution of 5.0 μL and a

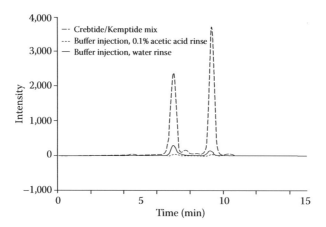

FIGURE 6.20 Chromatogram obtained during the injection of a CREBtide/Kemptide mixture (dashed line) compared to subsequent chromatograms obtained for injections of blank buffer. Solid line shows carryover resulting from rinsing needles with water between injections. Dotted line shows results when using 0.1% acetic acid to rinse needles between injections.

dispensing volume of 7.5 μL. After obtaining a stable baseline, three consecutive 0.5 μL injections (with a top-off-volume of 2.0 μL) of 2 mg/mL propyl paraben solution were run on a single column in three consecutive runs. After these three runs, a single blank injection of 0.5 μL of mobile phase was analyzed to determine percent carryover. Sample carryover, measured as the average over 24 columns, was consistently greater than 0.07%. Figure 6.19 is an example overlay of chromatograms obtained.

As expected, sample carryover can be significantly affected by the solvent used while rinsing the autosampler needles and the parameters employed for needle rinsing (i.e., number of iterations and aspirating and dispensing volumes). Figure 6.20 shows the results of a comparison study of carryover under different needle rinsing conditions. The dashed line demonstrates the initial injection of a CREBtide/Kemptide mixture. The needles were then rinsed before a blank buffer was injected in a subsequent set of injections (solid and dotted lines). The solid line exhibits the carryover generated when needles were rinsed with water only, while the dotted line shows the results of rinsing with 0.1% acetic acid + 0.01% Brij. As exemplified by the results, the carryover for CREBtide can be reduced from 15.4% for the water rinse to 2.2% for the acetic acid rinse, while the Kemptide carryover can be reduced from 4.4% to 0.9%. The residual acid remaining on the needles did not affect the assay. If acidity level is a potential problem, the needles can be programmed to rinse in pure water during a secondary rinse. Based on the results, proper choice of solvent use for washing autosampler needles will help minimize potential carryover. For example, with certain lipids, the recommendation is to employ methanol.

6.3.4 AUTOSAMPLER ACCURACY

To evaluate autosampler accuracy, an experiment involves injection of increasing volumes of a standard mixture containing an analyte into a cartridge. A solution containing methyl paraben (0.10 mg/mL) was employed and consecutive injections of increasing volume (0.5, 1.0, 1.5, 2.0, 3.0, 3.5, 4.0, 4.5, and 5.0 μL) of the test mixture containing the analytes described above were carried out employing a cartridge packed with a C18 stationary phase and per-column dimensions of 0.5 mm circular cross section and 80 mm length. The signal for methyl paraben was monitored via UV detection

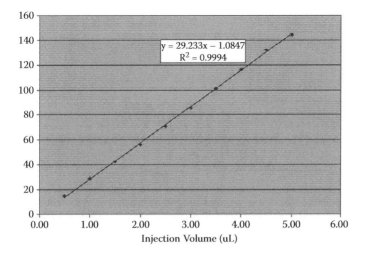

FIGURE 6.21 Results of evaluation of autosampler accuracy. Peak area values for methyl paraben are plotted against corresponding volume of injection. Values represent average peak areas obtained after triplicate injections. Error bars represent ± one standard deviation.

at 254 nm. The same gradient method as that depicted in Figure 6.16 was used. The compositions of mobile phases A and B were 5:95 $H_2O:CH_3CN$ with 0.1 HCOOH, and CH_3CN with 0.085% HCOOH, respectively, with a total flow rate of 300 $\mu L/min$ (corresponding to 12.5 $\mu L/min$ for each column). Triplicate injections of test mixture at each volume of injection were carried out and the results were evaluated as the linear regression obtained from the curve of peak area (for methyl paraben) versus the volume of injection.

As shown in Figure 6.21, excellent linearity was obtained, as represented by the high coefficient of correlation obtained for the least square linear regression. Similar results were obtained for the evaluation of autosampler accuracy when other analytes (propyl paraben and rhodamine 110 chloride) were employed in the determinations. Liu et al.[9] conducted similar evaluations for the samples employed in the evaluation of the drug release rate profile of OROS with similar results to those discussed above.

6.3.5 LINEARITY, LIMIT OF DETECTION, LIMIT OF QUANTITATION, AND SENSITIVITY

To evaluate linearity, limits of detection (LOD), limits of quantitation (LOQ), and sensitivity, an experiment assessed the responses for different concentrations of two analytes of interest. The analytes employed were methyl paraben and rhodamine 110 chloride. Consecutive 5.0 μL injections of a series of serial dilutions (four replicates) of this standard mixture containing the analytes described were carried out via a cartridge packed with C18 stationary phase and per-column dimensions of 0.5 mm circular cross section and 80 mm length.

Signals for methyl paraben were monitored with UV detection at 254 nm. The signal for rhodamine 110 chloride was monitored via fluorescence detection with an excitation filter of 482 nm (35 nm bandwidth) and emission filter of 535 nm (40 nm bandwidth). A gradient method (same as the one in Figure 6.16) was used. The compositions of mobile phases A and B were 5:95 H_2O: CH_3CN with 0.1 HCOOH and CH_3CN with 0.085% HCOOH, respectively, with a total flow rate of 300 $\mu L/$ min (corresponding to 12.5 $\mu L/min$ for each column).

TABLE 6.2
LOD and LOQ Values for Methyl Paraben and Rhodamine
110 Chloride

	LOD (S/N = 3)	LOQ (S/N = 10)
Methyl Paraben	0.2 μM	0.9 μM
Rhodamine 110 Chrloride	50 pM	300 pM

LOD is defined as the lowest concentration of an analyte that produces a signal above the background signal. LOQ is defined as the minimum amount of analyte that can be reported through quantitation. For these evaluations, a 3 × signal-to-noise ratio (S/N) value was employed for the LOD and a 10 × S/N was used to evaluate LOQ. The %RSD for the LOD had to be less than 20% and for LOQ had to be less than 10%. Table 6.2 lists the parameters for the LOD and LOQ for methyl paraben and rhodamine 110 chloride under the conditions employed. It is important to note that the LOD and LOQ values were dependent upon the physicochemical properties of the analytes (molar absorptivity, quantum yield, etc.), methods employed (wavelengths employed for detection, mobile phases, etc.), and instrumental parameters. For example, the molar absorptivity of methyl paraben at 254 nm was determined to be approximately 9000 mol/L/cm and a similar result could be expected for analytes with similar molar absorptivity values when the exact methods and instrumental parameters were used. In the case of fluorescence detection, for most applications in which the analytes of interest have been tagged with tetramethylrhodamine (TAMRA), the LOD is usually about 1 nM.

To evaluate linearity, calibration curves were generated for the values of peak areas obtained for analytes of interest against their concentrations in solutions prepared from serial dilutions of a standard mixture. In the case of methyl paraben, solutions with concentrations varying from 0.3 to

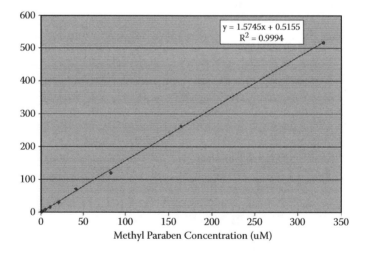

FIGURE 6.22 Standard calibration curve obtained for methyl paraben. Peak area values represent average value for four replicates. Error bars represent ± one standard deviation (%RSD is very small; error bars may not be visible at all concentration values).

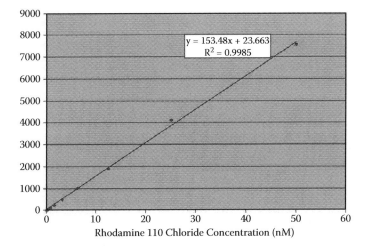

FIGURE 6.23 Standard calibration curve obtained for rhodamine 110 chloride. Peak area values represent average value for four replicates. Error bars represent ± one standard deviation (%RSD is very small; error bars may not be visible at all concentration values).

329 μM were employed. In the case of rhodamine 110 chloride, solutions with concentrations varying from 0.3 to 50 nM were employed. The results for methyl paraben are shown in Figure 6.22; those for rhodamine 110 chloride appear in Figure 6.23. Linearity was assessed by taking into consideration the value of the coefficient of regression encountered after least square linear regression analysis of the corresponding calibration curves. Values of R^2 greater than 0.99 were obtained in both cases. It is important to observe that these excellent values of R^2 were obtained even when different concentrations of analytes were injected in different columns, i.e., the evaluations were carried out across multiple columns (24 concentrations evaluated in a single run), again demonstrating the robustness of the analytical methodology.

The calibration sensitivity of the analytical method employed is simply determined as the slope of the calibration curve. For example, in the case of methyl paraben, the value of calibration sensitivity obtained was 1.6 mAU/min/μM (Figure 6.22). Analytical sensitivity is defined as the ratio between calibration sensitivity and the value of the standard deviation obtained at each concentration.[10] The value of the standard deviation encountered for a concentration of 0.6 μM was 0.1, resulting in an analytical sensitivity for methyl paraben at 0.6 μM of 16 mAU/min/μM. As indicated for LOD and LOQ, the values obtained for linearity and sensitivity depend on the analytes employed and the corresponding method and instrumental parameters. For example, Liu et al.[9] evaluated the LOD and LOQ for Drug A (released from OROS) for a particular analytical method employing μPLC to be 0.5 μg/mL and 2.0 μg/mL, respectively.

6.3.6 ANALYSIS TIME AND SOLVENT USAGE

The truly parallel approach employed in the design of the μPLC system described in this chapter produces a clear reduction in analysis time when compared with traditional HPLC techniques. For example, a separation method of 5 min duration in a μPLC system would allow simultaneous evaluation of 24 samples within that time—an average analysis time of 12.5 sec/sample. Similarly, the dimensions of the columns housed in the cartridge require smaller amounts amount of solvent (mobile phases) for the analysis.

Figure 6.24 compares the amounts of solvent required to analyze 24 compounds for solubility determination employing a four-point calibration and duplicate analysis when traditional HPLC and

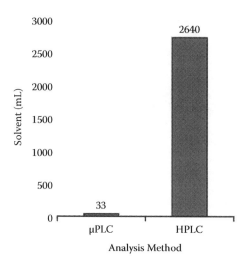

FIGURE 6.24 Amount of solvent needed for solubility determinations employing HPLC and μPLC (24 compounds, 4-point calibration curve, duplicate analysis).

μPLC approaches are employed. As illustrated, almost two orders of magnitude in solvent savings result with μPLC. These advantages—including reduced solvent consumption and reduced analysis time per sample—are clearly demonstrated for several applications.[1,4,9,11,12]

6.4 APPLICATIONS

This section describes examples of recent applications employing μPLC to evaluate relevant assays in the pharmaceutical industry. It covers three main types of applications: (1) those involving evaluations of ADMET and physicochemical properties; (2) those in which μPLC is used to evaluate separation-based enzymatic assays; and (3) those in which μPLC is coupled with time trigger fraction collection capabilities for the clean-up of complex samples prior to MS analysis. Depending on the specific application, instrumental capabilities, such as type of detection (UV, fluorescence, or both), plate exchanging ability, fraction collection capability, and other factors may be deemed necessary or advantageous.

6.4.1 High-Throughput Characterization of ADMET and Physicochemical Properties

The number of candidate drugs has increased drastically since the introduction of combinatorial chemistry.[13] One drawback is that the compounds generated often do not have favorable biopharmaceutical and pharmacokinetic properties.[13,14] For example, drug candidates may be poorly soluble in water, leading to low drug concentrations in GI fluids.[13,14] Forty percent of late stage failures are linked to poor pharmacokinetic properties.[15]

This is a significant failure rate, particularly if we consider that development, production, and marketing of a compound with low solubility are more expensive and time consuming than developing, producing, and selling one with more desirable properties.[16] It becomes obvious that physicochemical profiling at the early discovery stage is a very attractive proposition in the pharmaceutical industry.[17,18]

Solubility is one of the most important properties to be considered in selecting drugs to absorbed effectively after oral dosage.[16,19] Furthermore, these analytical determinations are often considered bottlenecks in the discovery process.[1,20,21] The following sections discuss μPLC evaluations of some of these properties..

6.4.1.1 Solubility and Purity

Compound solubility (at the usual pH range present in the GI tract) should be determined during drug lead optimization to aid in the selection of promising candidates prior to biological testing to ensure that screening results will be meaningful.[22,23] Aqueous solubility data can then be used to estimate dissolution, absorption, and bioavailability.[18]

The evaluations of compound solubility employing conventional approaches (such as the shake flask method) are often inadequate in modern high-throughput screening environments because they require large amounts of samples in solid form, time, and manpower in addition to samples at 10mM in DMSO. During early drug discovery, large numbers of compounds are generated in 1- to 5-mg quantities, making the testing of solubility challenging, especially when compound supplies is limited while many other properties such as potency, metabolism, toxicology, and permeability must also be evaluated. Furthermore, compound stability issues can often compromise the results of methods that require solutions to stand for several hours.

Because samples are routinely supplied at 10mM in DMSO solution for activity screening, the use of these same solutions for physicochemical property screening may save labor and time. Thus, it is highly desirable to establish methodologies that can adapt to these constraints and be integrated within the activities of a screening laboratory.[16,24] Some techniques used for solubility evaluations such as nephelometry and flow cytometry provide adequate throughput, but are not sensitive to compound purity and identity. Conventional HPLC and LC/MS methods overcome these drawbacks but suffer from their intrinsic low throughput when used in serial mode. High-throughput μPLC offers the advantages of a separation-based approach (such as HPLC and LC/MS) and allows 10 times more throughput than conventional HPLC while reducing sample consumption, solvent use, and waste generation.

Figure 6.25 is a diagram (similar to Figure 6.1) showing throughput and sensitivity for some of the common techniques used for solubility evaluations. As shown, μPLC systems can be employed to increase the throughput of thermodynamic and kinetic solubility evaluations without compromising sensitivity. Thermodynamic (equilibrium) solubility is usually determined by shaking the compound of interest with the buffer of choice for at least 24 hr or until the remaining solid does not dissolve. The solution is then filtered, and the concentration of the dissolved compound determined by a suitable analytical method. Of course, alternative methods can also determine thermodynamic

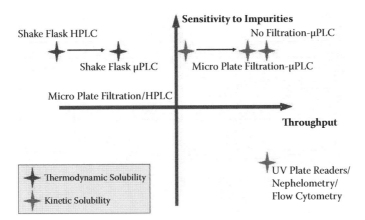

FIGURE 6.25 Comparison of techniques employed for solubility determination.

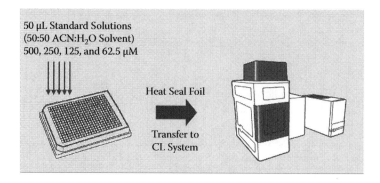

FIGURE 6.26 Overall view of processes involved in preparation and analysis of standard for solubility evaluations employing a μPLC system.

solubility, such as following the same approach but reducing the amount of sample needed by miniaturizing the device used.[23,24]

Other methods employ a microplate format followed by fast HPLC. Some researchers approach the determination from a different perspective. For example, an alternative method for ionizable substances is the pSol determination based on an acid–base titration.[25,26] Kinetic solubility determinations involve determining the concentration of the compound in the buffer of interest when an induced precipitate first appears.

To evaluate compound solubility, a μPLC system equipped with a cartridge containing 24 parallel columns (80 × 0.5 mm (inner diameter equivalent)) was employed. Sets of calibration standards were prepared for 24 compounds at different concentrations (in a 50:50 CH_3CN:H_2O solvent). A maximum standard concentration of 500 μM was selected to maintain the amount of DMSO co-solvent in all samples and standards below 5% v/v to minimize possible solubility enhancements due to the presence of DMSO when working with stock solutions provided at 10 mM in DMSO. Standards were added to the appropriate wells of a 384-well plate. The plate was covered with a heat seal foil and transferred to the μPLC system for analysis. Figure 6.26 depicts the process for preparation of standards; 95 μL of a buffer of desired pH were added to the appropriate wells. An additional 5 μL of each compound at a concentration of 10mM (in DMSO) was added to the corresponding wells. The plate was shaken for 90 min and centrifuged at 4000 rpm for 3 min.

Figure 6.27 presents the sample preparation process. Accurate and reproducible control of the depth of the autosampler needles permitted the sampling of the supernatant solution without perturbing the precipitate, thus avoiding the need for sample filtration, as shown in Figure 6.28. Of course, a step could be added to sample preparation to filter the solutions prior to analysis.

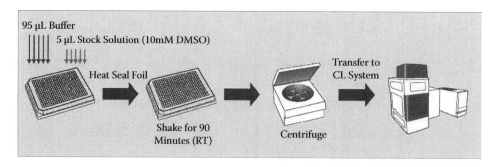

FIGURE 6.27 Overall view of processes involved in preparation and analysis of samples for solubility evaluations employing a μPLC system.

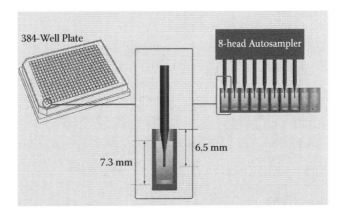

FIGURE 6.28 Precise control of depth of sampling can be achieved with an autosampler employed in a μPLC system. The dimensions indicated correspond to those used during the evaluations described.

Compound solubility values were evaluated from interpolation of the corresponding compound peak area obtained from a solution prepared with appropriate buffer within the corresponding external standard calibration curves of compound standards (peak area versus concentration). Sample purity was assessed by two methods: (1) evaluating area percentage at 254 nm, and (2) evaluating area percentage at 214 nm (without considering the DMSO peak). Figure 6.29 shows sample chromatograms obtained simultaneously for the 24 samples.

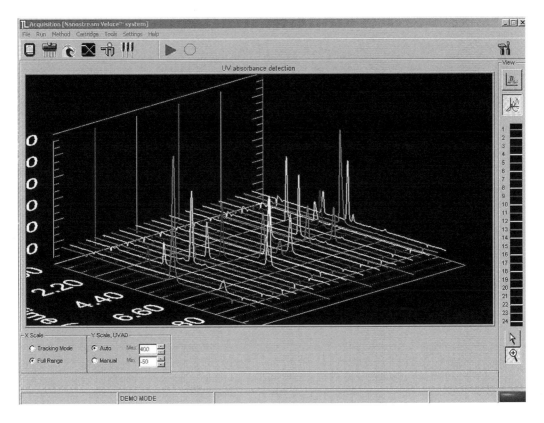

FIGURE 6.29 Example chromatograms (24) obtained simultaneously during evaluation of compound solubility employing a μPLC system.

TABLE 6.3
Coefficient of Correlation Values for Least
Squares Regression of Standard Curves

Standard Compound	Standard Curve (R^2)
Amiodarone HCl	1
Benzanthrone	0.997
Buspirone	0.998
Caffeine	0.999
Chlorpheniramine	0.999
Clozapine	0.998
Colchicine	0.999
4,5 Diphenylimidazole	0.999
Haloperidol	1
Hydrocortisone	0.997
Indomethacin	0.995
Ketoprofen	1
2-Naphthoic acid	0.993
Naproxen	0.999
Nifedipine	0.989
Nortriptyline HCl	0.998
Phenazopyridene HCl	0.999
Piroxicam	0.992
Prednisone	0.999
Probenecid	0.998
Quinine HCl	0.995
Reserpine	0.983
Thioridazine HCl	0.999
Warfarin	1

Data analysis was performed using Nanostream's advanced software to automate the analysis of samples and generation of calibration curves. Linear regressions were evaluated for all standard curves yielding R^2 values between 0.98 and 1.0, as shown in Table 6.3. Figure 6.30 shows an example of a calibration curve obtained for one analyte (clozapine). Linear regression analysis by the least squares method yielded a high value for the corresponding coefficient of correlation.

Aqueous solubility values for the samples analyzed compared favorably with results obtained by traditional methods. The solubility values for amiodarone HCl, reserpine, and benzanthrone were lower than the LOQ of the μPLC system used for the evaluation. Results of the evaluation of compound solubility employing no-filtration μPLC were compared with those obtained by two traditional methods: (1) multiscreen filtration followed by a UV plate reader, and (2) the shake flask method followed by a UV plate reader. As shown in Figure 6.31, the solubility values determined by the different methods are comparable for most compounds examined. Figure 6.32 shows the results of evaluations of aqueous solubility at four different pH levels for phenazopyridine and piroxicam samples.

Compound impurities can lead to biased results when plate readers are used for detection since the total absorbance at a particular wavelength is employed for the determinations. The μPLC system employed in these determinations can easily compensate for this problem since only the value for the peak area due to the compound (at the corresponding retention time) is considered for the calculations. For example, in the case of a nifedipine sample, purity was determined to be

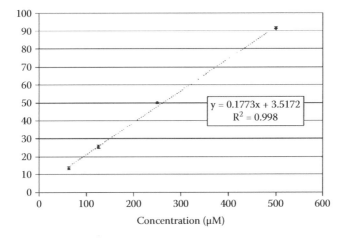

FIGURE 6.30 External standard calibration curve for clozapine obtained with a μPLC system. Error bars represent $\pm$ one standard deviation.

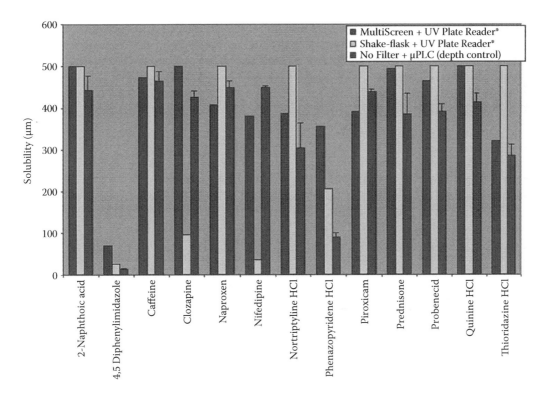

FIGURE 6.31 Comparison of solubility results obtained by no-filtration method followed by PLC detection and those obtained by two traditional methods: multiscreen filtration followed by UV plate reader, and shake flask method followed by UV plate reader. (Data provided by Steven Hobbs, Courtney Coyne, and Gregory Kazan.)

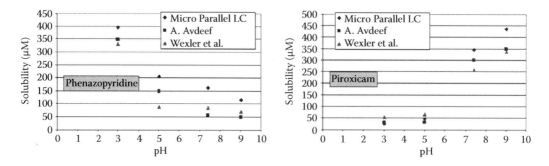

FIGURE 6.32 Solubility values at different pH values for phenazopyridine and piroxicam samples obtained with a Nanostream PLC system (error bars represent ± one standard deviation). Values are compared with reported values. Literature values adapted from cited manuscripts.

significantly lower. As presented in Figure 6.33, the chromatogram obtained for the nifedipine sample demonstrated three small peaks due most likely to impurities in the sample. Purity evaluations carried out by μPLC were compared with those obtained by traditional HPLC; results of the comparison studies are presented in Figure 6.34. As shown in the figure, results obtained by both techniques compared favorably.

The results clearly demonstrate the utility of μPLC for high-throughput determinations of compound solubility. μPLC required significantly lower volumes of mobile phase than traditional HPLC methods. For example, to evaluate 24 compounds using a four-point external standard calibration curve and duplicate analysis, a total of 33 mL of mobile phase was consumed by μPLC. A traditional HPLC system required 2.6 L for the same determinations. The high-throughput capabilities of μPLC allowed the generation of calibration curves for 24 compounds (four-point calibration curve, duplicate analysis) and duplicate solubility measurements for each sample (240 separations) in approximately 2 hr (not including incubation time). Employing similar conditions, approximately 1900 samples could be evaluated weekly (four-point calibration curve, singlet, 1152 separations

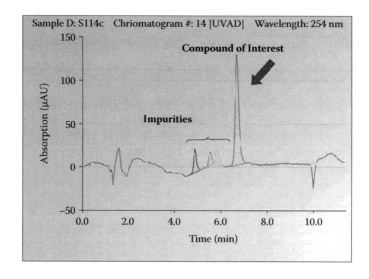

FIGURE 6.33 Impurities may be taken into consideration when evaluating solubility with a μPLC system. For solubility calculations, only the peak area of the analyte of interest is used for interpolation within the corresponding external standard calibration curve.

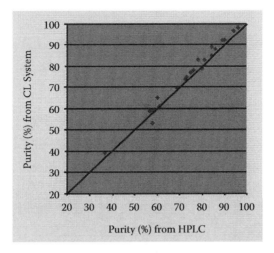

FIGURE 6.34 Comparison of primary peak area results for 24 samples analyzed using PLC and HPLC. The primary peak purity (%) obtained using conventional HPLC and PLC differed by an average of 2%.

per day). Minimal sample quantities (5 μl of 10mM solutions in DMSO) were required for all these determinations.

Another approach can be used for purity and solubility determinations when compound confirmation is sought; it involves connecting the system to an existing MS system. For example, a μPLC system's UV detectors were employed for all columns and the 24th column was also connected to a mass spectrometer (Agilent 1100 API-ES) for mass confirmation (employing 90 cm of a 100 μm inner diameter fused silica capillary tube to minimize extra-column broadening). Figure 6.34 depicts the general instrumental set-up. In all experiments, the last column of the 384-well plate was dedicated to sample identity, mass confirmation, and purity determination by MS (i.e., the last column in the plate contained individual samples for mass confirmation). The configurations employed allowed for solubility determination at 254 nm and purity determinations at 214 and 254 nm. In addition, compound purity and identity were also assessed by MS.

With the μPLC system connected to a MS, sample identity could easily be confirmed. If the sample chromatograms shown in Figure 6.29 were to be considered, MS spectra for all compounds could be obtained after all runs were completed, as shown in Figure 6.36. The MS capability

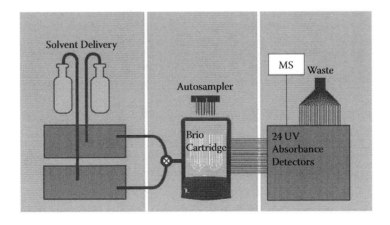

FIGURE 6.35 Configuration of a microparallel liquid chromatography (PLC) system with MS capability.

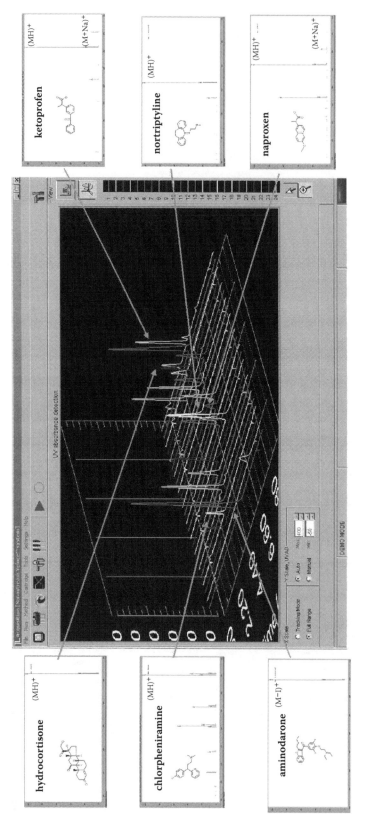

FIGURE 6.36 Example chromatograms and mass spectra obtained from a μPLC system with MS capabilities.

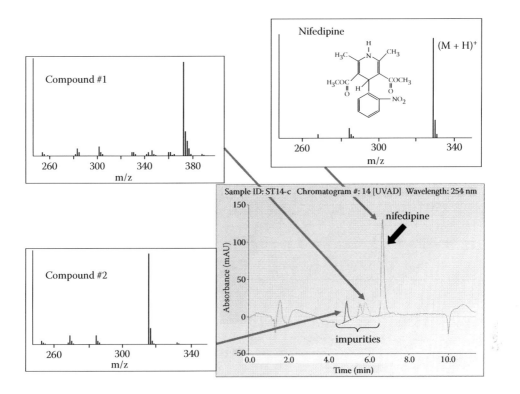

FIGURE 6.37 Chromatogram and mass spectra acquired from a μPLC/MS system for a sample containing nifedipine. Impurities present in the sample displayed differences in their corresponding mass spectra, clearly indicating different chemical compositions for each compound present in the sample.

of the system allowed the mass spectra of all compounds suspected of being impurities to be obtained. As shown in Figure 6.37, in the case of nifedipine, the mass spectra of the impurities present in the sample could be easily determined. Comparison with an appropriate MS library or detailed study of the mass spectra acquired could lead to the proper identification of these impurities, if desired. This example demonstrates the utility of μPLC MS for purity evaluation and identity confirmation.

6.4.1.2   Dissolution and Drug Release Rate Profile

Dissolution indicates the rate-limiting step for compound absorption when drugs are administered orally. The solubility of a pharmaceutical compound represents its maximum concentration in an aqueous buffer. Additional compound will not dissolve above this concentration. The solubility value is often heavily dependent upon pH and temperature and is typically measured at physiologically important pH levels and body temperature. The standards for dissolution testing are determined by the *United States Pharmacopoeia* (USP). Testing typically requires sampling of a solution at 15, 30, 45, and 60 min for immediate-release products. μPLC is ideally suited for use in conjunction with USP apparatus types I or II and can rapidly analyze multiple time points or replicate samples.

The approach used for the evaluation of dissolution via μPLC is very similar to the one described for evaluating solubility (see Section 6.4.1.1). External standard calibration curves are generated for each compound to be analyzed. Figure 6.38 shows an example of a calibration curve obtained for acetaminophen. As was the case during the evaluation of compound solubility, linear regressions were evaluated for all standard curves generated for the compounds to be characterized, yielding R^2 values higher than 0.9995 in all cases studied. Peak areas attained from the sample runs at regular

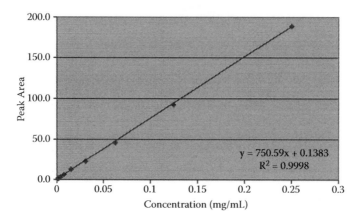

FIGURE 6.38 External standard calibration curve for acetaminophen (error bars represent ± one standard deviation).

time intervals were used to evaluate compound concentrations by interpolation within the corresponding external standard calibration curve. Results are commonly reported as percentages of drug released at each time period used for sampling. Figure 6.39 presents results acquired for dissolution for acetaminophen. Figure 6.40 compares results obtained for six different vessels from one USP apparatus type II system employed for the determination of dissolution. It shows excellent agreement among the replicates used.

A similar approach to the one described here for the evaluation of dissolution can be applied to the characterization of drug release rate profile. For example, Liu et al.[9] demonstrated the application of μPLC to characterize an OROS drug release rate profile. They used a USP type VII apparatus to monitor drug released at specific time intervals (2-hr intervals for 24 hr) in modified artificial gastric fluid at pH 1.0 and 37°C, with analysis performed both by traditional HPLC and μPLC. Figure 6.41 illustrates the results of a comparison of HPLC and by μPLC. Excellent agreement of results obtained by HPLC and μPLC was found, with the added advantages for the μPLC system of analysis time reduction and solvent savings.

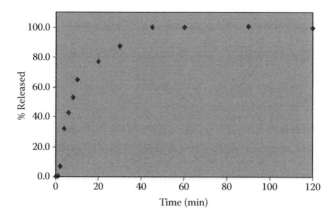

FIGURE 6.39 Dissolution profile of acetaminophen obtained from the Nanostream CL System.

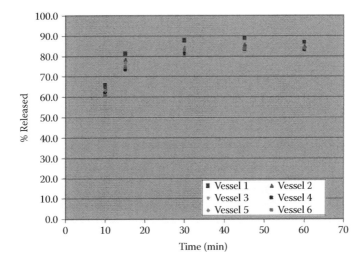

FIGURE 6.40 Dissolution profile of acetaminophen obtained from the Nanostream CL System.

6.4.1.3 Log P and Log D Determinations

Lipophilicity is an important property of molecules in relation to their biological activities. It is one of the key physiochemical parameters that determine the distribution and transport of drugs into the body and target organs. Measurements of lipophilicity, expressed as the logarithm of the

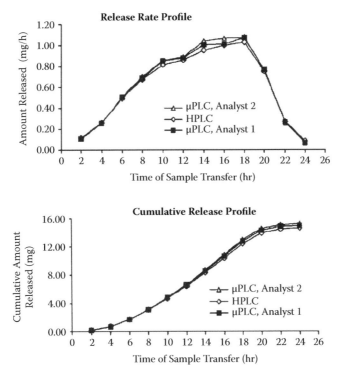

FIGURE 6.41 Comparison of drug release profiles obtained by HPLC and by μPLC, as reported by Liu et al., 2007.[9]

octanol–water partition ratio (log P) of a neutral molecule, are indicative of the tendency of a compound to associate with a lipid-like environment. This value correlates with biological transport processes.[27] The traditional shake flask method used to obtain partition data is not suitable for modern drug discovery environments; it is time-consuming, labor-intensive, and suffers from inaccuracies from several sources. Among recent advances in column technology, reverse-phase HPLC has become one of the most popular techniques for indirect estimations of Log P.[28–30] Most HPLC methods utilize extrapolation of retention indices such as k′ (retention factor) to 100% water (0% organic) conditions.

The logarithm for the capacity factor correlates well with known log P values obtained by the shake flask method. In practice, the k′ values are determined isocratically from 70 to 30% organic mobile phase and then extrapolated to 0%. Prior to determining the log P for an unknown compound, a set of structurally related molecules (standards) are analyzed to construct a correlation model between the logarithm of the retention factor and known log P values. The process is then repeated for the test compounds and their log P values determined from the mathematical relationship established for the standard compounds.

One drawback of this approach is the relatively low sample throughput of traditional HPLC systems. The primary reason for this low throughput is that each standard and sample must be assayed under a minimum of three different isocratic conditions. Assuming a run time (injection to injection) of 10 min, it would take 50 hr to analyze 4 standards and 96 unknowns. μPLC is ideally suited for determination of log P since it facilitates parallel analysis of a large number of compounds under identical chromatographic conditions (Table 6.4).

To exemplify the determination of log P employing μPLC, a single cartridge was used to generate a standard curve for four known compounds and analyze a fifth (unknown) compound.[31] Retention times were determined for the four known compounds (acetanilide, benzophenone, naphthalene, and diphenylamine) and standard (uracil) at five different isocratic mobile phase compositions (Table 6.5).

Ten columns of the 24 available in a cartridge were employed to analyze all compounds in duplicate. Uracil, was employed as a dead volume marker (t0) needed for the evaluation of retention factor [k′ = (tr – t0)/t0]. Two additional columns were used for simultaneous analysis of the unknown. Values for the log of the capacity factor k′ were calculated for every compound at each percent organic content of the mobile phase: log k′ = log [(tr – t0)/t0. For each compound, a plot of log k′ versus percent acetonitrile was used to calculate log k′w (log k′ at 0% acetonitrile).

A standard curve was then generated by plotting the log k′w data against log P values obtained for all compounds cited in the literature. The standard curve was then used to determine log P for the unknown compound. For comparison, the procedure was repeated using HPLC instrumentation with

TABLE 6.4
Mobile Phase Compositions Employed
for μPLC Determination of Log P

Acetonitrile (%)	Run Time (min)
40	10
50	6
60	3
70	3
80	3

TABLE 6.5

Average Retention Times from Isocratic Separations at Five Mobile Phase Concentrations

Compound	Retention Time, t_R (min)				
	80% ACN	70% ACN	60% ACN	50% ACN	40% ACN
Uracil	0.509	0.493	0.532	0.525	0.525
Acetanilide	0.627	0.660	0.742	0.824	0.988
Benzophenone	0.958	1.239	1.768	2.969	6.268
Naphthalene	1.128	1.525	2.253	3.978	8.787
Diphenylamine	0.989	1.340	2.069	3.877	9.408
Unknown	0.671	0.756	0.883	1.154	1.754

a 50×4.6 mm (inner diameter) column packed with 5 μm C18 stationary phase. Table 6.5 shows the average retention times (tR) for each compound at various mobile phase compositions. Figure 6.42 shows an example of the determination of log k'w for benzophenone. Data were extrapolated to determine the y intercept corresponding to a value of log k'_w = 2.05 for this compound. Log k'_w values for the four compounds based on data obtained from the μPLC system were comparable to results from the same study performed with HPLC, as shown in Table 6.6.

These values were plotted against known log P values to generate standard curves such as those shown in Figure 6.43. The standard curves were then used to predict log P of the unknown compound. Results appear in Table 6.7. The slight variation (2%) between the values predicted by the two methods is reasonable within experimental error limits.

Using a μPLC system, log P for one unknown compound was determined in less than 1 hr. It is important to note that the excess capacity provided by the system (24 columns are available for simultaneous analysis) allows simultaneous determination of log P for six additional compounds. The same study required 5 hr using conventional HPLC, and consumed 300 mL of solvent, equivalent to 15 times the volume of solvent used for the evaluations via μPLC. A similar approach can be used to evaluate log D, the octanol–water distribution coefficient—a measure of the distribution ratios of all combinations (ionized and unionized) of octanol and pH-buffered water.

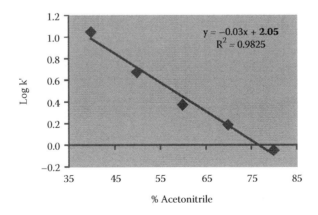

FIGURE 6.42 Log k' versus percent acetonitrile for benzophenone. Linear regression was used to determine the y-intercept corresponding to log k'_w, i.e., log k' at 0% acetonitrile.

TABLE 6.6
Log k′$_w$ Values for Compound Analyzed via μPLC and HPLC

Compound	Log k′$_w$ CL System	Log k′$_w$ HPLC
Acetanilide	0.48	0.47
Benzophenone	2.05	2.11
Naphthalene	2.22	2.30
Diphenylamine	2.30	2.48

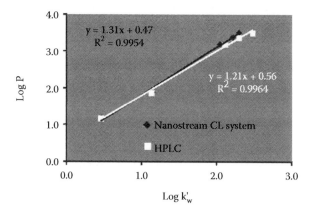

FIGURE 6.43 Standard curves obtained for log P versus log k′$_w$ employing μPLC and HPLC systems.

TABLE 6.7
Comparison of Log P Values of Unknown Compound Determined by μPLC and Traditional HPLC

	Log k′$_w$	Log P, Predicted
CL System	1.15	1.97
HPLC	1.14	1.94

6.4.2 SEPARATION-BASED ENZYMATIC ASSAYS

Target characterization and assay development are commonly considered bottlenecks during drug discovery. These areas often involve many "pain points" such as working with challenging enzymes, dealing with assay feasibility, developing multiple assays in a short time, and facing concerns about the robustness of the screening assays to be implemented. Work in screening areas often involves issues related to hit identification and lead optimization, which become challenging in the presence of assay artifacts, false positives, false negatives, and concerns about true activators.

Separation-based assays are preferred in many applications because they allow discrimination of signals due to substrate, product, and interference. When assays that involve fluorescence detection are developed, they are typically carried out by employing plate readers. When separation-based methods are employed for these applications, the influences of interferences (quenchers and other fluorescent compounds) on the final results are minimized because both substrate and product are quantified. With a separation-based approach, the label employed does not need to be placed in close proximity to the site of action of the enzyme, therefore minimizing the effect of the label on the mode of action of the enzyme. Of course, it is often desirable to develop assays that employ substrates free of labels.

Figure 6.44 presents an example highlighting the separation-based assay approach—results of a kinase assay. A substrate is incubated with kinase and ATP for a specific time. The formation of product and the consumption of substrate are followed by their corresponding fluorescence signals obtained after separation by liquid chromatography in which the non-phosphorylated substrate peptide is separated from the phosphorylated product. Enzyme activity is determined by measuring product formation. Since the separation via liquid chromatography allows the isolation of interferences, the numbers of artifacts and false positives decrease significantly.

As was the case for the applications described for the characterization of ADMET and physicochemical properties (see Section 6.4.1), the throughput limitations imposed by traditional HPLC often result in the adoption of a different technology for certain biochemical assays although the advantages of a separation-based technique are well understood. The ability of μPLC to allow parallel simultaneous analyses of samples also allows a non-compromise approach to assays because it provides all the advantages of separation-based techniques along with the very desirable advantage of more than adequate throughputs for many applications. Other advantages resulting from use of μPLC for assay design include carrying out assays at low enzyme conversion, separation and detection of multiple phosphorylations, and the use of label-free substrates.

This section describes recent applications of μPLC methodologies for separation-based enzymatic assays. It covers the most common applications: (1) those involving the development and optimization of assays; (2) those in which μPLC is use to evaluate real-time enzyme kinetics; and (3) those in which μPLC is used to determine substrate specificity.

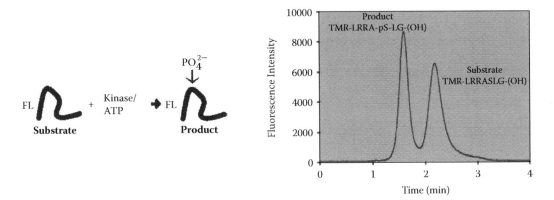

FIGURE 6.44 Separation of substrate and product employing a μPLC system.

6.4.2.1 Streamlined Assay Development and Optimization

When assays are first developed and when optimization is sought, four key parameters must usually be considered: enzyme concentration, substrate concentration, ATP concentration, and time needed for running the assay. The ability of μPLC to analyze 24 samples simultaneously easily translates into the possibility of evaluating 24 assay conditions in a single matrix experiment, making this approach very attractive because of the significant time saving benefits.

To demonstrate the ability of the system to perform a matrix experiment as described above, concentrations of enzyme, substrate, and ATP were varied across the 24 wells in a row of an SBS 384-well microtiter plate. Results of these types of evaluations for the optimization of an assay for a protein kinase A and Kemptide system were presented by Wu et al.[12] All the reactions were carried out in 100mM HEPES, pH 7.4, 10mM $MgCl_2$, 10mM DTT, and 0.015% Brij-35. No quenching agent was used. A sample from each of the 24 wells was analyzed in parallel every 6.5 min as the 24 enzymatic reactions progressed.

The first data point was taken 3 min after the enzyme was added. For each analysis, a 1-μL sample of each reaction was injected via an integrated 8-channel autosampler onto a μPLC cartridge loaded with 24 parallel columns (30 mm × 0.5 mm) containing C18 stationary phase. Product and substrate peaks were monitored simultaneously by fluorescence detection (λex = 525 nm and λem = 585 nm). The mobile phase flow rate was 600 μL/min, equivalent to 25 μL/min for each column. Mobile phase A was 50mM NH_4OH:HCl, pH 9.0; mobile phase B was acetonitrile. The substrate and product peaks were separated using a 4-min gradient method starting with 26% B and held steady for 2.5 min, ramped from 26 to 28% B from 2.5 to 3.0 min, and re-equilibrated at 26% B from 3.0 to 4.0 min.

Figure 6.45 presents results. The percentage conversion [product/(substrate + product) × 100] obtained under different assay conditions is plotted against the corresponding values for the parameters varied during the evaluations. Other types of matrix experiments can also be designed based on the initial results from a first set of experiments such as those described above to optimize for a desired level of substrate conversion in a desired time. Figure 6.46 presents an example of the results acquired from such an experiment. Overlays of multiple chromatograms obtained during these evaluations are presented in Figure 6.47. The concentration of the substrate (Kemptide) was kept constant (1.32 μM). As shown in the figure, some assay conditions resulted in low product formation, suggesting insufficient enzyme or non-optimal conditions for the enzyme.

Other overlays demonstrate a very fast rate of conversion indicated by the onset of a predominant product peak in a very short time. Some assay conditions were deemed optimal for a particular application because they provided the desired interchange of decreasing substrate concentration and increasing product concentration in a desired time. Once conditions are optimized, an assay can be easily adapted for screening via a μPLC system.

Jezequel-Sur et al.[6] described successful evaluations using μPLC as a platform for small molecule secondary screening on kinase drug targets. Their results with μPLC permitted the identification of false positive hits previously obtained with FP and TR-FRET. As described by Wu et al.,[12] by employing a μPLC system to screen compound inhibitors against the target of a protein A kinase, a total of 4400 compounds could be easily screened in only 20 hr.

6.4.2.2 Real-Time Enzyme Kinetics

The evaluation of results of assay optimization experiments such as those described above (see Section 6.4.2.1) also provides valuable information about enzymatic kinetic behavior. For example, the results shown in Figures 6.45 and 6.46 already provide information on enzymatic activity at each time point. In general, when evaluating enzyme kinetics, assays are designed to yield a measured conversion close to initial velocity.[32]

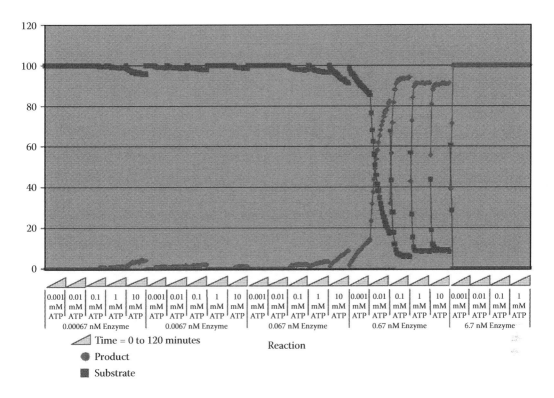

FIGURE 6.45 Profiles of percentage conversion obtained from matrix experiments described in text.

The problem is that substrate conversions are frequently lower than 10% and thus difficult to detect in most traditional formats. As previously discussed in this chapter, μPLC allows optimal operation even when working with low substrate conversion (e.g., as low as 1% as described by Wu et al.[12]). Data obtained allow the calculation of the Michaelis-Menten constant (K_m) for ATP. An example of such an evaluation is presented in Figure 6.48 in which the obtained reaction velocities

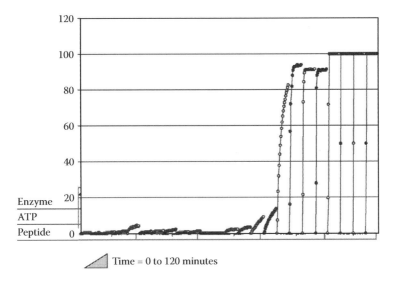

FIGURE 6.46 Profiles of percentage conversion for follow-up matrix experiments described in text.

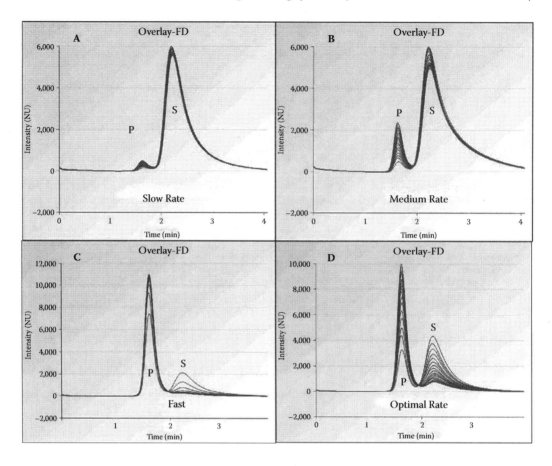

FIGURE 6.47 Chromatogram overlays obtained for 19 time points during evaluations described in text. S denotes substrate. P denotes product. Observed levels of enzymatic conversion are indicated as slow, medium fast, and optimal for a particular application.

are plotted against the ATP concentration. The value of K_m can be calculated by carrying out a non-linear regression to the equation proposed by Michelis-Menten and by a least squares linear regression to a double reciprocal plot such as the one proposed by Lineweaver-Burk. For the example in Figure 6.48, the calculated K_m (for the protein kinase A and Kemptide system) was 28.6 μM. Results of time course studies can also be obtained by evaluating changes in peak area values for a peak of interest (e.g., a product peak) with respect to time. Figure 6.49 presents an example of this evaluation for the formation of product during a study performed for the protein kinase A (PKA) and Kemptide system (ATP= 0.01 nM, PKA = 0.67 nM, Kemptide = 1.32 μM).

Another important kinetic parameter usually determined is the value of IC_{50}, defined as the half maximal inhibitory concentration that provides an indication of the potency of an inhibitor under certain the specific conditions of an evaluation. This parameter is typically determined from dose–response plots in which the percentage conversation or inhibition is plotted against the logarithmic concentration of inhibitor employed in each evaluation. The parallel nature of the μPLC system described in this chapter is ideal for the evaluation of this parameter.

Figure 6.50 shows the results obtained for a dose–response curve for H-89 inhibitor at a substrate conversion of 1%. The IC_{50} value for this inhibitor under the conditions employed was determined to be 19 nM. Jezequel-Sur et al.[6] also showed the application of μPLC to generate dose–response curves for a reference compound.

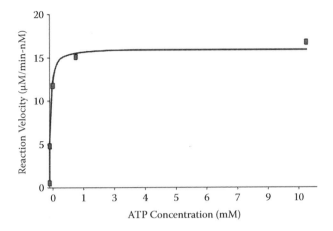

FIGURE 6.48 Plots of reaction velocity versus ATP concentration. (Adapted from Wu, J. et al., *Assay Drug Dev. Technol.*, 4, 653, 2006.)

In many cases, it is desirable, after an enzymatic inhibitor is identified, to understand the mechanism by which the inhibitor interacts with the enzyme of interest. For example, the inhibitor may interact with the primary substrate binding pocket, which is usually preferred, or it may interact with a secondary site on the enzyme such as the binding pocket for a secondary substrate (e.g., ATP for kinases). For example, the mode of enzyme inhibition of H-89 (described for the evaluation of IC_{50} above) may be investigated easily with μPLC. Matrix experiments similar to those described in Section 6.4.2.1 are performed with varying concentrations of ATP and H-89 inhibitor. V_{max} and K_m values obtained under different conditions help explain the mode of inhibition. For example, when competitive inhibition is present, the addition of inhibitor will increase the apparent value of K_m, leaving the value of V_{max} unchanged as shown in Figure 6.51 for H-89 and ATP. Figure 6.52

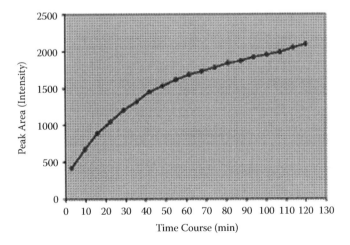

FIGURE 6.49 Time course obtained for evaluation of product formation in a protein A kinase and Kemptide system. Values of peak areas obtained for product are plotted against corresponding time points used during evaluation.

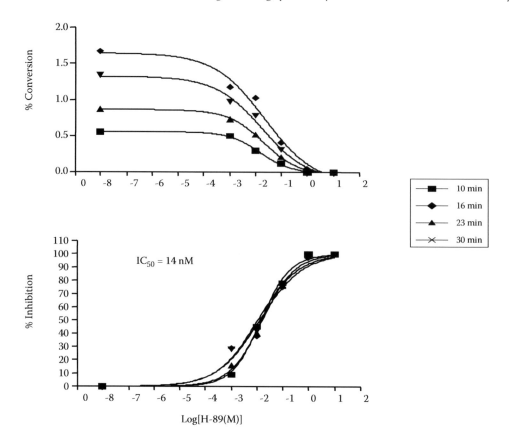

FIGURE 6.50 Dose–response plots obtained for H-89 inhibitor. (Adapted from Wu, J. et al., *Assay Drug Dev. Technol.*, 4, 653, 2006.)

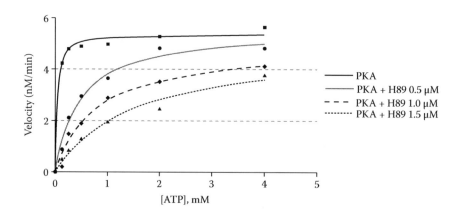

FIGURE 6.51 Determination of V_{max} and K_m for ATP. A Michaelis-Menten plot of PKA phosphorylation of labeled peptide substrate was used to determine V_{max} and K_m in the presence and absence of three concentrations of inhibitor H-89.

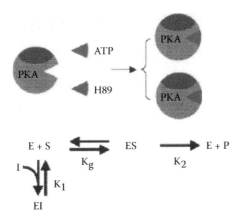

FIGURE 6.52 H-89 mode of enzyme inhibition.

represents the mode of enzyme inhibition expected for H-89 based on the results obtained by μPLC.

Another advantage of employing a separation-based approach for these studies is the possibility of simultaneously monitoring several related substrate–enzyme systems—multiplex assays. Figure 6.53 shows the results obtained from a multiplex P450 substrate assay. Since P450 enzymes contribute to the metabolism of more than 90% of marketed drugs,[33] this type of study is very relevant. Four different substrates for each CYP450 were incubated in the absence or presence of human liver microsome at room temperature for 20 min. The reactions were quenched and the samples analyzed using a μPLC system with UV detection, thus eliminating the need for fluorescence tags. Label-free substrates were used. The results presented in Figure 6.53 clearly demonstrate the ability of μPLC to separate all the substrates and product of interest monitored. This approach also demonstrates the feasibility of using μPLC to study potential drug–drug interactions.

Principle

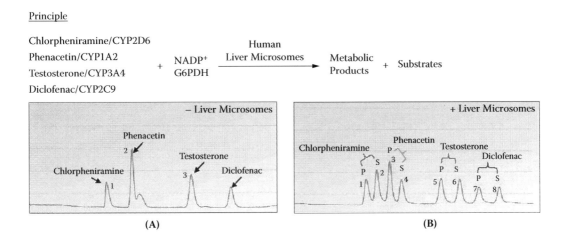

FIGURE 6.53 Multiplexing P450 substrate analysis. Four substrates for each CYP450 were incubated in the absence (A) or presence (B) of human liver microsomes at room temperature for 20 min. The reactions were then quenched and samples were analyzed using μPLC. The principle of this experiment is shown in the top panel.

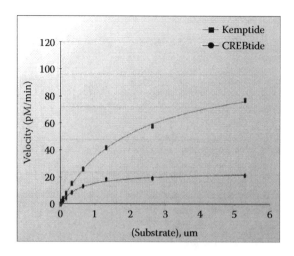

FIGURE 6.54 Velocity versus substrate concentration for Kemptide and CREBtide (substrates co-located in same assay buffer).

6.4.2.3   Evaluation of Substrate Specificity

Understanding the complex interconnectivity between the interactions of enzymes with their primary and secondary substrates is critical. Additionally, understanding enzyme selectivity is essential for novel assay development in drug discovery, especially if targeted inhibitors are desired. For example, the specificity of PKA against two different substrate targets (Kemptide and CREBtide) was evaluated with a μPLC system.[34] Relevant kinetic parameters for both the substrate binding and ATP binding were simultaneously determined. μPLC allowed simultaneous monitoring of multiple reaction conditions to ensure that differences in reaction rates truly arose from the natures of the substrates and not from minor changes in reaction conditions.[34]

The phosphorylation of each substrate was monitored via a one- or two-substrate reaction in real time and the kinetic parameters (V_{max}, K_m, k_{cat}, and k_{cat}/K_m) were determined. Figure 6.54 shows the results of the evaluations of velocity with respect to substrate (Kemptide and CREBtide) concentrations. The data were fitted using the Michaelis-Menten equation to determine the kinetic parameters shown in Table 6.8. The V_{max} of phosphor-Kemptide (105.57 pM/min) was approximately 4.3-fold larger than the V_{max} for CREBtide (24.33 pM/min) although both peptide substrates had similar

TABLE 6.8
Kinetic Constants of Kemptide and CREBtide

Parameter	Kemptide	CREBtide
V_{max} (pM·min^{-1})	105.57	24.33
K_m (μM)	2.05	0.59
k_{cat} (S^{-1})	0.0065	0.0015
k_{cat}/K_m (nM^{-1}·S^{-1})	3.17	2.54

Substrates monitored in same assay buffer; ATP concentration held constant.

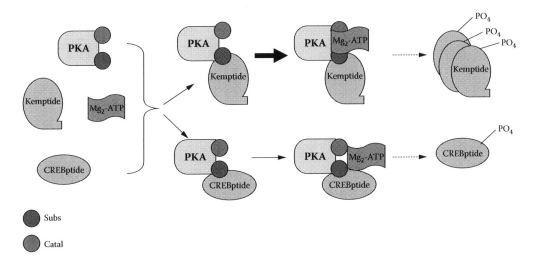

FIGURE 6.55 Preferential phosphorylation of Kemptide by PKA. The binding efficiency of Kemptide and CREBtide substrates to PKA is relatively equal while the binding of the ATP to subsequent PKA–substrate complexes is substantially greater for Kemptide, resulting in higher overall phosphorylation.

substrate specificity values (k_{cat}/K_m). These results suggest that both substrates had relatively similar affinities and substrate specificities for PKA. The obtained K_m value for CREBtide (0.59 μM) was much lower than the K_m value of Kemptide (2.05 μM)—a 3.5-fold difference—suggesting that the preferential phosphorylation of Kemptide in a competitive situation with CREBtide was not the result of preferential affinity. Figure 6.55 represents the mechanism of preferential phosphorylation of Kemptide by PKA.

Employing this approach, μPLC allows the evaluation of multiple peptide substrates of similar length in one reaction in real time with significant analysis time optimization. Several kinetic parameters including V_{max}, k_{cat}, K_m, and k_{cat}/K_m can be calculated from a single experiment in which all the products can be simultaneously analyzed and quantified. This approach has broad relevance for enzyme substrate specificity studies, target characterization, and enzyme kinetic studies for novel therapeutics. Table 6.9 highlights the main advantages of μPLC when compared with other common techniques such as radioassays and HPLC.

TABLE 6.9
Comparison of Common Evaluation Techniques

	Radioassay	HPLC	MS or ESI-MS	μPLC System
Workflow	Multistep	Step by step	Step by step	Real-time monitoring
Detection	Radioisotope	UV/fluorescent	MS	UV/fluorescent
Throughput	Low	Low	Low	High
Monitor matrix experiment in real time	Hard	Not practical	Not practical	Easy
Real-time kinetic analysis	Hard	Hard	Hard	Easy
Data analysis for multiple samples	Slow	Slow	Slow	Fast

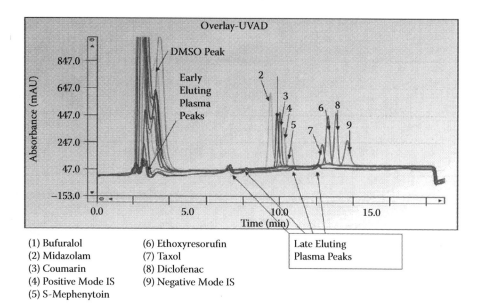

(1) Bufuralol   (6) Ethoxyresorufin
(2) Midazolam   (7) Taxol
(3) Coumarin   (8) Diclofenac
(4) Positive Mode IS   (9) Negative Mode IS
(5) S-Mephenytoin

FIGURE 6.56 Example chromatograms obtained with long gradient method employed for the isolation of analytes of interest from indigenous plasma components.

6.4.3 SAMPLE PREPARATION

6.4.3.1 High-Throughput Clean-Up of Complex Samples

Analysis of complex mixtures is often carried out via traditional LC/MS. Although this approach is widely accepted and effective, rapid gradients commonly employed to improve throughput often result in suboptimal separations of analytes of interest from compounds that can influence ionization, therefore affecting sensitivity. The use of μPLC with automatic time triggered fraction collection (as described in Section 6.2.6) for the clean-up of complex samples prior to MS analysis provides a significant increase in overall throughput. This increase can be achieved without sacrificing separations since longer gradient methods can be employed, as shown by Lloyd et al.[8]

Figure 6.56 depicts an example of μPLC separation obtained with a generic gradient method utilized in the experiments described by Lloyd et al. The endogenous plasma peaks were separated from the active compounds present in the samples. Stock solutions and blank plasma solutions were used to define the windows for fraction collection for each sample analyzed (individual drugs and seven compound cocktail samples). Figure 6.57 shows the overlaid chromatograms (24) obtained for a sample containing midazolam, a positive mode internal standard, and plasma, with the selected time window employed for fraction collection highlighted. Collected fractions were then analyzed by MS, thus carrying out flow injection analysis (FIA). In all cases evaluated, excellent reproducibility and linearity values were obtained after FIA of collected fractions.[8]

Figure 6.58 presents results obtained from these evaluations. Average peak areas obtained after FIA analysis employing an Agilent 1100 (API-ES) MS are shown. When relative peak area values (peak area due to analyte and internal standard) were used for reproducibility evaluation, percentage CV values ranged from 2.7 to 4.0% for the two lowest concentrations analyzed.

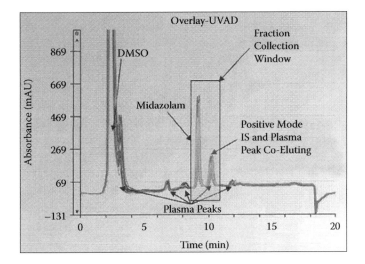

FIGURE 6.57 Overlaid chromatograms (24) obtained for blank plasma spiked with midazolam and positive mode internal standard.

Peak area data were employed to generate the external standard calibration curve shown in the figure. A least squares method was used for linear regression, yielding an r^2 value of 0.995, demonstrating that the dilution series remained linear after purification on the μPLC system. Overall, the utilization of μPLC to isolate active compounds from ion-suppressing interferences resulted in a significant increase in signal compared with no-separation, fast gradient methods that do not allow complete separation of ion suppression compounds, and solid phase extraction methods. These results demonstrate the utility of a fully automated platform with fraction collection to maximize the efficiency and throughput of complex biological sample clean-up that allowed the MS systems to operate on MS time scales (e.g., 30-sec flow injection analysis for this study) and not on LC time scales.

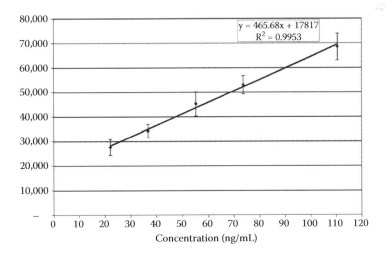

FIGURE 6.58 External standard curve obtained by FIA of fractionated samples (error bars represent ± one standard deviation). Concentration values represent analyte concentrations injected onto MS system for FIA.

6.5 CONCLUSIONS

This chapter discussed the utilization of microparallel liquid chromatography for high-throughput analyses in the pharmaceutical industry. It first described the instrumentation components and capabilities. Results of standardized performance characterization studies performed on the instrumentation were then discussed. The chapter concluded with discussions of several applications of μPLC for the characterization of ADMET and physicochemical properties, the evaluation of separation-based enzymatic assays, and the clean-up of complex samples. The advantages and applications provided by μPLC technology were also described. Overall, the system allows complex chemical and biological reactions to be analyzed with microliters (and submicroliters) of samples while operating in a truly parallel fashion (24 columns running simultaneously), with consequent significant savings in analysis time, solvent consumption, and mixed waste generation, without compromising data quality.

ACKNOWLEDGMENTS

The author would like to thank Sheila Bilbao, Chris Karp, Jun Wu, Paren Patel, Elizabeth Rietz, and Kristy Tantillo for all their help in producing this chapter.

REFERENCES

1. Lemmo, A.V., Hobbs, S., and Patel, P. *Assay Drug Dev. Technol.*, 2004, 2, 389.
2. Kilpatrick, P., *Nat. Rev. Drug Discov.*, 2003, 2, 337.
3. Mullins, R., *Chem. Eng. News,* July 26, 2004, 82, 23, http.//pubs.acs.org/cen/coverstory/8230/8230drug discovery.html.
4. Wielgos, T. and Havel, K., *LCGC*, August 2006.
5. Yasgar, A. et al., presented at SBS 12th Annual Conf., September 2006.
6. Jezequel-Sur, S. et al., presented at SBS 12th Annual Conf., April 2007.
7. Mehl, J.T., Trainor, N., and King, R.C., presented at 54th ASMS Conf. Mass Spectrom., June 2006.
8. Lloyd, T. et al., presented at 55th ASMS Conf. Mass Spectrom., June 2007.
9. Liu, Y. et al., *J. Pharm. Biomed. Anal.*, 2007, doi.10.1016/j.jpba.2006.12.026.
10. Mandel, J., Stiehler, R.D., *J. Res. Nat. Bur. Std.*, 1964, A53, 155–158.
11. Guazzotti, S. and Bilbao, S., *Lab. Equip.*, August 2006.
12. Wu, J. et al., *Assay Drug Dev. Technol.*, 2006, 4, 653.
13. Lipinski, C.A. et al., *Adv. Drug Disc. Rev.*, 1997, 23, 3.
14. Bergstrom, C.A.S. et al., *Pharm. Res.*, 2002, 19, 182.
15. Kilpatrick, P., *Nat. Rev. Drug Discov.*, 2004, 2, 337.
16. Bevan, C.D. and Lloyd, R.S., *Anal. Chem.*, 2000, 72, 1781.
17. Lomme, A., Marz, J., and Dressman, J.B., *J. Pharm. Sci.*, 2005, 94, 1.
18. Dehring, K.A. et al., *J. Pharm. Biomed. Anal.*, 2004, 36, 447.
19. Tan, H. et al., *JALA*, December 2005, 364.
20. Fang, L. et al., 2002, *Rapid Commun. Mass Spectrom.*, 16, 1440.
21. Greig, M., *Am. Lab.*, 1999, 31, 28.
22. Dressman, J. and Kramer, J., Eds., *Pharmaceutical Dissolution Testing*, Boca Raton, FL, Taylor & Francis, 2005.
23. Chen, X.Q. and Venkatesh, S., *Pharm. Res.*, 2004, 21, 1758.
24. Kassel, D.B., *High-Throughput Strategies for in Vitro ADME Assays. How Far Can We Go?* Korfmacher, W.A., Ed., Boca Raton, FL, CRC Press, 2005.
25. Avdeef, A. and Berger, C.M., *Eur. J. Pharm. Sci.*, 2001, 14, 271.
26. McFarland, J.W., Du, C.M., and Avdeef, A., *Drug Bioavailability: Estimation of Solubility, Permeability, Absorption and Bioavailability*, Weinheim, Wiley–VCH, 2003, p. 232.
27. Lambert, W.J., *J. Chromatogr.*, 1993. 656, 469.

28. Valko, K., *J. Liq. Chromatogr.*, 1984, 7, 1405.
29. Mirrles, M.S. et al., *J. Med. Chem.*, 1976, 19, 615.
30. Lombardo, F. et al., *J. Med. Chem.*, 2000, 43, 2922.
31. Patel, P., Nanostream Application Note AD1002060512, 2006.
32. Wu G., Yuan, Y., and Hodge, C.N., *J. Biomol. Screen.*, 2003, 8, 694.
33. Cohen P., *Nat. Rev. Drug Dis.*, 2002, 1, 309.
34. Wu J., Vajjhala, S., and O'Connor, S., *Assay Drug Dev. Technol.*, 2007, 5, doi.10.1089/adt.2007.072.

Walter Korfmacher

CONTENTS

ABSTRACT

This chapter describes strategies that can be used with mass spectrometry to support new drug discovery in a drug metabolism environment. The chapter discusses *in vitro* and *in vivo* tests that can be used to determine whether a compound is a potential candidate for development as a new drug. The chapter covers recent techniques used to increase throughput in the area of drug discovery bioanalytical assays. It also explains potential problems that can occur when using HPLC/MS/MS for bioanalytical assays and provides suggestions on avoiding the problems.

7.1 INTRODUCTION

It is important to understand the need for the multiple assays that are now routinely performed by most pharmaceutical companies to measure various absorption distribution metabolism and excretion (ADME) parameters to determine the pharmacokinetic (PK) properties of new chemical entities (NCEs). The goal of new drug discovery is to find NCEs that have the appropriate

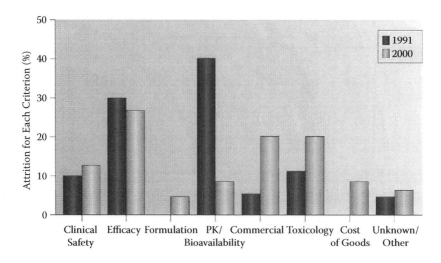

FIGURE 7.1 Major reasons for compound attrition during clinical studies (1991–2000). (*Source:* Adapted from Kola, I. and Landis, *J. Nat. Rev. Drug Dis.,* 2004, 3, 711. With permission.)

physical and chemical properties for drugs and show efficacy in preclinical models of the disease or ailment to be treated. The problem is that many compounds tested in the clinic fail and do not become drugs.

A landmark study reported that 40% of failures in clinical studies were due to PK problems.[1–3] This led to the need to develop drug metabolism studies that could be performed on a compound before it was recommended for development. The early drug metabolism and pharmacokinetic (DMPK) studies were used to assess the ADME/PK properties of NCEs. The major pharmaceutical companies were very successful at setting up exploratory drug metabolism departments using various models. This led to an explosion of new higher throughput ADME/PK assays that provided medicinal chemists with the necessary tools to improve the ADME/PK properties of NCEs.

The effort was very successful and the results were documented by Kola and Landis[1] who stated that the situation changed and that fewer than 10% of current clinical failures arose from PK problems. Figure 7.1 portrays this shift in reasons for compound attrition. The increased emphasis on early ADME/PK screening resulted in a significant change in reasons for compound failure from Phase I to FDA approval—PK is no longer a major reason.

The dramatic shift shown in Figure 7.1 serves as strong evidence that the build-up and use of exploratory drug metabolism as an integral aspect of new drug discovery is a successful strategy. This chapter will focus on some of the strategies and techniques used to implement the early ADME/PK studies. Specifically, it focuses on the analytical challenges overcome and discusses recent technologies that are available for this purpose.

7.2   CHALLENGE AND VISION

One of the first challenges was perception. During lead optimization, a discovery team uses ADME/PK data to improve NCEs so that the final compound recommended has acceptable ADME/PK characteristics.[3–11] Most of the "rules" for bioanalytical assays were covered by the good laboratory practices (GLP) guidelines.[12–15] The initial attempts to set up assays for these non-GLP ADME/PK studies often included only minor changes of the rules used for GLP assays. In the early days of exploratory drug metabolism, assay development, testing, and utilization took 2 to 4 weeks, leading to assay backlog and the perception that drug metabolism studies would not meet the challenges of participating in new drug discovery.

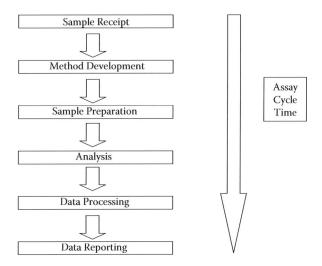

FIGURE 7.2 Discovery assay cycle, showing major steps from sample receipt to data reporting.

One way to view the issue of sample throughput is to define the steps required for the process. If we add an in-life portion to the assay time for an ADME/PK study, the term *drug metabolism discovery cycle time* (DMDCT) applies to the amount of time from the beginning to the end of a study. Thus, DMDCT = in-life cycle time + assay cycle time.

DMDCT is an important parameter because it governs how quickly one can deliver ADME/PK data for certain parameters to a discovery team. The shorter the DMDCT, the more discovery cycles can be completed. For example, if the DMDCT is 1 month, a laboratory could handle 12 discovery studies per year at most. If the DMDCT is 2 weeks, 24 discovery cycles per year are possible. In the early days of exploratory drug metabolism, the DMDCT was typically 4 to 6 weeks, with at least 50% of the time dedicated to assay cycle time. Clearly, a new vision was required for the assay cycle.

As shown in Figure 7.2, most assays involve a common series of steps that must be completed in order to report results. These steps include sample receipt, method development, sample preparation, analysis, data processing, and data reporting. While most researchers focus on speeding the analysis step, any of these steps can become bottlenecks. Thus it is important to optimize the whole process.

In the new vision, assay cycle time is dramatically reduced and the criteria used to measure assay acceptability are matched to sample type. Early screening samples may be assayed using simple methods and minimum numbers of standards. Samples from early preclinical PK studies in rats and other species may require additional standards. Finally, for PK studies performed in the lead characterization phase, one might add quality control (QC) samples. One set of "rules" for non-GLP assays has been codified in a recent publication.[16] These rules make it possible to match the assay cycle time with the in-life cycle time in order to minimize the total discovery cycle time.

This chapter will describe various strategies for higher throughput *in vitro* and *in vivo* assays. I will also discuss matrix effects that can lead to misleading results if one is unaware of their negative potential. Finally, I will highlight newer technologies that help to increase throughput.

7.3   *IN VITRO* ASSAYS

Several higher throughput *in vitro* assays may be used to assess various DMPK properties of NCEs. One common parameter is that HPLC/MS/MS is the method of choice for the analytical step.[11,17–26] These higher throughput assays include the Caco-2 assay, p450 enzyme inhibition assay, and *in vitro* stability assay. Each assay has different requirements and solutions and they will be described individually.

7.3.1 CACO-2 ASSAY

The human colon adenocarcinoma cell line (Caco-2) assay is still commonly used to measure the potential for a compound to be absorbed. It measures the permeability potential because permeability is a component of the absorption process,[3,27–31] Multiple reports discuss the use of HPLC/MS/MS to support the Caco-2 assay.[32–38]

Wang et al.[38] demonstrated the utility of HPLC/MS/MS technology to assay Caco-2 samples. Subsequent authors cite the use of newer technology in their efforts to better support the need for higher throughput Caco-2 assays. Hakala et al.[35] show that atmospheric pressure photoionization (APPI) and the more common electrospray ionization (ESI) may be utilized in HPLC/MS/MS analysis of Caco-2 samples. Van Pelt et al.[39] described the utility of a nanospray interface for assaying Caco-2 samples on a MS/MS system. Fung et al.[34] described the use of a four-way multiplexed electrospray interface (MUX) to increase assay throughput. The MUX interface allows interface of four parallel HPLC columns to a single MS/MS system. The MUX HPLC/MS/MS system allowed the authors to assay as many as 100 compounds per week.

7.3.2 ENZYME INHIBITION ASSAY

The high-throughput enzyme inhibition assay is important for measuring the drug–drug interaction potential of a NCE dosed in humans.[10,11,40–44] Various cytochrome p450 (CYP450) isozymes have been targeted by these methods.[19] In some cases, the major obstacle for a series of NCEs is CYP450 inhibition; for example, Berlin et al.[45] described how the CYP450 problem for a series of H3 receptor antagonists was resolved by using data from a CYP450 inhibition assay to guide the structural modifications needed to solve the issue. The data produced may be put into perspective as part of an overall assessment of a compound's potential for success. Generally one would consider the potency and projected human PK parameters for a compound along with the intended use for the NCE when trying to understand the significance of the results obtained via this assay.[40,46]

Multiple reports have focused on setting up higher throughput assays for measuring the inhibition potential of NCEs for various CYP450 isozymes.[19,47–54] One of the common features is that these assays utilize HPLC/MS/MS.[17] For example, both Testino et al.[48] and Li et al.[53] reported on high-throughput CYP450 inhibition screening assays for the five major human cytochrome P450 (CYP) enzymes (CYP1A2, CYP2C9, CYP2C19, CYP2D6, and CYP3A4) and performed the analytical step using HPLC/MS/MS. Peng et al.[49] described the use of a monolithic HPLC column to assay the same five major human CYP450 isozymes; their HPLC/MS/MS assay run times were shorter than 30 sec per sample. Kim et al.[54] showed that a single assay could be run in a high-throughput manner for a total of nine human CYP450 cytochrome enzymes (CYP1A2, CYP2A6, CYP2C8, CYP2C9, CYP2C19, CYP2B6, CYP2D6, CYP2E1, and CYP3A4) using HPLC/MS/MS for the analytical step.

Another concern is the possibility for mechanism-based inactivation of the CYP450 isozymes.[55] Typically, this can happen when a metabolite of a compound being tested binds with the CYP450 isozyme and inactivates it.[56] The common procedure for testing for mechanism-based inactivation of CYP450 isozymes is to perform the enzyme inhibition assay using human liver microsomes and compare the results of the test compound with and without a pre-incubation step.[57] A change of measured IC_{50} values is the test. If the value is significantly lower after preincubation, the test compound is likely to cause mechanism-based inactivation of the CYP450 being tested.[57,58] Lim et al.[56] described an alternative procedure for testing mechanism-based inactivation of CYP1A2, CYP3A4, CYP2C19, CYP2C9, and CYP2D6 based on an apparent partition ratio screen that uses HPLC/MS/MS for the analytical step.

7.3.3 *In Vitro* Metabolic Stability Assay

An important DMPK property of a NCE is oral bioavailability (F) of the compound in various pre-clinical species.[3] The oral bioavailability of a compound is dependent on several factors including intestinal permeability (estimated by the Caco-2 assay) and hepatic clearance (estimated with an *in vitro* metabolic stability assay).[3,30] The metabolic stability assay is typically performed by incubating test compounds in liver microsomes or hepatocytes. The results can provide estimates of *in vivo* stability in terms of metabolic liabilities.[3,8,59–62] Several authors described this assay as an important tool for the rapid assessment of the DMPK properties of NCEs.[3,6,8,11,18,19,26,44,59,62–65]

Multiple reports discuss various methods of conducting *in vitro* metabolic stability assays; what these reports have in common is their use of LC/MS for the analytical step.[17,18,66–73] In an earlier report, Korfmacher et al.[66] described an automated system that could accept 96-well plates containing samples from liver microsomal studies; the assay centered on a simple LC/MS system with an automated control procedure. With this system, the authors were able to assay 75 compounds per week with a single LC/MS system.[66] Di et al.[70] later described a higher throughput microsomal stability assay that uses robotic sample preparation along with a rapid LC/MS/MS assay. Jenkins et al.[71] described the utility of using robotic sample handling systems as part of their strategy for increasing throughput of their metabolic stability assay. Xu et al.[73] described one of the highest throughput microsomal stability assays to date. Their system makes use of robotics for sample preparation and includes a unique set-up of eight parallel HPLC columns connected to a single mass spectrometer. Xu's system can handle microsomal stability assays of 176 compounds per day.

One issue related to supporting a metabolic stability assay with HPLC/MS/MS is the need to set up an MS/MS method for each compound. While it may only take 10 min to infuse a compound solution and find the corresponding precursor and product ions (along with minimal optimization of the collision energy), the processes of MS/MS development would require 4 hr per day if one wanted to assay 25 compounds per day. MS vendors have responded to this need by providing software tools that can perform the MS/MS method development step in an automated fashion. Chovan et al.[68] described the use of the Automaton software package supplied by PE Sciex (Toronto, Canada) as a tool for the automated MS/MS method development for a series of compounds. The Automaton software was able to select the correct precursor and product ions for the various compounds and optimize the collision energy used for the MS/MS assays of each compound. They found that the Automaton software provided similar sensitivity to methods that would have been developed by manual MS/MS procedures. Chovan et al. also reported that the MS/MS method development for 25 compounds could be performed in about an hour with the Automaton software and required minimal human intervention.

Some authors used reduced numbers of sample time points and sample pooling strategies as ways to increase the throughput of *in vitro* stability assays. Zhao et al.[74] state that the standard microsomal stability study has a total of 5 time points per compound and describe a pooling strategy in which pool 1 is a mixture of the 0- and 5-min samples and pool 2 is a mixture of 15-, 30-, and 45-min time points. By calculating the theoretical percent remaining for each time point against the $t_{1/2}$ of the compound, the ratio of pool 2 to pool 1 (pooling ratio R) was related to $t_{1/2}$ in a systematic manner such that the $t_{1/2}$ for a compound could be calculated by determining the R value for the compound. The authors showed that this two-sample assay provided data comparable to that obtained from a standard five-time point assay.[74] In a simpler approach, Di et al.[69] suggested that a single time point of 15 min was sufficient for a metabolic stability assay in the course of new drug discovery screening.

Some authors searched for common oxidative metabolites as part of metabolic stability assays. Tong et al.[75] described a highly automated microsomal metabolic stability assay that achieved a throughput of 50 compounds per day with each compound tested in rats, dogs, monkeys, and humans. In addition to assaying the test compound, they monitored M+16 metabolites by using the predicted selected reaction monitoring (SRM) transitions for the addition of O on either side of a molecule. In another approach, Shou et al.[76] reported on the use of a QTRAP-based LC/MS/MS system that measure metabolic stability of NCEs in microsomal incubations while searching for metabolites of the NCEs. The results for the metabolic stability assay were the same whether they added or did not add the extra automated metabolite search to the assay. This is a good example of obtaining additional information from a set of samples that would normally yield only compound stability data.

7.4   *IN VIVO* ASSAYS

As a general rule, *in vivo* assays are more challenging than *in vitro* assays because the matrices for the samples are more complex. The most common use for *in vivo* assays is to measure the concentration of NCE dosed into a laboratory animal; by collecting multiple sample time points, one can use the analytical results to plot the PK profile of the NCE and also obtain various PK parameters that help determine a test compound's PK properties. Preclinical PK parameters of a test compound are then used to predict its human PK parameters. Another use of *in vivo* assays is combining the results with pharmacodynamic (PD) observations to perform PK/PD modeling.[77–82] PK/PD modeling is an important aspect of new drug discovery because it can be used to predict the exposures and durations required to determine clinical efficacy of a NCE.

7.4.1   Discovery PK Screening Assays: Sample Reduction

PK assays are generally not considered high throughput due to the need to dose laboratory animals, collect plasma samples, and subject the samples to a bioanalytical procedure to determine desired concentration values for the NCE dosed. Significant efforts have focused on developing PK screening assays with the goal to increase sample throughput.[83] As shown in Figure 7.2, the first step in a discovery study is setting up the in-life segment of the study to determine the number of samples to be delivered for assay.

For traditional PK studies in laboratory animals, one would dose three animals via an intravenous (IV) route and three animals via oral (PO) dosing, with a minimum of 8 time points per animal, resulting in a total of 48 samples for a PK study of one compound. For a comparison of six NCEs, using this standard PK procedure would result in the need to assay 288 samples. A large pharmaceutical company may have to assay 40 to 60 or more NCEs weekly via *in vivo* PK screening. This led to methods for reducing the number of samples that need to be assayed.

Two early examples of sample reduction used sample pooling to mix time points into one sample for assay. Hop et al.[84] described a sample pooling procedure that mixed various amounts of plasma samples into a single pooled sample that could be used to obtain an area-under-the-curve (AUC) estimate for the dosed compound. Cox et al.[85] used a simpler sample pooling process that also produced a pooled sample suitable for making an AUC estimate for NCEs dosed in rats. Both procedures expedited the assay process by reducing the number of samples to be assayed, but neither method provided concentration–time curve data. Han et al.[86] described a three-time point (1, 4, and 8 hr) approach for estimating the oral rat AUC and the Cmax for the NCE; they showed that their "rapid rat" approach could be used as a screening paradigm for early drug discovery support.

Sample pooling is also used in a process called cassette assay in which samples are pooled from multiple (typically 5 or 6) dosing experiments.[24,83,87,88] Hsieh et al.[89] showed that one could pool the plasma from six NCEs into one sample per time point to reduce sample assay time. Kuo et al.[88] used a similar sample pooling approach for NCEs dosed into rats. The advantage is that cassette assay requires fewer samples. Two disadvantages are the need to dilute samples and the difficult set-up.

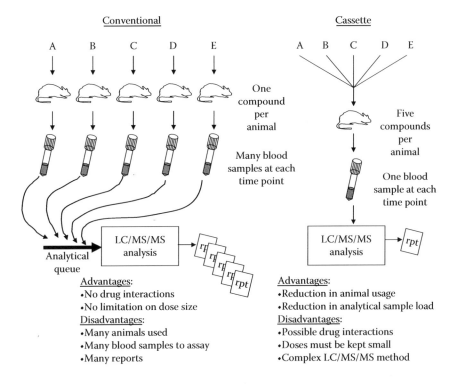

FIGURE 7.3 Conventional dosing versus cassette dosing. With conventional dosing, only one compound is dosed into each rat. Cassette dosing involves multiple compounds dosed in each rat. (*Source:* Adapted from Manitpisitkul, P. and White, R.E., *Drug Dis. Today*, 2004, 9, 652. With permission of Elsevier.)

Another sample reduction method still employed by some companies is the technique called cassette dosing.[90–99] As shown in Figure 7.3,[92] a group of compounds (typically 5 to 10) are dosed together in one laboratory animal to reduce the number of samples to be assayed and the number of animals to be dosed. An early report about it was from Olah et al.[98] who noted that about 400 NCEs were evaluated by this procedure in 24 weeks. Zhang et al.[97] used the cassette dosing approach to compare plasma and brain exposures in rats for NCEs that were candidates as possible central nervous system (CNS) drugs. Shaffer et al.[99] applied cassette dosing to dogs (5 to 20 NCEs per dog) and reported that it was a useful screening tool. Manitpisitkul and White[92,96] described the drug metabolism theory behind cassette dosing and showed that it produce false positive and false negative PK parameters due to drug–drug interactions. A second report by Manitpisitkul and White[92] discusses the reduced interest in cassette dosing due to drug–drug interactions and practical disadvantages in the dosing and assay steps.

An alternative to cassette dosing for rats is the use of the cassette-accelerated rapid rat screen (CARRS).[100] As described by Korfmacher et al., CARRS is a systematic approach for testing NCEs via oral rat PK screening. The CARRS model uses cassettes of six NCEs as a unit for dosing and assay. Each compound in the set of six NCEs is dosed individually into two rats (PO at 10 mg/kg), for a total of 12 rats dosed for each cassette of six compounds. The rats are sampled for six time points (0.5, 1, 2, 3, 4, and 6 hr post-dose). The samples are pooled across two rats at each time point, resulting in a total of 6 samples per compound dosed or 36 samples for each 6-compound cassette. The samples are assayed using a simple three-point standard curve in duplicate. All the standards, samples, and blanks for one cassette can fit onto one 96-well plate. This leads to efficiencies in sample preparation and assay[100] and also data reporting. The CARRS procedure recently provided *in vivo* PK screening of more than 7000 NCEs in 4 years.[92]

7.4.2 DISCOVERY PK STUDIES: SAMPLE PREPARATION

For discovery PK assays, the most common sample preparation procedure is protein precipitation[16,17,20–24] because it is fast, easy to automate, and requires no method development. While protein precipitation typically will not provide as clean a sample as will alternative procedures, it is sufficient for most discovery PK samples that use HPLC/MS/MS for the analytical step.[21,101]

Typical protein precipitation procedures use one volume of plasma plus three to six volumes of acetonitrile or methanol (or a mixture) with the internal standard at an appropriate concentration for the assay. Polson et al.[102] reported that protein precipitation using acetonitrile eliminates at least 95% of the proteins; after filtration or centrifugation, the supernatant can often be directly injected into the HPLC/MS/MS system. Usually this step is performed using 96-well plates that are ideal for semi-automation of sample preparation. Briem et al.[103] reported on a robotic sample preparation system for plasma based on a protein precipitation step and a robotic liquid handling system that increased throughput by a factor of four compared to a manual system.

In some cases, so called direct plasma injection techniques may be used[23,83,104–108] instead of protein precipitation for loading plasma samples onto an HPLC/MS/MS system. Some direct plasma injection systems use a column switching technique in which the plasma is loaded onto an extraction column that retains the small molecules. The other plasma components are sent to waste and the flow is switched so that the small molecules are eluted onto an analytical column that connects to the MS/MS.[23,83,108] One variation of the column switching method is turbulent flow chromatography commercialized by Cohesive Technologies (now part of Thermo, San Jose, CA).[23]

Turbulent flow chromatography uses large particle packing materials and high flow rates to separate small molecules from proteins and other matrix components in plasma. In one example, Herman et al.[109] reported that turbulent flow chromatography was useful for a series of discovery compounds as the online extraction step in LC/MS/MS analysis. As an alternative, Hsieh et al.[89,104–107] described the use of a single mixed function column as a simpler process for direct plasma injection applications.

The most common (off-line) sample preparation procedures after protein precipitation are solid phase extraction and liquid–liquid extraction. Multiple vendors and available chemistries utilize 96-well plates for solid phase extraction systems and liquid–liquid extraction procedures. Both extraction process can prepare samples for HPLC/MS/MS assay. Jemal et al.[110] compared liquid–liquid extraction in a 96-well plate to semi-automated solid phase extraction in a 96-well plate for a carboxylic acid containing analyte in a human plasma matrix and reported that both clean-up procedures worked well. Yang et al.[111,112] described two validated methods for compounds in plasma using semi-automated 96-well plate solid phase extraction procedures. Zimmer et al.[113] compared solid phase extraction and liquid–liquid extraction to a turbulent flow chromatography clean-up for two test compounds in plasma; all three clean-up approaches led to HPLC/MS/MS assays that met GLP requirements.

7.4.3 DISCOVERY PK STUDIES: RAPID METHOD DEVELOPMENT

For discovery PK samples, rapid method development is required. For HPLC/MS/MS assays, method development can be achieved within 2 hr if no unusual problems are encountered. Xu et al.[101] described a process for rapid method development as part of the discovery PK paradigm. As shown in Figure 7.4, the systematic process is based on using protein precipitation as the sample clean-up step and generic HPLC conditions for the HPLC/MS/MS assay.

The systematic procedure has checkpoints along the way to ensure that the final method is suitable as a discovery PK assay. For example, if protein precipitation plus a fast gradient HPLC method lead to a method that suffers from matrix effect issues (*vide infra*), possible solutions include revising the chromatography to a longer gradient or switching to solid phase extraction. In a similar scenario, the assay would be tested at the likely limit of quantitation (LOQ). An interfering

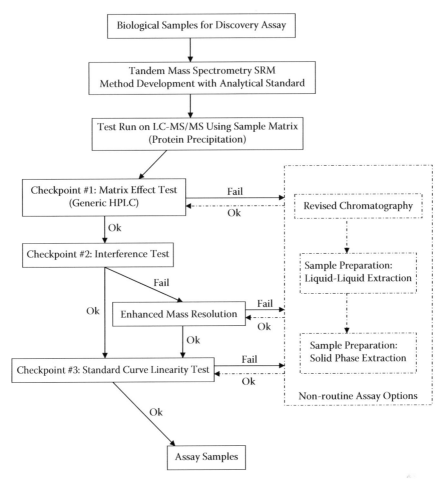

FIGURE 7.4 Steps in development of rapid HPLC/MS/MS methods for discovery PK assays. (*Source:* Adapted from Xu, X. et al., *Anal. Chem.,* 2005, 77, 389. With permission of the American Chemical Society.)

endogenous peak could be resolved with higher mass resolution (when possible) or changing to a different HPLC column or gradient. The point of this process is to test the HPLC/MS/MS method before the samples are assayed rather than waiting until the samples are assayed by a failed method.

7.4.4 Discovery PK Studies: Rules for Acceptance Criteria for Discovery Assays

While clear rules apply to the acceptance criteria for GLP assays,[12,14,15,114] little agreement surrounds what should be included in the acceptance criteria for discovery PK (non-GLP) assays. Korfmacher[16] published a set of rules for discovery PK assays based on the simple concept that the rules should become more rigorous as one moves from early PK screening (Level I) of many compounds to rapid PK studies for lead compounds (Lead Optimization—Level II), and finally special PK studies for compounds that are likely to be recommended for development (Lead Qualification—Level III).

Table 7.1 summarizes the major rules. Table 7.2 shows Level I rules in detail.[16] Table 7.3 covers Level II rules and Table 7.4, Level III rules.[16] These rules have been used for thousands of discovery compounds and ensure that the reported results are scientifically correct while minimizing the time required to develop the assay and report the results to the discovery project team in a timely manner.[115]

TABLE 7.1

Rules for Discovery (Non-GLP) Assays

Drug Stage	Assay Type (Level)	Summary of Major Rules	GLP?
Screening	I	Use two-point standard curve	No
Lead optimization	II	Use multipoint standard curve and no QCs	No
Lead qualification	III	Use multipoint standard curve plus QCs	No
Development	IV	GLP rules	Yes

Source: Adapted from Korfmacher, W., *in Using Mass Spectrometry for Drug Metabolism Studies,* Korfmacher, W., Ed., CRC Press, Boca Raton, FL, p. 1. With permission.

7.4.5 DISCOVERY PK STUDIES: METABOLITE PROFILING

As part of new drug discovery, the trend toward screening for metabolites in plasma samples being assayed for the dosed compound (NCE) is increasing. This effort is sometimes called metabolite profiling[116] and it is important for two reasons: (1) for compounds with low bioavailability due to extensive metabolism, metabolites may help medicinal chemists learn to modify the NCE to block

TABLE 7.2

Rules for Discovery (Non-GLP) Screening Assays (Level I)

1. Samples should be assayed using HPLC/MS/MS technology.
2. Sample preparation should consist of protein precipitation using appropriate internal standard (IS).
3. Samples should be assayed along with a standard curve in duplicate (at beginning and end of sample set).
4. The zero standard is prepared and assayed, but is not included in calibration curve regression.
5. Standard curve stock solutions are prepared after correcting standard for the salt factor.
6. The standard curve should be three levels, typically ranging from 25 to 2,500 ng/mL (may be lower or higher as needed). Each standard is 10 times the one below (a typical set is 25, 250, and 2500 ng/mL). The matrix of the calibration curve should be from the same animal species and matrix type as the samples.
7. QC samples are not used and the assay is not validated.
8. After the assay, the proper standard curve range for the samples is selected; this must include only two concentrations in the range that covers the samples. A one-order-of-magnitude range is preferred, but two orders of magnitude are acceptable if needed to cover the samples.
9. After the range is selected, at least three of the four assayed standards in the range must be included in the regression analysis. Regression is performed using unweighted linear regression (not forced through zero).
10. All standards included in the regression set must be back-calculated to within 27.5% of their nominal values.
11. The limit of quantitation (LOQ) may be set as the lowest standard in the selected range or as 0.4 times the lowest standard in the selected range, but must be greater than three times the mean value for the back-calculated value of the two zero standards.
12. Samples below the LOQ are reported as zero.
13. If the LOQ is 0.4 times the lowest standard in the selected range, samples with back-calculated values between the LOQ and the lowest standard in the selected range may be reported as their calculated value provided the S/N for the analyte is at least three.
14. Samples with back-calculated values between 1.0 and 2.0 times the highest standard in the selected range are reportable by extending the calibration line up to 2 times the high standard.
15. Samples found to have analyte concentrations more than 2 times the highest standard in the regression set are not reportable; these samples must be reassayed after dilution or along with a standard curve that has higher concentrations so that the sample is within 2 times the highest standard.

TABLE 7.3
Rules for Discovery (Non-GLP) Full PK Assays (Level II)

1. Samples should be assayed using HPLC/MS/MS technology.
2. Sample preparation should consist of protein precipitation using an appropriate internal standard (IS).
3. Samples should be assayed along with a standard curve in duplicate (at beginning and end of the sample set).
4. Zero standard is prepared and assayed, but is not included in calibration curve regression.
5. Standard curve stock solutions are prepared after correcting standard for salt factor.
6. Standard curve should be 10 to 15 levels, typically ranging from 1 to 5000 or 10,000 (or higher as needed) ng/mL. The matrix of the calibration curve should be from the same animal species and matrix type as the samples.
7. QC samples are not used.
8. After assay, the proper standard curve range for the samples is selected; this must include at least five (consecutive) concentrations.
9. Once the range is selected, at least 75% of the assayed standards in the range must be included in regression analysis.
10. Regression can be performed using weighted or unweighted linear or smooth curve fitting (e.g., power curve or quadratic), but is not forced through zero.
11. All standards included in regression set must be back-calculated to within 27.5% of their nominal values.
12. The regression r^2 must be 0.94 or larger.
13. The LOQ may be set as the lowest standard in the selected range or 0.4 times the lowest standard in the selected range, but the LOQ must be greater than three times the mean value for the back-calculated value of the two zero standards.
14. Samples below the LOQ are reported as zero.
15. If the LOQ is 0.4 times the lowest standard in the selected range, samples with back-calculated values between the LOQ and the lowest standard in the selected range may be reported as their calculated value provided the S/N for the analyte is at least three.
16. Samples with back-calculated values between 1.0 times and 2.0 times the highest standard in the selected range are reportable by extending the calibration curve up to 2 times the high standard as long as the calibration curve regression was not performed using quadratic regression.
17. Samples found to have analyte concentrations more than 2 times the highest standard in the regression set are not reportable; they must be reassayed after dilution or along with a standard curve that has higher concentrations so that the sample is within 2 times the highest standard.
18. Assay is not validated.
19. Final data does not need QA approval; assay is an exploratory non-GLP study.

TABLE 7.4
Additional Rules for Discovery (Non-GLP) PK Assays Requiring QC Samples (Level III)

1. Use all rules (except Rule 7) for full PK Level II assays plus Rules 2 to 6 below.
2. QC standards are required; a minimum of six QCs at three concentrations (low, middle, high) must be used. QC standards should be frozen at the same temperature as samples to be assayed.
3. QC standards must be traceable to a separate analyte weighing from the one used for standard curve standards.
4. Standard curve standards should be prepared the same day as samples are prepared for assay; standard curve solutions required may be stored in a refrigerator up to 6 mo until needed.
5. At least 2/3 of QC samples must be within 25% of their prepared (nominal) values.
6. If dilution of one or more samples is required, an additional QC at a higher level must be prepared, diluted, and assayed along with the sample(s) needing dilution. This QC should be run in duplicate and at least one of the two assay results must meet the 25% criteria.

the sites of metabolism; (2) in some cases, a metabolite is active and has a better PK profile than the dosed NCE. In this case, the metabolite may become the new lead compound.

Table 7.5 provides a complete list of common metabolites and their mass shifts relative to parent compounds.[117] While the concept of metabolite profiling is not new,[20] multiple advances in MS hardware and software allow researchers to look more easily for metabolites and include them in PK assays.[118]

One of the best tools for metabolite profiling is the hybrid QTRAP MS/MS system (Applied Biosystems).[119–121] While the hybrid QTRAP MS/MS was initially considered a premier tool for metabolite identification, it has more recently been seen as a tool for quantitation and metabolite profiling. Li et al.[122] described the use of a hybrid QTRAP MS/MS system for discovery PK assays plus metabolite profiling in the same analytical procedure. Because QTRAP MS/MS may be used as a triple quadrupole MS system, it can be used as part of a quantitative HPLC/MS/MS system. Because QTRAP MS/MS also has linear ion trap capabilities, it can be used for metabolite screening and characterization—essentially it combines the capabilities of a triple quadrupole mass spectrometer and a linear ion trap mass spectrometer.

The software tools accompanying the QTRAP MS/MS allow set-up of multiple selected reaction monitoring (SRM) transitions for all likely metabolites after the major product ion transitions for the dosed compound are known. Because QTRAP MS/MS can monitor up to 100 SRM transitions during a single assay, the SRM transitions required for quantitation of the dosed compound and internal standard are obtained along with the possible metabolite transitions. During sample analysis, when a possible metabolite transition exceeds a preset threshold value, the QTRAP MS/MS performs an enhanced product ion (EPI) scan. When the assay is complete, the EPI scans can be used to determine whether the "hits" are metabolites, and if they are metabolites, what part of the molecule has changed. Thus, one analytical run provides both quantitative and metabolite information.

Our laboratory used the QTRAP MS/MS system for both quantitative analysis and metabolite profiling when assaying plasma samples. We found it useful to use three rules for deciding when to report a possible metabolite:

1. Each metabolite must be chromatographically separated from the dosed compound and from other reportable metabolites.
2. Each metabolite must produce an SRM response of at least 1% of the dosed compound.
3. A product ion mass spectrum for each reportable metabolite must be obtained and must provide sufficient information to demonstrate that it is a metabolite of the dosed compound.

These rules help to avoid incorrect reporting of false metabolites and unnecessary reporting of minor metabolites. Typically, we report metabolites by showing the relative responses of the metabolite and dosed compound on the same graph; because the y axis of this graph is labeled relative response (as opposed to concentration units), we alert the recipient that the concentration responses of dosed compound and metabolite may vary. Figure 7.5 is an example of this type of report.

A good example of metabolite screening as part of a discovery PK study was reported by Tiller and Romanyshyn.[123] They described a dog PK study in which a monohydroxylated metabolite was found at much higher levels than the dosed NCE; it was active and fit the pharmacodynamic (PD) profile better than the dosed NCE.

Wainhaus et al.[116] described a decision tree process for a "targeted metabolite screening procedure" as shown in Figure 7.6. This procedure can be used to decide when a possible metabolite found during a screening step can be reported. They set a response equivalent to 25 ng/ml for the parent (dosed compound) as one threshold that to be met before reporting a compound as a metabolite. While the initial screening is performed using atmospheric pressure chemical ionization (APCI), further tests are performed with electrospray ionization (ESI) when potential for glucuronidation as the metabolic pathway is indicated. For acyl glucuronide metabolites, special

TABLE 7.5

Mass Shifts of Common Metabolites Relative to Initial (Parent) Compound Dosed

Reaction	Example	Metabolism/ Enzyme Phase	Mass Shift Da (NL, Parent Ion/Ion Mode)
Nitro reduction	R-NO$_2$> R-NH$_2$	I/amine oxidase	−30
N–, O– or S–demethylation	R-NH-CH$_3$ >R-NH$_2$	I/CYP	−14
N–, O– or S–dealkylation	R-NH-alkyl> R-NH$_2$	I/CYP	Depends on alkyl chain length
Dehydrogenation	R-CH$_2$-OH> R-CHO	I/dehydrogenase	−2
Hydroxylation	R-CH$_2$> R-CH-OH Ar-H> AR-OH	I/CYP	+16
di-Hydroxylation		I/CYP	+32
Oxidation	R$_1$-CH$_2$-R$_2$> R$_1$-CO-R$_2$	I/CYP	+14
N-oxidation	R-NH> R-N-OH	I/CYP and/or FMO	+16
Sulfoxidation	R-S-R> R-SO-R	I	+16
Aldehyde oxidation	R-CHO> R-COOH	I/alcohol dehydrogenase	+16
Alcohol oxidation	R-CH$_2$-OH> R-COOH	I/alcohol dehydrogenase	+16
Oxidation of CH$_3$–group to carboxylic acid	R-CH$_3$> R-COOH	I/CYP	+30
Epoxide hydroxylation	R-CH(O)-R> R-CH(OH)-CH(OH)-R	I/epoxide hydratase	+18
Epoxide formation and hydroxylation		I/epoxide hydratase and CYP	+34 (+18, +16)
Sulfation, aromatic	Ar-OH> Ar-O-SO$_3$H	II/sulfotransferase	+80 (Precursor m/z 97/−)
Sulfation, aliphatic	R-OH> R-O-SO$_3$H		
Glucuronidation	R-OH> R-O-GlcA	II/ UDP-transferase	+176 (NL 176/+ or −)
Carbamoyl–glucuronide	primary & secondary amines	II/UDP-transferase	+220 (NL 176/+)
Glycosylation hexose (Glc)		II/GDP-transferase	+162
Glutathione conjugation	R-CH=CH$_2$> R-CH$_2$-CH$_2$-SG R-CH$_2$-CH$_2$- CysOAc	II/glutathione transferase	+307 (305) Aliphatic (NL 129/+), Aromatic (NL 273/+)
N-acetylcysteines Mercapturic acid	R-CH$_2$-CH$_2$- CysOAc		+163 (NL 129/+)

(Continued)

TABLE 7.5 (CONTINUED)

Mass Shifts of Common Metabolites Relative to Initial (Parent) Compound Dosed

Reaction	Example	Metabolism/ Enzyme Phase	Mass Shift Da (NL, Parent Ion/Ion Mode)
Glutathione conjugation	Epoxide + GSH-H$_2$O>GS-parent		+305
Acetylated GSH			+347 (305 + 42)
Glutathione conjugation	Epoxide+GSH> R-CHOH-HCSG-		+323 (129)
GSH	R-CH$_2$-CH$_2$-CysGly		178 (176)
GSH	R-CH$_2$-CH$_2$-CysGlu		250 (248)
Cysteine conjugation	R-CH=CH$_2$> R-CH$_2$-CH$_2$-Cys		+121 (119)
Mercapturic acid (from GSH conjugation)			+161
Reduction of NO$_2$	R-CH$_2$-NO$_2$> R-CH$_2$-SG	II/GSH transferase	+160
Gly-conjugation	R-COOH> R-CO-Gly	I or II	+57
Ala conjugation	R-COOH> R-CO-Ala	I or II	+71
Methylation	R-OH> R-O-CH$_3$	I/methyl transferase	+14
Acetylation (1°, 2° amines)	R-NH$_2$> R-NH-CO-CH$_3$	I/N-acetyltransferase	+42
Phosphorylation	R-OHR-O-PO$_3$H		+79 Precursor *m/z* 63/- Precursor *m/z* 79/-

NL = neutral loss scanning.

Source: Adapted from Baranczewski, P. et al., *Pharmacol Rep.*, 2006, 58, 341. With permission.

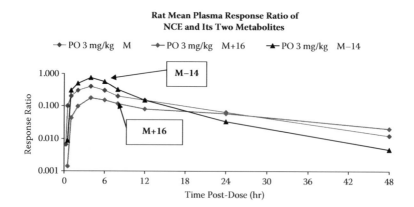

FIGURE 7.5 Graph showing dosed compound plus M+16 and M-14 metabolites. The M-14 metabolite showed a higher response than the dosed compound, suggesting that it was a major metabolite.

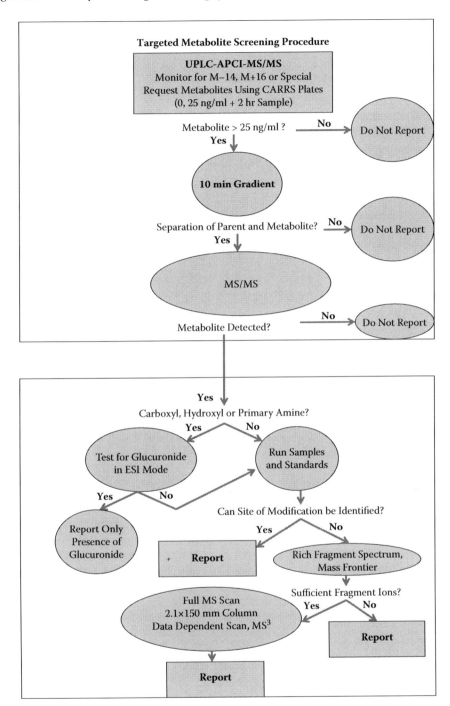

FIGURE 7.6 Targeted metabolite screening procedure, showing a flowchart that could be followed to determine whether to report a potential metabolite observed in a sample assay. (*Source:* Adapted from Wainhaus, S. et al., *Am. Drug Dis.,* 2007, 2, 6. With permission.)

precautions are needed to stabilize the acyl glucuronides and unique procedures can be used to estimate their concentrations.[124,125]

7.4.6 DISCOVERY PK STUDIES: MATRIX EFFECTS

Matrix effect is a phrase normally used to describe the effect of some portion of a sample matrix that causes erroneous assay results if care is not taken to avoid the problem or correct for it by some mechanism. The most common matrix effects are those that result in ion suppression and subsequent false negative results. Ion enhancement may lead to false positive results.[126,127] Several reports about matrix effects include suggestions on what can cause them and how to avoid them.[126–147]

While various ways to detect matrix effects have been reported, Matuszewski et al.[140] described a clear way to measure the matrix effect (ME) for an analyte, recovery (RE) from the extraction procedure, and overall process efficiency (PE) of a procedure. Their method is to prepare three sets of samples and assay them using the planned HPLC/MS/MS method. The first set is the neat solution standards diluted into the mobile phase before injection to obtain the A results. The second set is the analyte spiked into the blank plasma extract (after extraction) to obtain the B results. The third set is the analyte spiked into the blank plasma before the extraction step (C results); these samples are extracted and assayed along with the two other sets. The three data sets allow for the following calculations:

$$ME\ (\%) = B/A \times 100$$

$$RE\ (\%) = C/B \times 100$$

$$PE\ (\%) = C/A \times 100 = (ME \times RE)/100$$

This procedure allows one to identify the source of a problem if an assay shows poor process efficiency. The main disadvantage is the effort required that may be justified for a GLP validated assay but not for most discovery PK assays except as a tool for finding the source of a problem when an assay fails to work properly.

For PK assays, it is generally believed that most matrix effects are due to the sample matrix (typically plasma). While this is correct in many cases, this assumption has some exceptions (*vide infra*). One of the most useful tools for avoiding matrix effects is studying the sample matrix and proposed assay by using the post-column infusion technique described by Bonfiglio et al.[148] This technique allows visualization of the portion of the chromatographic step affected by ion suppression.[16,17,21] Xu et al.[101] recommended inclusion of this step in the method development process for drug discovery PK assays.

Another important consideration is chromatography. As we strive for faster assays by using shorter columns or faster HPLC gradients, the possibility for matrix effects is likely to increase. De Nardi and Bonelli[149] investigated the matrix effect issue when converting to a faster assay and found that by carefully selecting the sample precipitation process and performing proper matrix effect evaluations, one could increase assay speed without adverse matrix effects. Chambers et al.[145] described an extensive study of matrix effects in rat plasma samples. They compared various sample extraction procedures and tested chromatography conditions. One of their conclusions was that faster chromatography could be utilized as long as one took precautions to avoid matrix effects as part of method optimization.

As a general rule, APCI is less likely to demonstrate matrix effects and ESI is more likely to be affected by matrix effects. Sample clean-up is another important factor—protein precipitation is more likely to result in matrix effects than is solid phase extraction. Matrix effects may be caused by sample constituents that are not parts of the biological matrix. Mei et al.[126,129] showed that certain brands of sample tube containers can produce matrix effects. They also demonstrated that Li-heparin, a common anticoagulant for plasma samples, can produce significant matrix effects (for this reason, we avoid Li-heparin in our laboratory). Leverence et al.[150] reported on the possibility of having matrix effects from common co-administered (concomitant) medications (naproxen and ibuprofen). While this is not likely to become an issue for preclinical PK studies, it serves as a cautionary note for assaying multidrug dosing studies.

A few reports discuss the possibility matrix effects arising from dosing formulations.[126,127,151–153] This can be especially problematic because it can lead to time-dependent matrix effects if the amount of the formulation excipient varies over the time course of sample collection in a PK study.[127] As shown in Figure 7.7, a major matrix effect is caused by PEG 400, a common formulation additive in discovery PK studies.[127] Larger et al.[153] also reported that dosing formulations can lead to significant matrix effect errors in discovery PK studies. They found that diluting the early time points from an IV-dosed laboratory animal study provided an easy way to check the assay for matrix effects. Figure 7.8 shows a decision tree they used as a way to rapidly test PK samples for matrix effects.

7.4.7   DISCOVERY PK STUDIES: FASTER ASSAYS WITH FASTER CHROMATOGRAPHY

In 1995, when HPLC/MS/MS was becoming the premier tool for PK assays, chromatographic sample cycle times were typically 10 to 12 min. At 10 min per sample, 16 hr were required to process 96 samples. By 2000, scientists used shorter HPLC columns and per-sample cycle times decreased to 5 to 6 min. At 5 min per sample, it takes about 8 hr to assay one 96-well plate of samples. As a result, parallel HPLC became popular; Korfmacher et al.[154] described a two-column system and an MS vendor produced a triple quadrupole system designed to work with four HPLC columns.[16,155–158] Advances in fast chromatography continued and by 2005, sample cycle times of 1 to 2 min became common.[21,87,159–161] At 2 min per sample, 3 hr are required to assay one 96-well plate of samples.

A very dramatic change occurred when Waters (Bedford, MA) introduced the first commercial ultrahigh pressure liquid chromatography (UPLC) system. UPLC changed the pressure limits on HPLC systems, so that smaller particle size (1.7 µ diameter) packing could be used in columns.[162] UPLC systems allow scientists to perform sample assays in time frames of 0.5 to 1 min.[116] At this speed, a single 96-well plate can be assayed in less than 2 hr. This short assay time has changed thinking about timelines; our laboratory can assay high priority PK studies within a day of sample receipt. For small sample sets, we can frequently provide same-day ("rapid dog" assay) results.

A number of researchers described the utility of UPLC for bioanalytical efforts.[116,145,163–175] As show in Figure 7.9, Yu et al.[174] demonstrated that a 3-min run time might be needed using HPLC/MS/MS for five common drugs spiked in rat plasma. UPLC/MS/MS was able to assay the compounds with a total run time shorter than 1 min. UPLC allows faster run times and also achieves better chromatography with sharper peaks that typically lead to more sensitive assays. As shown in Figure 7.10, two test compounds were spiked into rat plasma at a 10 ng/ml concentration and the same extract was injected into an HPLC/MS/MS system. Another aliquot was injected into a UPLC/MS/MS system (in each assay comparison, the same ionization mode and SRM transition were utilized). The results were impressive. UPLC/MS/MS system yielded very sharp chromatographic peaks with a 3- to 10-fold signal increase as compared to HPLC/MS/MS. As noted by Yu et al.,[174] the sharp peaks (typical peak width was 2 sec) require a fast scanning mass spectrometer to produce sufficient data points. For the example in Figure 7.9, the MS dwell time was set to 5 msec for the UPLC/MS/MS assay.

Wainhaus et al.[116] provided a good overview of UPLC/MS/MS and showed how it could be used to enhance discovery PK assays. Their report indicated that PK screening assays be performed more quickly because of shorter run times with UPLC for the separation mode, and assay LOQs were often lower with UPLC. In one example, the LOQ for the fast HPLC/MS/MS assay was 1.5 ng/ml. The LOQ for the same compound using UPLC/MS/MS was 0.04 ng/ml. Wainhaus et al. also reported improved separation of metabolites of a test compound when using UPLC/MS/MS.

MS vendors continue to improve their triple quadrupole mass spectrometers. Typically, the latest models of these MS/MS systems are more sensitive than previous systems and scan faster than the

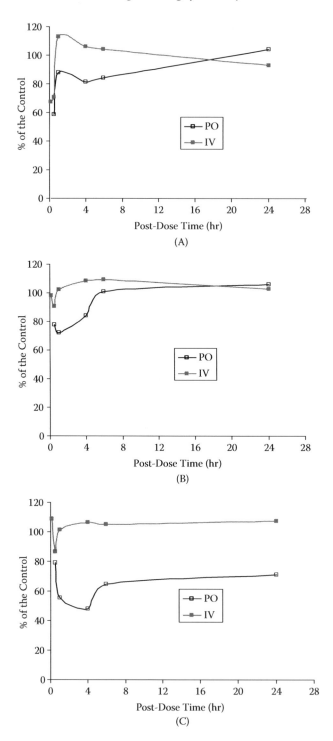

FIGURE 7.7 Time-dependent MS response for pseudoephedrine when PEG 400 was used in the formulation dosed to rats. HPLC/MS/MS was performed in the ESI mode using: (A) Thermo-Finnigan Quantum MS; (B) AB Sciex 3000 MS; (C) Waters-Micromass Quattro Ultima MS. The PK samples were spiked with pseudo-ephedrine after collection from rats. The dip (below 100%) in the profiles shows the time-dependent nature of this type of matrix effect. (*Source:* Xu, X. et al., *Rapid Commun. Mass Spectrom.*, 2005, 19, 2643. With permission of John Wiley & Sons.)

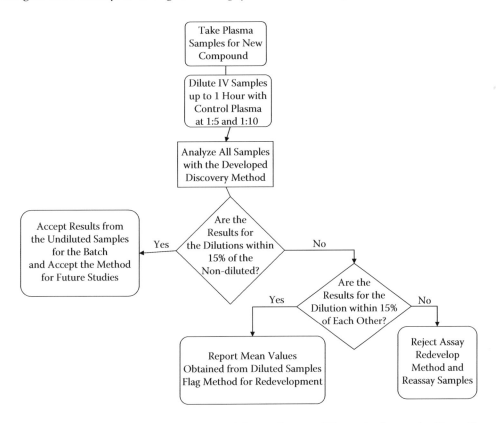

FIGURE 7.8 Decision tree that could be used to rapidly test discovery PK samples for matrix effects. (*Source:* Adapted from Larger, P.J. et al., *J. Pharm. Biomed. Anal.,* 2005, 29, 206. With permission.)

older models. Ionization sources are also becoming easier to use and simpler to switch between APCI and ESI. At least one vendor designed a combined ESI–APCI source that can be switched during a single assay via software tools. This type of source will be of benefit in performing discovery PK assays because it is common to want to analyze sample sets with multiple compounds during one chromatographic event. To show how well this approach works, Yu et al.[171] reported on a 30-sec assay of dapsone, sulfadimethoxine, tolbutamide and ibuprofen. As shown in Figure 7.11, each drug was chromatographically separated from the others, then assayed using a separate ionization mode (ESI+, APCI+, APCI–, ESI–).

7.4.8 METABOLITE IDENTIFICATION

While metabolite identification remains a lower throughput effort in most cases, it is a very important support procedure for new drug discovery. A thorough review of metabolite identification is beyond the scope of this chapter, so I will provide a brief overview and refer the reader to several recent reviews that will provide a more complete picture.[176–184] Ma et al.[176] reviewed the application of MS for metabolite identification; this comprehensive review describes how different types of MS equipment can be used for metabolite identification.

One of the more powerful techniques is a new software tool called mass defect filtering.[176,185–188] A mass defect can be defined as the difference between the exact mass and nominal mass of a compound.[189] Typically, drug-like molecules (and their metabolites) will have mass defects that differ from those of endogenous matrix materials. While a mass spectrometer that has unit mass resolution cannot differentiate a test compound from an isobaric matrix compound, a high mass resolution MS may be able to differentiate many isobaric matrix compounds from test compounds.

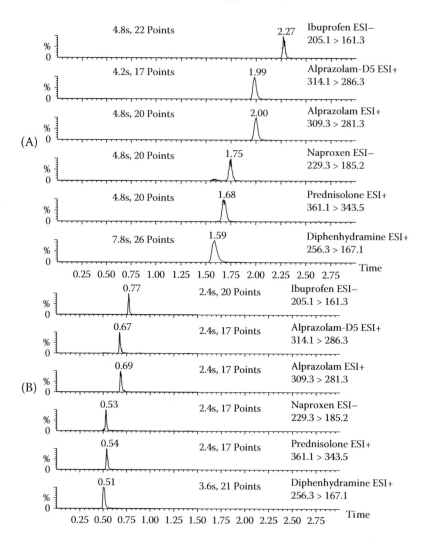

FIGURE 7.9 (A) Upper traces show HPLC/MS/MS mass chromatograms for a set of six test compounds (total HPLC assay time ca. 3 min). (B) Lower traces show UPLC/MS/MS mass chromatograms for the same set of six test compounds (total UPLC assay time ca. 1 min). (*Source:* Adapted from Yu, K. et al., *Rapid Commun. Mass Spectrom.*, 2006, 20, 544. With permission of John Wiley & Sons.)

Zhang et al.[186] generated an early report of mass defect filtering. They demonstrated its utility by using the software filter on a bile sample from a dog dosed at 20 mg/kg with ^{14}C-labelled compound A, as shown in Figure 7.12. The top trace shows the total ion chromatogram (TIC) from the HPLC/MS assay (the same result one would obtain using a quadrupole MS set at unit mass resolution). The middle trace shows the same assay after using the mass defect filter and the bottom trace shows the radioactivity chromatogram. It is clear that the mass defect filtered result shows very good agreement with the radio chromatogram.

Figure 7.13 illustrates the utility of mass defect filtering (also known as exact mass filtering) and UPLC. It shows results of UPLC/MS assay of a bile sample containing buspirone and its metabolites.[184] The top trace shows the (unfiltered) TIC for the sample; the middle trace is the result of an exact mass filter; the bottom trace is an extracted ion chromatogram for the M+16 or hydroxylated metabolites based on their exact masses. It is readily apparent that this new software tool may be very helpful for metabolite identification studies.

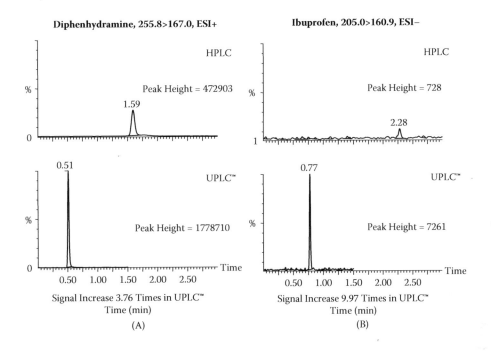

FIGURE 7.10 Sensitivity comparison of HPLC and UPLC. The same sample was injected twice; once using HPLC/MS/MS and again on a UPLC/MS/MS system: (A) response for diphenhydramine on each system; (B) response of ibuprofen. (*Source:* Adapted from Yu, K. et al., *Rapid Commun. Mass Spectrom.,* 2006, 20, 544. With permission of John Wiley & Sons.)

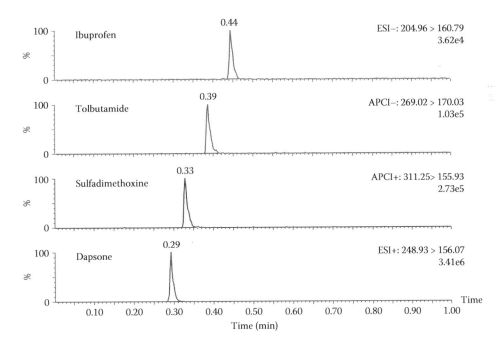

FIGURE 7.11 Multimode assay based on UPLC/MS/MS system. A four-compound mixture was assayed in one injection. The mass spectrometer switched among the four ionization modes quickly. The ionization modes for the analytes are shown. (*Source:* Adapted from Yu, K. et al., *Rapid Commun. Mass Spectrom.,* 2007, 21, 893. With permission of John Wiley & Sons.)

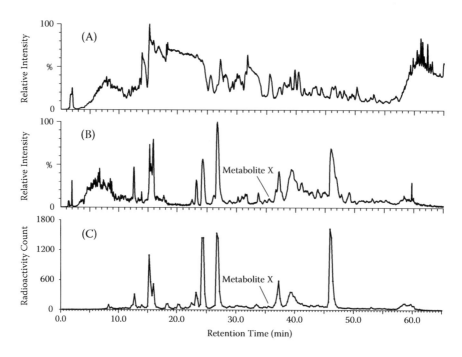

FIGURE 7.12 Metabolite profile of bile sample obtained from a dog dosed with 20 mg/kg of a ^{14}C-labeled test compound: (A) total mass chromatogram of unprocessed LC/MS data; (B) total mass chromatogram after mass defect filter processing; (C) radioactivity chromatogram. (*Source:* Adapted from Zhang, H.D. and Ray, K., *J. Mass Spectrom.*, 2003, 38, 1110. With permission of John Wiley & Sons.)

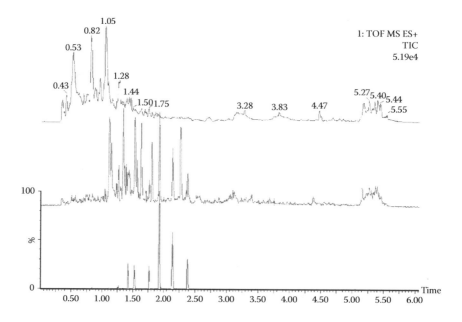

FIGURE 7.13 Results of UPLC/MS assay of a bile sample containing buspirone and its metabolites showing the utility of exact mass filtering and UPLC. Top: total mass chromatogram of unprocessed LC/MS data. Middle: total mass chromatogram after mass defect filter processing. Bottom: extracted mass chromatogram for M+16 (M + OH) metabolites. (*Source:* Adapted from Castro-Perez, J.M., *Drug Dis. Today,* 2007, 12, 249. With permission.)

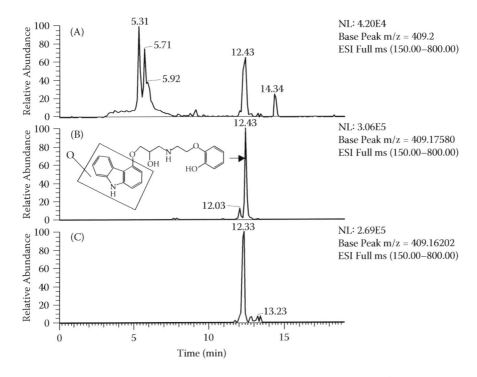

FIGURE 7.14 High mass resolution provided by Orbitrap MS allows separation of a metabolite and a co-eluting isobaric matrix component. (A) extracted mass chromatogram using nominal mass resolution (*m/z* 409). (B) chromatogram for the metabolite (*m/z* 409.17580). (C) chromatogram for co-eluting endogenous compound (*m/z* 409.16202). (*Source:* Adapted from Lim, H.K. et al., *Rapid Commun. Mass Spectrom.*, 2007, 21, 1821. With permission of John Wiley & Sons.)

One recent advance in MS hardware that has been found to be useful for metabolite identification studies is the Orbitrap. This MS has a mass resolution of 30,000 to 100,000 (two models). For many applications, 30,000 mass resolution capability is sufficient. While only a few current literature references cite the Orbitrap MS for metabolite identification, it is safe to predict that the Orbitrap will be the subject of many references in the future. Two references related to its use for metabolite identification are Peterman et al.[190] and Lim et al.[182] Lim's group related an an impressive example of the use of high mass resolution to differentiate a metabolite from a co-eluting isobaric matrix component, as shown in Figure 7.14.

Estimating the amount of a metabolite when an authentic reference standard is not available is still a challenge. Yu et al.[191] described a procedure that uses the results of an *in vitro* metabolite identification based on a test compound that produces ^{14}C-labelled metabolites; essentially the ^{14}C-labelled metabolites are used to provide a correction factor for the MS response when assaying samples that contain the same metabolite in a study that did not use the ^{14}C-labelled test compound. Hop[192] described another novel approach for metabolite quantitation based on the observation that the MS responses for most compounds are very similar to responses from nanospray ESI. Valaskovic et al.[193] also reported equimolar MS responses for multiple compounds when the flow rate to the nanospray ESI source was set to about 10 nl/min. It is too soon to know whether these intriguing findings can be readily applied to discovery metabolite identification studies.

7.5   CONCLUSIONS

It is certainly an exciting time to be a scientist involved in new drug discovery. As pharmaceutical companies continue to look for ways to efficiently screen thousands of compounds each year through various *in vitro* and *in vivo* ADME/PK screens, they face continuing pressures to obtain data more efficiently at a faster pace. The ongoing advances in the fields of HPLC and MS are helping scientists meet these demands.

REFERENCES

1. Kola, I. and J. Landis, *Nat Rev Drug Discov*, 2004. 3: 711.
2. Caldwell, G.W., *Curr Opin Drug Discov*, 2000. 3: 30.
3. Thompson, T.N., in *Using Mass Spectrometry for Drug Metabolism Studies*, Korfmacher, W., Ed., 2005, CRC Press: Boca Raton, FL. p. 35.
4. Zhang, L. and M. Banks, *Amer Drug Discov*, 2006. 1: 6.
5. Eddershaw, J., A. Beresford, and M.K. Bayliss, *Drug Discov Today*, 2000. 5: 409.
6. Kassel, D.B., *Curr Opin Chem Biol*, 2004. 8: 339.
7. Kerns, E.H. and L. Di, *Drug Discov Today*, 2003. 8: 316.
8. Thompson, T.N., *Curr Drug Metab*, 2000. 1: 215.
9. Roberts, S.A., *Xenobiotica*, 2001. 31: 557.
10. White, R.E., in *Pharmaceutical Profiling in Drug Discovery for Lead Selection*, Borchardt, R.T., Ed., 2004, American Association of Pharmaceutical Scientists Press, p. 431.
11. Korfmacher, W.A., *Curr Opin Drug Discov Devel*, 2003. 6: 481.
12. Bajpai, M. and J.D. Esmay, *Drug Metab Rev*, 2002. 34: 679.
13. Yang, L., N. Wu, and J. Rudewicz, *J Chromatogr A*, 2001. 926: 43.
14. Shabir, G.A., *J Chromatogr A*, 2003. 987: 57.
15. Shah, V. et al., *Pharm Res*, 2000. 17: 1551.
16. Korfmacher, W., in *Using Mass Spectrometry for Drug Metabolism Studies*, Korfmacher, W., Ed., CRC Press: Boca Raton, FL. p. 1.
17. Korfmacher, W.A., *Drug Discov Today*, 2005. 10: 1357.
18. Kassel, D.B., in *Using Mass Spectrometry for Drug Metabolism Studies*, Korfmacher, W., Ed., 2005, CRC Press: Boca Raton, FL.
19. Chu, I. and A.A. Nomeir, *Curr Drug Metab*, 2006. 7: 467.
20. Korfmacher, W.A. et al., *Drug Discov Today*, 1997. 2: 532.
21. Korfmacher, W., in *Mass Spectrometry in Medicinal Chemistry*, Wanner, K. and Hofner, G.., Eds., 2007, Wiley-VCH: Weinheim, p. 401.
22. Hopfgartner, G., C. Husser, and M. Zell, *Ther Drug Monit*, 2002. 24: 134.
23. Hopfgartner, G. and E. Bourgogne, *Mass Spectrom Rev*, 2003. 22: 195.
24. Ackermann, B.L., M.J. Berna, and A.T. Murphy, *Curr Top Med Chem*, 2002. 2: 53.
25. Briem, S., B. Pettersson, and E. Skoglund, *Anal Chem*, 2005. 77: 1905.
26. Korfmacher, W., *Identification and Quantification of Drugs, Metabolites and Metabolizing Enzymes by LC-MS*, Chowdhury, S., Ed., 2005, Elsevier: Amsterdam, p. 7.
27. Balimane, V. and S. Chong, *Drug Discov Today*, 2005. 10: 335.
28. Kerns, E.H. et al., *J Pharm Sci*, 2004. 93: 1440.
29. Markowska, M. et al., *J Pharmacol Toxicol Meth*, 2001. 46: 51.
30. Mandagere, A.K., T.N. Thompson, and K.K. Hwang, *J Med Chem*, 2002. 45: 304.
31. Balimane, V. et al., *Eur J Pharm Biopharm*, 2004. 58: 99.
32. Larger, P. et al., *Anal Chem*, 2002. 74: 5273.
33. Bu, H.Z. et al., *Rapid Commun Mass Spectrom*, 2000. 14: 523.
34. Fung, E.N. et al., *Rapid Commun Mass Spectrom*, 2003. 17: 2147.
35. Hakala, K.S. et al., *Anal Chem*, 2003. 75: 5969.
36. Li, Y. et al., *Comb Chem High Throughput Screen*, 2003. 6: 757.
37. Stevenson, C.L., F. Augustijns, and R.W. Hendren, *Int J Pharm*, 1999. 177: 103.
38. Wang, Z. et al., *J Mass Spectrom*, 2000. 35: 71.
39. Van Pelt, C.K. et al., *Rapid Commun Mass Spectrom*, 2003. 17: 1573.
40. Obach, R.S. et al., *J Pharmacol Exp Ther*, 2005.

41. Obach, R.S., *Drugs Today*, 2003. 39: 301.
42. Obach, R.S. et al., *J Pharmacol Exp Ther*, 2006. 316: 336.
43. Smith, D. et al., *J Chromatogr B*, 2007. 850: 455.
44. Singh, S.S., *Curr Drug Metab*, 2006. 7: 165.
45. Berlin, M. et al., *Bioorg Med Chem Lett*, 2006. 16: 989.
46. Yan, Z. and G.W. Caldwell, *Curr Top Med Chem*, 2001. 1: 403.
47. Vengurlekar, S.S. et al., *J Pharm Biomed Anal*, 2002. 30: 113.
48. Testino, S.A., Jr. and G. Patonay, *J Pharm Biomed Anal*, 2003. 30: 1459.
49. Peng, S.X., A.G. Barbone, and D.M. Ritchie, *Rapid Commun Mass Spectrom*, 2003. 17: 509.
50. Chu, I. et al., *Rapid Commun Mass Spectrom*, 2000. 14: 207.
51. Bu, H.Z. et al., *Rapid Commun Mass Spectrom*, 2000. 14: 1619.
52. Bu, H.Z. et al., *J Chromatogr B*, 2001. 753: 321.
53. Li, X. et al., *J Chromatogr B*, 2007.
54. Kim, M.J. et al., *Rapid Commun Mass Spectrom*, 2005. 19: 2651.
55. Obach, R.S., R.L. Walsky, and K. Venkatakrishnan, *Drug Metab Dispos*, 2007. 35: 246.
56. Lim, H.K. et al., *Drug Metab Dispos*, 2005. 33: 1211.
57. Favreau, L.V. et al., *Drug Metab Dispos*, 1999. 27: 436.
58. Ghanbari, F. et al., Curr Drug Metab, 2006. 7: 315.
59. Thompson, T.N., *Med Res Rev*, 2001. 21: 412.
60. Obach, R.S., *Drug Metab Dispos*, 1999. 27: 1350.
61. Obach, R.S. et al., *J Pharmacol Exp Ther*, 1997. 283: 46.
62. Rostami-Hodjegan, A. and G.T. Tucker, *Nat Rev Drug Discov*, 2007. 6: 140.
63. Di, L. and E.H. Kerns, *Curr Opin Chem Biol*, 2003. 7: 402.
64. Janiszewski, J.S. et al., *Anal Chem*, 2001. 73: 1495.
65. Pelkonen, O. et al., *Basic Clin Pharmacol Toxicol*, 2005. 96: 167.
66. Korfmacher, W.A. et al., *Rapid Commun Mass Spectrom*, 1999. 13: 901.
67. Inman, B.L. et al., *J Pharm Sci*, 2007. 96: 1619.
68. Chovan, L.E. et al., *Rapid Commun Mass Spectrom*, 2004. 18: 3105.
69. Di, L. et al., *J Pharm Sci*, 2004. 93: 1537.
70. Di, L. et al., *J Biomol Screen*, 2003. 8: 453.
71. Jenkins, K.M. et al., *J Pharm Biomed Anal*, 2004. 34: 989.
72. Lindqvist, A., S. Hilke, and E. Skoglund, *J Chromatogr A*, 2004. 1058: 121.
73. Xu, R. et al., *J Soc Mass Spectrom*, 2002. 13: 155.
74. Zhao, S.X. et al., *J Pharm Sci*, 2005. 94: 38.
75. Tong, X.S. et al., *J Chromatogr B*, 2006. 833: 165.
76. Shou, W.Z. et al., *J Mass Spectrom*, 2005. 40: 1347.
77. Derendorf, H. and B. Meibohm, *Pharm Res*, 1999. 16: 176.
78. Hickey, E., *Curr Opin Drug Discov Devel*, 2007. 10: 49.
79. Derendorf, H. et al., *J Clin Pharmacol*, 2000. 40: 1399.
80. Csajka, C. and D. Verotta, *J Pharmacokinet Pharmacodyn*, 2006. 33: 227.
81. Chien, J.Y. et al., *AAPS J*, 2005. 7: E544.
82. Dickinson, G.L. et al., *Br J Clin Pharmacol*, 2007.
83. Cox, K.A., R.E. White, and W.A. Korfmacher, *Comb Chem High Throughput Screen*, 2002. 5: 29.
84. Hop, C.E. et al., *J Pharm Sci*, 1998. 87: 901.
85. Cox, K. et al., *Drug Discov Today*, 1999. 4: 232.
86. Han, H.K. et al., *J Pharm Sci*, 2006. 95: 1684.
87. Hsieh, Y. and W.A. Korfmacher, *Curr Drug Metab*, 2006. 7: 479.
88. Kuo, B.S. et al., *J Pharm Biomed Anal*, 1998. 16: 837.
89. Hsieh, Y. et al., *J Chromatogr B*, 2002. 767: 353.
90. Ackermann, B.L., *J Am Soc Mass Spectrom*, 2004. 15: 1374.
91. Beaudry, F. et al., *Rapid Commun Mass Spectrom*, 1998. 12: 1216.
92. Manitpisitkul, P. and R.E. White, *Drug Discov Today*, 2004. 9: 652.
93. Ohkawa, T. et al., *J Pharm Biomed Anal*, 2003. 31: 1089.
94. Sadagopan, N., B. Pabst, and L. Cohen, *J Chromatogr B*, 2005. 820: 59.
95. Smith, N.F. et al., *Cancer Chemother Pharmacol*, 2004.
96. White, R.E. and Manitpisitkul, P., *Drug Metab Dispos*, 2001. 29: 957.
97. Zhang, M.Y. et al., *J Pharm Biomed Anal*, 2004. 34: 359.

98. Olah, T.V., D.A. McLoughlin, and J.D. Gilbert, *Rapid Commun Mass Spectrom*, 1997. 11: 17.
99. Shaffer, J.E. et al., *J Pharm Sci*, 1999. 88: 313.
100. Korfmacher, W.A. et al., *Rapid Commun Mass Spectrom*, 2001. 15: 335.
101. Xu, X., J. Lan, and W. Korfmacher, *Anal Chem*, 2005. 77: 389A.
102. Polson, C. et al., *J Chromatogr B*, 2003. 785: 263.
103. Briem, S. et al., *Rapid Commun Mass Spectrom*, 2007. 21: 1965.
104. Hsieh, Y., in *Using Mass Spectrometry for Drug Metabolism Studies*, Korfmacher, W., Ed., 2005, CRC Press: Boca Raton, FL. p. 151.
105. Hsieh, Y. et al., *J Pharm Biomed Anal*, 2002. 27: 285.
106. Hsieh, Y. et al., *Analyst*, 2001. 126: 2139.
107. Hsieh, Y. et al., *Rapid Commun Mass Spectrom*, 2000. 14: 1384.
108. Ackermann, B.L., A.T. Murphy, and M.J. Berna, *Amer Pharm Rev*, 2002. 5: 54.
109. Herman, J.L., *Rapid Commun Mass Spectrom*, 2002. 16: 421.
110. Jemal, M. et al., *J Chromatogr B*, 1999. 732: 501.
111. Yang, L. et al., *J Chromatogr B*, 2003. 792: 229.
112. Yang, L. et al., *J Chromatogr B*, 2004. 809: 75.
113. Zimmer, D. et al., *J Chromatogr A*, 1999. 854: 23.
114. Shah, V. et al., *Eur J Drug Metab Pharmacokinet*, 1991. 16: 249.
115. Xu, X. et al., *Amer Drug Discov*, 2007. 2: 6.
116. Wainhaus, S. et al., *Amer Drug Discov*, 2007. 2: 6.
117. Baranczewski, J. et al., *Pharmacol Rep*, 2006. 58: 341.
118. Triolo, A. et al., *J Mass Spectrom*, 2005. 40: 1572.
119. King, R. and C. Fernandez-Metzler, *Curr Drug Metab*, 2006. 7: 541.
120. Hopfgartner, G. et al., *J Mass Spectrom*, 2004. 39: 845.
121. Hopfgartner, G. and M. Zell, in *Using Mass Spectrometry for Drug Metabolism Studies,* Korfmacher, W., Ed., 2005, CRC Press: Boca Raton, FL.
122. Li, A.C. et al., *Rapid Commun Mass Spectrom*, 2005. 19: 1943.
123. Tiller, R. and L.A. Romanyshyn, *Rapid Commun Mass Spectrom*, 2002. 16: 1225.
124. Wainhaus, S.B. et al., *Amer Pharm Rev*, 2002. 5: 86.
125. Wainhaus, S.B., in *Using Mass Spectrometry for Drug Metabolism Studies*, Korfmacher, W., Ed., 2005, CRC Press: Boca Raton, FL. p. 175.
126. Mei, H., in *Using Mass Spectrometry for Drug Metabolism Studies*, Korfmacher, W., Ed., 2005, CRC Press: Boca Raton, FL. p. 103.
127. Xu, X. et al., *Rapid Commun Mass Spectrom*, 2005. 19: 2643.
128. Zheng, J.J., E.D. Lynch, and S.E. Unger, *J Pharm Biomed Anal*, 2002. 28: 279.
129. Mei, H. et al., *Rapid Commun Mass Spectrom*, 2003. 17: 97.
130. Schuhmacher, J. et al., *Rapid Commun Mass Spectrom*, 2003. 17: 1950.
131. Seliniotakis, E. et al. in *ASMS Conference on Mass Spectrometry and Allied Topics*. 2003. Montreal, Canada.
132. Souverain, S., S. Rudaz, and J.L. Veuthey, *J Chromatogr A*, 2004. 1058: 61.
133. Taylor, J., *Clin Biochem*, 2005. 38: 328.
134. Tiller, R. and L.A. Romanyshyn, *Rapid Commun Mass Spectrom*, 2002. 16: 92.
135. Chen, J. et al., *J Chromatogr B*, 2004. 809: 205.
136. Dams, R. et al., *J Am Soc Mass Spectrom*, 2003. 14: 1290.
137. Hsieh, Y. et al., *Rapid Commun Mass Spectrom*, 2001. 15: 2481.
138. Jemal, M., A. Schuster, and D.B. Whigan, *Rapid Commun Mass Spectrom*, 2003. 17: 1723.
139. Matuszewski, B.K., M.L. Constanzer, and C.M. Chavez-Eng, *Anal Chem*, 1998. 70: 882.
140. Matuszewski, B.K., M.L. Constanzer, and C.M. Chavez-Eng, *Anal Chem*, 2003. 75: 3019.
141. Avery, M.J., *Rapid Commun Mass Spectrom*, 2003. 17: 197.
142. King, R. et al., *J Am Soc Mass Spectrom*, 2000. 11: 942.
143. Matuszewski, B.K., *J Chromatogr B*, 2006. 830: 293.
144. Shen, J.X. et al., *J Pharm Biomed Anal*, 2005. 37: 359.
145. Chambers, E. et al., *J Chromatogr B*, 2007. 852: 22.
146. Heller, D.N., *Rapid Commun Mass Spectrom*, 2007. 21: 644.
147. Chen, J., W.A. Korfmacher, and Y. Hsieh, *J Chromatogr B*, 2005. 820: 1.
148. Bonfiglio, R. et al., *Rapid Commun Mass Spectrom*, 1999. 13: 1175.
149. De Nardi, C. and F. Bonelli, *Rapid Commun Mass Spectrom*, 2006. 20: 2709.

150. Leverence, R. et al., *Biomed Chromatogr*, 2007.
151. Tong, X.S. et al., *Anal Chem*, 2002. 74: 6305.
152. Shou, W.Z. and W. Naidong, *Rapid Commun Mass Spectrom*, 2003. 17: 589.
153. Larger, J. et al., *J Pharm Biomed Anal*, 2005. 39: 206.
154. Korfmacher, W.A. et al., *Rapid Commun Mass Spectrom*, 1999. 13: 1991.
155. Yang, L. et al., *Anal Chem*, 2001. 73: 1740.
156. Van Pelt, C.K. et al., *Anal Chem*, 2001. 73: 582.
157. Bayliss, M.K. et al., *Rapid Commun Mass Spectrom*, 2000. 14: 2039.
158. Jemal, M. et al., *Rapid Commun Mass Spectrom*, 2001. 15: 994.
159. Hsieh, Y. et al., *Comb Chem High Throughput Screen,* 2006. 9: 3.
160. Dunn-Meynell, K.W., S. Wainhaus, and W.A. Korfmacher, *Rapid Commun Mass Spectrom*, 2005. 19: 2905.
161. Tiller, R., L.A. Romanyshyn, and U.D. Neue, *Anal Bioanal Chem*, 2003. 377: 788.
162. Mazzeo, J.R. et al., *Anal Chem*, 2005. 77: 460A.
163. Castro-Perez, J. et al., *Rapid Commun Mass Spectrom*, 2005. 19: 843.
164. Churchwell, M.I. et al., *J Chromatogr B*, 2005.
165. Johnson, K.A. and R. Plumb, *J Pharm Biomed Anal*, 2005.
166. Plumb, R. et al., *Rapid Commun Mass Spectrom*, 2004. 18: 2331.
167. Plumb, R.S. et al., *Analyst*, 2005. 130: 844.
168. Wang, G. et al., *Rapid Commun Mass Spectrom*, 2006. 20: 2215.
169. Wang, G. et al., *J Chromatogr B*, 2007. 852: 92.
170. Wren, S.A., *J Pharm Biomed Anal*, 2005. 38: 337.
171. Yu, K. et al., *Rapid Commun Mass Spectrom*, 2007. 21: 893.
172. Hsieh, Y. et al., *J Pharm Biomed Anal*, 2007. 44: 492.
173. Xu, R.N. et al., *J Pharm Biomed Anal*, 2007. 44: 342.
174. Yu, K. et al., *Rapid Commun Mass Spectrom*, 2006. 20: 544.
175. Shen, J.X. et al., *J Pharm Biomed Anal*, 2006. 40: 689.
176. Ma, S., S.K. Chowdhury, and K.B. Alton, *Curr Drug Metab*, 2006. 7: 503.
177. Kostiainen, R. et al., *J Mass Spectrom*, 2003. 38: 357.
178. Jemal, M. et al., *Rapid Commun Mass Spectrom*, 2003. 17: 2732.
179. Hopfgartner, G., C. Husser, and M. Zell, *J Mass Spectrom*, 2003. 38: 138.
180. Cox, K., in *Using Mass Spectrometry for Drug Metabolism Studies*, Korfmacher, W., Ed., 2005, CRC Press: Boca Raton, FL. p. 229.
181. Nassar, A.E. and D.Y. Lee, *J Chromatogr Sci*, 2007. 45: 113.
182. Lim, H.K. et al., *Rapid Commun Mass Spectrom*, 2007. 21: 1821.
183. Chen, Y., M. Monshouwer, and W.L. Fitch, *Pharm Res*, 2007. 24: 248.
184. Castro-Perez, J.M., *Drug Discov Today*, 2007. 12: 249.
185. Bateman, K. et al., *Rapid Commun Mass Spectrom*, 2007. 21: 1485.
186. Zhang, H., D. Zhang, and K. Ray, *J Mass Spectrom*, 2003. 38: 1110.
187. Zhu, M. et al., *Drug Metab Dispos*, 2006. 34: 1722.
188. Sanders, M. et al., *Curr Drug Metab*, 2006. 7: 547.
189. Leslie, A.D. and D.A. Volmer, *Spectroscopy,* 2007: 24.
190. Peterman, S.M. et al., *J Am Soc Mass Spectrom*, 2006. 17: 363.
191. Yu, C., et al., *Rapid Commun Mass Spectrom*, 2007. 21: 497.
192. Hop, C.E., *Curr Drug Metab*, 2006. 7: 557.
193. Valaskovic, G.A., et al., *Rapid Commun Mass Spectrom*, 2006. 20: 1087.

8 High-Throughput Analysis in Drug Metabolism during Early Drug Discovery

Yau Yi Lau

CONTENTS

8.1 INTRODUCTION

Pharmaceutical profiling assays provide early assessments of drug-like properties such as solubility, permeability, metabolism, stability, and drug–drug interactions. This information can be used to alert project teams to potential property issues, predict and diagnose *in vitro* and *in vivo* assay results, guide structure–property relationships, provide insight into structure modification, and help drug discovery teams make informed decisions. Successful drugs can be developed when biological activities of interest and pharmaceutical properties are optimized in parallel.

8.2 LABORATORY AUTOMATION, INFORMATION, AND DATA MANAGEMENT

Laboratory automation, information, and data management are important parts in implementing high-throughput analysis in drug metabolism. These processes help avoid errors, reduce turnaround times and costs, and aid in integrating information. The type and extent of automation varies greatly among organizations and laboratories. Figure 8.1 illustrates important components that can be incorporated into laboratory automation and data management. Typically an automated system consists of different components such as pipettors, washers, plate holders, tip storage areas, readers, incubators, and robotic arms for moving plates. A typical automated system can handle plates containing wells in multiples of 96 (96-, 384-, and even 1536-well plates). Highly powerful detection instruments such as plate readers and fast LC/MS equipment are also essential.

The advent of automation techniques moved high-throughput ADME screening from individual test tube to multiwell plates. The use of 96- and 384-well plates produced a data explosion and the need to capture, store, and mine data so that it can be used effectively. A database for storing and

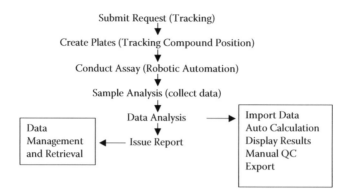

FIGURE 8.1 Laboratory automation, information and data management system flowchart.

processing rapidly generated data is crucial to high-throughput approaches. A database should be user-friendly, manageable, and open to data input and retrieval and should also be capable of handling *in vivo* and *in vitro* receptor binding and PK/ADME data.

Lloyd et al.[1] described automation processes for compound optimization and simultaneous implementation of (1) a LIMS system to automate and track the flow of sample information, data analysis, and reporting; (2) an automated data archiving system to handle a large number of LC/MS/MS data files; (3) custom software to track a large number of protocol flows; and (4) workstation automation.

Herbst et al.[2] described automated high-throughput ADME/Tox profiling for optimization of preclinical candidates, a "Profiling Toolkit" enables researchers to request profiles, track progress, receive notification when the requests have been filled, and view data. Automated systems and bar coding improve efficiency, help track compound and plate locations, and create audits trail of the profiling process.

In early 2005, Thermo introduced its fully automated ADME/Tox LeadStream[3] platform consisting of four distinct elements: three integrated instrumentation modules plus software that manages the flow of samples. The components include (1) LeadStream Orchestrator software that collects and reports data (LIMS connectivity) and optimizes ADME/Tox screening across multiple assay types; (2) the LeadStream Reformatter that provides online preparation of plates for the work cell; parallel processing dramatically reduces turn-around time; (3) the LeadStream WorkCell, a fully automated modular platform for conducting ADME/Tox assays; and (4) the LeadStream LC/MS(TM) analysis and quantification system. LeadStream has been integrated with Galileo, an ADME/Tox LIMS designed to enhance data analysis, review, and approval.

Other commercial high-throughput informatics systems developed to address this challenge include the Assay Explorer[4] (MDL Information Systems), ActivityBase[5] (IDBS), BioAssay Manager[6] (CambridgeSoft), CBIS[7] (ChemInnovation Software), and DS Accord Enterprise Informatics Suite[8] (Accelrys Inc).

MDL's Assay Explorer is a data management system for capturing, calculating, and analyzing high-throughput screening, and *in vivo* data. It has the capability to analyze data in real time as it streams off an automated workstation or robotics solution. It can also apply complex statistical analyses including Zprime, principal components analysis (PCA) and analysis of variance (ANOVA) to determine edge effects and visually validate large amounts of screening data to ensure quality results. Assay Explorer can capture all data types including images and documents as experimental results. Assay results are securely stored in a central database and project teams can access them to make informed decisions. Assay Explorer also has the flexibility of allowing customization.

ActivityBase enables the capture, validation, and visualization of high-throughput screening data. Integration with Microsoft Excel provides flexibility analysis template design. The chemically and biologically aware environment of ActivityBase allows users to relate results back to experimental conditions, protocols, and even chemical structures. The system can generate interactive SAR reports on stored data and display both biological and chemical data to support decision making.

Accelrys' Accord is a database for storage and management of chemical, biological screening, and inventory data within life science organizations. It has the ability to set up assay templates, create assay plate records, and handle plate tracking.[6] It can store reports that are retrievable and searchable. It enables integration of chemical and biological data into a unified and scalable system.

BioAssay manages both high- and low-throughput biological screening data. It is designed for complex lead optimization experiments. The software supports the quick set-up of biological models, integrates chemical and biological data, allows queries by structure or text with ChemFinder, and sets up Excel templates for reporting and graphing.

8.3 FAST HPLC/MS/MS ANALYSES

HPLC/MS/MS is a very important technique in high-throughput drug discovery. It provides excellent sensitivity and selectivity and short analysis times. MS/MS detection is based on a combination of the unique parent and fragment mass of each compound, eliminating the need for baseline separation, and achieving fast analyses. The highly selective nature of this method is well suited for high-throughput screening of compounds with diverse structures in the discovery phase and has been used extensively to support high-throughput metabolic screening.[9–17] The most time-consuming issues in the process are method development and the need to analyze large numbers of *in vitro* samples. A number of techniques have been developed to increase sample throughput including online sample preparation,[18,19] cassette dosing and compound analysis,[20–22] staggered parallel HPLC/tandem MS,[23] and multiple inlet electrospray interface (MUX) developed by Micromass.[24–29]

Staggered parallel analysis is used to reduce turn around time during analysis. During an HPLC run, much of the time spent during the generic LC run is not used for collecting valuable information. It is spent on equilibrating and washing columns to maintain good reproducibility. One resolution to the problem of inefficient use of MS is to offset multiple HPLC systems with a time delay. King et al.[23] described a four fully independent HPLC systems fed by two injection syringes. Samples were introduced into the MS interface via a selection valve. A single computer program handled timing and triggering of injections, gradient starts, and collection of data. Figure 8.2 illustrates the operating principle.

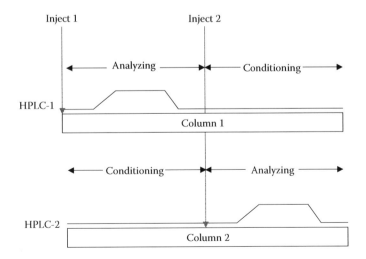

FIGURE 8.2 Staggered parallel analysis scheme showing two systems run with staggered start times.

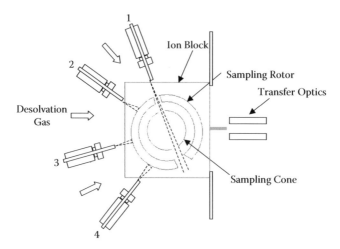

FIGURE 8.3 MUX system.

Parallel analysis was made possible by the development of the multiplexed ion source (MUX) technique (Figure 8.3). The MUX ion source consists of four electrospray needles around a rotating dish containing an orifice that allows independent sampling from each sprayer. Four samples using the same HPLC gradient are sequentially introduced into the MS in a sampling cycle that is fast relative to LC peak width. The disadvantages of MUX are interchannel variability and cross-talk. To eliminate interchannel variability, samples of the same compound are analyzed via the same channel. Compounds with different m/z values are introduced into different channels to reduce cross-talk. Fang et al. described eight-channel[28] and nine-channel[29] multiplexed electrospray systems that further increase sample throughput.

Chovan et al.[30] described a system that integrates different components of bioanalysis including automatic *in vitro* incubation, automatic method development (mainly SRM transitions for LC/MS/MS analysis), and a generic LC method for sample analysis to minimize human intervention and streamline information flow. Automaton software (Applied Biosystems) was used for automatic MS method development. Flow injection was used instead of a HPLC column to decrease run time to 0.8 min per injection. Two injections were performed. The first was performed to locate the precursor ion and optimal declustering potential (DP). The second injection was performed to locate the product ion and optimal collision energy (CE).

Compounds were optimized in positive ionization mode and in negative mode if necessary. Automaton can also perform automatic MS method development from solutions containing multiple compounds to increase throughput. When mixture solutions are used, Automaton injects a mixture once to determine all precursor ions and DP values and then injects once per compound to determine product ion and CE value. This approach allows automatic and unattended optimization of MS parameters for hundreds of compounds. The optimized parameters are stored in a compound database that permits fast and efficient retrieval of information about a specific compound and allows a compound to be used in multiple assays, eliminating the need to re-optimize the LC/MS/MS conditions.

QuanOptimize from Micromass also allows automated method development for quantitative LC/MS/MS. It automatically identifies the best method for each compound, then runs batches of samples for quantitative analyses and report results in a QuanLynx browser. Thermo recently launched a similar product for automatic MS tuning. Known as QuickQuan, it generates data and stores it in a central Microsoft Access or Oracle database for future access. The infusion-based valve switching auto-tuning device allows individual compounds to be fully and automatically optimized in about 1 min.

8.4 HIGH-THROUGHPUT METABOLISM SCREENING

Advances in techniques for chemical synthesis allow medicinal chemists to synthesize hundreds to thousands of compounds per month. Metabolic stability screening in liver microsomes is used extensively in early discovery to select the analogs or compounds most likely to have favorable pharmacokinetic parameters. This provides information on the relation of structure to stability, thus guiding synthesis strategies.

High-throughput laboratories have turned to assay automation, N-in-one (sample pooling) analysis strategies, and elaborate set-ups for parallel chromatography[30–33] to increase capacity and decrease turn-around time. Despite the relatively fast speed of HPLC/MS, this step still creates a bottleneck in ADME work flow. Xu et al.[32] reported a fast method for microsomal sample analysis that yields 231 data points per hour using a complex eight-column HPLC/MS set-up.

Solid phase extraction (SPE) is fast and can work as a clean-up method when combined with capillary electrophoresis or used prior to HPLC/MS.[34–37] Kerns et al.[35] described an online alternating parallel SPE column with MS/MS detection and a turn-around time of 1.1 min.

Inman et al.[38] demonstrated elution of metabolic stability samples directly from SPE cards into an MS/MS set-up that can circumvent the lengthy HPLC runs used in routine ADME analysis. The SPE card consists of a C18 resin bed between two non-woven polypropylene filter layers arranged in a standard 96-well format. Each zone is delineated by an O-ring seal. The samples from a 96-well plate are loaded onto individual zones of SPE cards using an array of 96 needles on a Harvex component of the SPExpress system. The loaded cards are processed on an Elutrix component that has the capacity to elute samples from 35 cards by automatic in-feeding of cards and elution of individual zones into the MS. Two opposing pistons on the Elutrix seal the O-rings of individual zones on the card. The mobile phase is then passed through the zone to elute all material from the resin; elution time is 24 sec. This system achieved the shortest injection-to-injection time (33 sec) reported in the literature. The authors also combined this technique with sample pooling and achieved an acquisition rate of 480 data points in 1 hr on a single MS.

8.5 SOLUBILITY

Physicochemical profiling at the early discovery stage is important in the pharmaceutical industry because poor bioavailability is a leading factor in compound attrition. The ability to rapidly measure absorption properties such as solubility, log P, and log D allows promising compounds to quickly pass into exploratory development.

Aqueous solubility is a critical characteristic simply because an oral drug must dissolve in the gastrointestinal tract before it can be absorbed. In the discovery phase, it is important to determine whether promising compounds show sufficient solubility in the solvents used in various biological screening assays and during dosing. Solubility parameters also aid in formulating potential candidates in later stages. Lipinski[39] described the causes of poor drug solubility and Kerns[40] detailed high-throughput screening for drug-like properties.

Traditionally, thermodynamic solubility has been determined by shaking a compound in generic pH buffers for at least 24 hr followed by filtering and/or centrifuging. The concentration of dissolved compound is measured by a suitable analytical assay using UV, nephelometry, or other high capacity plate reading equipment. The concentrations of compounds in the buffer are determined against a calibration curve. The throughput of this approach is no longer high enough to meet the demands of modern drug discovery.

Laser nephelometry is based on measuring the turbidity of an aqueous medium after adding a fixed amount of solution of a compound in dimethyl sulfoxide (DMSO).[41] Dehring et al.[42] described an automated robotic system with laser-based nephelometry for high-throughput kinetic aqueous solubility measurements. The limitation of this method is the inability to measure solubility in pure aqueous media without DMSO. Pan et al.[43] compared the solubility measuring capabilities of chromatographic, UV/vis, and nephelometer plate readers. The solubilities determined by the three methods correlated well, suggesting that UV/vis and nephelometric plate readers could replace HPLC for high-throughput determinations of solubility.

HPLC coupled with an UV detector[44] constitutes the conventional method for measuring solubility. It provides moderate sample throughput for evaluation of lead compounds. To increase the throughput, a multi-wavelength UV plate reader and disposable 96-well UV plates are used for fast solubility determinations. This system analyzes 96 samples in a single step, thus significantly increasing sample throughput. Chen et al. and Pan et al. demonstrated that this method has the sensitivity and reproducibility to effectively determine solubilities as low as 1 µM. In addition to excellent sensitivity and reproducibility, the UV plate reader method also offers the flexibility of determining thermodynamic solubility with or without DMSO—a solvent widely used for high-throughput screening of combinatorial compounds.

Chemiluminescent nitrogen detection (CLND) has been applied for high-throughput solubility measurements. CLND measures the nitrogen contents of samples of interest (Figure 8.4). The sample enters the nitrogen detector by direct flow injection or as an eluent from an HPLC column, then enters the pyrolysis tube through a nebulizer where it mixes with a blend of helium and oxygen to form a fine aerosol spray. The sample is completely pyrolyzed at 1050°C and the nitrogen in the sample is converted to nitric oxide that reacts with ozone to produce nitrogen dioxide in an excited state (NO_2^*), then decays to the ground state with the release of photons. The photons are captured and amplified in a photomultiplier tube. The number of photons released (chemiluminescent response) is proportional to the nitrogen content of the sample. However, compounds that contain adjacent nitrogen atoms such as N–N, N=N, and N≡N are converted to molecular nitrogen upon combustion; molecular nitrogen is not measured by CLND. Therefore, with the exception of compounds containing adjacent nitrogens, the response of the detector is equimolar with respect to nitrogen. Using a CLND signal and a known number of nitrogens per molecule, the concentration of a compound can be determined via a generic nitrogen calibration curve.

Bhattachar et al.[45] used CLND for solubility determinations and compared results to those obtained from UV spectrophotometry and HPLC. CLND has a throughput of 96 compounds per day with a reduced compound consumption of approximately 3 mg. The sensitivity of the instrument is approximately 6.25 µg/mL for a compound with a molecular weight of 350 and 4 nitrogens per molecule.

Typically, MS does not play an important role in solubility measurement because UV and nephelometric plate readers provide high throughput and are sufficiently sensitive for most compounds.

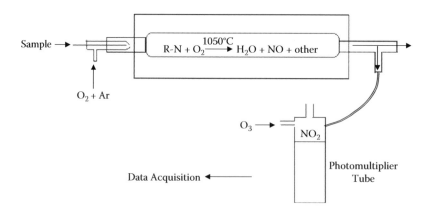

FIGURE 8.4 Chemiluminescent nitrogen detector.

MS has recently been used to measure compounds with significant levels of impurities and solubilities below the quantitation limits of other methods. Guo et al.[46] described the use of LC/MS for solubility measurements in buffer solutions in a 96-well plate. Fligge et al.[47] discussed an automated high-throughput method for classification of compound solubility. They integrated a Tecan robotic system for sample preparation in 384-well plates and fast LC/MS for concentration measurement. This approach is limited by LC/MS throughput.

8.6 HIGH-THROUGHPUT *IN VITRO* DRUG–DRUG INTERACTIONS

8.6.1 Inhibition

Drug–drug interactions have always been major concerns to the pharmaceutical industry. Several prominent drugs were withdrawn from the market because of serious adverse events related to drug–drug interactions. These interactions create problems for clinicians and patients and economic losses for pharmaceutical manufacturers. For this reason, pharmaceutical companies screen for enzyme inhibition and induction at the discovery stage.

Human liver microsomes (HLMs) are the most common *in vitro* sources of enzymes for inhibition studies, and selective probe substrates are required. Recombinant human P450 enzymes have become commercially available. They are widely used for screening, and less selective probe substrate can be used. Hepatocytes and liver slices[48] have also been used for P450 inhibition screening to a lesser extent.

Four high-throughput inhibition assays are common in the pharmaceutical industry: fluorescence,[49–52] bioluminescence,[53] radiometry,[54–57] and LC-MS/MS. The extensive use of microtiter plate-based fluorometric assays[49–52] in early drug discovery is based on their high throughput (hundreds of compounds per day), capacity, speed, sensitivity, and low maintenance. Advances in microtiter plate reader equipment allows a 96-well plate to be read in about 1 min. Naritomi et al.[52] reported an inhibition assay that can detect quasi-irreversible and irreversible inhibitors using fluorometric substrates. Jenkins et al.[58] described a high-throughput inhibition assay for CYP3A4 and CYP2D6 by combining liquid handlers and an integrated fluorescence plate reader. However, the fluorescent probe substrates are not specific for each CYP isozyme. Expressed enzymes are used for screening instead of human liver microsomes. Trubetskoy et al.[59] described an ultra-high-throughput assay using a 1536-well plate and Vivid® fluorescent substrates with recombinant human cytochrome P450.

Bioluminescence methods[53] are based on substrates that release luciferin as a metabolite. The addition of luciferase and ATP converts the freed luciferin to des-carboxyluciferin with light emission and the signal is detected with a luminescence plate reader. This assay requires the use of recombinant P450 enzymes because the probes are not specific. Compounds that interfere with light generation (luciferase enzymatic activity) may result in false positives. High precision liquid handling allows this assay to be scaled to low volume 384- and 1536-well formats.

A fully automated inhibition screen for the major human hepatic cytochromes P450 3A4, 2D6, 2C9, and 2C19 using radiometric analysis as described by Moddy et al.[57] Radiometric assays involve the use of a ^{14}C-labelled substrate and are based on CYP-catalyzed dealkylation with subsequent measurement of the radioactivity of the formaldehyde formed. Internal standard and HPLC separation are not needed. To increase throughput, a two-point IC_{50} estimate was used for initial screening instead of a full seven-point assay. The radiometric assay is relatively simple and sensitive (5.0 pmol HCHO/hr/mg microsomal protein).[54] The disadvantages of this assay relate to safety when using radioactive material for HTS. An extraction step before analysis is required before reactivity measurement slows the process, and only a limited number of CYPs can be screened due to the need for dealkylation. Also, the less specific probes used for CYP2C9 and CYP2C19 inhibition screening render the use of expressed enzymes.

High-throughput LC/MS/MS has high specificity and sensitivity. Each substrate is specific for each enzyme. Therefore, different mixed isozyme sources such as HLMs can be used. The specific metabolite generated is selectively detected. This allows a "cocktail" approach with multiple probe substrates. To increase throughput, cassette incubation[60–62] approaches were reported recently. A cocktail of specific drug-probe substrates is used with HLM and the signal due to each substrate metabolite is independently monitored using the specificity of LC/MS/MS. This approach increases throughput by determining inhibition of several isozymes simultaneously.

Di et al.[63] compared inhibition assays using HLM with LC/MS, a cocktail with LC/MS, and recombinant CYP450 with fluorescence. Higher IC_{50} values were observed with HLM/LC/MS as compared to fluorescent assay. Data from the cocktail approach correlate better with the fluorescent assay. These differences in assays are due to detection techniques, substrates used, enzyme sources, composition, and concentration. Based on recent progress in computational methods for prediction of P450 compound interactions, *in silico* screening[64,65] has become a desirable tool. A single stroke of a keypad can start screening of a large number of virtual compounds for P450 liabilities.

8.6.2 INDUCTION

Induction of cytochrome P450 is undesirable because of its link to tumor formation and poor drug exposure due to autoinduction. The most accepted method for studying P450 induction uses primary human hepatocytes (liver cells) in which mRNA levels or P450 activity can be measured. Cryopreserved human hepatocytes that can be plated are convenient and now available. However, induction assays are very expensive, time consuming, and subject to the availability of donor organs for obtaining primary human hepatocytes.

A medium-throughput 96-well reporter assay relies on immortalized human hepatoma cells (HepG2) that are transiently transfected with two plasmids, one carrying the PXR gene and the other containing CYP3A4 xenobiotic response modules upstream of the luciferase gene. PXR activation is measured by luminescence and compared to the fold vehicle (DMSO) response. If a compound is active, an EC_{50} is calculated. The responses to rifampicin and efavirenz, two positive controls that are also clinically relevant inducers, are measured on each plate.

8.7 BIOACTIVATION SCREENING

Liver injuries induced by drugs now constitute the major causes of acute liver failure. If liver transplant is not possible, deaths result. Liver injury is also the leading reason ($>50\%$)[66,67] that drugs are withdrawn from the market. Alternatively, their use is restricted and special monitoring of patients is required. Bioactivation of a drug to electrophiles and free radicals and subsequent covalent binding of the drug to proteins and nucleic acids[68,69] is one mechanism that produces liver injuries. However, drugs possessing functionalities susceptible to bioactivation are not always bioactivated and bioactivation does not always cause hepatotoxicity.[67,70] Because of complexity, adverse drug reactions cannot be predicted from preclinical toxicological assessments. The pharmaceutical industry is trying to implement higher throughput methods to screen for possible formation of reactive metabolites.

The current method for identifying drug candidates that form reactive intermediates is incubation with liver microsomes in the presence of a nucleophile (trapping agent) such as N-acetylcysteine,[71] glutathione,[71–74] and its glutathione ethyl ester[75] and dansyl glutathione[76] derivatives. The resulting conjugates may be detected and quantitated by fluorescence and their structures confirmed by MS after fluorescence detection[76] (long analysis time of 50 min). More commonly, GSH conjugates are detected by LC/tandem MS[77–80] or ion trapping[81] because of their high sensitivity and selectivity and relatively short run times.

Soglia et al.[75] described a more sensitive method using glutathione ethyl ester as an *in vitro* conjugating agent and microbore LC/microelectrospray ionization/tandem MS. This method requires knowledge of biotransformation mechanisms and the structures of GSH conjugates formed; total cycle time is 15 min per sample. Castro-Perez et al.[82] also described the use of exact mass neutral loss for screening glutathione conjugates provides better selectivity and confidence in compound confirmation and identification.

Dieckhaus et al.[83] reported a more general method of detecting unknown GSH conjugates by using negative precursor ion survey scans. This approach detected GSH conjugates that could not have been detected by neutral loss experiments. Another approach for rapid detection and characterization of minor reactive metabolites was described by Yan et al.[84] They used stable isotope-labeled glutathione (γ-glutamyl-cysteinyl-glycine-$^{13}C_2$-^{15}N) as a trapping agent in combination with

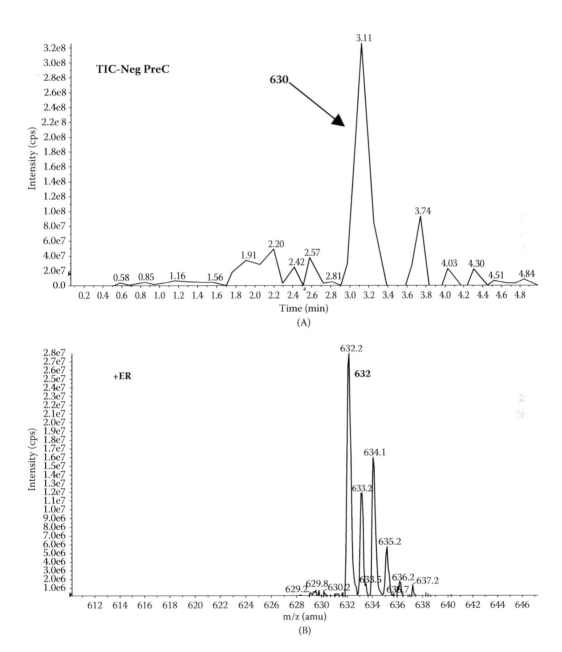

FIGURE 8.5 Simultaneous detection and identification of clozapine–GSH adduct in one single injection: (A) mass spectrum from negative precursor ion scan of 272 detected the clozapine–GSH adduct at m/z 630; (B) mass spectrum from positive enhanced resolution scan finds the clozapine–GSH adduct at m/z 632; (C) enhanced product ion spectrum of clozapine–GSH adduct at m/z 632.

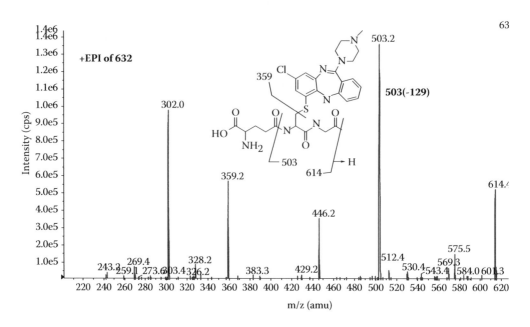

FIGURE 8.5 (Continued).

tandem MS. Reactive metabolites exhibited isotopic doublets (3 amu difference) and were detected rapidly.

Another high-throughput screening technique was presented by Lau[85] (data not published) at the 2007 Pittsburgh Conference. This technique has a turn-around time of 5 min per sample. It uses a negative precursor ion scans of 272 to detect the GSH adducts followed by an enhanced resolution scan for charge state and isotope confirmation. Information-dependent enhanced product ion scan is then used for structural identification. All these functions are handled within a single run. Figure 8.5 shows the MS information collected to detect GSH adducts of clozapine in a 5-min run.

REFERENCES

1. Lloyd, T.L. et al. 2006. Laboratory automation for compound optimization and early development drug metabolism: A Wyeth case study. *Am. Drug Discov.* 1: 27.
2. Herbst, J.J. and Dickinson, K. 2005. Automated high-throughput ADME/Tox profiling for optimization of preclinical candidate success. *Am. Pharm. Rev.* 8: 96.
3. DeWitte, R.S. and Robins, R.H. 2006. A hierarchical screening methodology for physicochemical/ ADME/Tox profiling. *Expert Opin. Drug Metab. Toxicol.* 2: 805.
4. MDL Information Systems, San Leandro, CA. www.mdli.com.
5. IDBS, Guilford, UK. www.idbs.co.uk.
6. CambridgeSoft.Com, Cambridge, MA. www.camsoft.com.
7. Cheminnovation Software Inc., San Diego, CA. www.cheminnovation.com.
8. Accelrys Inc., San Diego, CA. www.accelrys.com.
9. Lau, Y.Y. et al. 2004. Evaluation of a novel *in vitro* Caco-2 hepatocyte hybrid system for predicting *in vivo* oral bioavailability. *Drug Met. Disp.* 32: 937.
10. Lau, Y.Y. et al. 2002. The use of *in vitro* metabolic stability for rapid selection of compounds in early discovery based on their expected hepatic extraction ratios. *Pharm. Res.* 19: 1606.
11. Lau, Y.Y. et al. 2002. Development of a novel *in vitro* model to predict hepatic clearance using fresh, cryopreserved, and sandwich-cultured hepatocytes. *Drug Met. Disp.* 30: 1446.
12. Chu, I. et al. 2000. Validation of higher throughput high-performance liquid chromatography/atmospheric pressure chemical ionization tandem mass spectrometry assays to conduct cytochrome P450s

CYP2D6 and CYP3A4 enzyme inhibition studies in human liver microsomes. *Rapid Commun. Mass Spectrom.* 14: 207.

13. Fung, E.N., Chen, Y.H., and Lau, Y.Y. 2003. Semi-automatic high-throughput determination of plasma protein binding using a 96-well plate filtrate assembly and fast liquid chromatography/tandem mass spectrometry. *J. Chromgr. B.* 795: 187.

14. Korfmacher, W.A. et al. 1999. Development of an automated mass spectrometry system for the quantitative analysis of liver microsomal incubation samples: A tool for rapid screening of new compounds for metabolic stability. *Rapid Commun. Mass Spectrom.* 13: 901.

15. Deng, Y. et al. 2002. High-speed gradient parallel liquid chromatography/tandem mass spectrometry with fully automated sample preparation for bioanalysis: 30 seconds per sample from plasma. *Rapid Commun. Mass Spectrom.* 16: 1116.

16. Yee, S. 1997. *In vitro* permeability across Caco-2 cells (colonic) can predict *in vivo* (small intestinal) absorption in man: Fact or myth. *Pharm. Res.* 14: 763.

17. Bu, H.Z. et al. 2000. High-throughput caco-2 cell permeability screening by cassette dosing and sample pooling approaches using direct injection/on-line guard cartridge extraction/tandem mass spectrometry. *Rapid Commun. Mass Spectrom.* 14: 523.

18. McLoughlin, D.A., Olah, T.V., and Gilbert, J.D. 1997. A direct technique for the simultaneous determination of 10 drug candidates in plasma by liquid chromatography/atmospheric pressure chemical ionization mass spectrometry interfaced to a Prospekt solid-phase extraction system. *J. Pharm. Biomed. Anal.* 15: 1893.

19. Van der Hoeven, R.A. et al. 1997. Liquid chromatography/mass spectrometry with on-line solid-phase extraction by a restricted-access C18 precolumn for direct plasma and urine injection. *J. Chromatogr. A.* 762: 193-200.

20. Olah, T.V., McLoughlin, D.A., and Gilbert, J.D. 1997. Simultaneous determination of mixtures of drug candidates by liquid chromatography/atmospheric pressure chemical ionization mass spectrometry as an *in vivo* drug screening procedure. *Rapid Commun. Mass Spectrom.* 11: 17.

21. Berman, J. et al. 1997. Simultaneous pharmacokinetic screening of a mixture of compounds in the dog using API LC/MS/MS analysis for increased throughput. *J. Med. Chem.* 40: 827.

22. Cox, K.A. et al. 1999. Novel procedure for rapid pharmacokinetic screening of discovery compounds in rats. *Drug Discov. Today* 4: 232.

23. King, R.C. et al. 2002. Description and validation of a staggered parallel high performance liquid chromatography system for good laboratory practice level quantitative analysis by liquid chromatography/tandem mass spectrometry. *Rapid Commun. Mass Spectrom.* 16: 43.

24. Yang, L., Wu, N., and Rudewicz, P.J. 2001. Applications of new liquid chromatography/tandem mass spectrometry technologies for drug development support. *J. Chromatogr. A.* 926: 43.

25. Yang, L. et al. 2001. Evaluation of a four-channel multiplexed electrospray triple quadrupole mass spectrometer for the simultaneous validation of LC/MS/MS methods in four different preclinical matrixes. *Anal. Chem.* 73: 1740.

26. Morrison D., Davies, A.E., and Watt, A.P. 2002. An evaluation of a four-channel multiplexed electrospray tandem mass spectrometry for higher throughput quantitative analysis. *Anal. Chem.* 74: 1896.

27. Fung, E.N. et al. 2003. Higher-throughput screening for Caco-2 permeability utilizing a multiple sprayer liquid chromatography/tandem mass spectrometry system. *Rapid Commun. Mass Spectrom.* 17: 2147.

28. Fang, L. et al. 2002. High-temperature ultrafast liquid chromatography. *Rapid Commun. Mass Spectrom.* 16: 1440.

29. Fang, L. et al. 2003. Parallel high-throughput accurate mass measurement using a nine-channel multiplexed electrospray liquid chromatography ultraviolet time-of-flight mass spectrometry system. *Rapid Commun. Mass Spectrom.* 17: 1425.

30. Chovan, L.E. et al. 2004. Automatic mass spectrometry method development for drug discovery: Application in metabolic stability assays. *Rapid Commun. Mass Spectrom.* 18: 3105.

31. Jenkins, K.M. et al. 2004. Automated high throughput ADME assays for metabolic stability and cytochrome P450 inhibition profiling of combinatorial libraries. *J. Pharm. Biomed. Anal.* 34: 989.

32. Xu, R. et al. 2002. Application of parallel liquid chromatography/mass spectrometry for high throughput microsomal stability screening of compound libraries. *J. Am. Soc. Mass Spectrom.* 13: 155.

33. Wachs, T. and Henion, J. 2003. A device for automated direct sampling and quantitation from solid-phase sorbent extraction cards by electrospray tandem mass spectrometry. *Anal. Chem.* 75: 1769.

34. Visser, N.F. et al. 2003. On-line SPE-CE for the determination of insulin derivatives in biological fluids. *J. Pharm. Biomed. Anal.* 33: 451.

35. Kerns, E.H. et al. 2004. Integrated high capacity solid phase extraction-MS/MS system for pharmaceutical profiling in drug discovery. *J. Pharm. Biomed. Anal.* 34: 1.

36. Qi, L. and Danielson, N.D. 2005. Quantitative determination of pharmaceuticals using nano-electrospray ionization mass spectrometry after reversed phase mini-solid phase extraction. *J. Pharm. Biomed. Anal.* 37: 225.

37. Davies, I.D., Allanson, J.P., and Causon, R.C. 1999. Rapid determination of the anti-cancer drug chlorambucil (Leukeran) and its phenyl acetic acid mustard metabolite in human serum and plasma by automated solid-phase extraction and liquid chromatography-tandem mass spectrometry. *J. Chromatogr. B.* 732: 173.

38. Inman, B.L. et al. 2006. Solid phase extraction as a faster alternative to HPLC: Application to MS analysis of metabolic stability samples. *J. Pharm.Sci.* 2006, Nov. 8, Epub., ahead of print.

39. Lipinski, C.A. 2000. Drug-like properties and the causes of poor solubility and poor permeability. *J. Pharm. Tox. Meth.* 44: 235.

40. Kerns, E.H. 2001. High throughput physicochemical profiling for drug discovery. *J. Pharm. Sci.* 90: 1838.

41. Bevan, C.D. and Lloyd, R.S. 2000. A high-throughput screening method for the determination of aqueous drug solubility using laser nephelometry in microtiter plates. *Anal. Chem.* 72: 1781.

42. Dehring, K.A. et al. 2004. Automated robotic liquid handling/laser-based nephelometry system for high throughput measurement of kinetic aqueous solubility. *J. Pharm. Biomed. Anal.* 36: 447.

43. Pan, L. et al. 2001. Comparison of chromatographic and spectroscopic methods used to rank compounds for aqueous solubility. *J. Pharm. Sci.* 90: 521.

44. Li, P., Tabibi, S.E., and Yalkowsky, S.H. 1998. Combined effect of complexation and pH on solubilization. *J. Pharm. Sci.* 87: 1535.

45. Bhattachar, S.N., Wesley, J.A., and Seadeek, C. 2006. Evaluation of the chemiluminescent nitrogen detector for solubility determinations to support drug discovery. *J. Pharm. Biomed. Anal.* 41: 152.

46. Guo, Y. and Shen, H. 2004. In *Optimization in Drug Discovery*; Yan, S. and Caldwell, G., Eds., Humana Press, Totowa, NJ, p. 1.

47. Fligge, T.A. and Schuler, A. 2006. Integration of a rapid automated solubility classification into early validation of hits obtained by high throughput screening. *J. Pharm. Biomed. Anal.* 42: 449.

48. Lin, J.H. and Rodrigues, A.D. 2001. *In vitro* models for early studies of drug metabolism. In *Pharmacokinetic Optimization in Drug Research: Biological, Physicochemical and Computational Strategies*. Testa, B. et al., Eds., Wiley, New York, p. 217.

49. Kennedy, S.W. and Jones, S.P. 1994. Simultaneous measurement of cytochrome P4501A catalytic activity and total protein concentration with a fluorescence plate reader. *Anal. Biochem.* 222: 217.

50. Donateo, M.T., Gomez-Lechon, M.J., and Castell, J.V. 1993. A microassay for measuring cytochrome P450IA1 and P450IIB1 activities in intact human and rat hepatocytes cultured on 96-well plates. *Anal. Biochem.* 213: 29.

51. Crespi, C.L., Miller, V.P., and Penman, B.W. 1997. Microtiter plate assays for inhibition of human, drug-metabolizing cytochromes P450. *Anal. Biochem.* 248: 188.

52. Naritomi, Y. et al. Utility of microtiter plate assays for human cytochrome P450 inhibition studies in drug discovery. *Drug Metab. Pharmacokin.* 19: 55.

53. Cali, J.J. et al. 2006. Luminogenic cytochrome P450 assays. *Expert Opin. Drug Metab. Toxicol.* 2: 629.

54. Rodrigues, A.D. et al. 1994. Measurement of liver microsomal cytochrome p450 (CYP2D6) activity using [O-methyl-14C]dextromethorphan. *Anal. Biochem.* 219: 309.

55. Rodrigues, A.D. et al. 1996. [O-methyl 14C]naproxen O-demethylase activity in human liver microsomes: Evidence for the involvement of cytochrome P4501A2 and P4502C9/10. *Drug Metab. Dispos.* 24: 126.

56. Rodrigues, A.D. et al. 1997. [O-ethyl 14C]phenacetin O-deethylase activity in human liver microsomes. *Drug Metab. Dispos.* 25: 1097.

57. Moody, G.C. et al. 1999. Fully automated analysis of activities catalysed by the major human liver cytochrome P450 (CYP) enzymes. *Xenobiotica.* 29: 53.

58. Jenkins, K.M. et al. 2004. Automated high throughput ADME assays for metabolic stability and cytochrome P450 inhibition profiling of combinatorial libraries. *J. Pharm. Biomed. Anal.* 34: 989.

59. Trubetskoy, O.V., Gibson, J.R., and Marks, B.D. 2005. Highly miniaturized formats for *in vitro* drug metabolism assays using vivid fluorescent substrates and recombinant human cytochrome P450 enzymes. *J. Biomol. Screen.* 10: 56.

60. Testino, A.S., Jr. and Patonay, G. 2003. High-throughput inhibition screening of major human cytochrome P450 enzymes using an *in vitro* cocktail and liquid chromatography/tandem mass spectrometry. *J. Pharm. Biomed. Anal.* 30: 1459.

61. Weaver, R. et al. 2003. Cytochrome P450 inhibition using recombinant proteins and mass spectrometry/multiple reaction monitoring technology in a cassette incubation. *Drug Metab. Dispos.* 31: 955.

62. Kim, M.J. et al. 2005. High-throughput screening of inhibitory potential of nine cytochrome P450 enzymes *in vitro* using liquid chromatography/tandem mass spectrometry. *Rapid Commun. Mass Spectrom.* 19: 2651.

63. Di, L. et al. 2006. Comparison of cytochrome P450 inhibition assays for drug discovery using human liver microsomes with LC-MS, rhCYP450 isozymes with fluorescence, and double cocktail with LC/S. *Int. J. Pharmaceut.* http: //dx.doi.org/10.1016/j.ijpharm.2006.10.039.

64. De Groot, M.J., Kirton, S.B. and Sutcliffe, M.J. 2004. *In silico* methods for predicting ligand binding determinants of cytochromes P450. *Curr. Top. Med. Chem.* 4: 1803.

65. De Graaf, C., Vermeulen, N.P., and Feenstra, K.A. 2005. Cytochrome p450 *in silico*: An integrative modeling approach. *J. Med. Chem.* 48, 2725.

66. Lee, W.M. 2003. Drug-induced hepatotoxicity. *New Engl. J. Med.* 349: 474.

67. Park, B.K. et al. 2005. The role of metabolic activation in drug-induced hepatotoxicity. *Annu. Rev. Pharmacol. Toxicol.* 45: 177.

68. Brodie, B.B. et al. 1971. Possible mechanism of liver necrosis caused by aromatic organic compounds. *Proc. Natl. Acad. Sci. USA* 68: 160.

69. Gillette, J.R. 1994. Commentary. Perspective on the role of chemically reactive metabolites of foreign compounds in toxicity. I. Correlation of changes in covalent binding of reactivity metabolites with changes in the incidence and severity of toxicity. *Biochem. Pharmacol.* 23, 2785.

70. Kalgutkar, A.S. et al. 2005. A comprehensive listing of bioactivation pathways of organic functional groups. *Curr. Drug Metab.* 6: 161.

71. Ju, C. and Uetrecht, J.P. 1998. Oxidation of a metabolite of indomethacin (desmethyldeschlorobenzoyl indomethacin) to reactive intermediates by activated neutrophils, hypochlorous acid, and the myeloperoxidase system. *Drug Metab. Dispos.* 26: 676.

72. Mays, D.C. et al. 1995. Metabolism of phenytoin and covalent binding of reactive intermediates in activated human neutrophils. *Biochem. Pharmacol.* 50: 367-80.

73. Roy, D. and Snodgrass W.R. 1990. Covalent binding of phenytoin to protein and modulation of phenytoin metabolism by thiols in A/J mouse liver microsomes. *J. Pharmacol. Exp. Ther.* 252: 895.

74. Coles, B. and Ketterer, B. 1990. The role of glutathione and glutathione transferases in chemical carcinogenesis. *Crit. Rev. Biochem. Mol. Biol.* 25: 47.

75. Soglia, J.R. et al. 2004. The development of a higher throughput reactive intermediate screening assay incorporating microbore liquid chromatography–microelectrospray ionization–tandem mass spectrometry and glutathione ethyl ester as an *in vitro* conjugating agent. *J. Pharm. Biomed. Anal.* 36: 105.

76. Gan, J. et al. 2005. Dansyl glutathione as a trapping agent for the quantitative estimation and identification of reactive metabolites. *Chem. Res. Toxicol.* 18: 896.

77. Baillie, T.A. and Davis, M.R. 1993. Mass spectrometry in the analysis of glutathione conjugates. *Biol. Mass Spectrom.* 1993, 22, 319.

78. Chen, W.G. et al. 2001. Reactive metabolite screen for reducing candidate attrition in drug discovery. *Adv. Exp. Med. Biol.* 500: 521.

79. Tang, W. and Abbott, F.S. 1996. Characterization of thiol-conjugated metabolites of 2-propylpent-4-enoic acid (4-ene VPA), a toxic metabolite of valproic acid, by electrospray tandem mass spectrometry. *J. Mass Spectrom.* 31: 926.

81. Thomassen, D. et al. 1991. Partial characterization of biliary metabolites of pulegone by tandem mass spectrometry: Detection of glucuronide, glutathione, and glutathionyl glucuronide conjugates. *Drug Metab. Dispos.* 1991, 19, 997.

82. Castro-Perez, J. et al. 2005. A high-throughput liquid chromatography/tandem mass spectrometry method for screening glutathione conjugates using exact mass neutral loss acquisition. *Rapid Commun. Mass Spectrom.* 19: 798.

83. Dieckhaus, C.M. et al. 2005. Negative ion tandem mass spectrometry for the detection of glutathione conjugates. *Chem. Res. Toxicol.* 18: 630.

84. Yan, Z. et al. 2005. Rapid detection and characterization of minor reactive metabolites using stable-isotope trapping in combination with tandem mass spectrometry. *Rapid Commun. Mass Spectrom.* 19: 3322.

85. Lau Y.Y. 2007. High throughput simultaneous detection and identification of reactive intermediates using negative precursor ion scans combined with positive product ion scans. Presented at PittCon 2007.

9 High-Throughput Analysis in Support of Process Chemistry and Formulation Research and Development in the Pharmaceutical Industry

Zhong Li

CONTENTS

9.1 INTRODUCTION

Today's pharmaceutical industry faces the unprecedented challenge of surviving and succeeding in an increasingly complex, competitive, global business environment in which research and development (R&D) costs have skyrocketed and development and regulatory approval times have continuously lengthened despite flat revenue growth. According to analyses released by the Tufts Center for the Study of Drug Development, the fully capitalized cost to develop a single novel pharmaceutical (including studies conducted after receiving regulatory approval) averages $897 million (2003 dollars), and it takes 10 to 15 years to secure market approval of a new drug.[1] The situation has necessitated pharmaceutical companies to actively seek innovative ways to decrease costs, increase productivity, and enhance profitability in all phases of drug discovery and development and introduce new products that customers value.

In the small molecule drug discovery arena, the major pharmaceutical companies have invested heavily in genomics and proteomics, bioinformatics and computational modeling, combinatorial chemistry and high-speed synthesis, and high-throughput screening (HTS). These efforts led to a dramatic acceleration in the discovery of molecules as preclinical candidates (PCCs) for preclinical and clinical development. While the era of new technologies has contributed to the advancement of understanding of biology and medicinal chemistry in drug discovery, challenges remain in the drug development sector where speed and efficiency are primary strategic objectives due to the high direct cost of development and the substantial opportunity cost of delay in bringing a drug to market. It is estimated that a 1-day advantage typically saves $37,000 in out-of-pocket development costs and nets an additional $1.1 million in daily prescription revenue for an average performing drug according to a Tufts study.[2] The pressure is on industry to simultaneously cut costs, improve standards of quality, and shorten product development times required to get drugs on the market. The increased pressure to deliver improved returns to shareholders is driving various efficiency improvements related to all aspects of pharmaceutical development and manufacturing.

As vital components of drug development, chemical process R&D (CPR&D) and pharmaceutical process R&D (PhR&D) run in parallel from the preclinical phase forward and any successful shortening of drug development time must involve increased efficiency in both areas. The role of CPR&D is delivering a drug substance [active pharmaceutical ingredient (API)] suitable for preclinical and clinical studies in addition to designing practical, efficient, environmentally responsible, and economically viable chemical syntheses. PhR&D has the tasks of defining optimal stable and bioavailable formulations that allow evaluation of new chemical entities (NCEs) in humans and developing robust, efficient commercial manufacturing processes for drug products (formulated products or finished dosage forms). An increasingly important step in formulation development is the thorough physicochemical, mechanical, and biopharmaceutical characterization of PCCs conducted by a preformulation unit.

As the race to develop NCEs with novel pharmacological activities heats up, scientists in both CPR&D and PhR&D currently face the challenge of maintaining increased efficiency and productivity while contending with a deluge of new PCCs of increasing complexity. They have responded by developing innovative new technologies, such as high-throughput experimentation (HTE) techniques that dramatically increase the number of experiments by downscaling and parallelizing experiments for R&D work on a laboratory scale and allowing rapid and extensive investigation of far more parameters. Over the past 10 years, HTE has increasingly been applied to synthetic route exploration, optimization, and scale-up; polymorph screening and salt selection studies; solubility and pK_a measurements; forced degradation studies and stress testing of pharmaceuticals; and liquid and/or semisolid formulation screening for poorly soluble compounds. The application of these massive automated parallel approaches to experimentation in R&D has produced tremendously increased demand for analytical throughput and placed significant pressure on the analytical functions of CPR&D and PhR&D to provide timely decision-making information associated with the characterization of the processes under development. Analytical chemistry is now recognized as a potential bottleneck in HTE and analytical scientists must develop high-throughput analytical techniques and strategies to respond expeditiously to R&D process demands.

In addition to the support of process R&D, a key element in the role of pharmaceutical analysis in drug development is to ensure that the development and preparation of drug substances and drug products for preclinical and clinical studies meet the good manufacturing practice (GMP) requirements, i.e., they ensure the safety, identity, strength, quality, and purity of the drug product. These tasks include the testing and release of raw materials, the characterization and testing of synthetic intermediates and drug substances and controlling the chemical reactions leading to these chemical entities; physical, chemical, and microbiological characterization and testing of excipients, drug products, and packaging components; process cleaning validation; and the assessment of the stability and monitoring of the quality of product from release through shelf life.

Today's pharmaceutical analytical scientists are feeling the same economic pressure as their colleagues to generate accurate, reliable data of the highest possible quality for high volumes of samples while dealing with constantly decreasing time limits and issues of cost, safety, and efficiency. Accomplishment of these multiple goals can be facilitated only by innovations in technologies and methodologies to achieve faster, simpler, higher performance and more cost-effective analytical solutions. It should be emphasized that any new technology must also meet stringent regulatory standards for validation and documentation before its full benefits can be realized in this heavily regulated industry.

In the past two decades, new technologies in analytical chemistry have continually evolved to meet the demands for high-throughput analysis (HTA). Examples are fast chromatography and parallel separation, automation and robotics, novel detection systems, chemometrics and process analytic technologies (PAT), and miniaturization, among others. Although a vast amount of HTA information appears in the literature, it is difficult to find a systematic discussion of its application to CPR&D and PhR&D. In this chapter, the state of the art of high-throughput analysis in support of drug substance and drug product development and manufacturing for pre-IND, IND, and NDA requirements is critically reviewed with the aim of describing the most practical and effective approaches currently in use. The intention is not to cover the fundamental aspects and provide a detailed description of each HTA technique. Rather, the author will focus on practical applications by reviewing the most recent literature and citing references to books and review articles devoted to specific techniques when appropriate. The author strives to offer a sense of the general trends in high-throughput pharmaceutical analysis. The important elements of the various techniques are compared and the advantages and limitations of each technique will be discussed along with insight into future trends and developments.

9.2 HIGH-THROUGHPUT SEPARATION-BASED TECHNIQUES

Separation-based techniques, especially high-performance liquid chromatography (HPLC) and gas chromatography (GC), have long been the work horses of pharmaceutical analysis laboratories. They are among the most powerful and versatile tools for the detection and quantitation of analytes (chemical components) in complex matrices frequently encountered in the course of PhR&D.

The dominance of HPLC and GC in pharmaceutical laboratories is based on their excellent selectivity and sensitivity and to their ability to run automated and unattended analyses. However, one of drawbacks is that chromatographic analyses tend to have relatively long turn-around times, especially for GMP and GLP samples, for which a series of system suitability (SS) check injections must be performed. A typical full sequence consists of a blank, SS standards, working standards, and sample injections and can often include more than 10 injections. For a typical 30- to 60-min HPLC or GC procedure including column re-equilibration, the total cycle time easily exceeds 6 hr. Obtaining results even for a single sample during the same workday is a challenging task—and highly desirable for enabling process development groups to make decisions to fine-tune their experiments. A long sequence involving multiple samples analyzed with a long gradient HPLC or GC method can run for several days. Normally, analyses are set to run overnight to take advantage of autosamplers. However, it is not an uncommon scenario for an analyst to arrive the next morning and discover that the results are unacceptable due to a failure in SS that requires system adjustment. Because rapid analysis produces fewer disruptions in work flow since samples can be quickly re-run if an error occurs, it is not surprising that much effort has been devoted to the development of high-throughput chromatographic techniques.

9.2.1 Fast Liquid Chromatography

Today, fast liquid chromatography commonly refers to rapid HPLC analyses of 2- to 5-minute (fast LC) to sub-minute (ultra-fast LC) run times. The concepts, applications, and benefits of fast or ultra-fast LC are well documented.[3–5] Based on several obvious advantages including increased sample throughput and productivity, reduced solvent consumption, and enhanced mass sensitivity, the application of fast LC in PhR&D and quality control has grown rapidly. Fast LC also offers another benefit: rapid method development. Significantly shortened separation run times enable HPLC users to speed method optimization by simply conducting more trial runs in a given period. Different strategies can be implemented to perform fast HPLC analysis without sacrificing performance and reliability by balancing and optimizing the flow rate, column length, particle size, temperature, backpressure, and stationary phase.

9.2.1.1 Small Particle Liquid Chromatography

The standard formats of analytical HPLC columns in pharmaceutical R&D laboratories of 150 and 250 mm × 4.6 mm columns packed with 5 μm or 3.5 μm particles with plate numbers (N) of 17,000 to 20,000 (for well behaved small molecules) are sufficient for most separations.[6] The maximum resolving power of these columns is achieved when the run time is 30 to 60 min at an optimum flow rate of ~1 mL/min.[7] The simplest approach to fast HPLC analysis is to use shorter columns because separation time is proportional to column length. In addition, the reduction of column size allows a higher flow to be maintained without serious backpressure. However, significant loss of separation efficiency can result from the shortened column bed (N ≈ 5000 for 5 cm, 3.5 μm columns) and the increased flow rate above the optimum linear velocity for 5 or 3.5 μm particles. Therefore, this approach can be employed only for the analysis of samples with simple matrices that do not require high resolving power. In fact, when speed is a primary factor for analysis of major components, the desired resolution can still be achieved using a short (2 to 5 cm) column, resulting in significant reduction in run time and solvent consumption. For content uniformity and dissolution tests on conventional dosage forms for which a large number of samples must be handled and only the active ingredient must be quantitated, a run time of 1 to 2 min can be achieved with a 5 cm, 5 μm column.

When combined with the high resolving power of a mass spectrometry (MS) detector, short narrow-bore columns can be utilized to achieve HPLC cycle times of 1 min or less (ballistic gradient) for high-throughput LC/MS analyses that serve as valuable tools for synthetic chemists in CPR&D and for preformulation scientists in forced degradation studies.[8,9] Recent applications of fast LC in preformulation studies also include the high-throughput lipophilicity (logD) determination of drug candidates,[10,11] high-throughput drug excipient compatibility testing,[12] and high-throughput solubility measurements.[13] Novakova and Solich demonstrated the benefits of using a short C18 3.5 μm column for estradiol formulation testing.[14] The LC run time was shortened to 3.5 min compared to 12 min for a 250 × 3 mm, 5 μm column.

Although short columns with standard particle sizes have found many practical applications in pharmaceutical analysis, the compromised separation efficiency prohibits their use in situations

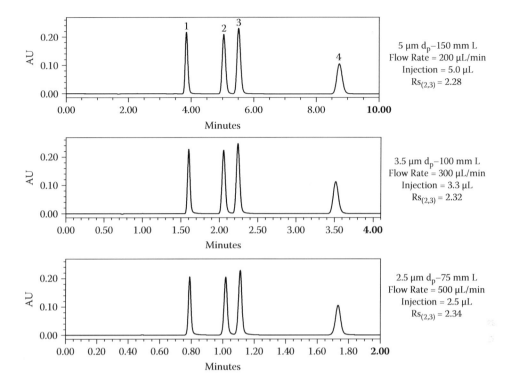

FIGURE 9.1 Comparison of resolution and selectivity with increased speed. L = column length. d_p = particle size. (Courtesy of Waters Corp.)

where high resolution separations are required for reliable quantitation of analytes in complex matrices. To maintain or minimize the loss of column efficiency while shortening column length (L), the particle size (d_p) of the packing material must be simultaneously decreased. According to chromatographic theory (van Deemter's plot, HETP versus flow rate), smaller particles in a packed column LC reduce eddy diffusion and mass transfer resistance in the mobile phase, produce higher separation efficiency per unit length of the column (lower HETP), and result in greater optimum mobile phase flow rate (for maximum efficiency or lowest HETP). With increased column efficiency, the column size can be decreased without a concomitant loss of resolving power. Consequently, shorter columns with smaller particles can provide the resolution of a longer column with larger particles as shown in Figure 9.1.

Most column manufacturers have developed short (2 to 5 cm) columns packed with 2- to 3-μm particles and various column diameters (2.1 to 4.6 mm) for fast LC on conventional HPLC instruments without exceeding the pressure limit of 400 bars (~6000 psi). These columns offer significantly increased separation efficiency compared to short columns packed with 3.5 or 5 μm particles and allow a pharmaceutical analyst to develop fast LC procedures for more complex samples. In formulated drug product development, methods using short columns with small particles are especially suited for content uniformity and dissolution testing of relatively complex formulations such as liquid-filled capsules (LFCs). Another important application of small particle LC is analytical support of safety assessment (GLP) studies that involve a large diversity of compounds entering the analysis stream. A generic HPLC method, through the use of short columns with small particles in typical runtimes from 5 (fast) to 2 min (ballistic), provides a means of high-throughput analysis of samples from preclinical studies.[15]

With the advance of column technology, HPLC columns packed with particle sizes smaller than 2 μm have been developed and are commercially available. Agilent Technologies introduced its Zorbax RRHT™ sub-2 μm (STM) column for ultra-fast LC in 2003. At Pittcon 2007, more than 10 column suppliers exhibited high performance columns (20 to 150 mm length and 0.075 to 4.6 mm internal diameter) packed with STM particles. One major drawback of stationary phases with small particle sizes is the dramatic increase of column backpressure arising from the reduction of particle size. As a result, many STM particles have been intentionally designed to have broader size distributions to produce lower backpressures. Shorter length (e.g., 50 mm) columns packed with these particles may be used on conventional HPLC equipment and usually allow run time reductions by a factor of two or three for modest separation complexity.

STM columns are increasingly employed in pharmaceutical laboratories. One potential area that may benefit greatly from STM technology is the effort to shorten the lengthy stability-indicating methods (SIMs) for determining impurity profiles for drug substances and products. To develop a stability-indicating method for early formulation development, one must demonstrate the method specificity against synthetic process impurities, potential drug degradation products, and excipients used in various prototype formulations. It is not uncommon to monitor more than a dozen peaks during a stability-indicating assay for a drug product, resulting in the need for a long gradient HPLC method with a conventional column to provide high resolving power. As candidate pharmaceutical compounds become more potent and are dosed at lower levels, more sensitive assays to detect and quantitate impurities are required. Low-throughput SIM can become the rate-limiting step in product release testing or process evaluation. Any further improvements in throughput and sensitivity would greatly benefit the processes of product release and identification of drug-related impurities.

Although a short (5 cm) STM column provides rapid separations, its column plate number is only about 50 to 60% of a 25 cm, 5 μm, or 15 cm, 3 μm column commonly used for SIM. The high resolution required for complex multicomponent samples may not always be achievable on 5 cm STM columns. A 10 cm STM column would generate comparable theoretical plates at the expense of greatly increased column pressure. Additionally, the van Deemter curve shows that a column packed with STM particles gives a flatter curve at high linear velocity than a 3.5 or 5 μm column.[5] Therefore, faster flow rates (linear velocities) can be employed with STM columns while maintaining separation efficiency, resulting in reduced analysis time. However, conventional HPLC systems are not capable of operating at optimal linear velocities for 10 cm STM columns without exceeding the instrument pressure limits of 350 to 400 bars. Therefore, to fully leverage the benefits of STM columns by maximizing separation efficiency and minimizing analysis time for the most chromatographically challenging SIMs, special HPLC instruments capable of handling pressure above 400 bars are required as discussed in the next section.

9.2.1.2   Ultra-High Pressure Liquid Chromatography

Only a few academic laboratories studied and used ultra-high pressure liquid chromatography (UHPLC) before 2003. Capillary columns packed with 1 to 1.5 μm non-porous particles were used with pressures up to 50,000 psi (4000 bars), one order of magnitude greater than those found in conventional HPLC (350 to 400 bars), on home-built ultra-high pressure instrumentation to generate plate numbers as high as 300,000.[5,16–18] Advancements in the development of LC columns packed with high quality STM porous particles and the necessary LC hardware able to withstand the associated increases in system pressure resulted in the introduction of the first commercial UHPLC instrumentation capable of handling pressure up to 15,000 psi, Waters' Acquity UltraPerformance LC (UPLC™) system, at the 2004 Pittsburgh Conference. Since then, several other companies introduced HPLC instruments and corresponding STM columns capable of operating at a pressure of 1000 bars.

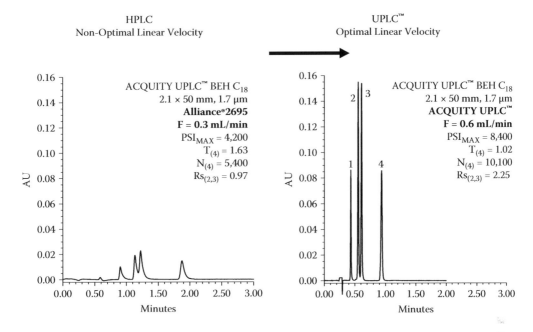

FIGURE 9.2 Comparison of Acquity UPLC column on HPLC and UPLC instruments. (Courtesy of Waters Corp.)

Agilent Technologies introduced a Rapid Resolution Liquid Chromatography (RRLC™) system with pressure capabilities up to 600 bars.

The commercialization of UHPLC promises significantly faster separations in addition to increased resolution and sensitivity. This technique allows the use of STM particle columns of lengths up to 150 mm to achieve high separation efficiencies. While 1000-bar is a modest pressure in comparison to the pressures of UHPLC systems used in academic laboratories, it is a significant increase in standard HPLC conditions, and is likely to offer considerable benefits in fast LC analysis since the higher optimal linear velocity required for STM columns can be realized under pressures above 400 bars.[5] Figure 9.2 shows that significant improvements in speed, resolution, and sensitivity may be seen with a 2.1 × 50 mm, 1.7 μm column operated under high pressure (8400 psi) compared to HPLC with a backpressure of 4200 psi. Nguyen et al. compared the chromatographic behaviors of various STM columns and showed that the best chromatographic performances were reached with high pressure systems (up to 1000 bars).[19]

Several recent articles discuss the applications of UPLC in pharmaceutical analysis and comparisons of UPLC and conventional HPLC.[20–22] Villiers et al. conducted a comparative study on the use of 1.7 μm particles at 1000 bars against conventional LC with 3.5 and 5 μm particles at 400 bars.[20] They concluded that UPLC offers advantages in terms of speed of analyses for required theoretical plate counts up to ~80,000. A gain in speed by factors of ~4.3 and 3.5 using 1.7 μm particles in comparison to 5 and 3.5 μm particles, respectively, can be realized without sacrificing efficiency. Wren and Tchelitcheff explored the potential of UPLC to improve the analysis of samples encountered during pharmaceutical development.[21] UPLC with a 2.1 × 100 mm, 1.7 μm column was compared to conventional HPLC with a 150 × 4.6 mm, 3 μm, or 3.5 μm column in terms of resolution, speed, and method development time for three developmental compounds in CPR&D. A speed reduction factor up to six was obtained. UPLC also showed benefits such as a flatter baseline, sharper peaks, and a simpler mobile phase and gradient program. The high linear velocities and rapid column re-equilibration made it possible to develop UPLC methods within 2 hr. Novakova et al. compared UPLC and HPLC analysis of four complex topical formulations.[22] The UPLC system with 1.7 μm particles showed a run time reduction up to nine times compared to conventional HPLC using 5 μm particles while demonstrating a run time reduction of ~3 times compared to a 3 μm column. Improvement in sensitivity for UPLC was also noted. The high resolving power, improved sensitivity, and faster processing offered by UHPLC with STM columns make it an excellent choice for impurity profile determination in pharmaceutical analysis. Jones and Plumb reported its use to develop an impurity profile method.[23] Ranitidine, an API, was forcefully degraded and used to test the performance of UPLC (2.1 × 100 mm, 1.7 μm) and HPLC (3.9 × 150 mm, 5 μm). UPLC gave rise to a factor of nearly six in reduction of analysis time (40 min to 7 min) while the resolution factor increased by more than five. In addition, UPLC detected 45 peaks with 0.05% area or greater compared to 34 on the HPLC chromatogram.

Figure 9.3 shows an impurity separation under conventional pressures with a 5 μm particle, 2.1 × 150 mm column, and the same separation performed via UPLC using a 2.1 × 50 mm column with 1.7 μm particles. The run time was improved by a factor of six, with overall resolution comparable to that of the original separation on the 5 μm column. The application of UHPLC technology to impurity profile analysis can exert a significant impact on laboratory productivity by achieving a

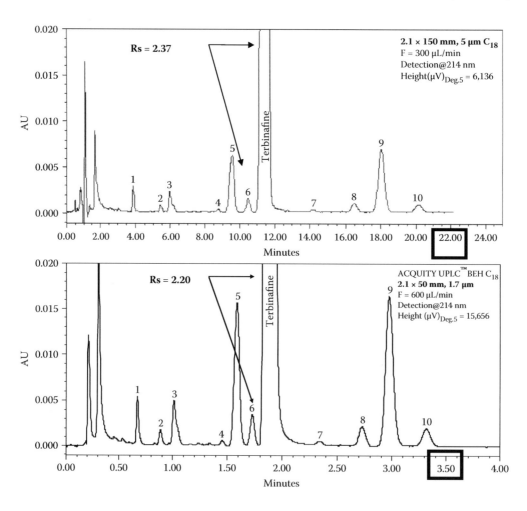

FIGURE 9.3 Impurity separation performance of UPLC and HPLC in an assay to determine stability. (Courtesy of Waters Corp.)

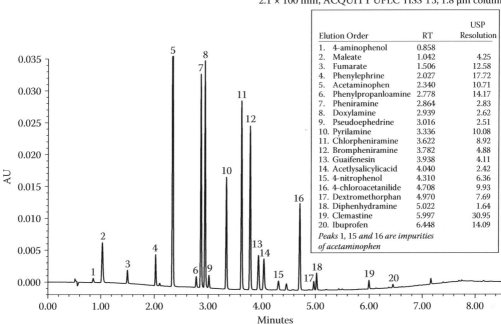

2.1 × 100 mm, ACQUITY UPLC®HSS T3, 1.8 μm column

Elution Order	RT	USP Resolution
1. 4-aminophenol	0.858	
2. Maleate	1.042	4.25
3. Fumarate	1.506	12.58
4. Phenylephrine	2.027	17.72
5. Acetaminophen	2.340	10.71
6. Phenylpropanloamine	2.778	14.17
7. Pheniramine	2.864	2.83
8. Doxylamine	2.939	2.62
9. Pseudoephedrine	3.016	2.51
10. Pyrilamine	3.336	10.08
11. Chlorpheniramine	3.622	8.92
12. Brompheniramine	3.782	4.88
13. Guaifenesin	3.938	4.11
14. Acetlysalicylicacid	4.040	2.42
15. 4-nitrophenol	4.310	6.36
16. 4-chloroacetanilide	4.708	9.93
17. Dextromethorphan	4.970	7.69
18. Diphenhydramine	5.022	1.64
19. Clemastine	5.997	30.95
20. Ibuprofen	6.448	14.09

Peaks 1, 15 and 16 are impurities of acetaminophen

FIGURE 9.4 Analysis of cold medicine formulations containing active ingredients, impurities, and counterions. (Courtesy of Waters Corp.)

3-5 fold cycle time reduction with improved resolution and sensitivity. The possibility of sub-5 min stability-indicating HPLC methods will minimize the need to conduct overnight runs, thus allowing much more to be accomplished in an 8-hr work day.

Another potential benefit of UHPLC is its capability of solving the most challenging separation tasks in pharmaceutical analysis. Figure 9.4 shows a UPLC method developed to analyze pharmaceutical formulations used to treat the common cold. Cold products often contain multiple active ingredients to treat different symptoms and can contain decongestants, antihistamines, pain relievers, cough suppressants, expectorants, and numerous excipients of various polarities. The analysis of a total of 20 components was achieved within 10 min.

Some potential issues of UHPLC have been cited in the literature. One concern is safety related to routine use of high pressure in chromatography laboratories. Another possible negative aspect of the high working pressures is the negative influence on column life expectancy. Also, the densely packed small particle columns are inherently more susceptible to premature plugging from fines present in packing materials and from particulates and contaminants present in a mobile phase or sample. The overall costs of UHPLC system ownership and return on investment are additional concerns of current HPLC users according to a 2007 *LCGC* survey. Despite these apparent concerns, more pharmaceutical laboratories are evaluating and are expected to purchase commercialized UHPLC systems. Several years of experience with practical experiments will be required to fully evaluate the disadvantages of UHPLC. Nevertheless, as interest in UHPLC continues to grow, it seems likely that it will find a broader spectrum of applications in pharmaceutical analysis.

9.2.1.3   High-Temperature Liquid Chromatography

It has long been recognized that LC separation times can be significantly reduced by operating the HPLC column at higher than ambient temperatures. Antia and Horvath predicted that a 20-fold reduction in analysis time would be realized if an HPLC column was operated at high temperature (150 to 200°C).[24] The new stationary phases and more robust columns designed for use with high temperatures have gained high temperature liquid chromatography (HTLC) increasing attention since the late 1990s.

The advantages of utilizing elevated temperatures in HPLC analyses are well documented.[25,26] One direct consequence of increased column temperature is a decrease of the viscosity of the mobile phase. This means that higher flow rates are possible with existing HPLC equipment without increasing backpressure. The lower backpressure in turn allows the use of smaller particles or longer columns that lead to increased separation efficiency. The rate of solute mass transfer within the stationary phase and the mobile phase increases with elevated temperatures. This gives rise to a flattened van Deemter curve and allows operation at flow rates that are many times the optimal velocity without the sacrifice in efficiency found at ambient temperatures. The kinetics of the interactions of the solutes and the stationary phase are accelerated at elevated temperatures. This often reduces or eliminates peak tailing. However, note that temperature can also affect retention and selectivity. Not all compounds have the same responses to temperature, so the selectivity of a separation can change dramatically when temperature is changed, resulting in improved or deteriorated resolution. High temperature can cause a less retained peak to be eluted so quickly that it can be eluted at or near the void time, making it difficult to quantify. This can be corrected by weakening the mobile phase with less organic (or even using water).

Most modern HPLC instruments include a column oven that can thermostat the column to at least 100°C. A typical HPLC analysis can be done in half the time by elevating the column temperature from ambient to 50 or 60°C. At temperatures above 100°C, it is not uncommon to decrease analysis time by a factor of 5.[26] Also, re-equilibration time for the column is much shorter, so it is possible to achieve ultra-fast gradient analysis with HTLC.

Despite its obvious advantages, HTLC was not considered a routine approach in the pharmaceutical industry until recently. Implementation of HTLC presents three main obstacles: (1) the thermal stability of the analytes, (2) the thermal stability of the stationary phases, and (3) the compatibility of the HPLC equipment. To address the primary concern of pharmaceutical scientists—the effect of high temperature on thermally labile compounds, Thompson and Carr conducted a study of the ability of a number of pharmaceuticals to withstand super-ambient temperatures (up to 190°C) with HTLC,[27] and found that as the exposure time of an analyte to high temperature decreases during a fast HTLC run, the likelihood and extent of on-column degradation greatly diminish. Therefore, the degradation of solutes may not be a major problem for fast HTLC because analytes are exposed to high temperatures for only short periods during the separation process and normally do not degrade significantly. The analytes must be thermally stable only for the duration of the chromatographic run of an HTLC analysis. Criteria for excluding thermally unstable analytes from measurements at high temperature have been proposed. Nevertheless, the fear of potential thermal degradation on the column will likely prevent widespread use of HTLC in the pharmaceutical laboratories in the near future.

The thermal stability of stationary phases can also be an obstacle to successful HTLC separations. This concern has been alleviated through recent advances in the development of stationary phases for LC separations at elevated temperatures.[28] Many high temperature HPLC columns packed with silica-based, metal oxide-based, and polymer stationary phases are now commercially available. Silica-based stationary phases are usually stable at temperatures up to 60°C and in some cases up to 90°C (Agilent Technologies' Zorbax StableBond C18 and Waters' XBridge BEH). Certain novel organic-bonded silica stationary phases may be stable at or above 100°C. Zirconia-based columns can withstand temperatures as high as 200°C. Thompson and Carr[29] provided column selection recommendations for fast HTLC analyses on conventional HPLC. They recommended that a highly retentive column be used to counteract the loss of retention with elevated temperature. With regard to column format, narrow-bore (2.1 mm inner diameter) columns are recommended to balance extra-column broadening, flow rate, and required heat-transfer tubing. Columns of 3 μm, 5 cm length or 5 μm, 10 cm length can be used for fast separations requiring low to moderate efficiency. Columns of 3 μm × 10 cm and 5 μm × 25 cm should be used at high temperatures for highly efficient sub-minute separations. Traditional silica-based packings are desirable for pharmaceutical analysis because of better peak shape and column efficiency. Thus, the benefits in analysis time using HTLC have not been fully explored because maximum temperature increases only 40 to 60°C above ambient conditions are normally used.

The third challenge to the routine use of HTLC is the thermal mismatch band broadening caused by a radial temperature gradient across the diameter of the column,[30] necessitating the modification of standard LC equipment. The elutant must be pre-heated to ensure complete thermal equilibration of the column. In addition, the column effluent must be cooled down to the temperature of the detector compartment to protect the detector hardware and reduce noise levels for improved sensitivity. Newly designed HPLC systems have column compartments capable of pre-heating mobile phases and cooling column effluents for temperatures up to 100°C. Column heaters providing mobile phase pre-heating (up to 200°C) and post-column cooling are commercially available for older instruments. The handling of organic solvents at elevated temperatures engenders some safety concerns. An after-detector backpressure regulator is required to maintain about 30 bars on the system to prevent the heated mobile phase from boiling as it exits the column and also to prevent the flashing of organic solvent in the absence of post-column cooling when the column temperature exceeds 80°C. The possibility of a leak in the column connection inside the heating chamber is another significant hazard concern.

HTLC is not routinely applied for rapid analysis of pharmaceutical samples at present, but its potential in this field certainly warrants further exploration. HTLC may offer unique efficiency advantages in some situations such as ultra-fast (seconds) HTLC analysis for online reaction monitoring. Another attractive application is in combination with UHPLC to achieve ultra-high resolution separation for very complex mixtures.[31] Lestremau et al.[32] incorporated temperature effects on mobile phase viscosity and analyte diffusion into experimental kinetic plots and demonstrated that high temperature LC allows faster separation for a given particle size, whereas higher pressure increases the efficiency attainable for a given particle size. Figure 9.5 shows that more than 100,000 plates were generated in under 12 min with three linked 2.1 × 150 mm, 1.7 μm columns operated at 90°C and 14,100 psi.

9.2.1.4 Monolithic Liquid Chromatography

One unique approach to fast HPLC analysis is increasing column permeability by using monolithic columns that consist of a single rod of very porous silica gel or polymer encased in a column package.[33–34] Silica-based monoliths are excellent tools for the separation of small molecules. Polymeric monoliths are better suited for large molecules such as proteins. The chromatographic features of a monolithic silica column result from its bimodal structure characterized by internal macropores and mesopores.[35] Macropores are on average 2 μm in diameter, dramatically reduce column backpressure, and allow faster flow rates. Mesopores of 13 nm diameter form a fine porous structure and create a large active surface area for high efficiency separations.

Current commercial silica-based columns have two important characteristics; (1) they can produce efficiency similar to that of columns packed with 3.5 μm particles and (2) they typically produce a pressure drop of half that caused by a column packed with 5 μm particles.[35] Monolithic columns have been shown to exhibit flat van Deemter curves, resulting in little loss of efficiency at high flow rates.[36] As a result, high-throughput separations on conventional HPLC instruments can be achieved by increasing flow rate up to nine times (up to 9 ml/min) the usual rate in a conventional packed column. Cycle times for HPLC analysis as short as 1 min (injection-to-injection) have been reported by users of monolithic columns. Additional benefits of monolithic columns cited include

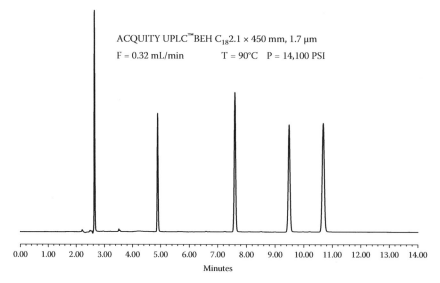

ACQUITY UPLC™ BEH C$_{18}$ 2.1 × 450 mm, 1.7 μm

F = 0.32 mL/min   T = 90°C   P = 14,100 PSI

Name	Retention Time	Area	Height	K Prime	USP Resolution	USP Tailing	USP Plate Count
thiourea	2.65	151654	93228	0.00		1.08	61789
toluene	4.89	119158	49674	0.85	42.28	1.06	94203
heptanophenone	7.60	203245	58588	1.87	34.71	1.02	108848
octanophenone	9.49	198872	45002	2.59	18.04	1.00	104645
amylbenzene	10.68	238837	45602	3.04	9.32	1.00	94419

FIGURE 9.5 Increasing efficiency HT UHPLC with three linked 2.1 × 150 mm 1.7 μm columns. (Courtesy of Waters Corp.)

fast column re-equilibration between runs, increased column lifetime based on high resistance of the macroporous structure to clogging, and reduced maintenance on pumps and injectors as a result of low operating pressures. Monolithic columns can also be arranged in series to generate very high efficiency separations at moderate pressures.

Monoliths also suffer drawbacks. Because silica monoliths are patented, widespread development has been hindered since the technology is not widely available. Currently, only Merck KGaA (Darmstadt, Germany) and Phenomenex (Torrance, CA) independently market equivalent silica-based monolithic columns under the trade names Chromolith™ and Onyx.™ The limited sources present a problem to pharmaceutical laboratories in regulated environments where an original second source of columns is required. The current choices of sizes and chemistries of monolithic columns are also limited. To date only three stationary phases available: C8, C18, and silica. The silica monoliths, until recently only available in 4.6 mm inner diameter columns, require higher flow rates than desirable, resulting in incompatibility with MS detectors and great consumption of solvents. Longer (1 to 2 mm inner diameter) columns are not yet commercially available although, this situation was alleviated to an extent when Phenomenex introduced the Onyx 3 mm monolithic HPLC column in 2006.

Reproducibility of monolithic columns has also been cited as a major concern because the monoliths are manufactured individually.[34,35] An extensive study by Kele and Guiochon indicates that the reproducibility results of Chromolith columns were almost comparable to those from different batches of particle-packed columns.[37] Other drawbacks of monolithic columns include weak retentivity for polar analytes,[38] efficiency loss at high flow rates for larger (800 MW) molecules,[39] and peak tailing, even for neutral non-ionizable compounds.[36–38,40] Furthermore, silica-based monolithic columns are not recommended for use above 45°C. They are made to operate at pressures up to 200 bars (3000 psi). This may become an issue when column coupling is employed. The recommended pH range is from 2.0 to 7.5. Another concern worthy of mention is that a strong dependence of baseline noise on flow rate was observed on some HPLC instruments.[39] A drastic increase of baseline noise for flow rates above 3 mL/min was observed, and this hindered the detection of impurities and degradates at low levels.

Despite these limitations, monolithic columns have found many applications in pharmaceutical laboratories. Several groups demonstrated their practical applications for pharmaceutical process development.[38,41,42] Wu et al. described the development of HPLC methods with a Chromolith column for impurity profiling of crude drug substances, reaction monitoring for impurity growth, and analysis of mother liquids for catalyst screening.[38] The analysis times were decreased three to seven times compared to a typical 5 μm particle-packed 250 mm × 4.6 mm column while comparable resolution, selectivity, and batch-to-batch reproducibility were maintained. Liu et al. demonstrated the enhanced throughput and speed of analysis of process R&D samples including column fraction screening and fast analysis of unstable analytes.[41] Higginson and Sach[42] demonstrated a rapid HPLC reaction analysis for high-throughput experimentation utilizing monolithic column technology. A cycle time of only 2.5 min enabled the analysis of a typical protocol of 100 reactions in hours permitting the capture of multiple time points for each reaction. Monolithic columns have also been proven advantageous in the analysis of formulated drug products.[43–45] A monolithic column-based method with a cycle time of 1 min using a 50 × 4.6 mm Chromolith column at a flow rate of 4 mL/min was developed and validated for a dissolution test of Lizepat® tablets.[43] Rapid impurity profiling of a Hagevir® cream formulation was accomplished within 3 min using a 100 × 4.6 mm monolithic column and a flow gradient to reduce post run re-equilibration.[44] Using a 100 × 4.6 mm column, Aboul-Enein et al. achieved a 2-min run time for the analysis of Plavix® tablets.[45]

9.2.1.5 Strategy for Implementing Fast LC

With several available techniques that can achieve rapid HPLC separations, pharmaceutical scientists are faced with the challenge of selecting the appropriate tools for their daily tasks. A number of researchers attempted to conduct head-to-head comparisons of various approaches to fast LC.[39,46–51] It is difficult to determine a sole winner among various technologies and associated products because many experimental factors are application- and user-dependent. For example, the overall separation efficiency of an LC system depends on the packing material and packing quality of the column and also the instrument bandwidth (IBW). Often, the advantages of a particular technology were demonstrated by comparative data obtained under experimental conditions that had been optimized only for the proposed approach. It is therefore not surprising that the literature reveals no agreement with regard to the best technology for performing rapid separations. It is evident that each technology has its own strengths and limitations, and should be used only when its individual advantages can be best leveraged. No single strategy fits all situations. From the perspective of pharmaceutical applications in a GMP environment, a need clearly exists to find a judicious balance of speed, resolution, sensitivity, reproducibility, and ruggedness.

Table 9.1 summarizes the advantages and limitations of different approaches for fast LC along with potential applications in pharmaceutical R&D and QC. When optimized, they are all potentially capable of achieving rapid HPLC analysis for various pharmaceutical analysis applications. The best separation time is the one that resolves all analytes of interest in the least amount of time. Both short columns packed with 2 to 3 μm particles and monolithic columns can be used with conventional LC equipment and require few or no instrument modifications. Fast separations of low or modest complexity such as content uniformity and dissolution test for conventional dosage forms can be obtained with these columns. Monolithic columns offer the additional advantage of better suitability for more challenging samples such as separating and characterizing potential unknown side products in crude or

TABLE 9.1
Comparison of Fast LC Approaches

	Advantages	Limitations	Applications
Short column with small particles (2 to 3 µm)	Use of conventional HPLC with minimal or no equipment modifications Greatest diversity of column chemistry Easy method transfer Best reproducibility and ruggedness	Limited separation efficiency Low sensitivity	High throughput LC/MS In-process testing for API manufacturing Potency assay of API and drug product CU/diss test of conventional dosages Fast generic method for GLP samples
Monoliths	Low backpressure, suited for conventional HPLC Higher separation efficiency by column coupling Rugged against delay volume and extra-column band broadening; fast column re-equilibration Reduced maintenance on pumps and injector seals Reduced need for sample pre-treatment	Single source of column Very limited column chemistry Method transfer difficult High solvent consumption Peak asymmetry pH 2 to 7; <200 bars Noisy baseline at high flow rate	"Dirty" samples Process impurities and degradates of APIs CU/diss for non-conventional formulations SIM with modest complexity Ultra-fast generic method for GLP samples New 3 mm inner diameter column for HT LC/MS
Short column with STM particles (< 400 bar)	Potential use with conventional HPLC Significant reduction in analysis time Easy method transfer Higher efficiency Relatively good variety of column chemistry Two-fold increase in speed for SIM	Modification of conventional LC required High back-pressure High column back-pressure limits speed of analysis and column length or N Reduced column life due to clogging	SIM with modest to high complexity Method development Ultra-fast high throughput LC/MS Ultra-fast generic method for GLP samples
UHPLC (600 to 1000 bar)	Significant runtime reduction for ultra-fast separation; minimal solvent consumption Five-fold increase in speed for SIM Significantly higher efficiency for most complex separations Higher mass sensitivity Rapid method development	Special equipment required Column chemistry limited Carry-over Viscous heating Detection at low wavelength (blending noise) Increased care due to high pressure	Rapid SIM method development Isomer separation SIM with highest complexity Separations with challenging matrices, e.g. LFCs
HTLC	Use of conventional HPLC Low backpressure Good peak symmetry Less organic in mobile phase Leverage of temperature for selectivity	Temperature mismatch broadening: special equipment required for heating and cooling Limited column chemistry due to thermal instability of stationary phases Thermal instability of analytes Method transfer difficult (selectivity change) Safety concerns	Ultra-fast LC for online reaction monitoring Combine with long STM column on conventional LC Combine with UHPLC and long STM column for most challenging separations

semi-purified reaction mixtures and formulations with complex matrices. For separations of modest to high complexity such as reaction monitoring, sub-2-μm particles packed into short columns allow high-throughput separation with minimal loss of resolution on upgraded conventional HPLC equipment. When packed into long columns, they provide high resolution separation of most complex mixtures using UHPLC. Rapid SIM methods can be developed with these approaches. In addition, STM columns are best suited for method development. HTLC can be combined with STM columns and UHPLC to further improve efficiency and speed of the most challenging separations. Based on current limitations, HTLC at temperatures above 100°C is less convenient for routine QC applications. However, ultra-fast (less than 1 min) LC with HTLC at very high temperatures (150 to 200°C) has potential for high-throughput applications such as those involving monitoring of rapidly changing systems such as online reaction monitoring.

Several important aspects relevant to the implementation of fast LC technologies in pharmaceutical laboratories should be mentioned. First, increases in speed should not compromise the quality of the analytical data or the robustness of the chromatography. All methods must be reproducible and validatable to meet the applicable GMP and GLP requirements. Instrumentation should be easily maintained and have minimal downtime.

To minimize unacceptable interruptions in highly regulated work flows, the smooth transfer of legacy methods from conventional to fast LC methods (via geometric transfer or method redevelopment) is a critical issue for implementing fast LC for pharmaceutical applications. Method transfer from HPLC to UPLC is discussed in detail in the literature.[52,53] Moreover, method transfer software that provides parameter conversion between UHPLC and conventional HPLC is available from instrument vendors.

Method development is one of the most time-consuming tasks in a pharmaceutical laboratory. Most method development is still performed manually via trial and error despite all the HPLC sophistication and automation. Fast LC offers the advantage of accelerating method development, but another approach that can be far more powerful is to work with selectivity by screening stationary phases and adjustments of solvent, pH, temperature, and gradient slope. Finally, although fast LC reduces analysis time to a few minutes, great attention is required to ensure that extra-column effects are minimized to achieve the expected efficiency of small particle columns, especially STM columns. Extra-column band broadening or IBW arising from injection valves, detectors, and connecting tubing must be minimized if the true advantages of STM columns are to be realized. Some modern liquid chromatographs have been designed or modified to minimize extra-column effects. Upgrade kits can modify older HPLC instruments to work satisfactorily with STM columns. Note that smaller flow cells often have shorter path lengths, leading to some compromises in sensitivity. Extra-column band broadening and similar effects for fast LC are detailed in the literature.[54,55]

9.2.2 Parallel HPLC

High-throughput experimentation implemented in PhR&D generates tens of thousands of samples that require qualitative and quantitative characterization. Although the fast LC techniques discussed above have significantly increased throughput, the number of samples that can be processed by serial LC analysis is still limited. The gradient can be ballistic, but an HPLC column requires a certain amount of time for reconditioning and equilibration. Parallel HPLC represents an interesting approach for high-throughput separation. In this mode, 8 to 24 columns and detectors are used in parallel and controlled by a single computer with a user-friendly software interface for instrument control, data acquisition, and sample tracking.[56] Parallel analysis offers many potential advantages over serial analytical methods, the most obvious of which is dramatically increasing throughput while maintaining chromatographic integrity.

For example, an 8-channel system with a modest separation time of 5 min can analyze 8 samples simultaneously, resulting in an average analysis time of 38 sec per sample. Another significant benefit of parallel HPLC is its application to procedures such as chiral column method development or optimization and testing selectivity and retentivity for different brands and types of reversed-phase columns. Using parallel HPLC in this manner means that the optimal column can be identified in a mere fraction of the time it would take a single HPLC to do the same task. An answer generated within 15 min instead of 2 hr or more has tremendous value for laboratories that need to rapidly develop separation methods for large numbers of projects in early drug development.

Several parallel HPLC systems are commercially available. The Sepmatrix™ (Sepiatec, www.sepiatec.com) can operate with up to eight columns at narrow-bore, analytical, and preparative flow rates.[57] The instrument uses a single standard HPLC pump to deliver an equal amount of flow to each column. Multiplexing fiberoptics collect full spectra for all eight channels. The Veloce™ microparallel liquid chromatography (μPLC) system (Nanostream, Pasadena, CA) allows 24 identical analyses to be performed in parallel using Brio™ prepackaged column cartridges.[58,59] The system produces chromatograms comparable to conventional HPLC instrumentation, and its use has been demonstrated for several high-throughput applications. Eksigent Technologies (Livermore, CA) introduced the ExpressLC-800™ eight-channel system powered by a proprietary microfluidic flow control system.[60] Each of the eight parallel separation channels is fully independent, allowing the analysis of eight different samples via eight different methods, further increasing the parallel analysis capabilities of the system. Sajonz et al. investigated the use of the ExpressLC-800™ system for multichannel screening and development of fast normal phase chiral separations.[61] The multiparallel approach was shown to provide "near real-time" method development, often affording an optimized method in less than an hour versus a next-day result offered by an automated sequential chiral method screen. Chromatographic data obtained with eight-channel microfluidic systems are comparable to those obtained with conventional HPLC systems. The authors also described the application of the ExpressLC-800 for high-throughput normal-phase chiral analysis in support of high-throughput pharmaceutical process research.[62] The system can carry out high-throughput analysis of enantiopurity to support HTS of asymmetric catalysis and other HTEs. The cycle time for a single 96-well microplate is typically on the order of an hour or two—a significant improvement over conventional analysis techniques that usually take several days.

A comparison of the ExpressLC-800 and the Veloce microparallel system for HTA in support of pharmaceutical process research was reported by Welch et al.[63] They concluded that the Eksigent system offers advantages including the ability to execute very fast gradient separations and use columns containing many different (highly efficient and chiral) stationary phases. The Veloce was noted to produce poor peak shape, leading to excessively long chromatograms to obtain baseline resolution. The Veloce may be better suited to single component analysis situations such as solubility studies, log P determinations, and dissolution tests. Liu et al. applied a high-throughput method developed with a 24-column Brio cartridge and the Veloce system to determine drug release profiles in OROS tablets.[64] The profiles generated via μPLC were comparable to those obtained by conventional HPLC; total analysis time was reduced from 20 to 2 hr. In addition, the μPLC system consumed only 36 mL of mobile phase compared to 1.6 L for conventional HPLC. Although not described in the literature, μPLC has potential for analytical support of GLP studies and preformulation studies.

9.2.3 Supercritical Fluid Chromatography

In recent years, packed-column supercritical-fluid chromatography (SFC) has begun to emerge as an alternative to conventional HPLC for high-throughput separations.[65–68] Unlike conventional HPLC from which SFC inherits particle columns, the mobile phase in SFC consists of carbon dioxide in supercritical condition and a modifier (typically an alcohol).[69] The low viscosity and high diffusivity of supercritical carbon dioxide allow fast separations at much higher linear velocities on longer columns and with faster column re-equilibration, leading to considerable reduction in analysis time without impractical pressure increases. Using carbon dioxide as the major eluent offers additional advantages of simplified evaporation, reduction of waste disposal costs and associated environmental impacts, and the ability to collect only one fraction per product.[70]

Despite many distinct advantages, the penetration of SFC in pharmaceutical laboratories appear limited based on several technological and commercial factors including unfamiliarity with the technique, hardware complexity, high capital cost, and disinterest of major instrumentation manufacturers. Enantiomeric separation is currently the most successful application of SFC. In general, SFC can be considered a normal phase chromatography ideally suited for the isolation of polar solutes that are challenging to separate by other chromatographic techniques such as enantiomeric separations.[70–72] SFC typically provides a three- to five-fold faster separation with better resolution than normal phase HPLC for enantioseparation.[71] It constitutes a powerful alternative in the area of chiral separations. Because of more stringent FDA guidelines for marketing chiral drugs, stereoselective syntheses have rapidly become common in the development of new drug candidates. The separation and the quantification of enantiomeric mixtures are among the great challenges of the past decade in pharmaceutical analysis. Increased sophistication of chiral HPLC separations has arrived at the forefront of pharmaceutical studies including analysis of atropisomeric (hindered rotation around a single bond) species. As commercial SFC instruments continue to improve, SFC is becoming the first choice of many pharmaceutical companies for high-throughput enantioseparation and purification.[73,74]

Many applications of SFC in pharmaceutical process research have been reported.[75–78] Toribio et al.[75] reported a comparison of HPLC and SFC for enantioselective separation of several antiulcer drug substances. SFC was faster, produced better resolution, and required less organic solvent than HPLC. Welch et al. developed a rapid method for analysis of the enantiopurity of the Soai reaction product by chiral SFC.[79] An analysis time of 0.6 min per sample was obtained. More recently, Alexander and Staab described the use of coupled achiral and chiral SFC/MS to achieve one-step isomeric profiling of a semi-purified reaction mixture to determine the diastereomeric and/or enantiomeric composition of the final product and identify remaining E/Z isomers present from the starting material.[68] Another novel SFC tandem column screening tool for solving multicomponent chiral separation challenges was developed by Welch et al.[80] A Berger analytical SFC system was modified to allow software-controlled selection of 25 different tandem column arrangements and 10 different single column arrangements, resulting in rapid development of chiral SFC methods in support of enantioselective catalysis screening. The enantiomer of interest had to be resolved from residual starting materials and reaction by-products. SFC is also recognized as a powerful technique for analysis of pharmaceutical compounds in various dosages forms—tablets and capsules,[67] emulsions and suspensions,[67] and liquid formulations.[76] Mukherjee developed a direct assay of an aqueous formulation of a drug compound by chiral SFC that eliminated the sample processing steps.[67]

9.2.4 Fast Gas Chromatography

Gas chromatography (GC) is a mature separation technique. Its use in industrial analytical laboratories has diminished somewhat as a result of the introduction and rapid spread of HPLC. Nevertheless, GC still plays an important role in areas of pharmaceutical analysis where the analytes are volatile and thermally stable. Applications of GC in drug development include testing of organic solvents for API synthesis, determination of extractables and leachables in pharmaceutical packing materials,[81] analysis of flavoring excipients,[82] and determination of moisture and headspace oxygen in pharmaceutical packages.[83] GC may also be employed for the analysis of compounds that are unsuitable for HPLC analysis, such as compounds that lack UV chromophores.[84,85] A new field of GC application of growing importance is the analysis of organic volatile impurities (OVIs) in pharmaceuticals including bulk drug substances, excipients, finished drug products, and packaging materials. Potential OVIs include the residues of solvents used during manufacturing[86] and trace amounts of reactive species such as aldehydes and acids.[87]

For most GC analyses using conventional capillary GC columns with internal diameters of 0.25 to 0.53 mm, analysis times range from 10 to 60 min. Like HPLC, GC since the 1990s has seen revived interest in developing fast technologies as a result of the desire for high-throughput analysis and the reducing operation costs in routine analysis.[88–91] Fast GC means peak widths of 1 to 3 sec, analysis times in minutes (5- to 30-fold reductions in run time), and equal or better separation efficiency compared to conventional capillary GC.[88] The many options to speed up GC analysis include short, narrow-bore columns, rapid heating, higher pressure drops, hydrogen as a carrier gas, and selective detection such as MS. Several references[88–91] detail strategies for implementing fast GC. Modern GC instruments are equipped with inlets and flow systems, column ovens, detectors, and data collection systems that are compatible with fast GC requirements. Columns of 100 μm inner diameter with a wide choice of stationary phases are commercially available. Ovens allow programming rates up to 50 to 100°C per min, while resistive heating enables rates up to 1200°C per min and cooling from 300 to 50°C in less than 30 sec. The latest technologies designed to reduce analysis cycle time and improve productivity include comprehensive flow modulated 2D GC,[92] parallel GC,[93] and column backflush.[94]

Fast GC has a number of applications in pharmaceutical analysis. Xu et al. described the use of micro-GC for the rapid, simultaneous determination of headspace oxygen and moisture in pharmaceutical packages.[83] The GC run is shorter than 90 sec. Residual solvent or OVI analysis has been one of most challenging tasks in pharmaceutical laboratories. The lengthy U.S., Europe, and Japan compendial methods normally require ~45 to 60 min injection-to-injection times. Many efforts have been devoted to shortening OVI analysis time.[95–99] Chen et al. first applied fast GC to the analysis of residual solvents in drug substances.[95] A 10 mm × 0.1 mm DB-624 column was employed along with direct injection to achieve fast separation of 38 commonly used ICH class 2 and class 3 organic solvents in less than 4.9 min with baseline resolutions for most analytes. Raghani described a high speed GC analysis of OVIs in APIs using solid phase microextraction and resistively heated column technology.[96] Separation of 13 solvents was achieved in less than 3 min while the total analysis time including extraction was ~6 to 9 min. David et al. employed a custom-made 20 mm × 180 μm × 1 μm DB-624 column mounted in a low thermal mass oven mounted on a standard GC instrument for high-throughput analysis of residual solvents in pharmaceutical products.[98] Complete separation of 20 solvents with headspace sampling was achieved in a total GC cycle time of less than 4 min. Pavon et al. reported use of headspace-programmed temperature vaporization (HSPTV) GC/MS for analysis of ICH class 1 residual solvents in formulated products.[99,100] Another effective technique to shorten GC run time is column backflush that allows removal of late eluting compounds by reversing the flow. The technique has been applied for fast OVI analysis using a commercial instrument.[101]

9.2.5 Capillary Electrophoresis

Since the 1980s, various modes of capillary electrophoresis (CE) such as capillary zone electrophoresis (CZE), micellar electrokinetic capillary chromatography (MECC), capillary gel electrophoresis (CGE), capillary electrochromatography (CEC), and chiral CE have attracted great interest in the pharmaceutical industry as possible alternatives or complements to HPLC.[102–107] Compared to other separation techniques, CE offers several distinct advantages including extremely high efficiency in liquid phase separation, fast analysis time (usually only a few minutes), relative simplicity (rapid and simple method development and optimization and ease of operation), low cost, and applicability for separating widely different compounds from small inorganic ions to large biomolecules. The high resolving power and speed make CE a potential candidate for high-throughput analysis. However, despite these advantages, CE has not yet become a real rival to HPLC in pharmaceutical analysis. Limited injection volume, low sensitivity with UV detection, and relatively poor reproducibility continue to be drawbacks many years after its introduction.[103,105] Additionally, CE is less robust than HPLC in handling pharmaceutical in-process samples containing catalysts[105] and dissolution samples containing surfactants.[108] Consequently, CE is primarily used when other separation techniques are limited or impractical, e.g., small inorganic and organic ion determinations,[109] studies of physiochemical properties,[104] and enantiomeric separations.[106]

An additional advantage of CE over HPLC (that can be used to enhance throughput) is that CE can be highly multiplexed for parallel analyses.[110,111] A 96-lane system is now commercially available (www.combisep.com) and boosts throughput up to 100-fold compared to a commercial single capillary system.[110] Multiplexed CE systems show great potential for efficient and simultaneous multisample analysis such as HTE. Microchip CE, the result of a marriage of the ultra-small sample volume (nL) capability of CE and microfabrication technology, is revolutionizing chemical and biochemical testing. It offers the integration of multiple steps of complex analytical procedures and furthers the potential of a lab-on-a-chip.[112] The integration of injection and detection on a microchip allows the use of much smaller volumes than is possible with standard CE. This means the separation channels can be very short (a few centimeters) for certain applications, resulting in faster analysis. The possibility of rapid, parallel separation of small amounts of samples on microchips will potentially afford high-throughput analysis of complex, multicomponent mixtures. Although the application of multiplexed and microchip CE techniques is still novel and modest, a growing number of research groups in different areas of the pharmaceutical industry (synthetic chemistry for rapid screening and microreactors for pharmaceutics) have shown interest in these techniques.

9.3 HIGH-THROUGHPUT SPECTROSCOPIC TECHNIQUES

Modern spectroscopy plays an important role in pharmaceutical analysis. Historically, spectroscopic techniques such as infrared (IR), nuclear magnetic resonance (NMR), and mass spectrometry (MS) were used primarily for characterization of drug substances and structure elucidation of synthetic impurities and degradation products. Because of the limitation in specificity (spectral and chemical interference) and sensitivity, spectroscopy alone has assumed a much less important role than chromatographic techniques in quantitative analytical applications. However, spectroscopy offers the significant advantages of simple sample preparation and expeditious operation.

Advances in instrumentation and chemometric techniques have made spectroscopic techniques effective alternatives to separation methods for high-throughput analysis in the pharmaceutical industry. In some industrial processes, spectroscopic techniques such as near-infrared (NIR) and Raman are gradually superseding chromatographic methods in online applications as evidenced by the renewed popularity of process analytical technology (PAT) in recent years. Information about the implementation of spectroscopic techniques in PAT for drug development is readily available[113] and is not discussed here. The discussion that follows focuses on applications of spectroscopic techniques in high-throughput off-line (in-laboratory) measurements in support of ChPR&D and PhR&D.

9.3.1 ULTRAVIOLET-VISIBLE SPECTROPHOTOMETRY

Although considered a basic technique, ultraviolet-visible (UV-vis) is perhaps the most widely used spectrophotometric technique for the quantitative analysis of pure chemical substances such as APIs in pharmaceutical analysis. For pharmaceutical dosage forms that do not present significant matrix interference, quantitative UV-vis measurements may also be made directly.[114,115] It is estimated that UV-vis-based methods account for ~10% of pharmacopoeia assays of drug substances and formulated products.[116]

The excipients present in pharmaceutical formulations can and often do interfere with quantitation of APIs, limiting the applications of direct UV-vis measurement for analyzing formulated products due to its lack of specificity. To minimize the interference of excipients, colorimetric methods based on chemical reactions have been used for rapid determination of drug substances in pharmaceutical formulations although their role in pharmacopoeias has been greatly reduced.[117–122]

Derivative UV spectroscopy has also been utilized for quantitative analysis of APIs in pharmaceutical dosage forms. The technique offers significant advantages over conventional absorption methods, allowing focus on specific details of the UV spectra and providing more accessible data useful for the selective quantitative analyses suitable for routine quality control of dosage forms.[123–126] For formulations containing more complex matrices or multiple components, multivariate calibration has become a useful chemometric tool for rapid and simultaneous UV-vis determination of each component in a mixture, with minimum sample preparation and without need for lengthy separations. A number of applications of multivariate calibration approaches in the determination of APIs in pharmaceutical dosage forms have been published.[127–132] UV-vis spectroscopy has been used extensively as a high-throughput measurement tool for automation such as flow-injection analysis and *in situ* dissolution testing (see Section 9.4).

9.3.2 INFRARED SPECTROSCOPY

9.3.2.1 FT-IR

The pharmaceutical industry comprises the largest segment, roughly 15 to 20%, of the infrared (IR) market. Modern mid-infrared instrumentation consists almost exclusively of Fourier transform (FT) instruments. Because of its ability to identify molecular species, FT-IR is routinely used as an identification assay for raw materials, intermediates, drug substances, and excipients. However, the traditional IR sample preparation techniques such as alkali halide disks, mulls, and thin films, are time-consuming and not always adequate for quantitative analysis.

The development of the FT-IR spectrometer with significantly improved signal-to-noise components and throughput brought new sampling techniques for pharmaceutical analysis including diffuse reflectance (DRIFTS) and attenuated total reflectance (ATR) that simplify sample handling and offer potential for high-throughput quantitation of drugs.[133–135] Chemometric approaches such as partial least squares (PLS) and principal component regression (PCR+) have also been used in data processing. Bunaciu et al. described the direct determination of bucillamine in Rimatil® tablets and dehydroepiandrosterone in capsule formulation using DRIFTS spectra processed with PLS and PCR+.[136,137] Quantitation can be performed in 5 to 10 min including the sample preparation and spectral acquisition. Boyer et al. reported the direct determination of niflumic acid in a pharmaceutical gel by ATR/FTIR spectroscopy and PLS calibration.[138] The method is rapid, non-destructive, easy to use, requires no sample pretreatment (reagent-free measurement), and constitutes a powerful alternative to separation methods. A simple ATR/FT-IR method was developed for the determination of residual acetic acid in a moisture-sensitive anhydride raw material.[139]

Other recent applications of FT-IR in pharmaceutical analysis include reaction monitoring by fiberoptic FT-IR/ATR spectroscopy[140] and stability studies of pharmaceutical emulsions using FT-IR microscopy.[141] A novel equipment cleaning verification procedure using grazing angle fiberoptic FT-IR reflection–absorption spectroscopy was described by Perston et al.[142]

APIs on a glass surface at loadings well below those visible to the naked eye can be quantitated simultaneously in significantly less than 1 min using a conventional laboratory spectrometer. Another important application of FT-IR in drug development is imaging for high-throughput analysis of pharmaceutical formulations.[143–146] About 100 samples can be analyzed simultaneously with a macro ATR/FT-IR spectroscopic imaging system. The proposed high-throughput methodology allows fast screening of many different formulations to identify those that exhibit drug recrystallization or polymorphism.

9.3.2.2 Near-Infrared Spectroscopy

The near-infrared (NIR) region (780 to 2500 nm) of the electromagnetic spectrum is situated between the visible light and mid-IR regions. With the introduction of efficient chemometric data processing techniques and novel spectrometer configurations based on fiberoptic probes, near-infrared spectroscopy (NIRS) has become one of the most rapidly growing methodologies in industrial laboratories. NIRS is a fast technique in which a spectrum can be recorded in only a few seconds. It is also non-destructive. A sample of virtually any matrix can be analyzed with little or no pretreatment. Due to the complex nature of NIR spectra, NIRS typically requires statistical manipulation of the spectra using chemometrics to extract relevant data for qualitative and quantitative analysis. A complete description of the underlying theory of NIRS can be obtained from the literature.[147,148]

Based on its distinct advantages over other analytical techniques for high-throughput, NIRS has found many applications in the pharmaceutical industry as documented in two extensive review articles with more than 400 references cited.[148,149] NIRS is considered the top technique in PAT for process monitoring and control. It has also gained wide acceptance in pharmaceutical laboratories for both qualitative and quantitative analyses. Its primary application is the rapid identification of raw materials, excipients, APIs, and finished products. The implementation of vendor qualification led many pharmaceutical companies to develop systems to accept vendors' certificates of analysis, so that the testing of raw materials and excipients after receipt is no longer required. However, the regulatory requirement that single container identification be performed for any lot of raw materials at any time of dispensing remains in place. Many companies use NIR techniques based on fiberoptic probes with a spectral library approach for fast and nondestructive identification of incoming materials in a GMP receiving area. Even direct identification of materials through protective polyethylene packing is possible. With an appropriate calibration set-up, data about physical state (crystallinity and powder size) of a material can be obtained. Identification of finished drug products can be performed by NIRS through blisters and vials to provide identity checks for clinical trial samples at clinic sites. Other important applications of NIRS in pharmaceutical analysis are the rapid quantitative analyses of intact dosage forms including content uniformity and water content. Detailed discussions can be found in the literature[148,149] where data are cited through early 2006.

Most recently, Lee et al. described the use of a novel NIR chemical imaging system for measuring the content uniformity of multiple drug tablets simultaneously.[150] A total of 20 tablets including 5 calibration and 15 unknown samples were measured simultaneously in less than 2 min with no sample preparation. Feng and Hu demonstrated the feasibility of building universal quantitative models that can be used in different instruments for rapid analysis of pharmaceutical products using NIR reflectance spectroscopy.[151] John and Pixley proposed a method for NIR identification of pharmaceutical finished products with emphasis on negative controls and data-driven threshold value selection.[152] Additional noteworthy studies include hardness testing of intact pharmaceutical tablets by NIRS,[153] NIR assay of low-dose tablets,[154] and use of short wavelength NIRS with an artificial neural network to determine drug substance in a powder.[155]

9.3.3 RAMAN SPECTROSCOPY

As a complementary tool to IR in vibrational spectroscopy, Raman spectroscopy is one of the fastest growing analytical techniques in use today due to its ability to reveal fundamental molecular structural information and immediate chemical environment through light-transparent materials and without sample preparation. It also offers submicron spatial resolution and very high sensitivity (when coupled with surface-enhanced Raman spectroscopy or SERS). IR spectroscopy is based on the absorption of electromagnetic radiation by a molecular system, whereas Raman spectroscopy relies upon inelastic scattering of the system. Strong IR bands are related to polar functional groups, whereas non-polar functional groups give rise to strong Raman bands. Raman is an extremely flexible technique and offers many advantages over IR spectroscopy including a wide variety of acceptable sample forms, flexible sample interfaces and sample size, no sample preparation, and high sampling rates. Furthermore, Raman spectra with sharp, well resolved bands with minimal water interference are easily interpreted. Simple univariate calibration models are often sufficient for quantitative work. Limitations include interference of fluorescence, unsuitability for black or highly colored materials, high cost, and sensitivity to the local molecular environment. Fundamentals of Raman spectroscopy can be found in relevant references and monographs.[156–158]

Raman spectroscopy is emerging as a powerful analytical tool in the pharmaceutical industry, both in PAT and in qualitative and quantitative analyses of pharmaceuticals. Reviews of analyses of pharmaceuticals by Raman spectroscopy have been published.[158,159] Applications include identification of raw materials, quantification of APIs in different formulations, polymorphic screening, and support of chemical development process scale-up. Recently published applications of Raman spectroscopy in high-throughput pharmaceutical analyses include determination of APIs in pharmaceutical liquids,[160,161] suspensions,[162,163] ointments,[164] gel and patch formulations,[165] and tablets and capsules.[166–172]

Kim et al. described a direct Raman measurement of an API in pharmaceutical liquids through a low-density polyethylene (LDPE) bottle with wide area illumination (WAI) Raman scheme.[160] NIR absorption measures had been difficult for aqueous samples in larger bottles. A rapid and simple approach for identification of multidose pharmaceutical products using Raman spectroscopy was reported by Cantu et al.[166] FT-Raman spectra of different tablet and capsule formulations displayed a rich array of bands in the fingerprint region to enable dose differentiation and identification by using a logarithmic ratio of peaks or peak heights of inner diameter marker bands. Raman microscopy has also been used for high-throughput polymorph screening of APIs.[173–175]

9.3.4 OTHER SPECTROSCOPIC TECHNIQUES

Both nuclear magnetic resonance (NMR) spectroscopy and mass spectroscopy (MS) share a long-standing tradition in the elucidation and confirmation of the structures of synthetic products, impurities, and degradation products in drug development and analysis. When coupled with chromatographic techniques for quantitation, MS provides high selectivity that enables fast separation without running long separations. Three recently introduced ionization techniques, desorption electrospray ionization (DESI), desorption atmospheric pressure ionization (DAPCI), and direct analysis in real time (DART), offer high-throughput analysis by rapidly analyzing complex mixtures with little or no sample preparation.[176–178] Their applications have been demonstrated for high-throughput analysis of counterfeit drug products in an ambient environment.[178–180] Potential applications in pharmaceutical analysis include in-line monitoring in PAT and identification of pharmaceuticals.

Although NMR is not often used in quantitative analysis of pharmaceuticals, it can be employed as a simple and rapid quantitation technique due to its high selectivity under appropriate acquisition conditions and the fact that analyte standards are not required.[181–187] Quantitative NMR analysis is usually based on the integration ratio between a specific NMR signal of the analyte and a selected signal in the NMR pattern of the internal standard. A unique application in reaction monitoring enables plant operators to run simple and fast NMR analyses to determine reaction completions without involving QC personnel. Solid state NMR spectroscopy (ssNMR) along with Raman spectroscopy, x-ray diffraction, and thermal methods represent the current choices for solid state characterization[188]—an assay of great importance in developing pharmaceutical formulations with optimal bioavailabilities. However, when compared to other analytical techniques for compound characterization, one disadvantage of ssNMR is its limited relative throughput. To increase its efficiency, a multiple-sample probe can now simultaneously acquire up to seven ssNMR spectra for analysis of pharmaceutical dosage forms.[189]

Atomic techniques such as atomic absorption spectrometry (AA), inductively coupled plasma-optical emission spectrometry (ICP-OES), and inductively coupled plasma-mass spectrometry (ICP-MS), have been widely used in the pharmaceutical industry for metal analysis.[190–192] A content uniformity analysis of a calcium salt API tablet formulation by ICP-AES exhibited significantly improved efficiency and fast analysis time (1 min per sample) compared to an HPLC method.[193]

Laser ablation ICP-AES and LA-ICP-MS were also proposed for rapid, direct analysis of tablets containing metallic species.[194] Compendial heavy metal tests based on wet chemistry are among the most labor-intensive tasks in pharmaceutical laboratories. Both ICP-MS and ICP-OES have been proposed as alternatives for compendial methods and have the advantages of smaller sample size, element-specific information, quantitation, rapid sample throughput, and significantly improved accuracy.[195–198]

Vukjovic et al.[199] recently proposed a simple, fast, sensitive, and low-cost procedure based on solid phase spectrophotometric (SPS) and multicomponent analysis by multiple linear regression (MA) to determine traces of heavy metals in pharmaceuticals. Other spectroscopic techniques employed for high-throughput pharmaceutical analysis include laser-induced breakdown spectroscopy (LIBS),[200,201] fluorescence spectroscopy,[202–204] diffusive reflectance spectroscopy,[205] laser-based nephelometry,[206] automated polarized light microscopy,[207] and laser diffraction and image analysis.[208]

9.4 AUTOMATION AND ROBOTICS

The application of technology in laboratories via automation and robotics (flexible automation) minimizes the need for human intervention in analytical processes, increases productivity, improves data quality, reduces costs, and enables experimentation that otherwise would be impossible. Pharmaceutical companies continuously look for ways to reduce the time and effort required for testing. To meet the ever-increasing demands for efficiency while providing consistent quality of analysis, more pharmaceutical R&D and QC laboratories have now automated their sampling, sample preparation, and analysis procedures.

Pharmaceutical laboratory automation has greatly increased over the past two decades. Using robotic systems to carry out tedious tasks such as weighing samples and making dilutions allows chemists to perform value-added projects and tasks, thus increasing productivity. Automation of instrumentation minimizes human handling and human error. The pharmaceutical industry has often been at the forefront of implementing new technologies and invested heavily in laboratory automation and robotics.

9.4.1 FLOW INJECTION ANALYSIS

Flow injection analysis (FIA)[209] and its next generation technology known as sequential injection analysis (SIA)[210] are well established automated techniques that serve widespread applications in quantitative chemical analysis. FIA provides a simple, precise, and versatile means of automating manual wet chemical analytical procedures to achieve high sampling rates by exploring transient rather than conventional steady-state signals. SIA offers advantages over FIA in terms of robustness, simplicity, and low reagent and sample consumption. The fundamental principles of high-throughput FIA and SIA are discussed in several review articles and books.[209–213]

FIA has also found wide application in pharmaceutical analysis.[214,215] Direct UV detection of active ingredients is the most popular pharmaceutical analysis application of FIA. For single component analysis of samples with little matrix interference such as dissolution and content uniformity of conventional dosage forms, many pharmaceutical chemists simply replace a column with suitable tubing between the injector and the detector to run FIA on standard HPLC instrumentation. When direct UV detection offers inadequate selectivity, simple online reaction schemes with more specific reagents including chemical, photochemical, and enzymatic reactions of derivatization are applied for flow injection determination of pharmaceuticals.[216]

Many other selective techniques such as MS, FT-IR, ICP-MS, and electrochemical detection have also been used and several reviews of FIA applications in pharmaceutical analysis appear in the literature.[214–216] Several articles are dedicated to pharmaceutical analysis using SIA.[217,218] FIA and SIA have been applied to high-throughput analysis (up to 200 samples per hour with good precision) of APIs in a wide variety of pharmaceutical formulations including solids (tablets, capsules), pastes (ointments, creams), and liquids (emulsions, suspensions, solutions). Other applications include continuous monitoring of synthetic processes and dissolution assays. SIA shows potential for high-throughput determination of reactive species in pharmaceutical excipients such as aldehydes and peroxides for which derivatization is required.

Neither FIA nor SIA is suitable for multicomponent analysis. Recently introduced sequential injection chromatography (SIC) formed by coupling of a short monolithic column with SIA offers the possibility of performing separations in flow analysis manifolds that would not withstand the backpressure from conventional packed columns.[213] SIC is considered a convenient alternative to HPLC for high-throughput analysis of simple samples containing two to five analytes while maintaining the positive attributes of FIA and SIA such as manifold versatility for sample pretreatment, speed of analysis, and portability. SIC has been used for multicomponent analyses of various pharmaceutical formulations including syrups, drops, creams, capsules, and tablets. However, it is still not possible to use SIC for low pressure impurity profiling of pharmaceuticals. Most recently, the introduction of lab-on-valve (LOV) technology offers possible miniaturization of SIA and portable instruments and will further expand the utilization of flow injection techniques in pharmaceutical analysis.[212]

9.4.2 AUTOMATED SAMPLE PREPARATION

Sample preparation represents a major challenge and a very important step in the development and application of an analytical method. It is still considered the most labor-intensive, time-consuming, and error-prone step.[219] Although the technology advancements discussed above have significantly boosted the throughputs of chromatographic and spectroscopic techniques for pharmaceutical analysis, sample preparation often remains a data generation bottleneck.

Robotic systems that enable processing of hundreds of samples with no compromises in precision and accuracy have been commercially available for the last two decades. New technologies integrate sampling, preparation, and analysis in a single analytical platform and include the Caliper TPW3 (http://www.caliper.com) and the Sotax CTS (http://www.sotax.com). These fully automated, multitasking sample preparation workstations are specifically designed to prepare and analyze pharmaceutical solid dosage forms (tablet and capsule) and intermediate granulations for analyses such as content uniformity, stability-indicating assays, and blend and granulation uniformity. Both systems are compatible with most commercial UV-vis spectrometers for online sample readings. Online HPLC analysis is also available. Proper implementation of these systems can result in improved quality by eliminating human error and increased productivity by allowing more time for more critical challenges. The CTS can process multiple samples simultaneously, producing a further increase in sample throughput.

While these workstations are specially designed for tablets and capsules, other types of samples such as suspensions and viscous liquids can be processed using more flexible robotic systems such as the Symyx (www.symyx.com) and Tecan (www.tecan.com) platforms that are completely modular and may be configured into a range of workflows for specific applications and various types of samples. Increasing demands for high-throughput profiling of physicochemical and biopharmaceutical properties of PCCs led to widespread use of these robotic tools in preformulation studies of solubility,[220,221] polymorph and salt selection, liquid formulation screening,[222] forced degradation,[223] and excipient compatibility.[223] The flexibility of these robotic systems can allow full automation of tedious, labor-intensive analytical and preformulation tasks in pharmaceutical laboratories, e.g., extractable and leachable assays, microbial testing, and GLP sample preparations.

9.4.3 AUTOMATED DISSOLUTION TESTING

Dissolution testing of pharmaceutical dosage forms, one of the most frequent tasks to be performed in a pharmaceutical laboratory, is another laborious and time-consuming process that generates a large number of samples.[224,225] The extensive work required often prevents skilled analysts from assuming more challenging responsibilities. As a result, a great deal of effort has been given to automate dissolution testing in the pharmaceutical industry. Commercial dissolution systems providing various degrees of automation are available.

Most manufacturers of dissolution testing devices offer semi-automated systems that can perform sampling, filtration, and UV reading or data collection. These systems automate only a single test at a time. Fully automated systems typically automate entire processes including media preparation, media dispensing, tablet or capsule drop, sample removal, filtration, sample collection or analysis (via direct connection to spectrophotometers or HPLCs), and wash cycles. A fully automated system allows automatic performance of a series of tests to fully utilize unused night and weekend instrument availability.

Advanced media handling capabilities provide the flexibility to address any method requiring media changes or modifications such as dissolution testing of extended-release dosage forms. Fully automated systems can be purchased off the shelf or fully customized. Examples of commercial products include Caliper's MultiDose G3™ (Maryland, USA) and Sotax' AT 7smart™ (Basel, Switzerland). These fully automated systems provide significant increases in testing capacity. For example, up to 20 dissolution tests can be completed in 24 hr with the AT 7smart compared to only four manual tests.[226] Methods requiring baskets (USP I) and sinkers can also be performed on the AT 7smart. It allows automated basket changes for up to 10 unattended basket tests.

Fully automated dissolution systems eliminate time-consuming media preparation and system cleaning, increase testing throughput by maximizing the number of dissolution tests per analyst per day, eliminate human error, and increase cost savings in QC laboratories. One disadvantage is that all working steps must be carried out sequentially, leading to long processing times. Dissolution profiles with short sampling intervals present particular difficulties. Furthermore, in early development projects, robotic systems are not suitable for processing small series of test samples for which frequent changes of method are required because of limited flexibility. Semi-automated dissolution systems with online or *in situ* measurement options are important for efficient formulation development. Online UV systems with spectrophotometers are widely used and significantly reduce data turn-around time and analyst effort. Online HPLC can process samples that are not suitable for UV measurement due to matrix interference. Continuous real-time dissolution profiles can be obtained with *in situ* dissolution apparatus using UV fiberoptic probes.[227] Fiberoptic measurement eliminates sample withdrawal and enables more frequent sampling. Additional benefits include fewer moving parts, less carryover and fewer leaks and blockages by air bubbles and particulates. However, the disadvantages are light scattering interference and higher UV cut-off wavelength compared to online UV systems. *In situ* dissolution measurements with ion selective sensors,[228] fiberoptic chemical sensors,[229] and multivariate chemometric approaches[230] to minimize matrix interference are reported in the literature.

9.4.4 IMPLEMENTATION OF AUTOMATION

Laboratory automation in pharmaceutical analysis attained maturity since robots first appeared in pharmaceutical laboratories more than 20 years ago. While automation offers great promise for improving sample throughput and reducing sample backlog, its implementation has not been without problems. The industry cannot invest heavily in tools that produce little return on investment. Strategies in key aspects of automation such as planning, vendor selection, personnel, and efficient use of systems can determine the success or failure of an automation project.

Any decision to establish automated or robotic systems must carefully consider prerequisites such as the annual numbers of samples to be processed to achieve an acceptable cost-to-benefit ratio. Late phase development stability studies may benefit from fully automated systems based on the enormous numbers of samples to be analyzed for each stability time point. The use of automated systems in manufacturing quality control is now required due to the sheer number of samples to be tested for each marketed product. More flexible robotic tools are better suited for early development projects such as GLP and preformulation studies. Selection and customization of a wide variety of automated systems should be conducted in a way to judiciously balance current needs with future requirements to ensure successful deployment and continued use.

A dedicated automation specialist or group with necessary skill sets (analytical chemistry, computer literacy, and instrumentation) able to devote sufficient time to automation implementation is critical to take a project to fruition. The success of a robotics program can be enhanced greatly by making it accessible to a large population. Thus, it is essential for automation specialists to work closely with every analytical area to develop and validate automated methods for developmental products. After the technology is well established, the specialists can return to their operating areas.

Another challenge of implementing automation in pharmaceutical analysis within a regulated environment is the validation of each new automated method, especially during early development when the need to analyze relatively small volumes of samples of many different products may discourage chemists from utilizing automation. Fortunato[231] discussed strategies for validating automated instrumentation.

9.5 CONCLUSIONS

As the pharmaceutical industry embraces high-throughput technologies in drug discovery, more drug candidates move along the development pipeline as results of unprecedented numbers of novel drug targets. Consequently, analysis costs may escalate at certain points in the development process. Analysts in ChR&D and PhR&D face the challenges of implementing efficient tools for high-throughput analysis to help meet aggressive deadlines. Considerable efforts have gone into developing new analytical tools to increase productivity, decease operational costs, facilitate product development, and increase revenue generation.

The past decade has witnessed a continued trend in the development and implementation of high-throughput techniques in all aspects of pharmaceutical analysis. Technological advances have provided analytical capabilities that were unimaginable only a few years ago. Chromatographic-based techniques maintain their dominance in pharmaceutical laboratories. Dramatic advances in sample analysis throughput incorporate various approaches for fast chromatography. The focus of using HPLC for drug analysis is on faster separations with comparable or improved separation capability, based on applications of monolithic columns and STM columns that may be operated at high column pressures and/or temperatures. Parallel chromatography has also gained popularity as an efficient way to improve sample throughput. HPLC use is no longer limited to reversed-phase chromatography; increased interest relates to SFC applications. The trend toward pursuing more reliable and faster separations will continue. On the other hand, spectroscopic-based techniques are gaining increasing popularity in high-throughput analysis in pharmaceutical laboratories. Direct sample analysis by spectroscopic techniques is more widespread because of its inherent simplicity and compatibility with various imaging techniques. Developments in the near future include increasing emphasis on new techniques such as DEI-MS and on the use of chemometric techniques to avoid complex sample preparation, e.g., *in situ* testing using NIR or Raman. Commercially available laboratory automation systems are capable of significantly increasing sample throughput in analysis of dosage forms and allowing complex sample preparations. Robotics use has also greatly increased for high-throughput preformulation studies. Automated dissolution increases productivity through increased capacity and more consistent, technique-independent results.

A few key aspects to successful implementation of high-throughput analytical tools should be mentioned. The large variety of analytical tools available for high-throughput analysis serve as a challenge to an analyst who must select correct tools for each situation. No single strategy can be viewed as a universal solution. Each methodology offers advantages and imposes significant limitations. A novel, state-of-the-art technology may appear to have important advantages over existing equipment, but a single significant disadvantage such as lack of ruggedness, unreasonably high cost, or poor sensitivity will prevent its acceptance for routine laboratory use. As more laboratories explore these new approaches and technologies, experience will demonstrate which ones deliver their promises. One of greatest challenges of pharmaceutical analysis in an era of high-throughput drug development has been to balance the need for high-throughput and maintain analytical standards of high quality required by GMPs and GLPs. As data production rates increase, more laboratories seek help through information technology. The increase in data generation via HT techniques has made automated data processing and information management essential. Despite the acceleration in data processing that has been brought about by advances in laboratory information management systems, further developments are required to keep up with the ever-increasing amounts of data generated and the efficient use of expensive high-throughput tools serves as a challenge in today's pharmaceutical laboratories.

Because of space limitations, this chapter cannot cover all existing and potential high-throughput techniques such as automated method development, rapid microbial methods, chemical sensing, multidimensional chromatography, and high-throughput microplate readers used in pharmaceutical analysis. High-throughput analysis is a very dynamic field. New and exciting technologies and developments constantly emerge in response to the needs and pressures of modern drug discovery and development. Many of what are now considered novel techniques available only to innovators will advance into more robust, easy-to-use, and more accurate high-throughput analytical instruments suitable for routine applications in pharmaceutical laboratories.

REFERENCES

1. Tufts Center for the Study of Drug Development, *Impact Report*, May/June 2003.
2. Tufts Center for the Study of Drug Development, *Impact Report*, September/October 2006.
3. Dong, M. *Today's Chem.* 2000, 9, 46.
4. Majors, R.E. *LCGC* 2006, 19, 352.
5. Wu, N. and Clausen, A.M. *J. Sep. Sci.* 2007, 30, 1167.
6. Snyder, L.R., Kirkland, J.J., and Glajch, J.L. *Practical HPLC Method Development*, 2nd ed., 1997, John Wiley & Sons: New York, Chap. 5.
7. Ahujia, S. and Scypinski, S. *Handbook of Modern Pharmaceutical Analysis*, 2001, Academic Press: New York, Chap. 4.
8. Gomis, D.B. et al. *Anal. Chim. Acta* 2005, 531, 105.
9. Zhu, L. et al. *Anal. Chim. Acta* 2007, 584, 370.
10. Kerns, E.H. et al. *J. Chromatogr. B.* 2003, 791, 381.
11. Wilson, D.M. et al. *Combinator. Chem. High Throughput Screen.* 2001, 4, 511.
12. Wyttenbath, N. et al. *Pharm. Dev. Technol.* 2005, 10, 499.
13. Alsenz, J. and Kansy, M. *Adv. Drug Deliv. Rev.* 2007, doi:10.1016/j.addr.2007.05.007.
14. Novakova, L. and Solich, P. *J. Chromatogr. A.* 2005, 1088, 24.
15. Nardi, C. and Bonelli, F. *Rapid Commun. Mass Spectrum.* 2006, 20, 2709.
16. Jerkovich, A.D., Mellors, J.S., and Jorgenson, J.W. *LCGC* 2003, 21, 600.
17. Thompson, J.W. et al. *LCGC* 2003, April, 16-20.
18. Wu, N., Liu, Y., and Lee, M. *J. Chromatogr. A.* 2006, 1131, 142.
19. Ngugen, D.T.T. et al. *J. Chromatogr. A.* 2006, 1128, 105.
20. Villiers, A.D. et al. *J. Chromatogr. A.* 2006, 1127, 60.
21. Wren, S.A.C. and Tchelitcheff, P. *J. Chromatogr. A.* 2006, 1119, 140.
22. Novakova, L., Matysova, L., and Solich, P. *Talanta* 2006, 68, 908.
23. Jones, M.D. and Plumb R.S. *J. Sep. Sci.* 2006, 29, 2409.
24. Antia, F.D. and Horvath, C. *J. Chromatogr.* 1988, 435, 1.
25. Yan, B. et al. *Anal. Chem.* 2000, 72, 1253.
26. Vanhoenacker, G. and Sandra, P. *J. Sep. Sci.* 2006, 29, 1822.
27. Thompson, J.D. and Carr, P.W. *Anal. Chem.* 2002, 74, 1017.
28. Yang, Y. *LCGC* April 2006, 53.
29. Thompson, J.D. and Carr, P.W. *Anal. Chem.* 2002, 74, 4150.
30. Thompson, J.D., Brown, J.S., and Carr, P.W. *Anal. Chem.* 2001, 73, 3340.

31. Xiang, Y., Liu, Y., and Lee, M.L. *J. Chromatogr. A.* 2006, 198.
32. Lestremau, F. et al. *J. Chromatogr. A.* 2007, 1138, 120.
33. Ikegami, T. and Tanaka, N. *Current Opinion in Chemical Biology* 2004, 8, 527.
34. Svec, F. and Geiser, L. *LCGC*, April 2006, 22.
35. Cabrera, K. *J. Sep. Sci.* 2004, 27, 843.
36. McCalley, D.V. *J. Sep. Sci.* 2003, 26, 187.
37. Kele, M. and Guiochon, G. *J. Chromatogr. A.* 2002, 960, 19.
38. Wu, N. et al. *Anal. Chim. Acta* 2004, 523, 149.
39. Gerber, F. et al. *J. Chromatogr. A.* 2004, 1036, 127.
40. McCally, D.V. *J. Chromatogr. A.* 2002, 965, 51.
41. Liu, Y. et al. *J. Liq. Chromatogr. Rel. Technol.* 2005, 28, 341.
42. Higginson, P.D. and Sach, N.W. *Org. Proc. Res. Dev.* 2004, 8, 1009.
43. Tzanavaras, P.D. and Themelis, D.G. *J. Pharm. Biomed. Anal.* 2007, 43, 1483.
44. Tzanavaras, P.D. and Themelis, D.G. *J. Pharm. Biomed. Anal.* 2007, 43, 1526.
45. Aboul-Enein, H.Y. et al. *J. Liq. Chromatogr. Rel. Technol.* 2005, 28, 1357.
46. Kirkland, J.J. *J. Chromatogr. Sci.* 2000, 38, 535.
47. Novakova, L., Solichova, D., and Solich, P. *J. Sep. Sci.* 2006, 29, 2433.
48. Guillarme, D. et al. *J. Chromatogr. A.,* 20007, 1149, 20.
49. Bones, J., Macka, M., and Paull, B. *Analyst*, 2007, 132, 208.
50. Kofman, J. et al. *Am. Pharm. Rev.* 2006, 9, 88.
51. Cunliffe, J.M., Adams-Hall, S.B., and Maloney, T.D. *J. Sep. Sci.* 2007, 30, DOI 10.1002/jssc.200600524.
52. Guillarme, D. et al. *Eur. J. Pharm. Biopharm.* 2007, 66, 475.
53. Yang, Y. and Hodeges, C.C. *LCGC* 2005, May, 3.
54. Ahujia, S. and Scypinski, S. *Handbook of Modern Pharmaceutical Analysis*, Academic Press: New York, 2001, Chap. 3.
55. Gomis, D.B. et al. *J. Liq. Chromatogr. Rel. Technol.* 2006, 29, 1861.
56. Yan, B. *Analysis and Purification Methods in Combinatorial Chemistry*, John Wiley & Sons: New York, 2004, Chap. 12.
57. www.sepmatic.com.
58. Lemmo, A.V., Hobbs, S., and Patel, P. *Assay Drug Dev. Technol.* 2004, 2, 389.
59. www.eksigent.com.
61. Sajonz, P. et al. *Chirality* 2006, 18, 803.
62. Sajonz, P. et al. *J. Chromatogr. A.* 2007, 1145, 149.
63. Welch, C.J. et al. *J. Liq. Chromatogr. Rel. Technol.* 2006, 29, 2185.
64. Liu, Y. et al. *J. Pharm. Biomed. Anal.* 2007, 43, 1654.
65. Yaku, K. and Morishita, F. *J. Biochem. Biophys. Meth.* 2000, 43, 59.
66. Zhang, Y. et al. *Drug Dis. Today* 2005, 10, 571.
67. Mukherjee, P.S. *J. Pharm. Biomed. Anal.* 2007, 43, 464.
68. Alexander, A.J. and Staab, A. *Anal. Chem.* 2006, 78, 3835.
69. Anton, K. and Berger, C. *Supercritical Fluid Chromatography with Packed Columns,* Vol. 75, 1998, Marcel Dekker: New York.
70. Villeneuve, M.S. and Anderegg, R.J. *J. Chromatogr. A.* 2007, 1998, 826, 217.
71. Phinney, K.W. *Anal. Chem.* 2000, 72, 204A.
72. Garzotti, M. and Hamdan, M. *J. Chromatogr. B.* 2002, 770, 53.
73. Matfouh, M. et al. *J. Chromatogr. A.* 2005, 1088, 67.
74. White, C. *J. Chromatogr. A.* 2005, 1074, 163.
75. Toribio, L. et al. *J. Chromatogr. A.* 2005, 1091, 118.
76. Gyllenhaal, O. *J. Chromatogr. A.* 2004, 1042, 173.
77. Toribio, L. et al. *J. Chromatogr. A.* 2001, 921, 305.
78. Biba, M. et al. *Am. Pharm. Rev.* 2005, 8, 68.
79. Welch, C.J., Biba, M., and Sajone, P. *Chirality* 2007, 19, 34.
80. Welch, C.J. et al. *Chirality* 19, 184.
81. Gudat, A.E. and Firor, R.L. *Agilent Application Note* 2006, Publication 5989–5494EN.
82. Harvey, B.A. and Barra, J. *Euro. J. Pharm. Biopharm.* 2003, 55, 261.
83. Xu, H. et al. *J. Pharm. Biomed. Anal.* 2005, 38, 225.
84. Nevado, J.J. et al. *J. Pharm. Biomed. Anal.* 2005, 38, 52.

85. Mitrevski, B. and Zdravkovski, Z. *Forensic Sci. Int.* 2005, 152, 199.
86. Li, Z., Han, Y.H., and Martin, G.P. *J. Pharm. Biomed. Anal.* 2002, 28, 673.
87. Li, Z. et al. *J. Chromatogr. A.* 2006, 1104, 1.
88. Snow, N. *J. Liq. Chromatogr. Rel. Technol.* 2004, 27, 1317.
89. Matisova, E. and Domotorova, M. *J. Chromatogr. A.* 2003, 1000, 199.
90. Korytar, P. et al. *Trends Anal. Chem.* 2002, 21, 558.
91. Mastovska, K. and Lehotay, S.J. *J. Chromatogr. A.* 2003, 1000, 153.
92. Firor, R.L. *Agilent Application Note* 2007, Publication 5989–6078EN.
93. Wang, C. *Agilent Application Note* 2007, Publication 5989–6103EN.
94. Meng, C.K. *Agilent Application Note* 2006, Publication 5989–6026EN.
95. Chen, T.K., Phillips, J.G., and Durr, W. *J. Chromatogr. A.* 1998, 811, 145.
96. Raghani, A.R. *J. Pharm. Biomed. Anal.* 2002, 29, 507.
97. Rocheleau, M.J., Titley, M., and Bolduc, J. *J. Chromatogr. B.* 2004, 805, 77–86.
98. David, F. et al. *J. Sep. Sci.* 2006, 29, 695.
99. Pavon, J.J.P. et al. *J. Chromatogr. A.* 2007, 1141, 123.
100. Pavon, J.J.P. et al. *Anal. Chem.* 2006, 78, 4901.
101. Gudat, A.E., Firor, R.L., and Bober, U. *Agilent Application Note* 2007, Publication 5989–6079EN.
102. Camilleri, P. *Capillary Electrophoresis: Theory and Practice*, 2nd Ed., CRC Press: Boca Raton, FL, 1997.
103. Issaq, H.J. *J. Liq. Chrom. Rel. Technol.* 2002, 25, 1153.
104. Wan, H. and Thompson, R.A. *Drug Dis. Today* 2005, 2, 171.
105. Natishan, T.K. *J. Liq. Chrom. Rel. Technol.* 2005, 28, 1115.
106. Ha, P.T.T., Hoogmartens, J., and Schepdael, A.V. *J. Pharm. Biomed. Anal.* 2006, 41, 1.
107. Macia, A. et al. *Trends Anal. Chem.* 2007, 26, 133.
108. Marin, A. and Barbas, C. *Pharm. Biomed. Anal.* 2004, 35, 769.
109. Sung, H.H. et al. *Eur. J. Pharm. Biopharm.* 2006, 64, 33.
110. Liu, S. et al. *Talanta* 2006, 70, 644.
111. Pang, H.M., Kenseth, J., Coldiron, S. *Drug Dis. Today* 2004, 9, 1072.
112. Henry, C.S. *Microchip Capillary Electrophoresis: Methods and Protocols*, Humana Press, Totowa, NJ, 2006.
113. Bakeev, K.A. *Process Analytical Technology*, Blackwell Publishing, 2005.
114. Busaranon, K., Suntornsuk, W., and Suntornsuk, L. *J. Pharm. Biomed. Anal.* 2006, 41, 158.
115. Sarkar, M., Khandavilli, S., and Panchagnula, R. *J. Chromtogr. B.* 2006, 830, 349.
116. Gorog, S. *Trends in Anal. Chem.* 2007, 26, 12.
117. Amin, A.S. et al. *Spectrochim. Acta A,* in press.
118. El-Shabrawy, Y. et al. *Il Farmaco* 2003, 58, 1033.
119. Sultan, M. *Il Farmaco* 2002, 57, 865.
120. Abdine, H., Belal, F., and Zoman, N. *Il Farmaco* 2002, 57, 267.
121. Al-Ghanam, S.M. and Belal, F. *Il Farmaco* 2001, 56, 677.
122. Belal, F., Al-Zaagi, I.A., and Abounassif, M.A. *Il Farmaco* 2000, 55, 425.
123. Erk, N. *Il Farmaco* 2003, 58, 1209.
124. Bonsái, D. et al. *J. Pharm. Biomed. Anal.* 1997, 16, 431.
125. Ozaltin, N. and Kocer, A. *J. Pharm. Biomed. Anal.* 1997, 16, 337.
126. Fattah, A. and Walily, M.E. *J. Pharm. Biomed. Anal.* 1997, 16, 21.
127. Vignaduzzo, S. et al. *Anal. Bioanal. Chem.* 2006, 386, 2239.
128. Ni, Y., Qi, Z., and Kokot, S. *Chemom. Intell. Lab. Syst.* 2006, 82, 241.
129. Boeris, M.S., Luco, J.M., and Olsina, R.A. *J. Pharm. Biomed. Anal.* 2000, 24, 259.
130. Canafa, M.J. et al. *Talanta* 1999, 49, 691.
131. Bernal, J.L. et al. *J. Chromatgr. A.* 1998, 823, 423.
132. Blanco, M. et al. *J. Pharm. Biomed. Anal.* 1996, 15, 329.
133. Salari, A., Young, R. E. *Int. J. Pharm.* 1998, 163, 157.
134. Helmy, R. et al. *Anal. Chem.* 2003, 75, 605.
135. Dohi, K., Kaneko, F., and Kawaguchi, T.J. *Crystal Growth* 2002, 237–239, 2227.
136. Bunaciu, A.A., Aboul-Enein, H.Y., and Fleschin, S. *Il Farmaco* 2005, 60, 685.
137. Bunaciu, A.A., Aboul-Enein, H.Y., and Fleschin, S. *Il Farmaco* 2005, 60, 33.
138. Boyer, C. et al. *J. Pharm. Biomed. Anal.* 2006, 40, 433.
139. Pan, L., LoBrutto, R., and Zhou, G. *Talanta* 2006, 70, 661.

140. Friebe, A. and Siesler, H.W. *Vibrational Spec.* 2007, 43, 217.
141. Masmoudi, H. et al. *Int. J. Pharm.* 2005, 289, 117.
142. Perston, B.B. et al. *Anal. Chem.* 2007, 79, 1231.
143. Chan, K.L.A. et al. *Vibrational Spec.* 2007, 43, 221.
144. Chan, K.L.A. and Kazarian, S.G. *Vibrational Spec.* 2006, 42, 130.
145. Chan, K.L.A. and Kazarian, S.G. *J. Comb. Chem.* 2006, 8, 26.
146. Chan, K.L.A. and Kazarian, S.G. *J. Comb. Chem.* 2005, 7, 185.
147. Ciurczak, E.W. and Drennen, J.K. III. *Pharmaceutical and Medical Applications of Near-Infrared Spectroscopy*, Marcel Dekker: New York, 2002.
148. Reich, G. *Adv. Drug. Deliv. Rev.* 2005, 57, 1109.
149. Luypaert, J., Massart, D.L., and Heyden, Y.V. *Talanta* 2007, 72, 865.
150. Lee, E. et al. *Spectroscopy* 2006, 21, 24.
151. Feng, Y.C. and Hu, C.Q. *J. Pharm. Biomed. Anal.* 2006, 41, 373.
152. John, C.T. and Pixley, N.C. *Am. Pharm. Rev.* 2007, January/February, 120.
153. Blanco, M. and Alcala, M. *Anal. Chim. Acta* 2006, 557, 353.
154. Bodson, C. et al. *J. Pharm. Biomed. Anal.* 2006, 41, 783.
155. Zhao, L. et al. *Spectrochim. Acta A* 2007, 66, 1327.
156. Chalmers, J. and Griffiths, P.R. *Handbook of Vibrational Spectroscopy*, Vols. 1–5, John Wiley & Sons: Chichester, 2001.
157. Schrader, B. *Infrared and Raman Spectroscopy: Methods and Applications*, VCH: Weinheim, 1995.
158. Waterwig, S. and Neubert, R.H.H. *Adv. Drug Delivery Rev.* 2005, 57, 1144.
159. Baeyens, W., Verpoort, F., and Vergote, G. *Trends Anal. Chem.* 2002, 21, 8697.
160. Kim, M. et al. *Anal. Chim. Acta* 2007, 587, 200.
161. Mazurek, S. and Szostak, R. *J. Pharm. Biomed. Anal.* 2006, 40, 1235.
162. DeBeer, T.R.M. et al. *Anal. Chim. Acta* 2007, 192.
163. DeBeer, T.R.M. et al. *Eur. J. Pharm. Sci.* 2004, 23, 355.
164. DeBeer, T.R.M. et al. *Eur. J. Pharm. Sci.* 2007, 30, 229.
165. Dennis, A.C. et al. *Int. J. Pharm.* 2004, 279, 43.
166. Cantu, R. et al. *Am. Pharm. Rev.* 2007, 96.
167. Karabas, I., Orkoula, M.G., and Kontoyannis, C.G. *Talanta* 2007, 71, 1382.
168. Kauffman, J.F., Dellibovi, M., and Cunningham, C.R. *J. Pharm. Biomed. Anal.* 2007, 43, 39.
169. Kim, M. et al. *Anal. Chim. Acta* 2006, 579, 209.
170. Mazurek, S. and Szostak, R. *J. Pharm. Biomed. Anal.* 2006, 40, 1225.
171. Johansson, J., Pettersson, S., and Folestad, S. *J. Pharm. Biomed. Anal.* 2005, 39, 510.
172. Szostak, R. and Mazurek, S. *J. Mol. Struct.* 2004, 704, 229.
173. Lowry, S. et al. *JALA* 2006, April, 75.
174. Kojima, T. et al. *Pharm. Res.* 2006, 23, 806.
175. Gamberini, M.C. et al. *J. Mol. Struct.* 2006, 785, 216.
176. Morlock, G. and Ueda, Y. *J. Chromatogr. A.* 2007, 1143, 243.
177. Ricci, C. et al. *Anal. Bioanal. Chem.* 2007, 387, 551.
178. Williams, J.P. et al. *Rapid Commun. Mass Spectrom.* 2006, 20, 1447.
179. Leuthod, L.A. et al. *Rapid Commun. Mass Spectrom.* 2006, 20, 103.
180. Chen, H. et al. *Anal. Chem.* 2005, 77, 6915.
181. Shamsipur, M. et al. *J. Pharm. Biomed. Anal.* 2007, 43, 1116.
182. Salem, A.A., Mossa, H.A., and Barsoum, B.N. *J. Pharm. Biomed. Anal.* 2006, 41, 654.
183. Wawer, I., Pisklak, M., and Chilmonczyk, Z. *J. Pharm. Biomed. Anal.* 2005, 38, 865.
184. Zoppi, A., Linares, M., and Longhi, M. *J. Pharm. Biomed. Anal.* 2005, 37, 627.
185. Fardella, G. et al. *Int. J. Pharm.* 1995, 121, 123.
186. Hanna, G.M. and Lau-Cam, C.A. *J. Pharm. Biomed. Anal.* 1993, 11, 855.
187. Kwakye, J.K. *Talanta* 1985, 32, 1069.
188. Berendt, R.T. et al. *Trends Anal. Chem.* 2006, 25, 977.
189. Nelson, B.N., Schieber, L.J., and Munson, E.J. *Solid State Nuclear Magnetic Resonance* 2006, 29, 204.
190. Rao, R.N. and Talluri, M. *J. Pharm. Biomed. Anal.* 2007, 43, 1.
191. Huang, J. et al. *J. Pharm. Biomed. Anal.* 2006, 40, 227.
192. Zachariadis, G.A. and Michos, C.E. *J. Pharm. Biomed. Anal.* 2007, 43, 951.
193. Wang, L., Marley, M., and Bahnck, C. *J. Pharm. Biomed. Anal.* 2003, 33, 955.

194. Lam, R. and Salin, E. *J. Anal. At. Spectrom.* 2004, 19, 938.
195. Wang, T., Wu, J., and Egan, R.S. *J. Pharm. Biomed. Anal.* 2000, 23, 867.
196. Lewen, N. et al. *J. Pharm. Biomed. Anal.* 2004, 35, 739.
197. Wang, T., Jia, X., and We, J. *J. Pharm. Biomed. Anal.* 2003, 33, 639.
198. Zachariadis, G.A. and Kapsimali, D.C. *J. Pharm. Biomed. Anal.* 2006, 41, 1212.
199. Vukovic, J., Matsuoka, S., and Zupanic, O. *Talanta* 2007, 71, 2085.
200. St-Onge, L. et al. *J. Pharm. Biomed. Anal.* 2004, 36, 277.
201. St-Onge, L. et al. *Spectrochim. Acta B* 2002, 57, 1131.
202. Moreira, A.B. et al. *Anal. Chim. Acta* 2005, 539, 257.
203. Moreira, A.B., Dias, I.L.T., and Kubota, L.T. *Anal. Chim.Acta* 2004, 523, 49.
204. Pulgarin, J.A.M., Bermejo, L.F.G., and Lara, J.L.G. *Anal. Chim. Acta* 2003, 495, 249.
205. Gotardo, M.A. et al. *Talanta* 2004, 64, 361.
206. Dehring, K.A. et al. *J. Pharm. Biomed. Anal.* 2004, 36, 447.
207. Sugano, K., Kato, T., and Mano, T. *J. Pharm. Sci.* 2006, 95, 2115.
208. Tinke, A.P., Vanhoutte, K., and Winter, H.D. *Int. J. Pharm.* 2005, 297, 80.
209. Hansen, E.H. and Wang, J. *Anal. Lett.* 2004, 37, 345.
210. Hansen, E.H. and Miro, M. *Trends in Anal. Chem.* 2007, 26, 18.
211. Economou, A. *Trends in Anal. Chem.* 2005, 24, 416.
212. Wang, J. and Hansen, E.H. *Trends Anal. Chem.* 2003, 22, 225.
213. Chocholous, P., Solich, P., and Satinsky, D. *Anal. Chim. Acta* 2007, doi:10.1016/j.aca.2007.02.018.
214. Tzanavaras, P.D. and Themelis, D.G. *Anal. Chim. Acta* 2007, 588, 1.
215. Calatayud, J.M. In *Automation in the Laboratory*, Taylor & Francis: London, 1996.
216. Evgenev, M.I., Garmonov, S.Y., and Shakirova, L.S. *J. Anal. Chem.* 2001, 56, 313.
217. Pimenta, A.M. et al. *J. Pharm. Biomed. Anal.* 2006, 40, 16.
218. Solich, P. et al. *Trends Anal. Chem.* 2003, 22, 116.
219. Valcarcel, M., de Castro, M.D.L., and Tena, M.T. *Anal. Proc.* 1993, 30, 276.
220. Fligge, T.A. and Schuler, A. *J. Pharm. Biomed. Anal.* 2006, 42, 449.
221. Tan, H. et al. *JALA* 2005, 10, 364.
222. Dai, W.G., Dong, L.C., and Eichenbaum, G. *Int. J. Pharm.* 2006, doi:10.1016/j.ijpharm.2006.11.034.
223. Carlson, E. et al. *JALA* 2005, 10, 374.
224. Banakar, U.V. *Pharmaceutical Dissolution Testing*, 1991, Taylor & Francis: London, Chap. 4.
225. Dressman, J. and Kramer, J. *Pharmaceutical Dissolution Testing*, 2005, Taylor & Francis: London, Chap. 13.
226. Rolli, R. *J. Automated Meth. Mgt. Chem.* 2003, 25, 7.
227. Lu, X., Lozano, R., and Shah, P. *Dis. Technol.* 2003, 11, 6.
228. Bohets, H., Vanhoutte, K., and Nagels, L.J. *Anal. Chim. Acta* 2007, 581, 181.
229. Wu, J., Yang, M., and Chen, J. *Spectros. Spec. Anal.* 2006, 26, 1761.
230. Wiberg, K.H. and Hultin, U.K. *Anal. Chem.* 2006, 78, 5076.
231. Fortunato, D. *Pharm. Tech.* 2006, 30, 116.

10 Online Solid Phase Extraction LC/MS/MS for High-Throughput Bioanalytical Analysis

Dong Wei and Liyu Yang

CONTENTS

10.1 INTRODUCTION

Since the late 1980s, the liquid chromatographic tandem mass spectrometric (LC/MS/MS) method has been applied to quantitative measurements of a broad range of analytes in the pharmaceutical, diagnostic, environmental, food science, and forensics areas. Its advantages include high sensitivity, good selectivity, and the ability to measure multiple analytes simultaneously. It has been applied to small organic molecules, peptides, and even proteins.

The LC/MS/MS method utilizes the principle of three-dimensional separation to achieve excellent selectivity based on chromatographic separation (reversed-phase, size-exclusive, ionic, etc.), the unique mass-to-charge ratio of the analyte's parent ion, and the fragment ion. A sample clean up step such as protein precipitation, solid phase extraction, or liquid–liquid extraction is often needed before analysis and can serve multiple purposes such as extracting analytes from solid samples, enriching analytes (e.g., from urine samples), reducing sample complexity (e.g., removing proteins in plasma), and removing ion suppression and/or enhancement background interferences (e.g., inorganic salts in biological samples). Although offline sample preparations have become less labor-intensive since the introduction of automated liquid handling systems and 96-well plates, they still create bottlenecks that analysts must deal with daily.

A variety of online solid extraction devices and applications have been developed for bioanalysis. Many are easy to build in laboratories or commercially available. Unlike offline methods, minimal operator intervention is needed for daily sample analysis after online applications are set up, so the approach is both labor- and cost-effective. The technique can also minimize errors arising from manual operations, eliminate potential inconsistencies caused by different operators, and provide accessibility of LC/MS/MS applications to laboratories that have minimal analytical expertise.

This chapter will review online solid phase extraction (SPE) LC/MS/MS applications published in recent years. According to instrumentation set up, the online SPE systems are divided into three categories: column switching online SPE LC/MS/MS systems; commercial online SPE/LC systems with disposable cartridges, and turbulence flow chromatography. The applications of these systems in the quantitative analysis of pharmaceutical agents in the pharmaceutical industry will be discussed. Quantitative analysis of pharmaceutical agents and other chemicals in the environmental area will be briefly summarized as well. Due to the explosion of publications in this area in recent years, complete coverage is impossible. In addition, several other major online SPE techniques and applications are not included in this chapter: for example, online SPE HPLC coupled with mass spectrometry and magnetic resonance spectroscopy (SPE/LC/MS/NMR), online SPE combined with capillary electrophoresis and mass spectrometry (SPE/CE/MS), etc.

10.2   COLUMN SWITCHING ONLINE SPE LC/MS/MS SYSTEMS

10.2.1   Basic Concepts

The feasibility of online SPE LC/MS/MS has been tested since the introduction of thermospray ionization. In an early research paper by Lant and Oxford (1987), a prototype online SPE LC/MS system was set up and successfully applied for the measurement of labetalol, a hypertension drug and α- and β-adrenergic receptor, in plasma. This system was set up by coupling an advanced automated sample processor (AASP, Varian, Walton-on-Thames, UK) with a reversed-phase column, a ten-port switching valve, and an MS equipped with thermospray interface (Vestec, Houston, Texas) (Blakley et al. 1980, Blakley and Vestal 1983).

After overcoming the instability of the thermospray ionization source, a sensitivity of 2 ng/mL was achieved with a calibration range of 10 to 103 ng/mL for human plasma samples. In recent years, the online SPE LC/MS/MS technology has matured and is now easy to build and use. It is used widely in the pharmaceutical industry (Jemal et al. 2000; Hsieh 2004; Hennion 1999).

Figure 10.1 shows a basic online SPE LC/MS/MS system, a column switching system (Kahlich et al. 2006) that includes a six-port switching valve, an injection valve, an SPE cartridge, an analytical column, SPE washing and HPLC pumps, and MS detector. The online process has three steps:

1. Injection: the sample solution is injected into the sample loop by an autosampler.
2. Online SPE: after conditioning of the SPE cartridge, the sample solution is flushed into the cartridge by the washing solvent after a positional switch of the injection valve. A large volume of washing solvent is used to clean up the sample.
3. Chromatographic separation and MS detection: after a positional switch of the second valve, the analyte is eluted from the SPE cartridge into the analytical column by a gradient pump using a higher organic content mobile phase. A gradient is used to separate

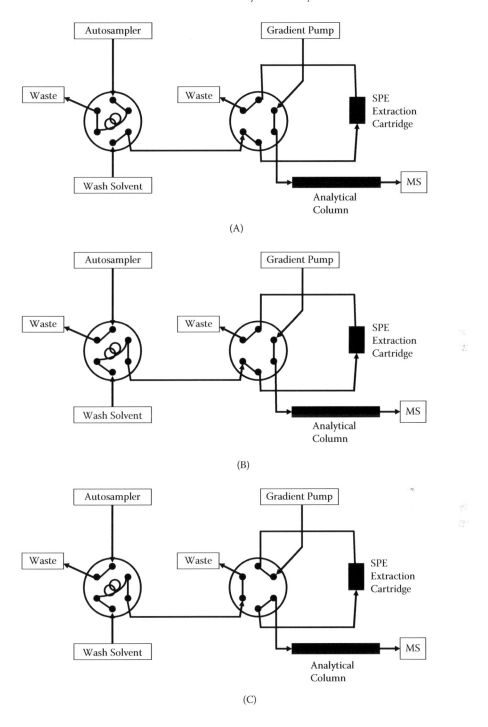

FIGURE 10.1 Online SPE-LC/MS/MS: (A) sample injection; (B) online SPE; (C) LC gradient elution and MS detection. (*Source:* Adapted from Kahlich. R. et al., *Rapid Commun Mass Spectrom* 2006, 20, 275. With permission.)

the analytes and elute them to the MS for detection. All components can be easily configured and the experimental process can be built up and controlled by a single piece of software.

A number of modifications have been adapted to improve online SPE: for example, online filters and precolumns to increase cartridge and column life; a protein precipitation step to increase cartridge life and shorten the analytical step; online aqueous dilution for peak focusing and pre-concentration of analytes from the SPE cartridge into analytical column; additional pumps to maintain pressure balance during valve switching; and time programming and use of dual SPE cartridges and/or a monolith analytical column with a fast flow to increase throughput.

A wide variety of SPE materials and cartridges are commercially available: for example, alkyl–diol silica-based restrictive access materials (RAMs) and a variety of silica- and polymer-based SPE materials of different binding abilities and capacities. Reversed-phase, size-exclusion, ion-exchange SPE, and turbulence flow methods will be discussed in this chapter related to real-world applications.

10.2.2 Bioanalytical Applications

RAMs packed in small cartridges are often used for online SPE of samples in biological matrixes. First developed by Boos et al. (1991), these alkyl-diol-silica (ADS) materials consist of large size particles (20 to 50 μm compared to 3.5 to 5 μm for analytical columns) with hydrophilic electron-neutral surfaces that do not retain proteins and hydrophobic internal pore surfaces that allow only small molecules to enter and bind.

Viehauer et al. (1995) used a cartridge (25 × 4 mm inner diameter, 25 μm) packed with ADS material for online SPE extraction of 8-methoxypsoralen in plasma. A LiChrosphere RP-18 column (125 × 4 mm inner diameter, 5 μM, Merck KGaA, Darmstadt, Germany) handled separation. Plasma samples (25 μL) were injected directly into the online cartridge and washed with a flow of HPLC-grade water (1 mL/min for 8 min). Via valve switching, the analyte was eluted from the cartridge with methanol:water (60:40 v/v) by backflush into the analytical column. Separation was carried out under isocratic conditions using the same mobile phase. The photoreactive drug was measured using fluorescence detection. The assay had a calibration range of 21.3 to 625.2 ng/mL and a total run time of 15 min. The lifetime of the cartridge exceeded 100 mL plasma.

Christiaens et al. (2004) measured cyproterone acetate (CPA), a drug used to treat prostate carcinoma, in human plasma. A LiChrosphere RP-4 (25 × 2 mm inner diameter, 25 μm, Merck KGaA) cartridge was coupled with an OminiSpher C18 column (100 × 2 mm inner diameter, 3 μm) and 30 μL plasma was injected directly. The lower limit of quantitation (LOQ) was 300 pg/mL. Recovery was 100%.

For applications in the diagnostics and biomarker area, 8-oxo-7,8-dihydro-2′-deoxyguanosine (8-oxoGuo) was measured as an oxidation stress biomarker in urine samples from smokers and non-smokers (Hu et al. 2006). When 100 μL of samples (10 times dilution) were used, a detection limit of 5.7 pg/mL (2.0 fmol) was achieved. The cycle time was 10 min per sample. The application was used for clinical scale. A similar approach was used for the detection of N7-methylguanine, another carcinogen exposure biomarker in human urine (Chao et al. 2005).

Online SPE LC/MS can also be used to measure peptides or small protein drugs. Dai et al. (2005) used a similar system to measure sifuvirtide, a 36-amino acid peptide with a molecular weight of 4727 Da in monkey plasma pre-spiked with protease inhibitors for pharmacokinetic studies. A [127]I-labeled peptide was used as an internal standard. Multiple charged parent and fragment pairs were used for selective reaction monitoring (SRM) with an ion trap instrument. Formic acid (FA) was added to plasma samples to adjust pH and overcome the zwitterionic properties of the peptides. A single sample run took 18 min. Spherical packing material was used for the SPE cartridges to improve reproducibility (a lifetime of more than 300 samples). Despite severe ion suppression and poor recovery, a calibration range of 4.88 to 5000 ng/mL was achieved.

10.2.3 Modifications of Online SPE LC/MS/MS

In addition to online filters and precolumns, a simple protein precipitation step often precedes online SPE LC/MS/MS to prolong cartridge life. Protein precipitation can also reduce analytical interference and shorten chromatographic separation time. Since an internal standard (IS) solution is often added to plasma samples and centrifugation is used to remove possible particles before loading into the autosampler, protein precipitation does not add labor to the process.

A common practice is to dissolve the IS in an organic solvent such as acetonitrile. Both the IS and organic solvent for protein precipitation can be added to the samples in a single step. Zell et al. (1997a and b) used a Supelcosil LC-ABZ online trapping column (20 × 4.6 mm inner diameter, Supelco, Gland, Switzerland) to detect a potassium channel opener and its metabolite in rat plasma. The plasma samples were first treated with ethanol and the organic contents evaporated. The remaining supernatant was diluted and injected into the trap for online SPE. A total of 500 μL plasma was used. The calibration range was 0.25 to 100 ng/mL with an 8-min run time.

Koal et al. (2004) measured four immunosuppressants (cyclosporine A, tacrolimus, sirolimus, and everolimus) in whole blood samples from transplant recipients. The samples were treated first with a protein precipitation step. The supernatant was extracted with a Poros R1/20 perfusion column (30 × 2.1 mm, 20 μm, Applied Biosystems, Darmstadt, Germany) online. A Luna phenyl hexyl column (2 × 50 mm, Phenomenex, Schaffenburg, Germany) was used for separation. The total run time was 2.5 min. The lower limit of quantitation was 10 ng/mL for cyclosporine A and 1 ng/mL for the other three analytes.

Sometimes orthogonal offline SPE steps were used prior to online SPE LC/MS/MS. These preparation steps were used to remove interference and concentrate samples. In an application to measure urinary N7-(benzo[α]pyren-6-yl)guanine (BP-6-N7Gua), a biomarker for exposure to polyaromatic hydrocarbons (PAHs), a two-step offline SPE was first performed using Sep-Pak C8 (Waters, Milford, Massachusetts) and Strata SCX (Phenomenex, Torrance, California) cartridges to obtain high sensitivity (Chen et al. 2005). The extracts were applied to an online reversed phase SPE LC/MS system. The lower limit of detection was 2.5 fmol/mL when 10 mL of urine was used.

Online dilution is used to improve performance parameters such as peak focusing. During online SPE, a mobile phase containing organic solvent is used to elute analytes from the cartridge into an analytical column after online SPE. The volume and organic content of the mobile phase needed for analyte elution is determined by the hydrophobicity of the analytes, SPE material, cartridge size, and mobile phase pH. An organic solvent may cause peak broadening on the analytical column. One solution is to use an additional pump to dilute the eluent with water using a T connection after the SPE cartridge.

Ye et al. (2005) used this technique for peak focusing during online SPE to measure nine environmental phenols in human urine. A LiChrosphere RP-18 ADS SPE cartridge (25 × 4 mm inner diameter, 25 μm, Merck KgaA) was used with two Chromolith Performance RP-18 columns (100 × 4.6 mm inner diameter, Merck KgaA) in tandem. After extraction, the analytes were flushed out of the SPE cartridge with organic solvent (50% methanol:50% water, 0.5 mL/min) and diluted with high aqueous mobile phase flow (25% methanol:75% water, 0.25 mL/min) before hitting the analytical column. A 10-port switching valve was used to accommodate these procedures. When 100 μL of urine was used, a lower limit of detection of 0.1 ng/mL or 0.4 ng/mL was reached for most analytes.

Gundersen and Blomhoff (1999) used online dilution with online SPE to measure vitamin A (retinol) and other active retinoids in animal plasma. The intention of online dilution in this application was on optimizing SPE extraction conditions rather than on peak focusing during analytical separation. An SPE cartridge packed with Bondapak C18 materials (37 to 53 μM, 300 A, Waters, Milford, Massachusetts) and a reversed-phase analytical column (250 × 2.1 mm inner diameter, Superlex pkb-100, Supelco, Bellefonte, Pennsylvania) were controlled by a six-port switching valve (Rheodyne, Cotati,

California). Plasma samples were first precipitated with acetonitrile. The supernatant was injected directly with a high organic carrying flow (acetonitrile:1-butanol:methanol:2% ammonium acetate: glacial acetic acid [69:2:10:16:3 v/v]) at 0.5 mL/min that was needed to prevent the analytes from precipitating out in the injection loop. An online dilution flow (doubly distilled deionized water, 2.2 mL/min) was T-connected in before the SPE cartridge, resulting in an increase of water content to 84.5% and a decrease of the ionic strength of the analyte solution when it hits the cartridge. After loading and washing for 5 min and the switching of a valve, a mixed flow of 38.7% water content was used for elution. With another valve switch, chromatographic separation was carried out on the analytical column while the cleaning and equilibration steps were performed on the SPE cartridge to ready it for the next injection.

An online filter was also used to protect the analytical column. A guard column was used before the T to prevent breakthrough. A restrictor was used to balance the pressure before, during, and after the valve switches. After sample transfer, the valve was switched back. With this method, both water-insoluble retinoids and water-soluble retinoic acid were extracted simultaneously. Because of the minimal light exposure of these light-sensitive analytes during the procedure, an extraction recovery of 97 to 100% was achieved with a quantitation range of 100 fmol to 3 nmol.

A variety of techniques can increase throughput. Since the detector (MS, diode array detector, etc.) must be switched online for detection only around the windows of analyte retention times, the entire procedure can be modified to maximize its output. The option of multiplexing SPE cartridges and analytical columns (as many as four) is discussed in the next sections. Another option is to optimize each step in the process, i.e. washing and re-equilibrating the SPE cartridge while separating and detecting analytes on the analytical column or re-equilibrating the analytical column while injecting and extracting the next sample on the cartridge.

The application from van der Hoeven et al. (1997) used an ADS cartridge online SPE to measure cortisol and prednisolone in plasma and arachidonic acid in urine. A precolumn packed with a C18 alkyl–diol support (LiChrosphere RP-18 ADS, 25 μm, Merck) was used. To reduce run time, column switching was programmed as "heart-cut", diverting only the analyte fraction into the analytical column. Another LiChrosphere column (125 × 4 mm inner diameter, Merck) handled separation. After the injection of 100 μL plasma, the lower limit of detection for prednisolone was 1 ng/mL while cortisol was readily quantitated at its endogenous level of ~100 ng/mL. The run time was 5 min. For arachidonic acid, a Hypersil ODS column (200 × 3.0 mm inner diameter, 5 μm) was used. The injection volume was 200 μL and run time was 9.5 min. The detection limit was 1 ng/mL and recovery was 77%.

The recent adaptation of a monolith column has accelerated analytical separation by using a high flow rate. Zang et al. (2005) incorporated a simple online SPE (Strata-X 20 × 2.1 mm inner diameter, 25 μM, Phenomenex) with a Chromolith Speed ROD RP-18e monolithic column (4.6 × 50 mm, Merck KgaA) to increase throughput in toxicokinetic and pharmacokinetic screening studies. A flow rate up to 4 mL/min was used and total run time was 2.8 min per sample. Up to eight analytes were separated and monitored simultaneously with a linear range of 1.95 to 1000 ng/mL.

In a similar application, Huang et al. (2006) used a monolith column to separate a drug compound from its metabolite in a shorter run time (5 min) when compared with a C18 column (10 min). Ye et al. (2005) measured nine environmental phenols in human urine using a LiChrosphere RP-18 ADS SPE cartridge (25 × 4 mm inner diameter, 25 μm, Merck KgaA). Two Chromolith Performance RP-18 columns (100 × 4.6 mm inner diameter, Merck KgaA) in tandem were used for analytical purposes. After extraction, the analytes were flushed from the SPE cartridge and diluted with a high aqueous mobile phase for peak focusing. The lower limit of detection was 0.1 to 0.4 ng/mL for most analytes when 100 μL of urine was used.

An interesting idea was to use a monolith column to perform dual functions of online SPE and chromatographic separation. Because of the porous structure of a monolith column and its very low backpressure, plasma or diluted plasma can be directly injected. Plumb et al. (2001) used this approach to quantitate an isoquinoline drug and 3'-azido-3'-deoxythymidine (AZT). Diluted plasma samples (plasma:water 1:1) were injected directly into a Chromolith Speed ROD RP-18e column

(4.6 × 50 mm, Merck KgaA). A 1.5 min gradient was used with a flow rate of 4 mL/min. The method achieved good separation of the analytes and their glucuronide metabolites. The linear range was 5 to 2000 ng/mL. Up to 300 samples were injected. However, significant decreases in column efficiency and resolution were observed.

Hsieh et al. (2003) and Zeng et al. (2003a), used direction injection to quantitate multiple drug discovery compounds simultaneously with good robustness. In one application, the flow rate of the HPLC gradient was programmed. The flow rate was first increased (4 to 8 mL/min) for fast protein clean-up, then decreased (1.2 mL/min) for better sensitivity. The column maintained similar efficiency after several hundred injections.

To improve chromatographic separation, another analytical column could be used in addition to the monolith (Xu et al. 2006). The monolith column served as an extraction column only. Hsieh et al. (2000, 2002) utilized a polymer-coated mixed function (PCMF) Capcell C8 column (4.6 × 50 mm, Phenomenex) to provide dual functions—online plasma extraction and analyte separation. The silica was coated with a polymer containing both hydrophilic polyoxythylene and hydrophobic groups. The diluted plasma samples (1:1 to 1:3) were injected directly. No column deterioration was observed after 200 injections.

Although the SPE cartridge is most commonly used for online SPE, other devices are also used, including SPE discs and solid phase microextraction (SPME). Wachs and Henion (2003) designed a 96-well SPE card to measure ritalin in human urine. Samples were loaded and extracted offline on the card and the card was loaded on an x–y–z positioner. Sample solutions in each disc were pneumatically drawn and electrosprayed directly into the detector. QC samples were measured at 24 to 3000 ng/mL. Altun et al. (2004) used microextraction in a packed syringe (MEPS) for online measurements of local anesthetics (ropivacaine, 3OH-ropivacaine, pipecoloxilidide, lidocaine, and bupivacaine) in human plasma samples. The calibration range was 2 to 2000 nM and sample loading volume was 25 μL. A 250 μL gas-tight syringe packed with 1 mg sorbent (ACX, 50 μm) was used for up to 100 extractions. The carry-over was below 0.5% and recovery was about 50%.

10.2.4 Environmental and Other Applications

Although this book focuses on high-throughput analyses in the pharmaceutical industry, applications in environmental analysis are closely related. The same technologies are applicable to both fields. Pharmaceuticals have been monitored as pollutants in surface water, soil, food, and human plasma. In environmental applications, as many as 30 to 40 analytes have been monitored simultaneously.

Water samples as large as 250 mL can be injected for preconcentration in online SPE and the resulting lower limits of quantitations may be as low as nanograms per liter. Online SPE was used for monitoring phenols (Papadopoulou-Mourkidou et al. 2001; Ye et al. 2005); phthalates (Kato et al. 2005); herbicides and pesticides (Sancho et al. 2004; Hernández et al. 1998, 2001; Koeber et al. 2001; Ibáñez et al. 2005); antibiotics (Pozo et al. 2006); polycyclic aromatic hydrocarbons (PAHs) and polycyclic aromatic sulfur heterocycles (PASHs) (Gimeno et al. 2002); flame retardants, plasticizers, organophosphorus triesters (Amini and Crescenzi 2003); and genistein, daidzein, and soy isoflavones (Doerge et al. 2000) in surface water, soil, human and animal plasma, and urine.

10.3 COMMERCIAL ONLINE SPE/LC SYSTEMS WITH DISPOSABLE CARTRIDGES

10.3.1 Basic Concepts

Commercially available fully automated online SPE systems include Prospekt, Prospekt-2, and Symbiosis systems (Spark Holland, the Netherlands) and Merck's OSP-2 (Darmstadt). Figure 10.2 shows a Symbiosis (Kuklenyik et al. 2005) two-cartridge, single analytical column online SPE

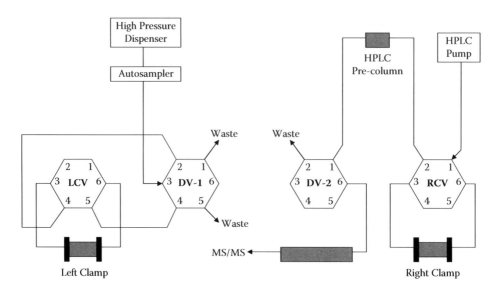

FIGURE 10.2 Tubing diagram for Symbiosis system (LCV = left clamp valve; DV-1 = divert valve; DV-2 = divert valve 2; RCV = right clamp valve). (*Source:* Adapted from Kuklenyik, Z. et al., *Anal Chem* 2005, 77, 6085. With permission.)

system. Unlike online SPE systems that use a single cartridge for multiple injections, this system uses a new cartridge for each injection. An automated cartridge exchange unit (ACE) is used to change cartridges automatically for each sample injection. Trays of 96 SPE cartridges are pre-loaded onto the system. After each analysis, a clamp drops the used SPE cartridge, picks up a new one, and connects it online. This one-time cartridge use eliminates potential carry-over and deterioration due to multiple injections.

10.3.2 BIOANALYTICAL APPLICATIONS

These systems have been used in many bioanalytical applications. A Prospekt system coupled with MS quantitated eserine N-oxide, a cholinesterase inhibitor, in human plasma for low level (4.5 mg) oral administration pharmacokinetic studies (Pruvost et al. 2000). After conditioning of the SPE cartridge (PLRP-S, Spark) with methanol (5 mL/min, 0.5 min) and water (5 mL/min, 0.5 min), a volume of 250 μL plasma plus internal standard was injected and washed (water, 1 mL/min, 3 min). The analytes were flushed out with 80:20 ammonium acetate (20 mM, pH 3.5 adjusted with formic acid) and acetonitrile (0.3 mL/min) and separated on a Zobax SB-CN column (150 × 2.1 mm inner diameter, 5 μm). A calibration range of 25 pg/mL to 12.5 ng/mL was achieved with a run time of 10.5 min.

Other applications include bioequivalent measurements of bromazepam, an anticonvulsant, in human plasma. The lower limit of quantitation (LLOQ) was 1 ng/mL (Gonçalves et al. 2005). Kuhlenbeck et al. (2005) studied antitussive agents (dextromethorphan, dextrophan, and guaifenesin) in human plasma; LLOQ values were 0.05, 0.05, and 5 ng/mL, respectively. Other compounds studied were nucleoside reverse transcriptase inhibitors, zidovudine (AZT) and lamivudine (3TC) (de Cassia et al. 2004) and stavudine (Raices et al. 2003) in human plasma, and paclitaxel, an anti-cancer agent, in human serum (Schellen et al. 2000).

Automated online SPE systems have been applied to various phases of drug discovery. McLoughlin et al. (1997) utilized the Prospekt system in pharmacokinetic animal studies for rapid drug candidate screening. Up to 10 compounds were simultaneously monitored. The lower limits of detection were

2.5 to 5 ng/mL and extraction recovery was 50 to 100%. Beaudry et al. (1998) utilized the Prospekt in a cassette dosing pharmacokinetic screening (n = 64). The LLOQ was 0.5 ng/mL.

Barrett et al. (2005) measured 6-beta-hydroxycortisol and cortisol in human urine. The ratio of the two compounds served as the indicator of CYP3A4 activity. A HySphere C18 HD cartridge (7 μm) was used in combination with a Symmetry Shield RP 18 analytical column (Waters, Morristown, New Jersey). The lower limits of quantitation were 1 and 0.2 ng/mL, for 6-beta-hydroxycortisol and cortisol, respectively. In a therapeutic drug monitoring (TDM) study, Nieder-länder et al. (2006) measured clozapine, a schizophrenia drug, and desmethyl-N-oxide metabolites in human serum. A HySphere-C18-HD cartridge (10 × 2 mm inner diameter, 7 μm), a Zorbax Eclipse XDB-C18 (30 × 2.1 mm, 3.5 μm, Agilent Technologies, Amstelveen, The Netherlands), and a guard column were used. The injection volume was 50 μL and total run time was 2.2 min. The lower limit of detection was 0.15 to 0.3 ng/mL.

10.3.3 SPE CARTRIDGE SELECTION

Cartridges with different SPE mechanisms (reversed-phase or ion exchange), binding strength (degree of hydrophobicity), and binding capacities are available for selection. The introduction of polymer-based SPE sorbents greatly increased the choices beyond conventional silica-based materials (Hennion 1999; Hsieh 2004). When choosing SPE cartridges, loading capacity along with the need to balance retention of and adsorption of all analytes are factors to consider.

An ideal SPE cartridge should have enough capacity and retain sufficient analytes to achieve good recovery while providing good adsorption so that chromatographic separation is not compromised. Other modifications such as online dilution may be needed to offset certain disadvantages.

Brostallicin, a synthetic DNA minor groove binding agent and anticancer drug candidate, was measured in human plasma using a Prospekt-2 system (Calderoli et al. 2003). With a pK$_a$ of 12, the compound was found not to retain well on C2, C8, C8-end capped, and C18 cartridges under neutral loading (water). A HySphere resin SH SPE cartridge (10 × 2 mm inner diameter, 15 to 25 μM, Spark) with a strong hydrophobic resin phase (modified polystyrene divinylbenzene) was used. Plasma samples (200 μL) were injected after the addition of an internal standard, then loaded onto a SPE cartridge with 500 μL of water (1 min/min) and washed with water (1 mL/min). The analytes were backflushed onto a precolumn and eluted from the analytical column (Platinum Cyano, 100 × 4.6 mm inner diameter, 3.6 μm, Alltech, Italy) with a 70:30 v/v acetonitrile–ammonium formate buffer (pH 3.5, 20mM) at a flow rate of 1 mL/min. The run time was 8 min and recovery was 40.2 to 57.9% for QC samples. The method was validated at a calibration range of 0.1 to 500 ng/mL with low carry-over (0.04%).

Rodriguez-Mozaz et al. (2004) measured eight estrogens and metabolites in natural and treated water. Four SPE cartridges were tested: PLRP-s (cross-linked styrene divinylbenzene polymer), two Hysphere resin GP (polydivinylbenzene, 10 to 12 μm and 8 μm) units, and a Hysphere C18 EC (end capped octadecyl-bonded silica cartridge, 10 × 2 mm, Spark). While other cartridges did not retain the polar compounds well (<70%), PLRP-s yielded good recovery (>74%) and chromatographic resolution with sample volumes as large as 250 mL. The LLOQ was 0.02 to 1.02 ng/L.

Wissiack et al. (2000) measured 12 phenols in surface water. Five polymer-based SPE materials were tested alone with a single silica-based compound: Hysphere SH (polydivinybenzene 15 to 25 μm, Spark), Hysphere GP (polydivinylbenzene, 5 to 15 μm, Spark), PRP-1 (cross-linked styrene divinylbenzene, 12 to 20 μm, Hamilton), PLRP-s (cross-linked styrene divinylbenzene, 15 to 25 μm, Polymer), Hysphere C18 HD (end-capped, C18 phase with high density of octadecyl chains, Spark), and Oasis (macroporous polydivinylbenzene-N-vinylpyrrolidone copolymer, Waters). Samples and SPE sorbents were acidified (sulfuric acid, pH 2.5) to achieve better retention. Hysphere GP and Waters Oasis cartridges yielded excellent recovery (>94%) for water samples up to 10 to 20 mL. The polymeric SPE cartridge retained analytes better than the analytical column. The mismatch was overcome by adapting an analytical column of larger dimension and stronger retention

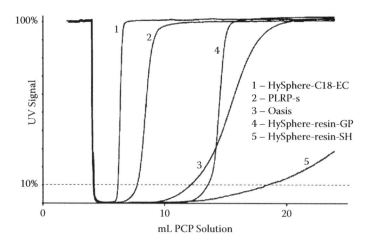

FIGURE 10.3 Breakthrough curves for pentachlorophenol (PCP) on various sorbents. PCP solution = 20 µg/L in acetonitrile:water (30:70), pH 2, 1 mL/min. (*Source:* Adapted from Schellen, A. et al., *J Chromatogr B* 2003, 788, 251. With permission.)

(Kromasil C18, 250 × 4 mm, Austrian Research Center, Seibersdorf) in connection with a smaller SPE cartridge (10 × 2 mm inner diameter). After online extraction, the LC elution gradient was used (40 to 100% organic, 2 mL/min) to flush the analytes out of the SPE cartridge and separate them on the analytical column. The larger flow rate applied to the smaller SPE cartridge allowed fast elution and focused peaks. Marchese et al. (1998) measured pranlukast, a peptidoleukotriene receptor antagonist, and its three oxidative metabolites for clinical studies. An end-capped phenyl (Ph-EC-IST) cartridge was used for online SPE. Calibration ranges were 40 to 2000 ng/mL for the parent and 1 to 200 ng/mL for metabolites; run time was 4.5 min.

A generic method was developed for 11 commercial drugs (sulfadiazine, taxol, propranol, and others) in porcine serum (Schellen et al. 2003). Five relatively strong hydrophobic sorbents were tested using a breakthrough measurement: silica-based HySphere C18 (EC), HySphere Resin GP, HySphere Resin SH (Spark), PLRP-s (Polymer), and Oasis HLB (Waters). The standard analytes (pH 2 or 7) were pumped directly into a UV detector to provide a constant signal. A cartridge was switched online to cause an initial UV signal drop to baseline when the analytes were retained, followed by an increase after maximum sorbent capacity was reached. An increase of UV signal to 10% of the original level was defined as the breakthrough point, at which the SPE cartridge capacity was measured. Figure 10.3 shows a typical example using pentachlorophenol. HySphere Resin GP (4) exhibited high retention capacity and desorption efficiency (steepness of breakthrough).

As a generic method, the SPE cartridge was conditioned with 1.5 mL methanol (5.0 mL/min) and 1.5 mL water (5.0 mL/min), after which 100 µL of spiked plasma was injected and washed with 3.0 mL water (2.0 mL/min). After switching online, the analytes were flushed and eluted with a fast gradient of mobile phase A (5:95 v/v acetonitrile:water, 0.1% formic acid, and 10 mM ammonium acetate) and B (95:5 v/v acetonitrile:water, 0.1% formic acid, and 10 mM ammonium acetate). The lower limit of quantitation was 0.2 to 2 ng/mL and linear range was 2 to 4 orders. Carry-over was 0.02 to 0.1 %.

Alnouti et al. (2005) used Symbiosis to measure propranolol and diclofenac in rat plasma. Twelve different SPE cartridges were screened. A C18 HD (2 × 10 mm inner diameter, Spark) was chosen because it provided the best recovery and peak shapes. When a Luna C18 (2.1 × 50 mm, 5 µm, Phenomenex) was used, the run time was 4 min; it was 2 min when a monolithic Chromolith C18 (50 × 2.1 mm, Merck KgaA) column was used.

Ion exchange online SPE provides options for ionic analytes. A two-step online SPE [strong anion exchange (SAX) and reversed-phase (RP)] method was developed to measure insulin derivatives (bovine, porcine, human, Arg-human, MW ~6000 Da) in aqueous and plasma samples (Visser et al. 2003). After the SAX cartridge (Isolute 10×4 mm inner diameter, 40 to 90 μm, Separations, The Netherlands) was conditioned with 3 mL of phosphate buffer (60 mM, pH 6.5, 1mL/min), the plasma sample was loaded using the same flow. At this pH, the insulin derivatives processed net negative charges and bound to the stationary phase. The SAX cartridge was washed with 500 μL of water (1mL/min) to remove phosphates. The analytes were eluted onto a second preconditioned RP-SPE cartridge (Luna C8, 4×2.0 mm, Phenomenex) with an acidic buffer (500 μL, 20% CAN:80% 0.1M perchloric acid, 0.2 mL/min). After a second wash with 1.5 mL of 0.05% trifluoroacetic acid (1 mL/min), the analytes were eluted from a RP-SPE cartridge and further separated on an analytical column. A detection limit of 200 nmol/L in spiked plasma could be achieved when UV detection was used and 100 nmol/L with MS detection.

Riediker et al. (2002) measured chlormequat and mepiquat herbicides in foods using an online strong cation exchange (SCX) SPE LC/MS/MS method. The four SCX resins tested were BondElut, Isolute, DVB (Spark), and LiChrolut (Merck). Peak broadening was caused by possible secondary polar interactions of the quaternary ammonium analyte with free silanol groups from BondElut and Isolute. LiChrolut provided the best balance of retention and elution. A GromSil SCX column (50×2 mm, 5 μm, Grom Analytik, Germany) was used for separation. Samples were homogenized and extracted with 1:1 v/v water:methanol. After filtration, the supernatant (2 mL) was injected into a preconditioned (2 mL methanol, 2 mL water, 4 mL 10mM HCl) SCX SPE cartridge. The cartridge was washed with acetonitrile (2 mL) and 1:1 v/v MeOH:H_2O (1 mL) at a flow rate of 4 mL/min. Elution was carried out with 160 mM ammonium formate in 1:1 v/v MeOH:H_2O at 0.3 mL/min. The calibration range was 5 to 195 μg/kg.

An SPE cartridge can be used multiple times, especially after the samples are pretreated with protein precipitation. Bourgogne et al. (2005) quantitated talinolol, a β1-adrenoceptor antagonist used to treat arterial hypertension and coronary heart disease, in human plasma. The sample was first precipitated with perchloric acid and the supernatant was injected directly. An Xterra MS analytical column (50×4.6 mm, 3.5 μm, Waters) with a C18 recolumn filter (4×2 mm, 3.5 μm, Phenomenex) and a C8 EC cartridge were chosen. The cycle time was 4.8 min and linear range was 2.5 to 200 ng/mL. Protein precipitation allowed the SPE cartridge to be used for more than 90 injections.

10.3.4 ONLINE INTERNAL STANDARD (IS) INTRODUCTION

Online IS introduction allows loading of samples in the biological matrix without preparation. ISs were introduced online in the quantitation of propranolol and diclofenac in plasma (Alnouti et al. 2006). Plasma samples were loaded into the autosampler without pretreatment. Both the plasma sample (10 μL) and IS (5 μL from an IS microreservoir) were aspirated into an injection needle sequentially and injected into the sample loop. After the switching of an injection valve, the mixed solution in the sample loop was loaded into a cartridge containing washing solution for online SPE. The accuracy and precision of the online IS method were comparable (85 to 119% and 2 to 12%, respectively) to values obtained offline (86 to 106% and 2 to 16%, respectively).

10.3.5 ENVIRONMENTAL APPLICATIONS

Automated online SPE LC systems are used extensively for environmental assays. Trays of SPE cartridges and autosampler can be used in the field. Water samples are preconcentrated; trays of SPE cartridges loaded with analytes are brought to the laboratory and mounted onto an online SPE LC/MS/MS system for analysis. Prospekt and Symbiosis systems were used for monitoring herbicides and transformation products (Hogenboom et al. 1998, 1999a and b; López-Roldán et al. 2004; Kato et al. 2003; Lacorte and Barceló 1995; Ferrer and Barceló 1999, 2001; Riediker et al. 2002), phenols

(Wissiack et al. 2000), estrogens and metabolites (Rodriguez-Mozaz et al. 2004), and perfluorinated organic acids and amides (Kuklenyik et al. 2005) in surface water, foods, and human sera.

10.4 TURBULENCE FLOW CHROMATOGRAPHY

10.4.1 BASIC CONCEPTS

Turbulence flow chromatography (TFC) allows direct injection of plasma samples and fast removal of proteins from small molecule analytes (Pretorius and Smuts 1966; Quinn and Takarewski 1997). A microbore extraction cartridge (typically 0.5 to 1 mm inner diameter × 50 mm) packed with large (30 to 50 μm) particles is typical. Plasma samples were injected with a high flow rate (4 to 6 mL/ min) of mobile phase. Under these conditions, the flow inside the cartridge deviates from laminar and becomes turbulence flow. Only small molecules are retained; large molecules like proteins are flushed out quickly. Since a TFC column provides minimal separation capacity, a second analytical column is often used to achieve good chromatographic separation (Jemal 2000). A TFC system can be set up with a turbulence flow cartridge in a valve switching system. Alternatively, commercial fully automated systems, TLX systems (Cohesive Technologies, Franklin, Massachusetts), are available.

10.4.2 DIRECT INJECTION TFC

TFC does not require an analytical column. The advantages of direct injection include ease of use, simplicity, very high throughput, and typical run times below 2 min per sample. This approach can be used for certain applications such as high-throughput drug discovery screening in which the analytes are well retained in high concentrations and present little biological matrix interference.

Ayrton et al. (1999) quantitated isoquinoline in plasma using TFC. The three cartridges used were the Oasis HLB (50 × 1 mm, Waters), the Prime C18 (50 × 1 mm, Capital HPLC, Broxburn, UK), and the Oasis HLB (50 × 0.18 mm) with a particle size of 30 to 50 μm. The plasma samples were mixed with an equal volume of aqueous IS solution. A 5 μL sample was injected into the 50 × 0.18 mm cartridge under a flow of 130 μL/min (0.1% formic acid and water). The analytes were washed for 0.2 min to remove proteins. The cartridge was switched online and eluted with a 0.6 min fast gradient (0 to 95% organic). The total run time was 1.2 min and calibration range was 0.5 to 100 ng/mL. With the 50 × 1 mm cartridge, the calibration range was 5 to 1000 ng/mL.

Manipulation of pH and organic content of washing solution can improve extraction efficiency and reduce or eliminate interferences. (Ding and Neve 1999) In one application, the Oasis HLB (50 × 1 mm, 30 μm, Waters) was coupled directly with a single quadruple MS. Antidepressant drugs (amitriptyline, nortriptyline, trimipramine) and narcotics (amphetamine, methamphetamine) served as model compounds. The injection volume was 50 μL. A systematic manipulation of pH levels and organic percentages of washing solvents were used to achieve maximal efficiency and ruggedness. A quantitation range of 5 to 500 ng/mL resulted for porcine plasma samples. The run time was 1.3 min.

A bioanalytical assay in compliance with GLPs was validated using direct TFC (Zimmer et al. 1999). The assay measured two drugs in animal plasma for toxicokinetic studies. The validated method employed HTLC C18 (50 × 1.0 mm inner diameter, 50 μm, Cohesive Technologies) and Oasis HLB (50 × 1.0 mm, 30 μm, Waters) extraction columns on a 2300 HTLC (Cohesive Technologies). Plasma samples were centrifuged before injection and 20 μL of plasma was injected onto a TFC cartridge and washed under a flow of 2 mM aqueous ammonium acetate (pH 6.8, 4.0 mL/min) for 60 sec. After a valve switch, a flow of fast gradient (0 to 95% organic, 1.5 mL/min) delivered the analytes into the detector (25 sec). The run time was 3.3 min per sample and lower the limit of detection was 1 μg/L. The precision and accuracy of the validated methods were similar or better than results with liquid–liquid extraction, SPE, and protein precipitation.

10.4.3 TFC with Analytical Column

TFC was used recently with an analytical column to obtain a desired chromatographic separation. The second column also reduced background ion suppression.

Both methods of direct TFC and TFC with an analytical column were developed for the quantitation of drug candidates in rat plasma (Jemal et al. 1998). Waters' Oasis HLB (50 × 1.0 mm, 30 μm) and Symmetry C18 (3.9 × 50 mm, 5 μm) were used. An aqueous IS was added to the samples. For direct TFC, 50 μL of sample was injected into a TFC cartridge and washed with a 20 mM formic acid mobile phase (4 mL/min, 1 min). After valve switching, the analytes were eluted directly into the detector with a fast gradient (0 to 100% organic, 0.8 mL/min, 0.5 min). In the second method, the plasma sample was injected and washed with 1 mM formic acid (4 mL/min, 1 min). After valve switching, the analytes were flushed from the TFC cartridge and separated on the analytical column with a slower gradient (0 to 62% organic, 0.5 mL/min, 2.8 min). The run times were 4 and 5 min. The calibration ranges were 1 to 1000 ng/mL for the first method and 0.5 to 100 ng/mL for the second.

Similarly, TFC analytical column methods were used to quantitate the β-lactam drug candidate and its metabolite in human plasma (LLOQ 0.980 ng/mL, run time 1.6 min, Jemal et al. 1999), cholesterol-lowering simvastatin and its *in vivo* metabolite (simvastatin acid) in human plasma (LLOQ 0.5 ng/mL, run time 2.5 min, Jemal et al. 2000); antifungal agent tebinafine (Lamisil®) in human and minipig plasma (LLOQ 0.0679 ng/mL, Brignol et al. 2000); ketoconazole, an antimycotic agent and cytochrome P450 3A4 inhibitor, in human plasma (LLOQ 2 ng/mL, Ramos et al. 2000); antidepressant fluoxetine and its norfluoxetine metabolite in human plasma (LLOQ 25 ng/mL, Souverian et al. 2003); antitussive dextromethorphan (DMP) and its two metabolites in rat plasma (LLOQ 0.5 ng/mL, run time 3.5 to 5 min, Ynddal and Hansen 2003); peroxisome proliferator-activated receptor (PPAR) α/γ agonist and potential diabetes II treatment agent in human plasma (LLOQ 4 ng/mL, Xu et al. 2005); and piritramide, a synthetic narcotic analgesic, in human plasma (LLOQ 0.5 ng/mL, Kahlich et al. 2006).

Automated online SPE systems were applied to various phases of drug discovery. Herman (2002) devised a generic method for high-throughput adsorption, distribution, metabolism and excretion (ADME) screening using TFC/LC/MS. They tested more than 1000 compounds and achieved a failure rate below 1%. A 2300 HTLC system was used with a dual column mode: a Cyclone polymeric TFC extraction column (Cohesive Technologies) and an Eclipse XDB C18 analytical column (4.6 × 15 mm, 3 μm, MacMod Analytical, Chadds Ford, Pennsylvania). Plasma samples were first subjected to protein precipitation with the addition of IS in acetonitrile (1:2 v/v). The supernatant (25 μL) was injected and washed with an aqueous mobile phase (0.05% formic acid, 4 mL/min, 0.5 min). The analytes were eluted (40% organic, 0.3 mL/min) and stored in a loop (200 uL). An aqueous mobile phase (1.2 mL/min) was used for online dilution that reduced the organic content of the analyte solution from 40 to 8% when it reached the analytical column. The analytes were eluted with a ballistic gradient (0 to 95% organic, 1 mL/min, 1.5 min); run time was 6 min.

Wu et al. (2000) used TFC to measure 10 and 14 compounds simultaneously in N-in-1 pharmacokinetic screening studies. A Symmetry C18 (150 × 2 mm, Waters) or a Develosil-MG C18 (150 ×2 mm, Phenomenex) was used in connection with an Oasis HLB (50 ×1 mm, 30 μm, Waters) TFC cartridge. The dynamic range for most compounds was 1 to 2500 ng/mL. Ceglarek et al. (2004) used TFC to quantitate cyclosporine A and tacrolimus for immunosuppressant TDM in post-transplant patients. The total run time was 3 min. The calibration ranges were 4.5 to 1500 ng/mL for cyclosporine A and 0.2 to 100 ng/mL for tacrolimus.

Another approach is increasing throughput via a monolith analytical column. Vintiloiu et al. (2005) used a self-made RAM online SPE under turbulent flow conditions to measure rofecoxib, a cyclooxygenase-2 inhibitor, in rat plasma. They constructed a cartridge (0.76 × 50 mm) packed with LiChrosphere 60 RP-18 ADS particles (40 to 63 μm, Merck KgaA). The analytical column was a Chromolith Speed ROD (RP-18, 50 × 4.6 mm, Merck KgaA). The injection volume was

50 μL and run time was 5 min. The calibration range was 0.3 to 30 μg/mL with ion trap MS. In another dual-column approach, Zhou et al. (2005) used a Cohesive Cyclone C18 (50 × 1.0 mm, 50 μm) for extraction and a Chromolith Speed ROD RP-18e (50 × 4.6 mm) column for separation to quantitate dextrorphan, dextromethorphan, and levallorphan; run time was 1.5 min.

10.4.4 PARALLEL SYSTEMS

Dual SPE cartridges allow shortening of sample run cycle time by starting the second injection and extraction on the second TFC column while the first remains online for elution. Xia et al. (2000) developed a ternary column online SPE/LC/MS/MS method to quantitate a drug candidate in rat plasma. Two Oasis HLB (50 × 1 mm, 30 μm, Waters) cartridges and a Symmetry C18 column (50 ×3.9 mm, 5 μm, Waters) were connected with a ten-port switching valve. The plasma sample was first injected into cartridge 1. After 0.3 min of washing with mobile phase A consisting of water;methanol:heptafluorobutyric acid (HFBA) 900:100:2.14 v/v, 4 mL/min), the analytes were backflushed from cartridge 1 and separated on the column with an isocratic mobile phase (30% A, 70% B, 1.3 min). Phase B consisted of water;methanol:HFBA) 100:900:2.14 v/v. Cartridge 2 was equilibrated and ready for injection at the same time. After detection of the first sample ended, injection of the second sample started with cartridge 2. The total run time was 1.6 min per sample; calibration range was 1 to 200 ng/mL.

Grant et al. (2002) designed a parallel system employing two HTLC columns (Cyclone, 50 × 1 mm, Cohesive Technologies) connected to one analytical column (Zorbax SB-C18, 50 × 2 mm, Hewlett Packard) on a 2300 HTLC. A polyarylethyl ketone (PAEK) six-port Valco (Valco Instruments, Texas) was used to increase switching speed and reduce carry-over. Peak focusing was used when the analyte was flushed from the TFC column into the analytical column by aqueous dilution. Compared to the dual column method, the overall time reduction was 1.5 to 4 min per sample with comparable data quality at the linear range of 0.1 to 100 ng/mL.

Hopfgartner et al. (2002) compared ternary column online SPE LC/MS and TFC with offline 96-well plate SPE LC/MS to quantitate three drug candidates in human plasma. A protein precipitation step was performed before the SPE LC/MS. Dual trapping columns (YMS AQ, 10 × 2.0 mm, 5 μm) were used with an analytical column (Intertsil Phenyl, 50 × 2.1 mm, 5 μm). The run cycle was 3 min; calibration range was 0.2 to 250 ng/mL. The run cycle was 2 min with a calibration range of 5 to 1000 ng/mL for TFC. Offline SPE LC/MS achieved the same calibration range with a run time of 2 min.

Because the instability of the N-oxide metabolite, which was subjected to decomposition during sample preparation (solvent evaporation during offline SPE), online SPE LC/MS became the method of choice for the application. Hsieh et al. (2004) built a system with two TFC cartridges and one analytical column, and another system with two TFC cartridges and two analytical columns for GLP quantitative bioanalysis of drug candidates. A Turbo C18 (50 × 1.0 mm, 5 μm, Cohesive Technologies), an Xterra MS C18 (30 × 2.0 mm, 2.5 μm), and a guard column were used. Protein precipitation preceded injection. The cycle times for the two systems were 0.8 and 0.4 min.

TFC has been extensively used in the diagnostics area. While the number of bioanalytical diagnostics assays may be limited compared to those in the pharmaceutical area, the sample numbers are overwhelming. The speed of TFC is a great advantage. Taylor et al. developed an assay to analyze 25-hydroxyvitamins D2 and D3 using a TFC column and a C18 column. The lower limits of detection were 4 and 2 ng/mL; run time was 2 min. Clarke and Goldman developed an assay to measure human steroids utilizing a Cohesive Aria TX-4 system with four TFC columns and four analytical columns. The lower limit of detection was 1 ng/L; run time was below 5 min.

10.4.5 ENVIRONMENTAL APPLICATIONS

TFC was used to measure 11 pesticides in water (Asperger et al. 2002). Five TFC columns (50 × 1 mm inner diameter) were tested. The columns were the silica-based Turbo C18 and Turbo

Phenyl (Cohesive Technologies), the polymer-based Oasis HLB (Waters), the Cyclone (Cohesive Technologies), and the porous graphitized carbon-based Hypercarb (ThermoHypersil, Cheshire, UK); Cohesive's 2300 system was the HTLC component. Merck's monolithic reversed-phased Chromolith Speed ROD (RP-C18 (50 × 4.6 mm) served as the analytical column. The Oasis HLB, Cyclone TFC, and Hypercarb yielded the best retention capacity and good elution efficiency and volume. Recovery was 42 to 94% with a sample volume of 10 mL. Run time was 14 min. LODs were 0.4 to 13 ng/L for most compounds.

10.5 CONCLUSIONS

Online SPE LC/MS/MS is commonly used for bioanalytical applications in the pharmaceutical industry. Column switching systems and TFC systems are easy to build and control. Sophisticated commercial systems and SPE cartridges are readily available. Compared to offline sample preparation, the online approach can save time and labor. However, the development of online SPE bioanalytical assays remains analyte-dependent. Generic methods can be applied to many analytes. For extremely hydrophobic, hydrophilic, and ionic analytes at normal pH range and analytes with a variety of hydrophobicity and pKa values, analyte-specific methods must be developed. An understanding of the chemistry of the analytes and SPE is critical.

REFERENCES

Alnouti Y. et al., 2005. Development and application of a new online SPE high throughput direct analysis of pharmaceutical compounds in plasma. *J Chromatogr A* 1080: 99.

Alnouti Y. et al., 2006. Method for internal standard introduction for quantitative analysis using online SPE LC/MS/MS. *Anal Chem* 78: 1331.

Altun Z., Abdel-Rehim M., and Blomberg L.G., 2004. New trends in sample preparation: Online microextraction in packed syringe (MEPS) for LC and GC applications Part III. *J Chromatogr B* 813: 129.

Amini N. and Crescenzi C., 2003. Feasibility of an online restricted access materal/liquid chromatography/tadem mass spectrometry method in the rapid and sensitive determination of organophosphorus trimesters in human blood plasma. *J Chromatogr B* 795: 245.

Asperger A. et al., 2002. Trace determination of priority pesticide in water by means of high-speed online solid-phase extraction-liquid chromatography-tandem mass spectrometry using turbulent-flow chromatography columns for enrichment and a short monolithic column for fast liquid chromatographic separation. *J Chromatogr A* 960: 109.

Ayrton J. et al., 1997. The use of turbulent flow chromatography for the rapid, direct analysis of a novel pharmaceutical compound in plasma. *Rapid Commun Mass Spectrom* 11: 1953.

Ayrton J. et al., 1998. Optimisation and routine use of generic ultra-high flow-rate liquid chromatography with mass spectrometric detection for the direct online analysis of pharmaceuticals in plasma. *J Chromatogr A* 8282: 199.

Ayrton J. et al., 1999. Ultra-high flow rate capillary liquid chromatography with mass spectrometric detection for the direct analysis of pharmaceuticals in plasma at sub-nanogram per millilitre concentrations. *Rapid Commun Mass Spectrom* 13: 1657.

Barrett Y.C. et al., 2005. Automated online SPE LC-MS/MS method to quantitate 6-beta-hydroxycortisol and cortisol in human urine. *J Chromatogr B* 821: 159.

Beaudry F. et al., 1998. *In vivo* pharmacokinetic screening in cassette dosing experiments: Use of online Prospekt liquid chromatography/atmospheric pressure chemical ionization tandem mass spectrometry technology in drug discovery. *Rapid Commun Mass Spectrom* 12: 1216.

Blakley C.R., Carmody J.J., and Vestal M.L., 1980. A new soft ionization technique for mass spectrometry of complex molecules. *J Am Chem Soc* 102: 5931.

Blakley C.R. and Vestal M.L., 1983. Themospray interface for liquid chromatography/mass spectrometry. *Anal Chem* 55: 750.

Boos K.S. et al., German Patent DE4130475A1. 1991.

Bourgogne E., Grivet C., and Hopfgartner G., 2005. Determination of talinolol in human plasma using automated online solid phase extraction combined with atmospheric pressure chemical ionization tandem mass spectrometry. *J Chromatogr B* 820: 103.

Brignol N. et al., 2000. Quantitatived analysis of terbinafine (Lamisil®) in human and minipig plasma by liquid chromatography tandem mass spectrometry. *Rapid Commun Mass Spectrom* 14: 141.

Calderoli S. et al., 2003. LC-MS-MS determination of brostallicin in human plasma following automated online SPE. *J Pharm Biomed Anal* 32: 601.

Ceglarek U. et al., 2004. Rapid simultaneous quantification of immunosuppressants in transplant patients by turbulent flow chromatography combined with tandem mass spectrometry. *Clin Chim Acta* 346: 181.

Chao M. et al., 2005. Rapid and sensitive quantification of urinary N7-mthelguanine by isotope-dilution liquid chromatography/electrospray ionization tandem mass spectrometry with online solid-phase extraction. *Rapid Commun Mass Spectrom* 19: 2427.

Chen Y., Wang C., and Wu K., 2005. Analysis of N7-(benzo[α]pyrene-6-yl)guanine in urine using two-step solid-phase extraction and isotope dilution with liquid chromatography/tandem mass spectrometry. *Rapid Commun Mass Spectrom* 19: 893.

Christiaens B. et al., 2004. Fully automated method for the liquid chromatographic–tandem mass spectrometric determination of cyproterone acetate in human plasma using restricted access material for online sample clean-up, *J Chromatogr A* 1056.

Clarke N. and Goldman M., Clinical applications of HTLC-MS/MS in the very high throughput diagnostic environment: LC-MS/MS on steroids. Application poster, Quest Diagnostics and Cohesive Technologies.

Dai S. et al., 2005. Quantitation of sifuvirtide in monkey plasma by an online solid-phase extraction procedure combined with liquid chromatography/electrospray ionization tandem mass spectrometry. *Rapid Commun Mass Spectrom* 19: 1273.

de Cassia R. et al., 2004. A rapid and sensitive method for simultaneous determination of lamivudine and zidovudine in human serum by online solid-phase extraction coupled to liquid chromatography/tandem mass spectrometry detection. *Rapid Commun Mass Spectrom* 18: 1147.

Ding J. and Neue U.D., 1999. A new approach to the effective preparation of plasma samples for rapid drug quantitation using online solid phase extraction mass spectrometry. *Rapid Commun Mass Spectrom* 13: 2151.

Doerge D.R., Churchwell M.I., and Delclos K.B., 2000. Online sample preparation using restricted-access media in the analysis of the soy isoflavones, genistein and daidzein, in rat serum using liquid chromatography electrospray mass spectrometry. *Rapid Commun Mass Spectrom* 14: 673.

Ferrer I. and Barceló, D., 1999. Simultaneous determination of antifouling herbicides in marina water samples by online solid-phase extraction followed by liquid chromatography-mass spectrometry. *J Chromatogr A* 854: 197.

Ferrer I. and Barceló, D., 2001. Identification of new degradation product of the antifouling agent irgarol 1051 in natural samples. *J Chromatogr A* 926: 221.

Gimeno R.A. et al., 2002. Determination of polycyclic aromatic hydrocarbons and polycyclic aromatic sulfur heterocycles by high-performance liquid chromatography with fluorescence and atmospheric pressure chemical ionization mass spectrometry detection in seawater and sediment samples. *J Chromatogr A* 958: 141.

Gonçalves J.C.S. et al., 2005. Online solid-phase extraction coupled with high-performance liquid chromatography and tandem mass spectrometry (SPE HPLC-MS-MS) for quantification of bromazepam in human plasma. *Ther Drug Monit* 27: 601.

Grant R.P., Cameron C., and Mackenzie-McMurter S., 2002. Generic serial and parallel online direct injection using turbulent flow liquid chromatography/tandem mass spectrometry. *Rapid Commun Mass Spectrom* 16: 1785.

Gundersen T.E. and Blomhoff R., 1999. Online solid-phase extraction and isocratic separation of retinoic acid isomers in microbore column switching system. *Meth Enzymol* 299: 430.

Hennion M., 1999. Solid-phase extraction: Method development, sorbents, and coupling with liquid chromatography. *J Chromatogr A* 856: 3.

Herman J.L., 2002. Generic method for online extraction of drug substances in the presence of biological matrices using turbulent flow chromatography. *Rapid Commun Mass Spectrom* 16: 421.

Hernández F. et al., 1998. Coupled-column liquid chromatography applied to the trace-level determination of triazine herbicides and some of their metabolites in water. *Anal Chem* 70: 3322.

Hernández F. et al., 2001. Rapid direct determination of pesticides and metabolites in environmental water samples at sub-μg/L level by online solid-phase extraction-liquid chromatography-electrospray tandem mass spectrometry. *J Chromatogr A* 939: 1.

Hogenboom A.C., Niessen W.M.A., and Brinkman U.A., 1998. Rapid target analysis of microcontaminanant in water by online single-short-column liquid chromatography combined with atomspheric pressure chemical ionization ion-trap mass spectrometry. *J Chromatogr A* 794: 201.

Hogenboom A.C., Niessen W.M.A., and Brinkman U.A., 1999a. Online solid-phase extraction-short-column liquid chromatography combined with various tandem mass spectrometric scanning strategies for the rapid study of transformation of pesticides in surface water. *J Chromatogr A* 841: 33.

Hogenboom A.C. et al., 1999b. Accurate mass determination for the confirmation and identification of organic microcontaminants in surface water using online solid-phase extraction liquid chromatography electrospray orthogonal-acceleration time-of-flight mass spectrometry. *Rapid Commun Mass Spectrom* 13: 125.

Hopfgartner G., Husser C., and Zell M., 2002. High-throughput quantification of drugs and their metabolites in biosamples by LC-MS/MS: Possibilities and limitations. *Ther Drug Monit* 24: 134.

Hsieh Y. et al., 2000. Direct analysis of plasma samples for drug discovery compounds using mixed-function column liquid chromatography tandem mass spectrometry. *Rapid Commun Mass Spectrom* 14: 1384.

Hsieh Y. et al., 2001. Quantitative screening and matrix effect studies of drug discovery compounds in monkey plasma using fast-gradient liquid chromatography/tandem mass spectrometry. *Rapid Commun Mass Spectrom* 15: 2481.

Hsieh Y. et al., 2002. Direct simultaneous determination of drug discovery compounds in monkey plasma using mixed-function column liquid chromatography/tandem mass spectrometry. *J Pharm Biomed Anal* 27: 285.

Hsieh Y. et al., 2003. Direct plasma analysis of drug compounds using monolithic column liquid chromatography and tandem mass spectrometry. *Anal Chem* 75: 1812.

Hsieh Y., 2004. Using mass spectrometry for drug metabolism studies, in *Direct Plasma Analysis Systems,* Korfmacher, W.A., Ed., Boca Raton, FL, CRC Press, Chap. 5.

Hsieh S. et al., 2004. Increased throughput of parallel online extraction liquid chromatography/electrospray ionization tandem mass spectrometry system for GLP quantitative bioanalysis in drug development. *Rapid Commun Mass Spectrom* 18: 285.

Hu C. et al., 2006. Clinical-scale high-throughput analysis of urinary 8-oxo-7,8-dihydro-2′-deoxyquanosine by isotope-dilution liquid chromatography-tandem mass spectrometry with online solid-phase extraction. *Clin Chem* 52: 7.

Huang M.Q. et al., 2006. Increased productivity in quantitative bioanalysis using a monolith column coupled with high-flow direct-injection liquid chromatography/tandem mass spectrometry. *Rapid Commun Mass Spectrom* 20: 1709.

Ibáñez M. et al., 2005. Residue determination of glyphosate, glufosinate and aminomethylphosphonic acid in water and soil samples by liquid chromatography coupled to electrospray tandem mass spectrometry. *J Chromatogr A* 1081: 145.

Jemal M., 2000. High-throughput quantitative bioanalysis by LC/MS/MS. *Biomed Chromatogr* 14: 422.

Jemal M., Xia Y., and Whigan D.B., 1998. The use of high-flow high performance liquid chromatography coupled with positive and negative ion electrospray tandem mass spectrometry for quantitative bioanalysis via direct injection of the plasma/serum samples. *Rapid Commun Mass Spectrom* 12: 1389.

Jemal M. et al., 1999. A versatile system of high-flow high performance liquid chromatography with tadem mass spectrometry for rapid direct-injection analysis of plasma samples for quantitation of a β-lactam drug candidate and its open-ring biotransformation product. *Rapid Commun Mass Spectrom* 13: 1462.

Jemal M., Ouyang Z., and Powell M.L., 2000. Direct-injection LC-MS-MS method for high-throughput simultaneous quantitation of simvastatin and simvastatin acid in human plasma. *J Pharm Biomed Anal* 23: 323.

Kahlich R. et al., 2006. Quantitative determination of piritramide in human plasma and urine by off- and online solid-phase extraction liquid chromatography coupled to tandem mass spectrometry. *Rapid Commun Mass Spectrom* 20: 275.

Kato K. et al., 2003. Determination of three phthalate metabolites in human urine using online solid-phase extraction–liquid chromatography–tandem mass spectrometry. *J Chromatogr B* 788: 407.

Kato K. et al., 2005. Determination of 16 phthalate metabolites in urine using automated sample preparation and online preconcentration/high-performance liquid chromatography/tandem mass spectrometry. *Anal Chem* 77: 2985.

Koal T. et al., 2004. Simultaneous determination of four immunosuppressant by means of high speed and robust online solid phase extraction-high performance liquid chromatography-tandem mass spectrometry. *J Chromatogr B* 805: 215.

Koeber R. et al., 2001. Evaluation of a multidimensional solid-phase extraction platform for highly selective online cleanup and high-throughput LC-MS analysis of triazines in river water samples using molecularly imprinted polymers. *Anal Chem* 73: 2437.

Kuhlenbeck D.L. et al., 2005. Online solid phase extraction using the Prospekt-2 coupled with a liquid chromatography/tandem mass spectrometer for the determination of dextromethorphan, detrophan and quaifenesin in human plasma. *Eur J Mass Spectrom* 11: 199.

Kuklenyik Z., Needham L.L., and Calafat A.M., 2005. Measurement of 18 perfluorinated organic acids and amides in human serum using online solid-phase extraction. *Anal Chem* 77: 6085.

Lacorte S. and Barceló D., 1995. Determination of organophosphorus pesticides and their transformation products in river waters by automated online solid-phase extraction followed by thermospray liquid chromatography. *J Chromatogr A* 712: 103.

Lant M.S. and Oxford J., 1987. Automated sample preparation online with thermospray high-performance liquid chromatography-mass spectrometry for the determination of drugs in plasma. *J Chromatogr A* 394: 223.

López-Roldán P., de Alda M.J.L., and Barceló D., 2004. Simultaneous determination of selected endocrine disrupters (pesticides, phenols and phthalates) in water by in-field solid-phase extraction (SPE) using the prototype PROFEXS followed by online SPE (PROSPEKT) and analysis by liquid chromatography-atomspheric pressure chemical ionization-mass spectrometry. *Anal Bioanal Chem* 378: 599.

Marchese A. et al., 1998. Determination of pranlukast and its metabolites in human plasma by LC/MS/MS with Prospekt™ online solid-phase extraction. *J Mass Spectrom* 33: 1071.

McLoughlin D.A., Olah T.V., and Gilbert J.D., 1997. A direct technique for the simultaneous determination of 10 drug candidates in plasma by liquid chromatography-atmospheric pressure chemical ionization mass spectrometry interfaced to a Prospekt solid-phase extraction system. *J Pharm Biomed Anal* 15: 1893.

Niederländer H.A.G. et al., 2006. High throughtput therapeutic drug monitoring of clozapine and metabolites in serum by online coupling of solid phase extraction with liquid chromatography-mass spectrometry. *J Chromatogr B* 834: 98.

Papadopouluo-Mourkidou E. et al., 2001. Use of an automated online SPE-HPLC method to monitor caffeine and selected aniline and phenol compounds in aquatic systems of Macedonia–Thrace, Greece. *Fresenius J Anal Chem* 371: 491.

Plumb R. et al., 2001. Direct analysis of pharmaceutical compounds in human plasma with chromatographic resolution using an alkyl-bonded silica rod column. *Rapid Commun Mass Spectrom* 15: 986.

Pozo O.J. et al., 2006. Efficient approach for the reliable quantification and confirmation of anitbiotice in water using online solid-phase extraction liquid chromatography/tandem mass spectrometry. *J Chromatogr A* 1103: 83.

Pretorius V. and Smuts T.W., 1966. Turbulent flow chromatography: A new approach to faster analysis. *Anal Chem* 38(2): 274.

Pruvost A. et al., 2000. Fully automated determination of eserine N-oxide in human plasma using online solid-phase extraction with liquid chromatography coupled with electrospray ionization tandem mass spectrometry. *J Mass Spectrom* 35: 625.

Quinn H.M. and Takarewski J.J., International Patent WO97/16724, 1997.

Raices R.S.L. et al., 2003. Determination of stavudine in human serum by online solid-phase extraction coupled to high-performance liquid chromatography with electrospray ionization tandem mass spectrometry: Application to a bioequivalence study. *Rapid Commun Mass Spectrom* 17: 1611.

Ramos L. et al., 2000. High-throughput approaches to the quantitative analysis of ketoconazole, a poten inhibitor of cytochrome P450 3A4, in huma plasma. *Rapid Commun Mass Spectrom* 14: 2282.

Riediker S. et al., 2002. Determination of chlormequat and mepiquat in pear, tomato, and wheat flour using online solid-phase extraction (Prospekt) coupled with liquid chromatography–electrospray ionization tandem mass spectrometry. *J Chromatogr A* 966: 15.

Rodriguez-Mozaz S., de Alda M.J.L., and Barceló D., 2004. Picogram per liter level determination of estrogens in natural waters and waterworks by a fully automated online solid-phase extraction-liquid chromatography-electrospray tandem mass spectrometry method. *Anal Chem* 76: 6998.

Sancho J.V., Pozo O.J., and Hernández F., 2004. Liquid chromatography and tandem mass spectrometry: A powerful approach for the sensitive and rapid multiclass determination of pesticides and transformation products in water. *Analyst* 129: 38.

Schellen A. et al., 2000. High throughput online solid phase extraction/tandem mass spectrometric determination of paclitaxel in human serum. *Rapid Commun Mass Spectrom* 14: 230.

Schellen A. et al., 2003. Generic solid phase extraction-liquid chromatography-tandem mass spectrometry method for fast determination of drugs in biological fluids. *J Chromatogr B* 788: 251.

Souverian S. et al., 2003. Rapid analysis of fluoxetine and its metabolite in plasma by LC-MS with column-switching approach. *Anal Bioanal Chem* 377: 880.

Taylor R.L., Grebe S.K., and Singh R., High throughput analysis of 25-hydroxyvitamins D2 and D3 by LC-MS/MS using automated online extraction. Application poster, Mayo Clinic and Cohesive Technologies.

van der Hoven R.A.M. et al., 1997. Liquid chromatography-mass spectrometry with online solid-phase extraction by a restricted-access C_{18} precolumn for direct plasma and urine injection. *J Chromatogr A* 762: 193.

Viehauer S. et al., 1995. Evaluation and routine application of the novel restricted-access precolumn packing material Alkyl-Diol Silica: Coupled-column high-performance liquid chromatographic analysis of the photoreactive drug 8-methoxypsoralen in plasma. *J Chromatogr B* 666: 315.

Vintiloiu A. et al., 2005. Combining restricted access material (RAM) and turbulent flow for the rapid online extraction of the cyclooxygenase-2 inhibitor rofecoxib in plasma samples. *J Chromatogr A* 1082: 150.

Visser N.F.C., Lingeman H., and Irth H., 2003. Online SPE-RP-LC for the determination of insulin derivatives in biological matrices. *J Pharm Biomed Anal* 32: 295.

Wachs T. and Henion J., 2003. A device for automated direct sampling and quantitation from solid-phase sorbent extraction cards by electrospray tandem mass spectrometry. *Anal Chem* 75: 1769.

Wissiack R., Rosenberg E., and Grasserbauer M., 2000. Comparison of different sorbent materials for online solid-phase extraction with liquid chromatography-atmospheric pressure chemical ionization mass spectrometry of phenols. *J Chromatogr B* 896: 159.

Wu J. et al., 2000. Direct plasma sample injection in multiple-component LC-MS-MS assays for high-throughput pharmacokinetic screening. *Anal Chem* 72: 61.

Wu J. et al., 2001. *Rapid Commun Mass Spectrom* 15: 1113.

Xia Y. et al., 2000. Ternary-column system for high-throughput direct-injection bioanalysis by liquid chromatography/tadem mass spectrometry. *Rapid Commun Mass Spectrom* 14: 105.

Xu R.N. et al., 2006. A monolithic-phase based online extraction approach for determination of pharmaceutical compounds in human plasma by HPLC-MS/MS and a comparison with liquid–liquid extraction. *J Pharm Biomed Anal* 40: 728.

Xu X. et al., 2005. Quantitative determination of a novel dual PPAR α/γ agonist using online turbulent flow extraction with liquid chromatography-tandem mass spectrometry. *J Chromatogr B* 814: 29.

Ye X. et al., 2005. Automated online column-switching HPLC-MS/MS method with peak focusing for the determnination of nine environmental phenols in urine. *Anal Chem* 77: 5407.

Ynddal L. and Hansen S.H., 2003. Online turbulent-flow chromatography-high-performance liquid chromatography-mass spectrometry for fast sample preparation and quantitation. *J Chromatogr A* 1020: 59.

Zang X. et al., 2005. A novel online solid-phase extraction approach integrated with a monolithic column and tandem mass spectrometry for direct plasma analysis of multiple drugs and metabolites. *Rapid Commun Mass Spectrom* 19: 3259.

Zell M., Erdin H.R., and Hopfgartner G., 1997a. Simultaneous dtermination of a potassium channel opener and its metabolite in rat plasma with column-switching liquid chromatography using atmospheric pressure chemical ionization. *J Chromatogr B* 694: 135.

Zell M., Husser C., and Hopfgartner G., 1997b. Column-switching high-performance liquid chromatography combined with ionspray tandem mass spectrometry for the simultaneous determination of the platelet inhibitor Ro 44-3888 and its pro-drug and precursor metabolite in plasma. *J Mass Spectrom* 32: 23.

Zeng H., Deng Y., and Wu J., 2003a. Fast analysis using monolithic columns coupled with high-flow online extraction and electrospray mass spectrometric detection for the direct and simultaneous quantitation of multiple components in plasma. *J Chromatogr B* 788: 331.

Zeng W. et al., 2003b. A direct injection high-throughput liquid chromatography tandem mass spectrometry method for the determination of orally active $\alpha_v\beta_3$ antagonist in huma urine and dialysate. *Rapid Commun Mass Spectrom* 17: 2475.

Zhang N. et al., 2000. Integrated sample collection and handling for drug discovery bioanalysis. *J Pharm Biomed Anal* 23: 551.

Zhou S. et al., 2005. High-throughput biological sample analysis using online turbulent flow extraction combined with monothic column liquid chromatography/tandem mass spectrometry. *Rapid Commun Mass Spectrom* 19: 2150.

Zimmer D. et al., 1999. Comparison of turbulent-flow chromatography with automated solid-phase extraction in 96-well plates and liquid-liquid extraction used as plasma sample preparation techniques for liquid chromatography-tandem mass spectrometry. *J Chromatogr A* 854: 1999.

Zwiener C. and Frimmel F., 2004. LC-MS analysis in the aquatic environment and in water treatment-a critical review. Par II: Applications for emerging contaminants and related pollutants, microorganisms and humic aicd. *Anal Bioanal Chem* 378: 862.

11 Applications of High-Throughput Analysis in Therapeutic Drug Monitoring

Quanyun A. Xu and Timothy L. Madden

CONTENTS

ABSTRACT

This chapter summarizes applications of high-throughput analysis in modern therapeutic drug monitoring. Today's medicine has become increasingly personalized. To maximize drug efficacy and minimize drug toxicity, optimize treatment, and reduce costs, it is critical to monitor the concentrations of drugs in biological fluids collected from patients under medical management. Many analytical procedures have been developed for therapeutic drug monitoring. To analyze a large number of patient samples in a short time, as required by modern therapeutic drug monitoring, high-throughput analytical methods have been developed and validated. These methods include immunoassay via auto-immunoanalyzer, high performance liquid chromatography (HPLC) coupled with a unique monolithic column, online sample clean-up (column switching), and automated solid phase extraction.

High performance liquid chromatography–tandem mass spectrometry (HPLC/MS/MS) is the current method of choice–a highly speedy, sensitive, and selective assay for complex biological samples. High-throughput analysis dramatically improves the efficiency of therapeutic drug monitoring and thus enhances drug performance. The combination of ultra-performance liquid chromatography (UPLC), providing ultra speed and ultra resolution power, and tandem spectrometry (MS/MS), furnishing high sensitivity and selectivity, is becoming increasingly popular for such monitoring and will probably replace other high-throughput analytical methods in the future.

11.1 INTRODUCTION

The previous chapters described various applications of high-throughput techniques for drug discovery and development in detail. This chapter will discuss applications of high-throughput technologies to therapeutic drug monitoring (TDM) in clinical settings.

Hundreds of new drugs are brought to the market annually by the pharmaceutical industry and they help fight many diseases. However, the absorption, distribution, metabolism, and excretion levels of a drug differ among individuals, resulting in differences in the relationship of plasma concentrations of a drug and its dosage. Furthermore, these inter-individual differences in pharmacokinetics can also be caused by drug–drug interactions, by differences in age or weight, or by co-morbid diseases such as renal failure and hepatic disease.[1–8] As a result of these differences, drug administration has become more individualized.

TDM plays a very important role in personalized medicine. In practice, TDM involves the assessment of the clinical indication for testing a drug, the analysis of samples, and the interpretation of assay results for possible dose adjustment. Among these three components, the measurement of drug concentrations in plasma or serum is the most important aspect. The goal of TDM is to ensure that drug concentration is within a defined optimal therapeutic range to maximize its efficacy, minimize its toxicity, and reduce costs. Not all drugs are candidates for TDM. A drug must meet the following criteria to be a TDM candidate: (1) a narrow therapeutic range; (2) significant inter-individual variability in systemic exposure at a given dose; (3) a clear relationship between blood exposure and clinical effect; and (4) a validated method to measure drug concentration in plasma or serum.

TDM has improved the performance of anticancer, antidementia, antidepressant, antiepileptic, anticonvulsant, antifungal, antimicrobial, antipsychotic, antiretroviral, anxiolytic, hypnotic, cardiac, addiction treatment, immunosuppressant, and mood stabilizer drugs for more than 30 years.[2–9] Many analytical procedures evolved as analytical techniques and instrumentation have advanced. This chapter briefly reviews the different types of analytical methods; the applications of high-throughput techniques in TDM are discussed in detail.

11.2 OVERVIEW: ANALYTICAL METHODS FOR TDM

TDM was first carried out on drugs in biological samples using ultraviolet (UV) light, fluorescence, and electrochemical detection, which measured physicochemical properties of drugs. Used alone, these detection methods had low sensitivity and selectivity and were soon obsolete.[10]

Since its invention in the late 1960s, gas chromatography (GC) has proven very useful for TDM.[1,11–15] First, a drug is extracted from a biological fluid with an organic solvent, derivatized before or after extraction, or derivitized online. The drug is then separated on a GC column at an elevated temperature, usually between 200 and 350°C. Flame ionization detector (FID), electron capture detector (ECD), nitrogen–phosphorus detector (NPD),[16] and mass spectrometer (MS)[17] handle detection. FID and ECD detectors are most commonly used for therapeutic drug analysis. The disadvantage of GC is that only a limited number of drugs can be safely volatized at temperatures over 200°, limiting its use for TDM.

Immunoassay is the sterile measurement of a drug molecule as an antigen using a specific antibody. Detection is performed by UV light absorption, radioactivity, or fluorescence polarization.

These methods have been used since the 1970s; they usually require little or no sample preparation and are rapid and easy to use. However, immunoassay has two limitations. First, it does not differentiate between active drugs and similar molecules such as metabolites or co-administered drugs.[9,11] Thus, cross-reactivity is a common problem. Second, its use is limited to only those drugs for which antibodies are available.

High performance liquid chromatography (HPLC) is by far the major method used to measure drugs in biological fluids.[9, 18,19] Conventional detectors are UV, fluorescence, refractive index, electrochemical, and photodiode array detector. A huge selection of columns and mobile phases provides HPLC methods with good sensitivity and selectivity. They have been used to analyze large numbers of drugs for TDM purposes. Thermally labile drugs can be analyzed by HPLC but it presents some limitations. One is the need for extensive sample preparation before injecting a sample onto a column. Preparation includes the precipitation of protein from the plasma, extraction of the drug with a specific solvent, extraction of the drug through a solid phase cartridge, or derivatization of the drug with a specific agent. Another disadvantage is the long average run time. HPLC with conventional detection has limited sample throughput capacity.

11.3   HIGH-THROUGHPUT ANALYTICAL METHODS

Today's personalized medicine requires analysis of a large number of biological samples in a short period on the day they are collected from patients so that a proper informed dose adjustment can be made before subsequent dosing. The high-throughput analytical procedures developed to meet this demand are reviewed in subsequent sections covering immunoassays, HPLC alone and combined with tandem mass spectrometry detection (HPLC-MS/MS), and ultra-performance liquid chromatography with MS/MS detection (UPLC-MS/MS).

11.3.1   Immunoassay

11.3.1.1   Fluorescence Polarization

Rao et al.[20] demonstrated a fluorescence polarization immunoassay for evaluating serum concentrations of tricyclic antidepressants (amitriptyline, imipramine, clomipramine, and doxepin) with respect to nonresponse, compliance, therapeutic window, and influences of age, sex, substance abuse, and toxicity. Abbott Laboratories' TDx/TDxFLx™ Toxicology Tricyclic Assay FPIA (fluorescence polarization immunoassay) was used. This assay of 50 μL samples contained tricyclic antidepressant antibodies raised in rabbits and fluorescein-labeled tricyclic antidepressant as a tracer. The assay was calibrated with imipramine in the range of 75 to 1000 μg/L (268 to 3571 nmol/L). Intra-assay and inter-assay coefficients of variation for internal quality control samples from the manufacturer were 4.2 and 4.7%, respectively. The limits of detection were 72, 71, 64, and 72 nmol/L for amitriptyline, imipramine, clomipramine, and doxepin, respectively. This high-throughput immunoassay was easy to use although amitriptyline, dosulepine, desipramine, and nortriptyline showed cross-reactivities ranging from 74 to 100%.

11.3.1.2   Homogeneous Enzyme

Pankey et al.[21] described a rapid, reliable, and specific enzyme multiplied immunoassay technique (EMIT®) for amitriptyline, nortriptyline, imipramine, and desipramine in sera. To overcome cross-reactivity, solid phase extraction was included in sample pretreatment. Disposable 1 mL columns packed with covalently labeled silica gel were conditioned with HPLC-grade methanol (1 mL) and then with de-ionized or distilled water (1 mL). Serum (calibrator, control, or patient sample, 500 μL) was applied onto the column, eluted to waste, washed with 900 μL of wash solution containing acetonitrile (236.1 g/L) and ion-pairing reagent in acetate buffer, pH 4.2, washed with 500 μL of mobile phase solution containing acetonitrile (393.5 g/L) in methanolic phosphate buffer, pH 7.0, followed by 1 mL of extraction diluent containing Tris HCl (55 mmol/L, pH 6.4) and preservatives to elute the metabolite-free drug.

Immunoassay analysis of the extracts was performed using Syva assay kits and an AutoLab system. The extract (50 μL) and assay buffer (250 μL) were delivered into a reaction cup, followed by 50 μL reagent A (antibody and substrate) plus another 250 μL assay buffer, incubated for 50 sec, and mixed with 50 μL reagent B (drug-labeled enzyme) plus 250 μL assay buffer. The change in absorbance of this final mixture was monitored at a wavelength of 340 nm by a spectrophotometer.

The dynamic range of the standard curve for amitriptyline, nortriptyline, and imipramine was 25 to 250 μg/mL; it was 50 to 500 μg/mL for desipramine. Within-run and between-run coefficients of variation for the assay were below 10%. Up to 40 patient samples could be analyzed in 1 hr.

11.3.2　High Performance Liquid Chromatography

HPLC has high-throughput capability when it can simultaneously determine multiple drugs and their metabolites or when coupled with a unique monolithic column or sample preparation technique. Some examples are summarized below.

11.3.2.1　Fluorescence Detection

Anthracyclines (daunorubicin, doxorubicin, idarubicin, and epirubicin) are anticancer drugs widely used to manage patients with acute leukemia or breast cancer.[22,23] To maximize therapeutic efficacy and minimize the acute myelosuppression and cumulative dose-related cardiotoxicity of these agents, several analytical methods were developed to measure anthracyclines and their metabolites in biologic fluids.[24–26]

Fogli et al. developed and validated an HPLC method with fluorescence detection for simultaneous routine TDM of anthracyclines and their metabolites.[27] They coupled a Waters LC Module I Plus system equipped with a WISP 416 autosampler with a Model 474 scanning fluorescence spectrophotometer. The stationary phase was a Supelcosil LC-CN column (250 × 4.6 mm, 5 μm particle size) with a μBondapak-CN guard column. The mobile phase consisted of $50mM$ monobasic sodium phosphate buffer and acetonitrile (65:35 v/v), adjusted to pH 4.0 with phosphoric acid. The flow rate was 1 mL/min. The fluorescence detection was set at excitation wavelengths of 233, 254, and 480 nm and at an emission wavelength of 560 nm.

Stock solutions of anthracyclines (1 mg/mL) were prepared in double distilled water and stored at 4°C in the dark. Standard working solutions were prepared by diluting stock solutions with double distilled water or 0.1M phosphoric acid. Aliquots of blank human plasma (0.5 mL) were spiked with working solutions of anthracyclines, mixed with 0.5 mL of 0.2M dibasic sodium phosphate buffer (pH 8.4), extracted with 4 mL of chloroform:1-heptane (9:1 v/v) by shaking for 15 min and centrifuged at 4000 rpm for 10 min. The lower organic layer was re-extracted with 0.25 mL of 0.1M phosphoric acid. The upper aqueous layer was collected and assayed. The injection volume was 50 μL. Retention times for daunorubicinol, daunorubicin, idarubicinol, idarubicin, doxorubicinol, doxorubicin, epirubicinol, and epirubicin were 6,7, 9.1, 8.0, 11.3, 5.1, 6.4, 5.5, and 7.0 min, respectively.

Calibration curves for anthracyclines were constructed for the concentration range of 0.4 to 10,000 ng/mL. Correlation coefficients exceeded 0.999. Within-day and between-day coefficients of variation were less than 10%. Recoveries ranged from 89 to 109%. Accuracies were 91 to 107%. Limit of detection and limit of quantification were both 0.4 ng/mL.

11.3.2.2　Monolithic Column

Carvedilol — This drug is a non-cardioselective β blocker used to manage hypertension and angina pectoris. Several methods have been developed to measure carvedilol in biological fluids.[28–30] Zarghi et al developed a simple, rapid, and sensitive HPLC method to analyze carvedilol in human plasma using a monolithic column coupled with fluorescence detection.[31] This method employed the unique properties of a monolithic column to improve the speed of the separation process and reduce column backpressure without sacrificing resolution.

The liquid chromatograph Zarghi et al. used consisted of a Wellchrom K-1001 pump, Rheodyne 7125 injector, Eurochrom 2000 integrator, and K2600 fluorescence detector. The stationary phase was a Merck Chromolith Performance RP-18e column (100 × 4.6 mm). The mobile phase consisted of 0.01M dibasic sodium phosphate buffer and acetonitrile (60:40 v/v) adjusted to pH 3.5. The flow rate was 2 mL/min. The detector operated at an excitation wavelength of 240 nm and an emission wavelength of 340 nm. Letrozole served as the internal standard (IS).

Stock solutions of carvedilol (8 mg/mL) and letrozole (10 mg/mL) were prepared in methanol and stored at 4°C. Standard solutions were prepared by spiking blank plasma with stock solutions. Aliquots (450 μL) of standard solutions and patient plasma samples were mixed with 50 μL of the IS (10 μg/mL) followed by 500 μL of acetonitrile, vortexed for 30 sec, and centrifuged at 8000 g for 10 min. Supernatants were collected and assayed. The injection volume was 20 μL. Figure 11.1 illustrates typical chromatograms of carvedilol and letrozole in plasma.

A linear calibration curve for carvedilol in plasma was constructed over a range of 1 to 80 ng/mL. The correlation coefficient exceeded 0.999. Intra-day and inter-day coefficients of variation were 1.93 and 1.88%, respectively. The average carvedilol recovery was 98.1%. The limit of quantification was 1 ng/mL. This high-throughput method enabled the analysis of more than 600 plasma samples without significant loss of column efficiency.

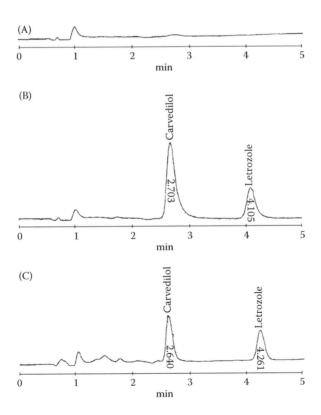

FIGURE 11.1 Chromatograms of (A) blank plasma; (B) blank plasma spiked with 60 ng/mL carvedilol and 1000 ng/mL letrozole (internal standard); and (C) plasma sample from a healthy volunteer 2 hr after oral administration of 25 mg of carvedilol. (*Source:* From Zarghi, A. et al. *J Pharm Biomed Anal.* 44, 252, 2007. With permission.)

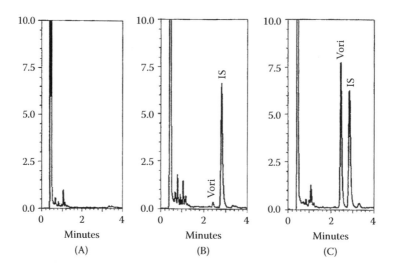

FIGURE 11.2 Chromatograms of (A) blank plasma, (B) patients' plasma containing 0.1 µg/mL voriconazole, and (C) 3.5 µg/mL voriconazole. Vori = voriconazole. IS = internal standard. (*Source:* From Wenk, M. et al. *J Chromatogr B*. 832, 315, 2006. With permission.)

Voriconazole — This novel wide-spectrum triazole antifungal agent combats *Aspergillus* and *Candida*. Due to its nonlinear pharmacokinetic behavior,[32,33] TDM of voriconazole in patients under medical treatment is important. Several analytical methods have been reported for determination of the drug in biological fluids.[34,35] Using a monolithic silica rod column, Wenk et al. developed a fast and reliable HPLC assay of voriconazole in human plasma.[36] They used a LaChrome Elite system that included a 2130 quaternary pump with online degassers, 2200 autosampler, 2300 UV detector, and 2400 column oven. The stationary phase was a Chromolith Performance RP-18e column (100 × 4.6 mm) with a guard column (5 × 4.6 mm). Column temperature was maintained at 32°C. The mobile phase consisted of 0.025M monobasic ammonium phosphate buffer (pH 5.8), acetonitrile, and tetrahydrofuran (74:25:1 v/v/v). The flow rate was 3.5 mL/min and total run time was 4 min.

Aliquots (0.25 mL) of plasma samples were mixed with 50 µL of the IS (Pfizer Global UK-115794, 20 µg/mL in water) followed by 0.5 mL 0.2M ammonium acetate buffer (pH 9.0), extracted with 7 mL ethylacetate:diethylether (1:1 v/v), vortexed for 90 sec, centrifuged at 1500 g for 3 min, and frozen. The organic layer was collected, evaporated to dryness at 40°C under a stream of nitrogen, reconstituted with 0.2 mL of mobile phase, and centrifuged at 10,000 g for 6 min. The supernatant was collected and assayed. The injection volume was 30 µL. Figure 11.2 shows chromatograms of voriconazole and the IS in plasma.

Calibration curves for voriconazole were constructed in concentration ranges of 0.1 to 10 µg/mL. Correlation coefficients exceeded 0.9998. Intra-day and inter-day coefficients of variation were less than 3.8 and 6.1%, respectively. The average extraction recovery was 94.6%. The limit of detection and the limit of quantification were 15 and 50 ng/mL, respectively.

11.3.2.3 96-Well Plate Solid Phase Extraction

Annerberg et al. reported a sensitive high-throughput assay to determine lumefantrine concentrations in plasma using 96-well plate solid phase extraction (SPE).[37] The LaChrom Elite system consisted of an L2130 LC pump, L2200 injector, L2300 column oven, and L2400 UV detector. The stationary phase was an Agilent SB-CN column (250 × 4.6 mm, 5 µm) coupled with a Phenomenex Security Guard CN column (4 × 3 mm). The mobile phase consisted of acetonitrile and 0.1M phosphate

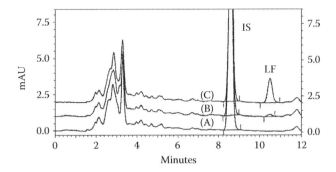

FIGURE 11.3 Overlay of chromatograms from (A) blank plasma, (B) spiked plasma sample at 25 ng/mL (LLOD), and (C) 300 ng/mL. (*Source:* From Annerberg, A. et al. *J Chromatogr B*. 822, 332 , 2005. With permission.)

buffer (pH 2.0) (58:42 v/v) containing $0.01M$ sodium perchlorate. The flow rate was 1.2 mL/min. UV detection was performed at a wavelength of 335 nm.

A stock solution of lumefantrine was prepared in methanol–acetic acid (99.8:0.2 v/v). Standard working solutions were prepared by diluting the stock solution with acidic methanol. Calibration standards in plasma were prepared by spiking blank plasma (4900 μL) with 100 μL of working solution.

Standards, controls, and samples (250 μL each) were treated with 500 μL acetonitrile–acetic acid (99:1 v/v) containing IS (2.50 μg/mL), vortexed for 10 sec, incubated for 5 min, and centrifuged at 15,000 g for 5 min. The supernatants (1650 μL) were loaded onto a polypropylene 96-well plate containing 900 μL HPLC water under low vacuum. The SPE plates were conditioned with 500 μL methanol followed by 300 μL acetonitrile–water–acetic acid (30:69.5:0.5 v/v/v) (solvent A), washed with 1000 μL solvent A, dried under full vacuum for 10 min, wiped dry with paper, eluted with 500 μL methanol–trifluoroacetic acid (99.9: 0.1 v/v) (solvent B) and then with 400 μL solvent B for 2 min, evaporated to dryness at 65°C under a gentle air stream, reconstituted with 200 μL methanol–hydrochloric acid ($0.1M$) (70:30 v/v) and assayed. The injection volume was 50 μL. Figure 11.3 shows chromatograms of blank plasma and spiked plasma with lumefantrine. A calibration curve was constructed in a concentration range of 25 to 20,000 ng/mL. Intra-assay and inter-assay coefficients of variation were below 5.2 and 4.0%, respectively. The limit of detection was 10 ng/mL. The limit of quantification was 25 ng/mL.

11.3.2.4 Automated Solid Phase Extraction

Kabra et al.[38] described a rapid, precise, cost-effective, and automated HPLC method for determining cyclosporine in whole blood. The liquid chromatograph was coupled with a Varian advanced automated sample processing (AASP) unit (Figure 11.4). The AASP sequentially routed the mobile phase through each octyl sorbent cartridge, transferring extracted cyclosporine and IS directly onto the guard and analytical columns for separation and quantitation.

Cyclosporine D (200 μg/L) in methanol and $ZnSO_4$ (50 g/L) aqueous solution (1:1 v/v) served as the IS. The IS (1.5 mL) was accurately transferred into disposable glass tubes (13 × 100 mm), mixed with 0.5 mL calibration standard, control, or patient sample by vortexing for 30 sec, and centrifuged at 500 g for 2 min. C8 cartridges (conditioned by 1 mL acetonitrile followed by 0.5 mL deionized water) were loaded with whole blood supernatants followed with 1 mL acetonitrile/water wash solution (2:3 v/v), then loaded into the AASP.

The liquid chromatograph was a Perkin Elmer Series 3 or Varian 5500 system equipped with a Perkin Elmer LC-100 or Varian 2080 column oven, Varian 2010 pump for backflushing the guard

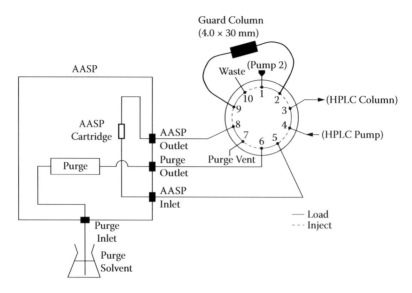

FIGURE 11.4 AASP fluidics showing guard and analytical column connection with the same valve. (*Source: From Kabra, P.M. et al. Clin Chem. 33, 2274, 1987. With permission.*)

column, Perkin Elmer LC 75 variable wavelength detector or Varian 9060 diode array detector, and Perkin Elmer LCI-100 integrator or Varian DS 604 data station. A Varian Micropak SP-C8-IP5 octyl column (150 × 4 mm, 5 μm particle size) with a Brownlee OSGU RP-8 guard column (30 × 4.0 mm) was maintained at a temperature of 70°C. The mobile phase consisted of 530 mL acetonitrile, 200 mL methanol, and 270 mL deionized water. The flow rate was 1.5 mL/min. The detector was set at a wavelength of 210 nm.

The calibration curve was linear up to 5000 μg/mL. Recoveries for cyclosporine ranged from 90 to 110 %. The limit of detection was below 30 μg/mL. Within-run and between-run coefficients of variation were less than 8%. About 100 whole blood samples could be analyzed within 3 hr with very high efficiency, sensitivity, and precision.

11.3.2.5 Single Quadrupole Mass Spectrometry

Zahlsen and colleagues[39] reported a high-volume, high-throughput liquid chromatography–mass spectrometry (LC-MS) TDM system for determining concentrations of clozapine and its metabolite, desmethylclozapine, in biological fluids.

An Agilent 1100 LC-MSD single-quadrupole instrument was equipped with a quaternary pump and electrospray. The stationary phase was a Zorbax SB-C18 column (30 × 4.6 mm, 3.5 μm particle size). The mobile phase consisted of methanol and 50mM ammonium acetate in water (60:40). The flow rate was 1 mL/min. A single quadrupole MS performed detection. The nebulizer was set at 25 psi at 350°C and a drying gas at 9 L/min. The instrument operated in selected ion monitoring (SIM) mode. Target: 327.0/frag 100V for clozapine, 313.0/frag 100V for desmethylclozapine, 388.1/ frag 100V for flurazepam, and 365.5/frag 100V for pericaizine; qualifier: 270.0/frag 150V for both clozapine and desmethylclozapine.

Internal standards were flurazepam for clozapine and pericaizine for desmethylclozapine, respectively. Aliquots (0.5 mL) of standards, controls, and patient samples were mixed with 50 μL of 10mM flurazepam and 50 μL of 10nM pericaizine, extracted with 4 mL of hexane/n-butanol/acetonitrile (93:5:2), shaken for 5 min, and centrifuged at 3000 rpm for 5 min. The organic layer was collected, evaporated to dryness at 40°C under an air stream, reconstituted in 50 μL of methanol,

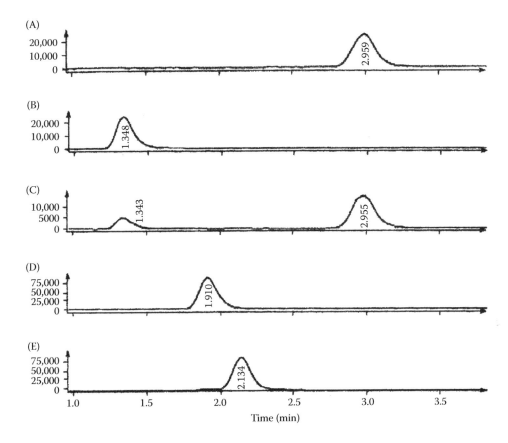

FIGURE 11.5 Extracted ion chromatogram of patient sample at typical low level. (*Source:* Zahlsen, K. et al. *LCGC N Amer.* 23, 390, 2005. With permission.)

and assayed. The injection volume was 2.5 μL. The retention times of desmethylclozapine, peric-aizine, flurazepam, and clozapine were 1.3, 1.9, 2.2, and 3.0 min, respectively. Figure 11.5 shows ion chromatograms of a patient sample. Calibration curves for clozapine and desmethylclozapine were constructed in a concentration range of 0 to 4000 nM. Correlation coefficients exceeded 0.999. Three-month rolling averages of the coefficient of variation were less than 9.1%.

11.3.3 HIGH PERFORMANCE LIQUID CHROMATOGRAPHY-MS/MS

Highly sensitive and specific determinations of drugs in biological samples can be achieved by coupling HPLC with tandem mass spectrometry (MS/MS). The high-throughput HPLC-MS/MS methodologies include several extraction procedures. Liquid–liquid extraction and cartridge-based SPE are the most common methods.

11.3.3.1 Liquid–Liquid Extraction

Sirolimus is a potent immunosuppressive agent. To prevent thrombocytopenia and hypercholester-olemia, optimize efficacy, and reduce organ rejection, assays were developed to monitor concentra-tions of sirolimus in the whole blood of patients under treatment.[40–42] Wallemacq et al.[43] developed and validated a simple high-throughput HPLC-MS/MS method to routinely monitor sirolimus concentrations in whole blood and applied the validated method to analyze more than 2000 clinical samples in daily TDM practice.

Their method included a Waters 2795 Alliance HT (high throughput) HPLC system with an integrated autosampler. The stationary phase was a Supelco C18 column (250×4.6 mm, 5 μm). The mobile phase consisted of solvent A (water containing $2mM$ ammonium acetate and 0.1% formic acid) and solvent B (methanol containing $2mM$ ammonium acetate and 0.1% formic acid). The mobile phase was delivered at a flow rate of 0.6 mL/min in a step gradient mode: 50% solvent B from 0 to 0.4 min and 100% solvent B from 0.4 to 0.8 min.

A Micromass Quattro Micro™ tandem MS was set in electrospray positive ionization mode. Settings were: capillary voltage 1.0 kV; cone voltage 25 V; source block temperature 140°C; desolvation temperature 350°C at a nitrogen flow of 600 L/hr; collision gas (argon) pressure 5×10^{-3} mbar; collision energy 18 eV; extractor 3 V; RF lens voltage 0.4 V; exit lens voltage –1 V; entrance lens voltage 1 V; and photomultiplier voltage 650 V. The multiple reaction monitoring (MRM) mode settings were m/z 931.5 → 864.6 for sirolimus and m/z 809.4 → 756.5 for ascomycin (IS).

A standard stock solution of sirolimus was prepared in methanol. Controls and standard working solutions were prepared by spiking blank whole blood with the stock solution. Standards, controls, and patient whole blood (10 μL) were transferred to 1.5 mL polypropylene tubes, mixed with 40 μL of $0.1M$ zinc sulfate solution, precipitated with 100 μL of methanol containing the IS (2 μg/L), vortexed vigorously for 5 sec, and centrifuged at 10,500 g for 5 min. Supernatants were collected and assayed. The injection volume was 20 μL. The retention times of sirolimus and ascomycin were 0.93 and 0.89 min, respectively. The total run time was 2.5 min. Representative MRM chromatograms of a patient sample are shown in Figure 11.6.

A calibration curve was constructed over a concentration range of 1 to 50 μg/mL with a correlation coefficient of 0.998. Intra-assay and inter-assay coefficients of variation were less than 7.9 and 9.5%, respectively. Mean absolute recoveries at 10 and 20 μg/mL were 72 and 76%, respectively. Limit of detection was 0.3 μg/mL. Limit of quantification was 1.0 μg/mL.

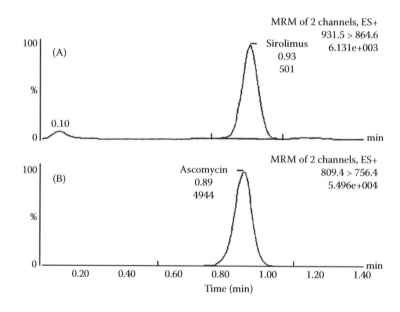

FIGURE 11.6 Representative multiple reaction monitoring (MRM) chromatogram of whole blood of patient treated with sirolimus: (A) m/z 931.5 → 864.6 represents transition of sirolimus at concentration of 10 μg/L eluted at 0.93 min; (B) m/z 809.4 → 756.4 represents transition of internal standard ascomycin eluted at 0.89 min. (*Source:* Wallemacq, P.E. et al., *Clin Chem Lab Med.* 41, 922, 2003. With permission.)

11.3.3.2 Solid Phase Extraction

Salm et al.[44] developed a high-throughput analytical method to measure cyclosporine in whole blood. They used a simple SPE procedure, followed by HPLC-MS/MS. An Agilent 1100 liquid chromatograph was coupled with an Agilent Zorbax Bonus C18 reversed-phase column (50 × 2.1 mm, 5 μm particle size). The column temperature was maintained at 70°C in a column oven. The mobile phase consisted of 80% methanol and 20% 40mM ammonium acetate buffer (pH 5.1) delivered isocratically at a flow of 0.4 mL/min. D_{12} cyclosporine was the IS.

A PE-Sciex API III triple quadrupole instrument was set in positive ionization mode using an electrospray interface. The orifice potential was 40 V and the interface heater was set at 40°C. The collision gas was argon at a thickness of 300×10^{12} molecules cm^{-2}. Cyclosporine and d_{12} cyclosporine were detected by MRM: m/z 1220 → 1203 for cyclosporine and m/z 1232 → 1215 for d_{12} cyclosporine, respectively.

Cyclosporine and d_{12} cyclosporine stock solutions (100 μg/mL) were prepared in methanol and remained stable for at least 12 mo at –80°C. Standard solutions were prepared by spiking blank whole blood with stock solutions. Standards, controls, and patient samples (50 μL), containing ethylenediaminetetraacetic acid as an anticoagulant were treated with 150 μL of precipitation reagent (mixture of acetonitrile, 0.1M zinc sulfate (70:30 v/v), and IS (200 μg/L), mixed, and centrifuged at 20,800 g for 2 min. Supernatants were loaded onto Waters Sep Pak C18 SPE cartridges (100 mg) conditioned with methanol (2 mL) followed by deionized water (2 mL). The loaded cartridges were washed with deionized water (4 mL), methanol/deionized water (65:35 v/v; 2 mL), and heptane (1 mL), placed under a vacuum for 15 min, and eluted with heptane/isopropanol (50:50 v/v; 1 mL). The resulting eluents were evaporated to dryness at 60°C under a gentle air stream, reconstituted in mobile phase (100 μL), centrifuged at 20,800 g for 1 min, and assayed. The injection volume was 10 μL. The retention times were 0.5 min for both cyclosporine and the IS. The total run time was 2 min. Figure 11.7 shows chromatograms of whole blood samples after extraction. A linear relationship was obtained at a concentration range of 10 to 2000 μg/mL, with a correlation coefficient exceeding 0.998. Intra-day and inter-day recoveries were greater than 96.0 and 94.9%, respectively. Intra-day and inter-day imprecisions were less than 4.2 and 7.6%, respectively.

11.3.3.3 Turbulent Flow Chromatography

Ceglarek et al.[45] reported the rapid simultaneous quantification of immmunosuppressants (cyclosporine A, tacrolimus, and sirolimus) in transplant patients by turbulent flow chromatography (TFC) coupled with HPLC-MS/MS. TFC is an online extraction technique involving the direct application of human plasma onto a turbulent flow column where protein is washed from the samples before the retained drug is backflushed onto an analytical column.

Cyclosporine D, ascomycin, and desmethoxyrapamycin were used as internal standards. A stock solution containing 200 μg/mL cyclosporine D, 20 μg/mL ascomycin, and 20 μg/mL desmethoxyrapamycin was prepared in methanol and diluted 1:1000 with methanol/zinc sulfate (50 g/L) aqueous solution (4:1 v/v) before use.

Blank, calibrator, control, and patient whole-blood samples (50 μL) were transferred into 1.5 mL conical test tubes, mixed with 100 μL of the IS, vortexed for 10 sec, and centrifuged at 13,000 g for 5 min. Twenty-five microliters of supernatant were injected onto a Cohesive Technologies Cyclone polymeric turbulent flow column (50 × 1 mm, 50 μm) and flushed with a mixture of methanol and water (10:90 v/v) at a flow of 5 mL/min. Column switching from the TFC to HPLC systems was via a Cohesive Technologies system. The analytical column was a Phenomenex Phenyl-Hexyl-RP (50 × 2.1 mm, 5 μm). The mobile phase consisted of methanol and ammonium acetate buffer (97:3 v/v). The buffer was 10mM ammonium acetate containing 0.1% v/v acetic acid. The flow rate was 0.6 mL/min.

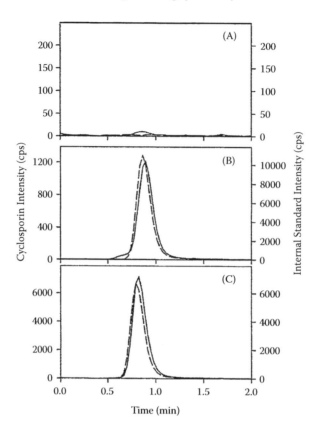

FIGURE 11.7 Representative chromatograms of blood obtained from (A) subject not receiving cyclosporine therapy, (B) patient's trough sample (84 μg/L), and (C) patient's 2-hr post-dose sample (818 μg/L). Peaks: solid line = cyclosporine (*m/z* 1220 → 1203). Dashed line = internal standard (*m/z* 1232 → 1215). (*Source:* From Salm, P. et al., *Clin Biochem.* 38, 669, 2005. With permission.)

A SCIEX API 3000 triple quadrupole mass spectrometer with Turbo ionspray was used in positive ion mode: nebulizer gas (10), auxiliary gas (71/min), curtain gas (8), collision gas (3), ionization voltage (4000 V), and source temperature (400°C). Analytes and ISs were detected by MRM: *m/z* 1220 → 1203 for cyclosporine A, *m/z* 1234 → 1217 for cyclosporine D, *m/z* 821 → 768 for tacrolimus, *m/z* 809 → 756 for ascomycin, *m/z* 931 → 864 for sirolimus, and *m/z* 901 → 834 for desmethoxyrapamycin.

Linear calibration curves were obtained over a range of 4.5 to 1500 ng/mL for cyclosporine A (*r* = 0.999), 0.2 to 100 ng/mL for tacrolimus (*r* = 0.998), and 0.4 to 100 ng/mL for sirolimus (*r* = 0.995). Within-run coefficients of variation were less than 8% for cyclosporine A, tacrolimus, and sirolimus. Between-run coefficients of variation were less than 2.7% for cyclosporine A, 8.4% for tacrolimus, and 9.3% for sirolimus. The total run time of an injection was 3 min including equilibrium time. The online extraction column worked for 800 injections without loss of cleaning capacity. The analytical column was good for at least 2000 injections.

11.3.3.4   Online Sample Clean-Up

Everolimus, a derivative of sirolimus, is a novel macrocyclic immunosuppressant. Risk of acute rejection increases when the everolimus trough level falls below 3 *μ*g/L in renal transplant patients.[46]

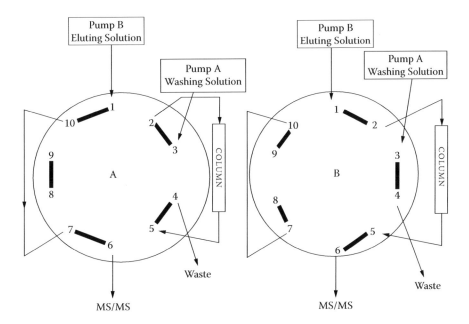

FIGURE 11.8 Switching valve set-up. (*Source:* From Korecka, M. et al., *Ther Drug Monit.* 28, 485, 2006. With permission.)

To improve long-term outcomes, it is critical to perform TDM of everolimus levels in patients under therapy.

Korecka et al.[47] developed a sensitive high-throughput HPLC-MS/MS method to measure everolimus in human whole blood using an online sample clean-up technique. An Agilent 1100 HPLC system was used. The stationary phase was a Waters Nova-Pak C18 column (150 × 2.1 mm, 4 μm) with a Waters C18 guard column (3 mm). The washing solution was a mixture of methanol and 30mM ammonium acetate buffer (pH 5.1) (80:20). The eluting solution was a mixture of methanol and 30mM ammonium acetate buffer (97:3). The flow rate was 0.8 mL/min. Figure 11.8 shows the switching valve set-up. The washing solution was delivered from 0 to 1.2 min for clean-up of the injected sample, then the switching valve was activated and the eluting solution was delivered through the column to elute analytes from 1.2 to 2.4 min. From 2.4 to 2.8 min, the washing solution was delivered for re-equilibration for the next injection.

An Applied Biosystems API 2000 mass spectrometer was coupled with HPLC and operated in positive ionization mode. The orifice potential was 60 V and focusing potential was 350 V. Collision energy was 20 eV, and ionization voltage was 4500 V. Detection was set in MRM: m/z 975.5 $\rightarrow$ 908.5 for everolimus, m/z 989.8 $\rightarrow$ 922.8 for SDZ RAD 223-756, and m/z 809.5 $\rightarrow$ 756.5 for ascomycin. Figure 11.9 is a typical mass chromatogram of everolimus in plasma.

Stock solutions of everolimus (600 μg/mL) and two ISs (ascomycin and SDZ RAD 223-756, 100 μg/mL) were prepared in methanol and stored at –70°C. The precipitation solution consisted of methanol and 0.1M zinc sulfate (70:30 v/v) containing an IS. The standards, controls, and patient samples (0.11 mL) were mixed with 0.2 mL of the precipitation solution, vortexed for 15 sec, and centrifuged at 9500 g for 15 min. Supernatants were collected and centrifuged again at 9500 g for 10 min. The injection volume was 90 μL.

A calibration curve for everolimus was obtained in a concentration range of 1 to 50 μg/mL with a correlation coefficient of 0.999. Between-day coefficient of variation was less than 8.6%. The limit of quantification was 1.0 μg/mL. Absolute recoveries averaged from 76.8 to 77.3%.

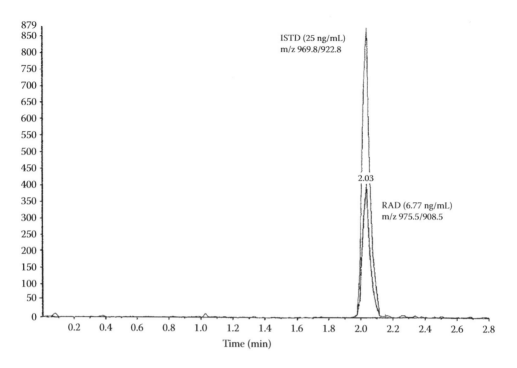

FIGURE 11.9 Representative mass chromatogram of everolimus (6.77 µg/L) and SDZ RAD 223-756 as internal standard (25 µg/L) in a blood sample from a renal transplant patient. (*Source:* From Korecka, M. et al., *Ther Drug Monit.* 28, 487, 2006. With permission.)

11.3.4 ULTRA-PERFORMANCE LIQUID CHROMATOGRAPHY (UPLC)-MS/MS

Based on its unique design and a column packed with smaller (1.7 µm) particles, UPLC can deliver a mobile phase at high linear velocities and it operates at high pressure (>10,000 psi). Thus, UPLC enables high-speed analysis, superior resolution, increased sensitivity, and reduced overall cost per analysis. Sub-microbore (<2.0 µm) HPLC is becoming increasingly popular because of its ultra speed and ultra resolution.[48,49] UPLC-MS/MS combines the high-throughput capability of UPLC with the identification power of MS/MS and is becoming very popular for TDM. Some examples of applications of high-throughput UPLC-MS/MS analytical methods in TDM are discussed below.

Amlodipine — Ma et al.[50] developed and validated a UPLC/MS/MS method for a pharmacokinetic study of amlodipine in human plasma after oral administration. Nimodipine 50 µg/mL in a mixture of methanol and water (50:50 v/v) served as the IS. Standard solutions of amlodipine were also prepared in a mixture of methanol and water (50:50 v/v).

A Waters Acquity™ UPLC system with a cooling autosampler and column oven was used. The stationary phase was a Waters Acquity BEH C18 column (50 × 2.1 mm, 1.7 µm particle size). The column was maintained at 40°C. The mobile phase consisted of water and acetonitrile, each containing 0.3% formic acid and was delivered at 0.35 mL/min in a gradient mode: at 60% water from 0 to 1.5 min, linearly decreased to 10% water in 0.5 min, and then returned to 60% water. Sample vials were maintained at 4°C.

A Waters Micromass triple quadrupole mass spectrometer was used with an electrospray ionization interface in positive ionization mode: desolvation gas (400), cone gas (70), collision gas (2.74 × 10⁻³ mbar), capillary (3.0 kV), cone (14 (kV), source temperature (105°C), and desolvation temperature (300°C). The detection and quantitation of amlodipine and nimodipine were performed by MRM: m/z 409 →238 for amlodipine and m/z 419 →343 for nimodipine, with a scan time of 0.05 sec per transition.

Twenty healthy male volunteers received two tablets containing 10 mg of amlodipine. Blood samples were collected before treatment and after 1, 2, 3, 4, 5, 6, 8, 10, 14, 24, 48, 72, 96, and 120 hours and centrifuged. Plasma was collected and stored at –20°C until analyzed. An aliquot (0.5 mL) of plasma sample was transferred into 10 mL glass tubes, mixed with 50 μL IS and then with 200 μL of 1M sodium hydroxide solution, vortexed for 60 sec, mixed with 3 mL of diethyl ether, vortexed for 60 sec, shaken for 10 min, and centrifuged at 3500 g for 10 min. The upper organic layer was collected, evaporated to dryness at 40°C under a gentle nitrogen stream, reconstituted in 100 μL of acetonitrile/water (70:30 v/v), and assayed. The injection volume was 5 μL in partial loop mode. The total run time was 3.0 min. Retention times of amlodipine and nimodipine were 0.75 and 1.38 min, respectively. No interference from metabolites or endogenous substances was observed. Linear calibration curves were constructed over a 0.15 to 16.0 ng/mL range. Correction coefficients exceeded 0.9984. Intra-day and inter-day coefficients of variation were less than 5.6 and 8.4%, respectively. The limit of quantitation was 0.15 ng/mL.

Doxazosine — This compound is a highly selective antagonist that effectively manages hypertension and benign prostatic hyperplasia.[51,52] Al-Dirbashi et al.[53] developed and validated a simple, sensitive, selective, and high-throughput UPLC-MS/MS assay for doxazosine in human plasma. A Waters Acquity UPLC system equipped with a thermostatted sampler and column oven was used. The stationary phase was a Waters BEH C18 column (50 × 2.1 mm, 1.7 μm). Solvent A was a mixture of 0.05 w/v pentadecafluorooctanoic acid in acetonitrile and solvent B was a mixture of 0.05 w/v pentadecafluorooctanoic acid in water. The mobile phase was delivered in a gradient mode. It was linearly increased from 10 to 99% solvent A from 0 to 1.45 min at 0.4 mL/min, then returned from 99 to 10% solvent A in 0.10 min at 1 mL/min, and held for another 0.05 min before the next injection.

A Micromass Quattro triple quadrupole mass spectrometer was used as the detector and set in the positive ionization mode at a capillary voltage of 4.5 kV. Cone voltages for the drug and IS were 50 and 35 V, respectively. Collision energies for the drug and IS were 32 and 23 eV, respectively. The ion source temperature was 125°C and the desolvation temperature was 400°C. Doxazosine and IS were detected by MRM: m/z 452 → 344 and 452 → 247 for the drug and m/z 409 → 228 and 409 → 271 for the IS.

Tamsulosin was the IS. Stock solutions of doxazosine and tamsulosin (1 mg/mL) were prepared in water/acetonitrile (50:50 v/v) and stored at 4°C in the dark. Aliquots (0.5 mL) of plasma samples were spiked with 50 μL of tamsulosin (250 ng/mL), mixed with 0.5 mL of 0.4M sodium borate buffer (pH 10) followed by 2 mL diethylether, vortexed for 5 min, and centrifuged at 3800 rpm for 5 min. The organic layer was collected, evaporated to dryness at 40°C under a gentle nitrogen stream, reconstituted with 100 μL of water/acetonitrile (50:50 v/v), filtered through a 0.45 μm Millipore PTFE hydrophilic filter, and assayed; injection volume was 5 μL.

A calibration curve for doxazosine was constructed in the range of 0 to 100 ng/mL with a correlation coefficient over 0.999. Intra-day and inter-day coefficients of variation were 5.7 and 8.0%, respectively. The limit of detection and the limit of quantification were 0.02 and 0.07 ng/mL, respectively. This validated method had a very short run time of 2 min compared with a 15-min run time for HPLC-fluorescence[54] and may be used for TDM of doxazosine.

Epirubicin — Anthracyclines have been used in cancer chemotherapy for more than 30 years and epirubicin (EPI) is one of the most widely used agents.[55] Li et al. developed a high-throughput method for the analysis of epirubicin in human plasma using UPLC-MS/MS.[56] A Waters Acquity UPLC system was coupled with a Micromass Quattro Premier MS. The stationary phase was a Waters Acquity BEH C18 column (50 × 2.1 mm, 1.7 μm). The column was maintained at 30°C. Solvent A was 0.1% formic acid in water and solvent B was acetonitrile. The mobile phase was delivered at a flow of 0.20 mL/min in a step gradient mode at 85% solvent A at 0 min, 70% solvent A at 1.00 min, and 85% solvent A from 2.50 to 4.00 min. Epidaunorubicin (EPR) was the IS.

The MS was set in the positive ionization mode and its source conditions were: capillary, 3.0 kV; source temperature, 110 °C; desolvation temperature, 180°C; cone gas, 50 L/hr; desolvation gas, 350 L/hr; collision gas, 2.34×10^{-3} mbar; multiplier, 650V; and dwell time, 0.1 sec. Detection was via MRM: m/z 544 → 130 and 544 → 397 for epirubicin, m/z 528 → 321 and 528 → 363 for epidaunorubicin, and m/z 546 → 399 for metabolite.

Stock solutions of epirubicin and the IS (1 mg/mL) were prepared in methanol/water (1:1 v/v) and stored at 4°C. Standard working solutions were prepared by diluting stock solutions with 20m*M*

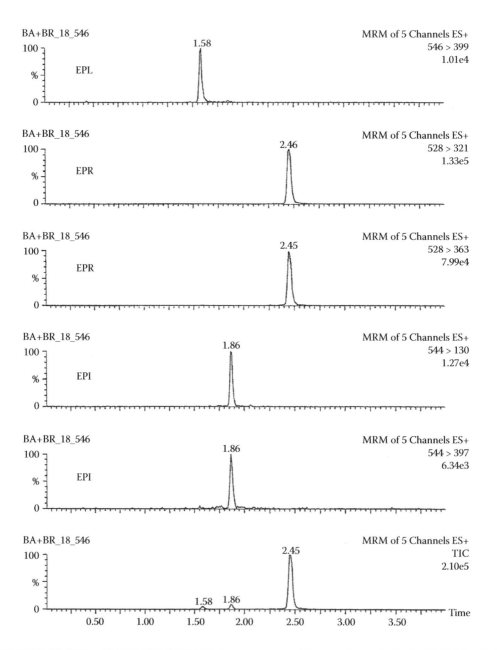

FIGURE 11.10 TIC and MRM UPLC-MS/MS chromatograms of human plasma (spiked with 100.0 ng/mL IS) from a patient obtained 8 hr after intravenous administration of 60 mg EPI. (*Source:* From Li, R. et al., *Anal Chim Acta.* 546, 171, 2005. With permission.)

formate buffer (pH 2.9). Aliquots (0.2 mL) of blank human plasma were spiked with 25 μL of IS and 25 μL of working solutions, loaded onto Oasis HLB cartridges preconditioned with 1 mL methanol followed by 1 mL deionized water, washed sequentially with 1 mL 5% v/v methanol and 1 mL 40% v/v methanol containing 2% ammonium hydroxide, eluted with 0.5 mL 0.5% formic acid in methanol, evaporated to dryness at 30°C under a gentle nitrogen stream, reconstituted with 200 μL of 15% acetonitrile in water, and assayed. The injection volume was 10 μL. Representative MRM chromatograms of epirubicin in plasma are shown in Figure 11.10.

A linear calibration curve for epirubicin ranged from 0.50 to 100.0 ng/mL with a correlation coefficient of 0.999. Intra-day and inter-day coefficients of variation were less than 5.2 and 11.7%, respectively. Limit of detection and limit of quantification were 0.1 and 0.5 ng/mL, respectively. The extraction recoveries ranged from 89.4 to 101.2%. The validated method was successfully applied to the routine analysis of plasma samples from patients treated with epirubicin.

Lercanidipine — Kalovidouris et al.[57] applied UPLC-MS/MS to the determination of lercanidipine in human plasma after oral administration of lercanidipine. A Waters Acquity UPLC system with cooling autosampler and column oven was coupled with a Waters BEH C18 column (50 × 2.1 mm, 1.7 μm). The mobile phase was composed of 70% acetonitrile in water containing 0.2% v/v formic acid, delivered at a flow of 0.30 mL/min. The column temperature was maintained at 40°C and sample vials at 5°C.

A Waters Quattro Micro API triple quadrupole mass spectrometer equipped with an ESI interface in positive mode was coupled to UPLC. Lercanidipine and nicardipine (IS) were detected and quantitated in MRM mode: m/z 612.2 →280.2 for lercanidipine, and 479.9 →315.1 for nicardipine. Other parameters were desolvation gas (400 L/hr), cone gas (10 L/hr), collision gas (0.0023 mbar), cone voltage (30.0 V), collision energy voltage (30.0 eV), transition dwell time (0.1 sec), source temperature (100°C), desolvation temperature (300°C), Stock standard solutions of lercanidipine and nicardipine were prepared in methanol. Working solutions were prepared by diluting stock standard solutions with methanol/water (50:50 v/v). Three patients (two male and one female, 50 to 60 years old) received 10 mg lercanidipine orally once daily in the morning. Blood samples were drawn 3 hr after dosing and centrifuged at 4000 rpm and 4°C for 10 min. Plasma was collected and stored at –20°C until analysis. Aliquots of calibrators, controls, and plasma samples were alkalined by adding 200 μL of 1.0M sodium hydroxide, vortexed for 30 sec, mixed with 4.0 mL *tert*-butyl methyl ether by gently shaking at 150 g for 30 min, and centrifuged at 2500 g for 10 min. The organic layer was transferred to a 10 mL glass tube after freezing the aqueous layer with dry ice for 5 min, evaporated to dryness at 30°C under a gentle stream of nitrogen, reconstituted in 100 μL of mobile phase, vortexed for 30 sec, and assayed. The sample injection volume was 20 μL. The total run time was 1.0 min. Retention times of lercanidipine and nicardipine were 0.41 and 0.38 minutes, respectively.

Linear $1/y^2$ regression analyses of the ratio of the peak area of lercanidipine to the concentration compared with the ratio of the IS were constructed over the range of 0.05 to 30.00 ng/mL. Correlation coefficients exceeded 0.995. Intra-assay and inter-assay coefficients of variation were less than 7.3 and 6.1%, respectively. The limit of detection was calculated to be 0.02 ng/mL, and the limit of quantitation was 0.05 ng/mL.

11.4 CONCLUSIONS

High-throughput analytical methods used for TDM use widely different technologies. Immunoassay is rapid and easy to use, but its disadvantages are cross-reactivity with other drugs or metabolites and the ability to analyze only a limited number of drugs. HPLC is very versatile and rapid when coupled with automated sample preparation or a monolithic column. It can simultaneously determine concentrations of multiple drugs and their metabolites in biological fluids. UPLC's high-throughput ability is due to the unique properties of its column and instrumentation, and it can be used with ultraviolet light, fluorescence, photodiode array detectors, and mass spectrometers.

The method of choice for TDM, however, is HPLC-MS/MS. Assay sensitivity is always an issue with potent drugs. The selectivity of the assay is also important when analyzing complex biological samples. HPLC-MS/MS can achieve high-sensitivity, high-selectivity, and high-throughput assays with proper sample preparation. The combination of powerful UPLC with tandem mass spectrometry (UPLC-MS/MS) is becoming increasingly popular and will probably replace many other analytical methods used for TDM.

REFERENCES

1. Baumann, P. et al. 2004. The AGNP-TDM expert group consensus guideline: Therapeutic drug monitoring in psychiatry. *Pharmacopsychiatry.* 37: 243.
2. Droste, J.A.H. et al. 2005. TDM: Therapeutic drug measuring or therapeutic drug monitoring? *Ther Drug Monit.* 27: 412.
3. Joerger, M., Schellens, J.H.M., and Beijnen, J.H. 2004. Therapeutic drug monitoring of non-anticancer drugs in cancer patients. *Meth Find Exp Clin Pharmacol.* 26: 531.
4. Clarke, W. and McMillin G. 2006. Application of TDM, pharmacogenomics and biomarkers for neurological disease pharmacotherapy: Focus on antiepileptic drugs. *Pers Med.* 3: 139.
5. Hendset, M. et al. 2006. The complexity of active metabolites in therapeutic drug monitoring of psychotropic drugs. *Pharmacopsychiatry.* 39: 121.
6. de Jonge, M.E. et al. 2005. Individualized cancer chemotherapy: Strategies and performance of prospective studies on therapeutic drug monitoring with dose adaptation. *Clin Pharmacokinet.* 44: 147.
7. Kuypers, D.R.J. 2005. Immunosuppressive drug monitoring: What to use in clinical practice today to improve renal graft outcome. *Transplant Int.* 18: 140.
8. Slish, J.C. et al. 2006. Update on the pharmacokinetic aspects of antiretroviral agents: Implications in therapeutic drug monitoring. *Curr Pharm Des.* 12: 1129.
9. Onal, A. 2006. Analysis of antiretroviral drugs in biological matrices for therapeutic drug monitoring. *J Food Drug Anal.* 14: 99.
10. Edholm, L.E. 1980. Specific methods for theophylline assay in biological samples. *Eur J Respir Dis Suppl.* 109: 45.
11. De la Torre, R. et al. 1998. Quantitative determination of tricyclic antidepressnts and their metabolites in plasma by solid phase extraction (bond-elut TCA) and separation by capillary gas chromatography with nitrogen-phosphorus detection. *Ther Drug Monit.* 20: 340.
12. Gupta, R. and Molnar, G. 1980. Plasma levels and tricyclic antidepressant therapy: Part I. A review of assay method. *Biopharmacol Drug Dispos.* 1: 259.
13. Scoggins, B. et al. 1980. Measurement of tricyclic antidepressants. Part I: Review of methodology. *Clin Chem.* 26: 5.
14. Rambeck, B. and Meijer, J.W. 1980. Gas chromatographic methods for the determination of antiepileptic drugs: A systematic review. *Ther Drug Monit.* 2: 385.
15. Varma, R. 1978. Therapeutic monitoring of anticonvulsant drugs in psychiatric patients: Rapid, simultaneous Gas chromatographic determination of six commonly used anticonvulsants without interference from other drugs. *Biochem Exp Biol.* 14: 311.
16. Least, C.J. Jr., Johnson, G.F., and Solomon, H.M. 1975. Therapeutic monitoring of anticonvulsant drugs: Gas chromatographic simultaneous determination of primidone, phenylethylmalonamide, carbamazepine, and diphenylhydantoin. *Clin Chem.* 21: 1658.
17. Lehane, D.P. et al. 1976. Therapeutic drug monitoring: Measurements of antiepileptic and barbiturate drug levels in blood by gas chromatography with nitrogen-selective detector. *Ann Clin Lab Sci.* 6: 404.
18. Wilson, J.M., Williamson, L.J., and Raisys, V.A. 1977. Simultaneous measurement of secondary and tertiary tricyclic antidepressants by GC/MS chemical ionization mass fragmentography. *Clin Chem.* 23: 1012.
19. Wong, S.H.Y. 1989. Advnces in liquid chromatography and related methodologies for therapeutic drug monitoring. *J Pharm Biomed Anal.* 7: 1011.
20. Rao, M.L. et al. 1994. Monitoring tricyclic antidepressant concentrations in serum by fluorescence polarization immunoassay compared with gas chromatography and HPLC. *Clin Chem.* 40: 929.
21. Pankey, S. et al. 1986. Quantitative homogeneous enzyme immunoassays for amitriptyline, nortriptyline, imipramine, and desipramine. *Clin Chem.* 32: 768.

22. Weiss, R.B. 1992. The anthracyclines: Will we ever find a better doxorubicin? *Semin Oncol.* 19: 670.

23. Vogler, W.R. et al. 1992. A phase III trial comparing idarubicin and daunorubicin in combination with cytarabine in acute myelogenous leukemia: A Southeastern Cancer Study Group study. *J Clin Oncol.* 10: 1103.

24. Coukell, A.J. and Faulds, D. 1997. Epirubicin: An updated review of its pharmacodynamic and pharmacokinetic properties and therapeutic efficacy in the management of breast cancer. *Drugs.* 53: 453.

25. de Jong, J. et al. 1990. Sensitive method for determination of daunorubicin and its known metabolites in plasma and heart by high performance liquid chromatography with fluorescence detection. *J Chromatogr A.* 529: 359.

26. Andersen, A., Warren, D.J., and Slordal, L. 1993. A sensitive and simple high performance liquid chromatographic method for the determination of doxorubicin and its metabolites in plasma. *Ther Drug Monit.* 15: 455.

27. Fogli, S. et al. 1999. An improved HPLC method for therapeutic drug monitoring of daunorubicin, idarubicin, doxorubicin, epirubicin, and their 13-dihydro metabolites in human plasma. *Ther Drug Monit.* 21: 367.

28. Varin, F., Cubeddu, L.X., and Powell, J.R., 1986. Liquid chromatographic assay and disposition of carvedilol in healthy volunteers. *J Pharm Sci.* 75: 1195.

29. Hokama, N. et al. 1999. Rapid and simple microdetermination of carvedilol in rat plasma by high performance liquid chromatography. *J Chromatogr B.* 732: 233.

30. Ptacek, P., Macek, J., and Klima, J. 2003. Liquid chromatographic determination of carvedilol in human plasma. *J. Chromatogr B.* 789: 405.

31. Zarghi, A. et al. 2007. Quantification of carvedilol in human plasma by liquid chromatography using fluorescence detection: Application in pharmacokinetic studies. *J Pharm Biomed Anal.* 44: 250.

32. Purkins, L. et al. 2002. Pharmacokinetics and safety of of voriconazole following intravenous to oral dose escalation regimens. *Antimicrob Agents Chemother.* 46: 2546.

33. Lazarus, H.M. et al. 2002. Safety and pharmacokinetics of oral voriconazole in patients at risk of fungal infection: A dose escalation study. *J Clin Pharmacol.* 42: 395.

34. Keevil, B.G. et al. 2004. Validation of an assay for voriconazole in serum samples using liquid chromatography–tandem mass spectrometry. *Ther Drug Monit.* 26: 650.

35. Egle, H. et al. 2005. Fast, fully automated analysis of voriconazole from serum by LC-LC-ESI-MS-MS with parallel column switching technique. *J Chromatogr B.* 814: 361.

36. Wenk, M., Droll, A., and Krahenbuhl, S. 2006. Fast and reliable determination of the antifungal drug voriconazole in plasma using monolithic silica rod liquid chromatography. *J Chromatogr B.* 832: 313.

37. Annerberg, A. et al. 2005. High throughput assay for the determination of lumefantrine in plasma. *J Chromatogr B.* 822: 330.

38. Kabra, P.M., Wall, J.H., and Dimson, P. 1987. Automated solid phase extraction and liquid chromatography for assay of cyclosporine in whole blood. *Clin Chem.* 33: 2272.

39. Zahlsen, K., Aamo, T., and Zweigenbaum, J. 2005. A high-volume, high-throughput LC-MS therapeutic drug monitoring system. *LCGC N Amer.* 23: 384.

40. Ren, B. et al. 2004. Determination of sirolimus in human whole blood by HPLC. *Zhongguo Yaoxue Zazhi.* 39: 52.

41. Taylor, P.J. et al. 2000. Simultaneous quantification of tacrolimus and sirolimus in human blood by high performance liquid chromatography–tandem mass spectrometry. *Ther Drug Monitor.* 22: 608.

42. Holt, D.W. et al. 2000. Validation of an assay for routine monitoring of sirolimus using HPLC with mass spectrometric detection. *Clin Chem.* 46: 1179.

43. Wallemacq, P.E. et al. 2003. High-throughput liquid chromatography–tandem mass spectrometric analysis of sirolimus in whole blood. *Clin Chem Lab Med.* 41: 921.

44. Salm, P. et al. 2005. A rapid HPLC mass spectrometry cyclosporine method suitable for current monitoring practices. *Clin Biochem.* 38: 667.

45. Ceglarek, U. et al. 2004. Rapid simultaneous quantification of immunosuppressants in transplant patients by turbulent flow chromatography combined with tandem mass spectrometry. *Clin Chem.* 346: 181.

46. Kovarik, J.M. et al. 2002. Exposure–response relationships for everolimus in *de novo* kidney transplantation: Defining a therapeutic range. *Transplantation.* 73: 920.

47. Korecka, M., Solari, S.G., and Shaw, L.M., 2006. Sensitive, high throughput HPLC/MS/MS method with online sample clean-up for everolimus measurement. *Ther Drug Monit.* 28: 484.

48. Grumbach, E.S. et al. 2005. Developing columns for UPLC: Design considerations and recent development. *LCGC Eur.* 37.
49. Grumbach, E. et al. 2006. Improving LC separations: Transferring methods from HPLC to UPLC. *LCGC N Amer Suppl.* 80.
50. Ma, Y. et al. 2007. Determination and pharmacokinetic study of amlodipine in human plasma by ultra performance liquid chromatography–electrospray ionization mass spectrometry. *J Pharm Biomed Anal.* 43: 1540.
51. Gillenwater, J.Y. et al. 1995. Doxazosin for the treatment of benign prostatic hyperplasia in patients with mild to moderate essential hypertension: A double-blind, placebo-controlled, dose–response multicenter study. *J Urol.* 154: 110.
52. Jyothirmayi, G.N., Alluru, I., and Reddi, A.S. 1996. Doxazosin prevents proteinuria and glomerular loss of heparin sulfate in diabetic rats. *Hypertension.* 27: 1108.
53. Al-Dirbashi, O.Y. et al. 2006. UPLC/MS/MS determination of doxazosine in human plasma. *Anal Bioanal Chem.* 385: 1439.
54. Zagotto, G. et al. 2001. Anthracyclines: Recent developments in their separation and quantitation. *J Chromatogr B.* 764: 161.
55. Sripalakit, P., Nermhom, P., and Saraphanchotiwitthaya, A. 2006. Validation and pharmacokinetic application of a method for determination of doxazosin in human plasma by high performance liquid chromatography. *Biomed Chromatogr.* 20: 729.
56. Li, R., Dong, L., and Huang, J. 2005. Ultra performance liquid chromatography – tandem mass spectrometry for the determination of epirubicin in human plasma. *Anal Chim Acta.* 546: 167.
57. Kalovidouris, M. et al. 2006. Ultra-performance liquid chromatography/tandem mass spectrometry method for the determination of lercanidipine in human plasma. *Rapid Commun. Mass Spectrom.* 20: 2939.

12 High-Throughput Quantitative Pharmaceutical Analysis in Drug Metabolism and Pharmacokinetics Using Liquid Chromatography–Mass Spectrometry

Xiaohui Xu

CONTENTS

12.1 INTRODUCTION

Drug metabolism and pharmacokinetics (DMPK)-related studies are among the most important phases of drug discovery and development and are increasingly involved in lead optimization. Recent advances in high-throughput screening, combinatorial chemistry, and genomics have significantly increased the numbers of samples requiring DMPK profiling. Hence, high-throughput pharmaceutical bioanalysis is necessary to support the increasing numbers of DMPK studies to expedite the drug discovery and development processes. High-throughput bioanalysis has been achieved by using a combination of high performance liquid chromatography and tandem mass spectrometry (LC/MS/MS).[1–11] This technique revolutionized bioanalysis by dramatically increasing throughput for the quantitative determination of drugs and metabolites in biological matrices due to its inherent specificity and sensitivity.

In recent years, new strategies for sample preparation to support high-throughput LC/MS/MS. The ultimate goal of sample preparation is to eliminate potential matrix interferences from biological matrices and other interfering compounds that may impact ionization during sample analysis. Various offline and online sample preparation strategies have been thoroughly evaluated. Although solid phase extraction (SPE) is considered a generic method, liquid–liquid extraction (LLE) is preferred for compounds requiring extensive sample clean-up. Direct injection is gaining more interest because of its apparent simplicity.

Fully automated sample preparations including direct injection, column switching extraction, and 96-well formats have been extensively applied in the pharmaceutical industry in recent years.[12,13] Although these approaches are capable of front-end high throughput, the efficient use of expensive MS equipment may be a limitation. Parallel HPLC online with a MUX® system reportedly improves the efficiency of the MS up to nine times in terms of accurate mass screening.[14] The MUX system involves multiple ion sprays on a single ion source on a triple quadrupole or time-of-flight (TOF) MS. The individual ion spray of the MUX is connected with an individual HPLC column.

Cassette dosing and plasma sample pooling are other widely applied alternatives for increasing the capacity of expensive LC/MS/MS equipment by decreasing the numbers of samples to be analyzed without compromising pharmacokinetics. Overall, high-throughput bioanalysis via LC/MS/MS is a very dynamic field and many novel approaches to improve bioanalytical processes and throughputs and solve new challenges continue to emerge at a very fast pace.

12.2 HIGH-THROUGHPUT QUANTITATIVE BIOANALYSIS: TRENDS AND GENERAL CONSIDERATIONS

Pharmaceutical development consists of four distinct stages: (1) discovery, (2) preclinical testing, (3) clinical phases, and (4) manufacture. Different approaches are usually required for the different stages to support their specific foci. Bioanalyses of both *in vitro* and *in vivo* samples are required to support DMPK-related studies during the first three stages. Typical LC/MS supports within DMPK[15–25] involve the screening for or evaluation of colon adenocarcinoma (Caco-2) cell line permeability, microsomal and/or hepatocyte metabolism, protein binding, cytochrome P450 isoforms (CYPs), enzyme inhibition, metabolic stability, bioavailability, animal mass balance studies, human absorption, distribution, metabolism and elimination (ADME) studies, metabolic profiling and structure elucidation, exposure and other pharmacokinetics parameters, and finally, formulation. Among typical DMPK screenings, pharmacokinetic sample analyses and metabolic screening and profiling are already conducted in a high-throughput fashion, due mainly to to the widespread use of LC/MS. A routine LC/MS screening of five major human cytochrome P450 compounds (CYP3A4, CYP2D6, CYP2C9, CYP2C19, and CYP1A2) was reported to support[26] the screening of new drugs showing potential for P450 inhibition. The analysis was achieved on a monolithic silica column within 24 sec via a generic gradient. Sample generation and analysis have always been important aspects of pharmaceutical development. Rapid, high-throughput, sensitive, and selective methods are critical for meeting the increasing needs created by dramatic improvements in sample generation. The high-throughput strategy is well established in both drug discovery and development but with different emphases. In drug discovery, bioanalytical methods support the selection of drug candidates. The challenge is to process large numbers of compounds and metabolites with relatively small numbers of samples per study. Analyses that provide quick data turn-around and require minimum method development are desirable. A generic method requiring minimal modification is preferable. As a lead candidate moves through the development process, analyses become more focused. In the development stage, especially in clinical development, the number of samples may vary from ten to several thousand per study. Robust, rugged, validated, and automated analytical methods that comply with strict regulatory guidelines are specifically needed for each compound.

12.2.1　Role of DMPK Screening in Lead Optimization

The three states of drug discovery are (1) target identification, (2) lead selection, and (3) lead optimization. Lead compounds identified by screening efforts are further optimized through the close collaboration of medicinal chemistry, exploratory drug metabolism, and drug safety assessment. As a segment of lead optimization, DMPK studies are becoming increasingly important based on the high attrition rate in later phases of drug development due to poor DMPK properties.[27,28] The identification of lead compounds that have desirable DMPK characteristics at an early discovery stage is expected to largely improve the success rate of drug candidates downstream. To achieve this, early DMPK screening of a larger number of samples is required, which in turn calls for higher throughput bioanalytical approaches.

At the drug discovery stage, the rapid quantification of leads and their metabolites remains a challenge that is often driven by the need for fast results from testing large numbers of samples. Therefore, many different LC/MS/MS methods have been developed to achieve higher throughputs. To streamline the process, generic sample preparation with a generic LC/MS/MS protocol applicable to most samples and requiring minimal modification is desirable.

Cassette dosing and sample pooling — To overcome the throughput limitation of classical DMPK supports, cassette dosing is widely applied to PK screening by use of LC/MS/MS[29,30] and has proven effective for improving throughput.[31–40] The prototype of cassette (N-in-one) dosing was initially reported in 1997.[34] The name arises from the ability to simultaneously dose laboratory animals with multiple compounds, typically to accelerate the pace of exposure screening. Combining with the extraordinary specificity of tandem MS detection, cassette dosing is seen as a way to maximize the use of expensive LC/MS/MS instrumentation and minimize the number of animals used. When the dosing of mixtures of several compounds (cassette dosing) is followed by sample analysis using LC/MS/MS, the possibilities of drug–drug interaction (DDI) are introduced. To reduce the risk of DDI, the doses of individual compounds must be reduced or a reference compound with a known PK profile should be included in the mixture. An alternative to circumvent the disadvantages of cassette dosing is sample pooling in which dosing an individual compound to each animal is followed by analyzing pooled plasma samples from different animals dosed with different compounds. This approach allows the use of one plasma sample per animal instead of eight to 10 to generate area-under-the-curve (AUC) data. Sample pooling[41–43] approaches have been reported to dramatically enhance efficiency during drug candidate screening. Both cassette dosing and sample pooling require development of methods for multiple components. Issues such as selecting a sample preparation procedure that fits all analytes, response factor balance, and the dynamic range of compounds are often involved. Not all compounds are suited for "cocktail" analysis, and special attention must be given to the selection of the cocktail, particularly for compounds that differ chemically.

12.2.2 PRECLINICAL DEVELOPMENT

The preclinical stage of drug development focuses on the activities required for filing an investigational new drug (IND) application. An IND application includes animal toxicity data, protocols for early phase clinical trials, and an outline of specific details and plans for clinical evaluation. Process research, formulation, metabolism and pharmacokinetics, and toxicology are the major areas of responsibility at this stage and analyses are targeted at obtaining specific and detailed information for evaluating drug properties.

This stage of drug development is also the first point at which regulatory issues are addressed; therefore, the use of validated analytical methods and compliance with U.S. Food & Drug Administration (FDA) guidelines[44] are critical. The all aspects of drug discovery and production must be conducted in accordance with FDA regulations and good laboratory practice (GLP) guidelines. Appropriate bioanalytical methods are normally developed for a series of toxicological studies typically focusing on ADME, system exposure, and metabolism.

12.2.3 CLINICAL DEVELOPMENT

Because DMPK properties vary among different species, *in vitro* human and animal data and *in vivo* animal data cannot always be extrapolated to human *in vivo* responses. The three main reasons that drugs fail during clinical trials are (1) lack of efficacy, (2) unacceptable adverse effects, and (3) unfavorable ADME properties. Hence, clinical development is necessary to establish solid experiment-based human exposure and safety data through both short- and long-term monitoring.

During clinical development, a lead candidate can be fully characterized in humans. Subsequent analyses must continue to be performed under strict protocols and regulatory compliance to register a new drug application (NDA) and supplementary NDA (sNDA) if required. Both quantitative and qualitative LC/MS/MS bioanalysis data must be obtained to prove adequate efficacy and favorable ADME properties of a compound in addition to achieving DMPK parameter optimization. Clinical development involves three phases of trials (I through III) and subsequent NDA filing. Each phase involves one or more pharmaceutical development processes such as scale-up, pharmacokinetics, drug delivery, and drug safety evaluations. High-throughput bioanalysis is even more critical at this phase because sample numbers increase dramatically. In clinical development, high-throughput approaches are required for sample analysis and other front-end operations such as tube labeling, centrifugation, decapping and recapping, and removing fibrinogen clots.

12.3 EXPERIMENTAL PERSPECTIVES IN DMPK RELATED HIGH-THROUGHPUT QUANTITATIVE BIOANALYSIS

The high-throughput concept for quantitative bioanalysis applies to steps such as assay development, sample collection and sorting, sample preparation, sample analysis, and data processing and reporting. Those processes are closely interlinked and improvement of process throughput is equally important.

Traditional quantitative high-throughput approaches such as 96-well format sample collection, automated or semi-automated sample extraction, automated LC/MS/MS method development, parallel concepts, and automated data processing have been widely applied to shorten turn-around times. Novel approaches such as monolithic columns,[45–47] turbulent flow chromatography,[37,48–49] fast gradient HPLC,[50–53] UPLC,[54,55] online SPE,[56–58], and MUX[14,17,59–61] show even more promising results for high-throughput quantitative analysis by further shortening turn-around.

12.3.1 HIGH-THROUGHPUT APPROACHES TO AUTOMATED SAMPLE PREPARATION

With decreasing LC run times, sample pretreatment, and the associated method development, sample analysis may be the rate-limiting step in bioanalytical assays. Both semi-automated and automated approaches are suitable for reducing the time necessary to remove interfering matrix components, but both have different advantages and foci.

Solid phase extraction (SPE), liquid–liquid extraction (LLE), and protein precipitation (PPT) in 96-well formats coupled with robotic liquid handling systems are straightforward for method development. Semi-automated and automated SPE, LLE, and PPT are the sample preparation techniques of choice for LC/MS/MS. However, for routine sample analysis, a fully online automated sample preparation is preferred to efficiently eliminate interfering endogenous components such as proteins, carbohydrates, salts, and lipids from biological samples. In particular, introducing plasma directly[62–64] may further reduce sample preparation time during routine analyses of large batches of plasma samples.

The recent trend of decreasing available sample volumes and requiring lower limits of quantitation (LLOQs) means better sample preparation procedures are under consideration. Further improvements MS sensitivity will eventually impact sample preparation strategies and sample throughput.

12.3.1.1 Automated Solid Phase Extraction (SPE)

SPE is the most common technique for sample pretreatment during pharmaceutical compound bioanalysis. Method development for SPE may follow a chromatographic procedure because it is based on the same principles as HPLC. Like HPLC column packing materials, SPE cartridge materials can utilize a wide range of silica-based and polymer-based sorbents. The extraction can follow generic protocols or be individually optimized if better sample clean-up is needed. Three types of extraction cartridges are commonly employed for SPE. The most common type is filled with 50 to 200 mg of SPE material, depending on size. The second type is a disk cartridge. The membrane contains chromatographic particles immobilized in an inert polytetrafluoroethylene matrix. The SPE bed thickness is less than 1 mm and therefore provides faster flow rates and smaller elution volumes (as low as 75 μL).

The third cartridge type is the novel 96-well HLB μElution SPE plate. The cartridge consists of 2 mg high-capacity SPE sorbents and a focusing tip. The unique design of the tip makes efficient use of the sorbents and allows elution of target compounds using as little as 25 μL of elution solvent.[65,66] The evaporation and reconstitution steps are not necessary mainly due to the concentrating ability of the plate. This type of plate helps to avoid common problems such as surface adsorption, thermal degradation, and re-solubilization associated with evaporation to dryness and reconstitution. The introduction of the 96-well format SPE in 1996 was a significant contribution to the application of high-throughput LC/MS/MS.[67] By using a pipetting stations where a 96-well plate vacuum manifold is integrated in the workstation, SPE is easy to automate and can be performed in either automated or semi-automated process. The use of a 384-well format[68] using a 96/384 multi-channel dispenser was also reported. The performance was equivalent to the 96-well plate format, whereas the sample preparation time was half compared to that of the equivalent 4 × 96 well plates with the existing robotic liquid handling system.

12.3.1.2 Automated Liquid–Liquid Extraction (LLE)

Although SPE is relatively simple to automate, it has a reputation for generating extracts that are less clean than those obtained by LLE—the second most common sample preparation based on simplicity and its ability to provide clean extracts. One advantage of LLE over other sample preparation methods is the easy removal of inorganic salts. They are not soluble in organic solvents and consequently remain in the aqueous phase. This makes LLE well suited for lipophilic compounds as the analyte transfers from the aqueous-based matrix to an apolar organic phase. In addition, proteins from plasma samples that often appear as interfering components are almost insoluble in the organic solvents used for LLE.

When extraction times and other conditions are fixed in LLE, the extraction efficiency is related to the partition coefficient that is controlled by the characteristics of the extraction solvent (e.g., viscosity, surface tension, solubility in water, etc.). In general, organic solvents with low solubility in water and high polarity and hydrogen-bonding properties enhance the recovery of pharmaceutical compounds in the organic phase, and are preferred for LLE. As a popular alternative to SPE, LLE in the 96-well format is relatively fast and simple, and allows extraction of a large number of samples. The application of liquid-handling workstations allows semi-automated operation. High-throughput LLE in the 96-well format was first reported by Zhang et al.[69] They used a Tomtec Quadra 96 pipetting station that can process 96 samples in parallel. The reported sample preparation time was approximately 1.5 hr for a 96-well plate. LLE's requirement for a large volume of extraction solvent can limit use of the 96-well format. To mitigate the impact of the volume limitation, a single-spot LLE method was introduced. As little as 50 μL of methyl-t-butyl ether (MTBE) can provide adequate recovery by simply adding acetonitrile to the samples prior to LLE.[70,71] Automated LLE combined with LC/MS/MS has been widely used in recent years[69–75] for the analysis of pharmaceutical moieties based on advantages in efficiency, cost, and throughput. Cartridge format solid–liquid extraction (SLE) is based on a diatomaceous earth plate, but its extraction mechanism is similar to that of LLE. Diatomaceous earth is a micro-amorphous silica that has many active sites on which polar sample molecules can adsorb strongly. It also contains small amounts of alumina and other metallic oxide impurities.[76] It has a very high surface area-to-weight ratio. The high surface area of this special synthetic inert material earth incorporated into cartridges ensures that the organic eluents remain uncontaminated by the aqueous matrix, eliminates emulsion problems, and facilitates efficient interactions between the sample and the organic solvent. Diatomaceous earth sorbents can easily be placed in a 96-well format as a fully automated alternative to LLE without the solvent volume limitation often observed with LLE. This methodology was applied to the quantitation of indolcarbazole in human plasma with a limit of quantification of 50 pg/mL, using an 0.25 mL plasma aliquot.[76]

12.3.1.3 Automated Protein Precipitation (PPT)

Due to its simplicity and wide applicability, PPT is important for sample pretreatment in early drug discovery when generic extraction of mixtures of candidates is more important than sensitivity. As a generic technique, PPT is attractive for high-throughput bioanalysis because it offers fast sample preparation and easy automation and requires minimal manual labor.

However, PPT cannot be considered a true sample preparation technique because it removes plasma proteins only through the addition of a precipitating solvent and subsequent homogenization and centrifugation.

Acids or water-miscible organic solvents are used to remove proteins by denaturation and precipitation. Acidic compounds such as trichloroacetic acid (TCA) and perchloric acid, efficiently precipitate proteins. Organic solvents such as acetonitrile, methanol, and ethanol, while relatively inefficient for removing plasma proteins, are widely used in bioanalysis because of their compatibility with HPLC mobile phases. Organic solvents allow the injection of the supernatant after an evaporation or dilution step. When analyzing a supernatant from a plasma or urine sample using PPT, salts and endogenous materials are present and can cause matrix effects that produce greater sample variation. In most cases, a divert valve is placed prior to the MS to remove the solvent front in order to enhance the robustness of the MS detector and prevent salt and other interferences from entering the MS source. Although the number of reported automated PPT procedures has been small in comparison with reports about automated SPE protocols, we expect this type of assay to become more routine. At times, PPT is a better choice for sample pretreatment than LLE. In one case,[77] the re-assay concentration of a highly protein binding compound (>99%) in rat and monkey plasmas was higher than nominal data generated by LLE. Data were, however, brought back to normal levels when PPT was used. One explanation is that typical LLE extraction solvents such as hexane did not disrupt all the protein bindings. While the temperature change of each freeze–thaw cycle gradually denatured proteins and released the analyte from the protein interactions, producing concentrations with a higher bias. PPT can disrupt protein binding completely and therefore provide unbiased data.

12.3.2   SEPARATION AND DETECTION

The advent of the atmospheric pressure ionization (API) source in the early 1990s allowed direct coupling of LC to MS. By the mid-1990s, this technology was a common in drug metabolism laboratories. The enhanced selectivity of tandem mass spectrometry (MS/MS) experiments reduced the need for exhaustive chromatographic separations prior to detection and this feature was exploited to significantly reduce analysis times.

12.3.2.1   HPLC Perspectives

The demand for analytical laboratories to increase sample throughput provided the impetus for HPLC column manufacturers to introduce new stationary phases and a range of column geometries to meet requirements for speed, high sensitivity, and reduced sample availability.

Fast gradient chromatography — Fast gradients and short columns are two typical chromatographic approaches for high throughput. In combination with narrow particle size distribution, they shortened analysis time without the loss of the chromatographic resolution. The concept of the rapid gradient elution with short columns and high flow rates was initially introduced in 1998[50] for the analysis of combinatorial chemical samples. Later, in combination with PPT, rapid gradient LC coupled with MS was used to screen P450 probe substrates.[23,78,79] An analytical run time of 2 min with sufficient resolution was achieved. The extra resolution provided by the gradient separations reduced both ion suppression and metabolite co-elution. Using short columns packed with small particles (3 μm) and flow rates of 1 to 2 mL/min, LC/MS run times of 15 sec were reported[80] for five analytes. The experiment demonstrated that 240 samples could be analyzed in 1 hr. Fast gradient is now common for bioanalysis and many papers discuss its direct and extended applications.

Monolithic column — The trend to use shorter columns in liquid chromatography means that the resultant lower separation efficiency is of concern. One way to improve HPLC separation efficiency on a shorter column is to reduce the size of the packing material, but at the cost of increased backpressure. Another approach to improve performance is increasing permeability with a monolithic column. Such a column consists of one solid piece with interconnected skeletons and flow paths. The single silica rod has a bimodal pore structure with macropores for through-pore flow and mesopores for nanopores within a silica rod[81,82] (Figure 12.1).

The higher performance at higher flow rates is presumably due to the combination of high porosities, small sizes of silica skeletons, and the resulting greater ration of through-pore size to skeleton size. The contribution of the mobile phase mass transfer for a monolith column is greater than that for a traditional particle-packed column, while the contribution of the stationary phase mass transfer is far less than than of a particle-packed column that produces a similar pressure drop.[83–85] The large theoretical plates-per-unit pressure drops[81] provide the unique properties of this material and allow the columns to operate at very high linear flow rates with little loss of the performance and very low backpressure. Monolithic columns were introduced for analyzing organic polymers in the late 1980s. Before the conventional-size columns became commercially available in 2000, only the polymer-based, low efficiency monoliths were used for analyzing biological macromolecules. The commercial silica rod column was claimed to produce efficiency similar to that of columns packed with 3.5 μm particles and typically yields a pressure drop half that caused by a column packed with 5 μm particles.[86] Monolithic columns have been successfully used with fast gradients and high flow rates for the direct analysis[87–89] of the pharmaceutical compounds in human plasma.

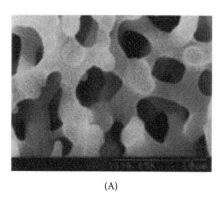
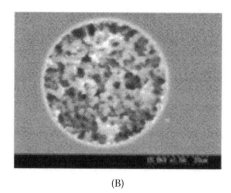

(A)   (B)

FIGURE 12.1 SEM photographs of monolithic silica columns: (A) monolithic silica prepared from TMSO in a test tube; (B) 50 μm inner diameter silica skeleton, size 2 μm, through-pore size 4.5 μm. (*Source:* From Ikegami, T. and Tanaka, N., *Curr. Opin. Chem. Biol.*, 2004, 8, 527. With permission from Elsevier Scientific Publishing.)

Ultra-performance liquid chromatography (UPLC) — This technique was introduced in 2004. Chromatographers no longer need to choose between speed and resolution because they can have both with UPLC. The technology retains the practicality and principles of traditional HPLC while offering significant advantages in resolution, speed, sensitivity, and efficiency for analytical determinations. Based on smaller particle size and column diameter, increases of column and system pressures is often a concern. However, it can be alleviated with use of a commercial UPLC system capable of providing liquid flow at pressures up to 1034 bar and columns packed with unique 1.7 μm pressure-tolerant particles that can withstand these pressures.[90,91] Optical and the mass detectors at high speeds contribute to the increased sensitivity and faster signal responses, and help to manage the increased speed and resolution requirements of UPLC. Coupling UPLC with mass spectrometers capable of high speed and integrated system software is claimed to increase unattended sample capacity up to ten times. According to the Van Deemter plot (Figure 12.2), the theoretical plate height (HETP) is proportional to the particle size (d_p). The move to smaller particles, specifically as particle size decreases to 1.7 μm, produced a significant gain in efficiency (N = theoretical plate number) resulting from the decreasing theoretical plate height (HETP) and the efficiency did not diminish at increased flow rates or linear velocities.[92] Therefore, using smaller particles allows extensions of speed and peak capacity to new limits. From a packed column LC perspective, the use of smaller particles to shorten the diffusion path of the analytes provides improved separation efficiencies. Ultimately, the flow rate at which optimal efficiency obtained is much wider on the 1.7 μm particle UPLC column. In summary, the UPLC column provided greater efficiency at high flow rates, leading to faster analysis and better sensitivity. Certain practical concerns still need improvement before UPLC is routinely used in laboratories. They include sample introduction, reproducibility, and detection. Another critical aspect is the possible formation of temperature gradients within UPLC columns at such high pressures.[93] The UPLC columns require extremely narrow sample plugs to minimize sample volume contributions to peak broadening. The commercial Acquity UPLC system is designed to ensure exceptionally low carryover, reduced cycle time, and minimal system volume. The first practical application of UPLC was carried out in connection with a TOF mass spectrometry detection[94] in the fields of metabonomics and genomics. The works showed explicit advantages of UPLC™ over HPLC in peak resolution together with the increased speed and sensitivity in these fields. Not until recently, the application of UPLC™ combining LC/MS/MS in the high-throughput quantitative analysis became widely applicable.[95-99]

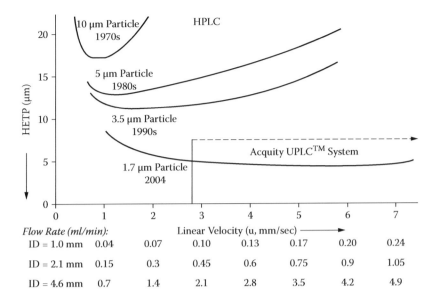

FIGURE 12.2 Van Deemter plot illustrating evolution of particle sizes and resulting changes of relationship of plate height and linear velocity. (*Source:* From Swartz, M., *J. Liq. Chromatogr. Rel. Technol.,* 2005, 28, 1253. With permission from Taylor & Francis Group.)

12.3.2.2 Mass Spectrometry Perspectives

Over the years, pharmaceutical developers have relied on improvements to MS hardware and software that allows users to benefit from lower levels of detection and ease of use. MS is designed to separate gas phase ions according to their m/z (mass-to-charge ratio) value. The combination of the separation power of HPLC and the detection power of MS represents a major advance in analytical chemistry. The sensitivity and high specificity of LC/MS/MS allows significant reductions of run times—1 to 3 min runs are typical. MS detection of ionized analytes can be carried out in a linear scan mode in which a range of m/z values is constantly monitored. In a more selective and sensitive mode called selective ion monitoring (SIM), a particular ion of a specific m/z value is selected for the monitoring. Another key detection technique for LC/MS/MS is multiple-reaction monitoring (MRM). With triple quadrupole MS, MRM adds another level of selectivity by isolating the precursor ion of the analyte in the first quadrupole (Q1) of the instrument, fragmenting this ion in a collision chamber (Q2), and isolating a selected product ion of the precursor in the third quadrupole (Q3). The triple quadrupole behaves better in MRM mode because only the analytes of interest undergo fragmentation. All other ions are excluded from the collision-induced dissociation (CID) step. This eliminates interference and suppression effects from other co-eluting species. A sequential series of precursor ion isolations, fragmentations, and product ion re-isolations along with a very rapid cycle time (10 to 50 msec) makes triple quadrupole MS particularly suitable for multi-component monitoring and quantitative high-throughput analysis.

With triple quadrupole MS, automatically obtaining sensitive full scan CID spectra can serve as a powerful tool in metabolite identification due to flexibility in performing post-acquisition data processing.

Another type of commercially available triple-quadrupole known as the TSQ Quantum was recently reported[100] to achieve significantly better resolution than a traditional triple quadrupole instrument without any significant loss of transmission. Based on the improved inherent resolution, assay development of an analyte on a classic TSQ that requires extensive sample preparation may be easily validated with enhanced sensitivity on the newer TSQ through a simple "dilute-and-shoot approach."[101] Unlike triple quadrupole MS that performs MS/MS in space, ion trap spectra are obtained by MS/MS in time. After the externally generated ions are trapped in the device, the analyte of interest is isolated in the trap and undergoes CID fragmentation in the same device. Due to its small footprint and high sensitivity, the ion trap was considered somewhat of an alternative to the triple quadrupole MS for quantitative analysis.[101] Despite the good sensitivity, the ion trap has not been used extensively for high-throughput analysis mainly because (1) the generation of MS/MS data requires a relatively long duty cycle (typically 100 msec or more) in the ion trap and consequently limits the numbers of analytes that can be quantified simultaneously because an insufficient number of data points may potentially be collected over the peak, and (2) the fact that only a certain number of ions can be stored simultaneously in an ion trap. With clean samples, most of the stored ions are analyte ions. With "dirty" biological extracts, most stored ions represent interference and will affect precision and accuracy. Thus, the ion trap instrument is less attractive for quantitative analysis of samples in biological matrices.

For high-throughput analysis, it is important to increase the specificity of each bioanalytical method. The enhancement of chromatographic resolution presents various limitations. Better selectivity can be obtained with TOF mass analyzers that routinely provide more than 5000 resolution (full width at half-mass or FWHM). The enhanced selectivity of a TOF MS is very attractive for problems such as matrix suppression and metabolite interference. In one report of quantitative analysis using SRM, TOF appeared less sensitive than triple quadrupole methods but exhibited comparable dynamic range with acceptable precision and accuracy.[102]

12.3.2.3   Data Acquisition

The increased data generation achieved via high-throughput techniques made automated data processing and information management essential. Despite the acceleration in data processing resulting from advances in laboratory information management systems (LIMSs), further developments such as electronic sample- and data-tracking systems are required to keep up with the increasing amounts of data generated. During data processing, ample details and results are typically linked by a sample identification number automatically generated by a LIMS, and stored in a searchable database. The database is designed to check the status of data acquisition, processing, and reporting, move data files from acquisition to the processing PC, and upload the information to the database for the storage and archiving. For metabolite ID and protein sequence definition, fully automated LC/MS packages could provide detailed interpreted results. Automated data acquisition in quantitative analysis is not a new concept. In order to ensure the collection of reliable *in situ* data, instrument detectors and associated data acquisition software have been continuously improved since commercial instruments first appeared. Equipment such as automated sample label readers, freezer temperature tracking systems, and automated data collection, processing and back-up have exponentially improved data integrity.

Quantitative analysis requires at least 12 data points for effective integration of the chromatographic peak. This means an efficient data capture rate detector is necessary to address the very narrow peaks produced by instruments such as UPLC. During parallel LC/MS injection, MS data acquisition is initiated upon receipt of a contact closure signal from an autosampler after the first injection valve is rotated from "load" to "inject." Data acquisition and processing software may require customization to meet specific needs such as high-throughput support.The autosampler injection cycle (wash, rinse, load, and re-initialize between runs) is time-consuming and can be an impediment to the high-throughput approach. When designed correctly, up to eight injections on an autosampler with sequential data collection on an MS may save time. A relatively fast new autosampler and data acquisition system led to a time lag shorter than 17 sec between consecutive injections.[45] Using the same autosampler, 768 protein-precipitated rat plasma samples (eight 96-well plates) containing both two different analytes were analyzed in 3 hr and 45 min.

12.3.3 OTHER METHODS

Additional gains in bioanalytical throughput have been realized by shortening sample preparation and LC analysis times through column switching, high flow rate analysis, and parallelism.[56,103–105] Online extraction is preferred to reduce labor required for sample preparation. Online extractions often employed polymer-based[37,106–108] or other restricted access media columns[109] in conjunction with high flow rates to wash away matrix interferences such as plasma proteins while retaining the analyte on the column and subsequently elution with a high organic mobile phase onto the MS. Throughput can be further improved by using multiple autosamplers feeding a single analytical column with detection by LC/MS/MS or using multiple inlets on a single MS source.

12.3.3.1 Direct Injection

Direct injection of plasma or supernatant after protein precipitation on a short column with a high liquid flow rate is a common method for reducing analysis time in the pharmaceutical industry. The direct injection of a sample matrix is also known as the "dilute-and-shoot" (DAS) approach.[62] DAS can be applied to all types of matrices and approaches and is the simplest sample preparation method with matrix dependency. Direct injection can also be approached through the extraction of eluent from PPT, SPE, and LLE onto a normal phase analytical column. The procedure is called hydrophilic interaction liquid chromatography (HILIC)[70,110,111] and it avoids the evaporation and reconstitution steps that may cause loss of samples from heat degradation and absorption.

SPE online with LC/MS/MS and resulting column switching — This combination has been applied to the direct analysis of pharmaceuticals since early 1980s, but only with fluorescence or UV detection. One early limitation was the poor selectivity of UV detection. Basically, a column switching platform consists of two HPLC systems connected by a six- or ten-port switching valve. In the first step, the analytes of interest are retained on column 1 (trapping column) and the matrix components are washed out. In a second step, the first column is switched in line with column 2 (analytical column), where the analytes are chromatographically separated. The procedures are fully automated, but require a relatively long cycle time. Online SPE with MS/MS detection offers speed, high sensitivity based on preconcentration, and relatively low extraction cost per sample. It requires only the installation of complex valves and column configurations at the LC end. For method development, the investigation of different packing materials on a trapping cartridge may be time-consuming. A column selector device allows automation of this procedure. An online SPE system with a dual cartridge column device developed by Spark Holland provides full flexibility and a cycle time of 80 sec including sample preparation and analysis. The time limiting step of a column switching set-up can be achieved by the trapping phase or the analysis phase. Throughput can be increased 50% by a dual trapping column or dual analytical column configuration. Xia et al.[112] demonstrated the dual trapping column approach by using two Oasis HLB extraction columns and a single C18 analytical column. The total analysis time to determine the drug candidate in rat plasma was 1.4 min.

Turbulent flow chromatography as online extraction coupled with LC/MS/MS — Turbulent flow chromatography (TFC) is another column switching technique that was introduced[113] for the direct injection of the biological fluids onto a column packed with large spherical porous particles. Sample preparation using TFC has proven to be fast, easy, and less labor-intensive than traditional offline and other online methods. In combination with API MS detection, TFC has been used wash for qualitative and quantitative analysis of biological matrices.[7] With TFC coupled to LC/MS/MS, plasma samples are injected directly onto the extraction column at very high LC flow rate (4 to 5 mL/min). The particle-packed TFC extraction columns often contained large pores and particles (30 to 60 μm)[106,114] that allowed proteins and salts to wash away while retaining small organic molecules on the column. TFC may be operated under single or dual column mode. Although the single

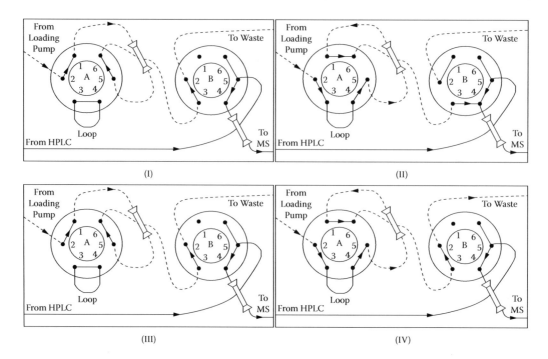

FIGURE 12.3 System configuration during (I) sample loading; (II) sample transfer; (III) cleaning and elution; and (IV) loop refilling and elution. Sample re-equilibration to aqueous conditions (V) followed the same flow scheme and flow rate as sample loading (I) and was therefore not included. Arrows denote direction of mobile phase flow. Dotted line indicates extraction column process. Solid line indicates analytic column process. (*Source:* From Xu, X. et al., *J. Chromatogr. B*, 2005, 814, 29. With permission from Elsevier Scientific Publishing.)

extraction column approach for both extraction and analysis provided high-throughput, it produced little or no chromatographic separation. To address this, an analytical column was added in line with the extraction column to achieve adequate LC separation in dual column mode. Figure 12.3 shows a typical configuration for dual column TFC coupled with a quadrupole tandem MS for quantification of biological samples.

In dual column mode, the first column that was often packed with 50 μm porous particles (60 Å) served as the extraction column to elute the plasma matrix components and retain the analytes. Chromatographic separation was achieved on the second (analytical) column. The loading, extraction, and elution of solvent can be optimized. Adding an analytical column allows the system to operate in a classic column switching configuration. Typical cycle times range from 2 to 4 min. Special washing conditions for autosampler, extraction columns, and switching valves were applied to minimize carry-over, but at the cost of operating cycle time. Due to the use of a single extraction column, both single and dual column modes require extra time to re-equilibrate the column between injections. The use of a ternary column system with two parallel extraction columns and a single analytical column addressed this issue. Sample purification takes place on one extraction column while the other extraction column is equilibrated. As a result, no additional equilibration time is included in the total time per sample. In addition to serving as a powerful tool for high-throughput quantitative analysis, TFC is also useful in qualitative analyses. In combination with an ion trap MS, TFC used to screen microsomal stability and metabolite profiles by direct injection.[108]

TABLE 12.1
Major Matrix Components Showing Potential for Dilute-and-Shoot (DAS) Approach

Matrix	Water	Proteins	Salts	Other Components
Urine	High	Trace amino acid	High	Hydrophilic only
Dialysate	High	No	High	Sugar, electrolytes, small waste products in blood, e.g., urea and creatine, potassium, extra fluid
Saliva	High	Amino acids and mucin	Minimal	Hydrophilic and hydrophobic components
Cerebrospinal fluid	High	Yes	Some	Close to blood component except low cholesterol and glucose
Plasma	90%	7%	Some	Fatty acid (lipid) 2%, amino acids, vitamins, hormones, waste products of metabolism

Dilute and Shoot (DAS) — Direct injection of plasma via DAS received little attention in the literature.[62] Most reports discussed it in combination with column switching. DAS has been used for matrices that do not require elaborate sample pretreatment, e.g., cerebrospinal fluid, tear fluid, and urine because they consist mainly of water. The selection of sample pretreatment for such fluids depends on the expected analyte concentrations and required limits. For example, urine samples with high concentrations allow simple dilution of samples prior to analysis, thus making DAS the method of choice. The major components that show potential for use with DAS are summarized in the Table 12.1. Issues such as carry-over often observed with other online sample preparation approaches are not of concern with DAS because other online sample pretreatments go through sample focusing or sample concentration steps. DAS requires only dilution of samples before detection. Its advantages over other online sample preparations include wide dynamic range, automation compatibility, and lack of heat-related sample destruction. Generic assays that minimize method development are particularly critical for drug discovery and development and therefore highly desirable. Methods such as DAS that require little or no sample preparation provide the answers. Intensive efforts are underway to develop robust solutions that will allow DAS to simply methods development and achieve high-throughput sample analysis. Direct plasma injection onto a standard HPLC column leads to rapid deterioration of analytical performance. The low injection volume required for DAS ensures better column performance. A divert valve is necessary to prevent solvent fronts from contaminating the MS source, especially when dialysates containing high salt contents are used. Interference components such as proteins and other endogenous compounds in matrices can also accumulate on components such as rotors, connecting tubing, injection needles, sample loops, and probes. Monitoring column pressure, peak shape, and carry-over will make performance relatively easy to control.

12.3.3.2   Parallel Approach to High Throughput

Parallel HPLC — Despite the selectivity of triple quadrupole MS, LC/MS/MS still presents relatively long run times because of the time required to separate analytes from matrix components to reduce ion suppression whether using isocratic or gradient elution. The inherent dead volume and required autosampler rinse times associated with LC devices also consume a fraction of run time. These steps add useless times to analytical runs, particularly during use of expensive MS instruments. Serial chromatography is expected to provide significant boosts in throughput without requiring modifications of existing methods. It also maintains chromatographic integrity and allows efficient use of expensive MS equipment. Two or more HPLC columns are run in parallel using a single MS source sprayer. Unfortunately, HPLC pumps are not designed for parallel chromatography because one pumping system is often necessary for each column in the established LC/MS/MS set-up. Combining four or more columns in parallel is a very complex and expensive endeavor. A parallel LC/MS/MS design utilizing multiple separate HPLC systems has the advantage of better flow control and thus consistency in retention time. Another advantage is the choice of HPLC conditions. When different compounds are analyzed in different channels, each compound may be optimized through choices of HPLC column, mobile phase. and gradient mode. The disadvantages of a multisystem arrangement include requirements for more laboratory space and consequent high costs.

A system with four pumps and selector valves achieved precision and accuracy results similar to a single column arrangement for four different compounds.[104] Each compound was analyzed on a separate channel. Alternatively, liquid flow split evenly into multiple channels from one HPLC system can be introduced into an MS detector using a Valco splitter. The flow rates are pressure-regulated. To maintain the same flow rate across the channels, care must be taken to ensure that backpressures in the four channels remain the same. Connection tubing, columns, and guard columns all generate backpressures. Columns and guard columns from the same manufacturer and even from the same batch can generate different backpressures. A modification of a conventional HPLC system was reported to analyze samples in parallel.[115] The modification involved the incorporation of three valves, four HPLC columns, and a binary pump. To achieve time-shifted gradient-separation on all four columns, a mobile phase delay tubing system was installed between the binary pump and autosampler.

Multiple Sprayer (MUX®) — Multiple flows may be introduced into a MS interface in serial mode with a valve selector connected to a single sprayer as in parallel chromatography or through parallel sprayers on the MS interface. For parallel sprays, the designated MS source is equipped with several API spray probes so that each analytical column in parallel will be connected to a separate spray. Each spray will be sampled in rapid successions for data acquisition by MS, with a separate data file for each spray. Thus, within a single chromatographic run, several samples can be injected simultaneously onto parallel columns, one sample per column. The MS must be optimized to record data at a specific time window for the chromatogram generated by each sprayer. An autosampler equipped with multiple injector ports can create or allow offset between injections, thus synchronizing an optimum and narrow window for the MS. The main drawback of multiple sprayers is the requirement for multiple pumps. A commercially available four-channel MUX system was interfaced to TOF/MS.[16] The fast acquisition capacity of TOF combined with a MUX interface containing up to nine channels was reported to maintain mass spectral integrity. The interface of MUX to TOF allows fast chromatography with narrow chromatographic peak widths. Although successful applications of MUX reported acceptable inter-column and inter-spray variabilities,[102] the disadvantages of MUX/TOF and MUX quadrupole MS for quantitative analysis include interferences from spray to spray, limited dynamic range, and less sensitivity than a single sprayer source. Concentration-dependent cross-talk exists between the sprayers because multiple LC effluents are sprayed simultaneously into the MS. The cross-talk effect therefore presented limitations of the MUX interface in the dynamic range of the assay and sensitivity up to three times lower than that of a single sprayer interface.

One reason for lower sensitivity is the lack of flexibility to optimize the positions of the sprayers on the MUX interface; another may be the lower electrospray desolvation efficiency on the MUX. The longer total cycle time on a MUX interface with a quadrupole MS in comparison to a single sprayer interface adds another concern. Assuming typical chromatographic peak widths appeared on average at 15 sec, 17 data points could be easily detected across the peak for each transition with a total cycle time of 0.88 sec on a conventional single sprayer set-up. With the MUX, only 12 data points could be detected across the same peak even with a total cycle time of 1.24 sec because of the introduction of additional interspray time on top of dwell time. Hence, when MUX is used with a quadrupole mass analyzer, it is important to consider dwell time and chromatographic peak width because these parameters will limit the numbers of analytes and internal standards monitored in an assay.

12.4 HIGH-THROUGHPUT QUANTITATIVE BIOANALYSIS IN LARGE MOLECULE DMPK

The DMPK parameter optimizations for large molecule drugs such as proteins and peptides are similar to those for small molecules except that metabolic profiling may not be involved in compound development unless unnatural amino acids are expected in the metabolic products or pathway. Using high throughput for large molecules quantitative analysis is not as well explored as its application for small molecules due to the analytical challenges associated with large peptides and proteins.

For small peptide compounds, a high-throughput quantitation follows the same concept as the assay for small molecules. All high-throughput strategies discussed earlier in this chapter may be applied. For large peptide and protein drugs, high-throughput quantitation is difficult because routine analysis protocols often involve dilution followed LC/UV or enzyme-linked immunosorbent assay (ELISA). High throughput is difficult to achieve with these techniques. ELISA is mainly used for quantifying biological molecules because of its accuracy and reliability. However, most immunoassays including ELISA are time- and labor-intensive because of needs for dilutions, reagent additions, and washes, and also the lack of universality, especially for drugs extensively metabolize and produce breakdown compounds that maintain conformation and are cross-reactive. ELISA and radioimmunoassay (RIA) have other disadvantages such as limited dynamic curve range, matrix interactions, and waste generation.

Quantitation and qualification of large molecules follow similar principles as procedures for small molecules but involve different foci and approaches. The quantitation of proteins and peptides includes relative and absolute amount evaluations. Most proteomic applications in drug discovery are concerned with relative abundances of proteins.

The comparison of diseased and control cell lines or tissues is generally necessary to suggest drug targets or predict the effects of drug candidates. In this type of analysis, the determination of relative amounts of protein or relative degrees of post-translational modifications associated with the proteins of interest is often necessary. In contrast, when absolute quantitation is needed, the use of internal standards and standard curves in a proteomic analysis is desirable to quantitate the absolute amount of protein in a biological sample. The creation of a standard curve can be based upon one or more peptides derived from the protein of interest to determine the molarity of or the absolute protein contained in a sample. The measurement of the exact protein or peptide amounts in a system often involves qualitative analysis prior to the quantitative analysis, so that the entity to be measured is already well defined. MS is one of the most sensitive and specific techniques for identifying and characterizing biomolecules. Matrix-assisted laser desorption/ionization mass spectrometry (MALDI/MS) and electrospray ionization mass spectrometry (ESI/MS) have been proven efficient for the analysis of large biomolecules. MALDI is more suitable for mixture analysis than ESI because the latter generates ions with many different charge states from each heavy compound. Recent improvements in MALDI/TOF sensitivity and versatility make this method highly competitive for the detection and the characterization of biopolymers.

The analyzing mass range has also increased significantly since the introduction of MALDI in 1988.[116,117] MALDI analysis of biopolymers with masses up to 500 kDa[118] has been reported. The simplicity and rapidity of sample preparation, automated sample introduction, fast data acquisition, and coupling of MS to intelligent interpretation software and protein databases make MALDI ideally suited for routine quantification and/or identification of proteins and peptides. The achieved sensitivity for analysis of peptides and proteins by MALDI/TOF/MS is typically in the low- to mid-femtomole range. However, the mass accuracy is highly dependent on the molecular mass. For every

30 kDa, the mass accuracy frequently drops by 0.1 to 0.2%. The analysis time is normally below 5 min for each sample. Although MALDI has been reported as providing adequate quantification, factors such as variable crystallization and laser ablation may lead to poor standard curves. Inadequate mass accuracy is always a major concern, and thus MALDI is not very useful for quantitative analysis. Bench-top instruments such as quadrupole ion trap analyzers and triple quadrupole instruments effectively generate qualitative and quantitative data. Routine analysis of large peptides and proteins by LC/MS/MS is possible because these molecules often possess high net charges that force their mass-to-charge ratios (m/z) into the relatively narrow mass analyzer ranges of bench-top instruments ($m/z \leq 3000$ Da).

Despite the significant numbers of publications detailing qualitative characterization of macromolecules using LC/MS/MS, limited numbers of reports discuss the use of this technique for quantitative bioanalysis.[119–121] Ion trap mass spectrometers generally offer better sensitivity than triple quadrupole instruments when using full-scan MS/MS operative mode. If differential analysis of two samples containing a mixture of peptidic species is required, ion traps are ideal. However, triple quadrupole instruments offer increased performance over ion trap analyzers for single reaction monitoring (SRM) experiments, for example, targeting and quantitating a single species, or even a simple mixture of peptides. Triple quadrupole instruments are superior for performing absolute quantitations or comparative quantitations of non-complex mixtures. The recent introduction of commercial electrochemiluminescence (ECL) detection instruments,[122] makes bioassay-based high-throughput quantitation for large molecules feasible. ECL is a well established process in which certain chemical compounds emit light when electrochemically stimulated. It is highly sensitive; reactive species are generated from stable precursors (i.e., the ECL-active label) at the surface of an electrode. This novel technology has many distinct advantages over other traditional large molecule detection systems: (1) extremely low detection limits for label (200 fmol/L); (2) wide dynamic range for label quantification that extends over six orders of magnitude; (3) less immunoreactivity because multiple labels can be coupled to proteins or oligonucleotides without affecting solubility or ability to hybridize; and (4) separation and nonseparation assays can be set up with simple and rapid measurements. ECL detection has replaced numerous immune, enzymatic activity, and binding assays for various types of analytes in large molecule quantitation.Overall, the high-throughput absolute quantitation of MS-based peptide and protein analysis may be achieved, particularly for small peptides because the sensitivity and accuracy of the instrumentation continue to improve. In the future, MS-based methods may replace immunological quantitation and LC/UV techniques and works as a complement to ECL detection since both techniques present different advantages. However, quantitative analysis of macromolecules is still performed principally via immunoassay and bioassay techniques[122] due to more precise quantitation than MS at very low protein concentrations provided by immunology-based assays such as ELISA, ECL, and RIA.

DEDICATION AND ACKNOWLEDGMENTS

This chapter is dedicated to Dr. James T. Stewart on the occasion of his 70th birthday. Formerly a professor at the University of Georgia's College of Pharmacy and my PhD advisor, Dr. Stewart is now retired. My thanks also to Dr. Wu Du of the Merck Research Laboratories and Dr. Mark Arnold of BMS for their reviews and comments.

REFERENCES

1. Niessen W.M.A. *J. Chromatogr. A* 2003, 1000, 413.
2. Korfmacher W.A. *Drug Discov. Today* 2005, 10, 1357.
3. Lee M.S. and Kerns E.H. *Mass Spectrom. Rev.* 1999, 18, 187.

4. Ackermann B.L., Berna M.J., and Murphy A.T. *Curr. Topics Med. Chem.* 2002, 2, 53.
5. Hopfgartner G. and Bourgogne E. *Mass Spectrom. Rev.* 2003, 22, 195.
6. Oliveira E.J. and Watson D.G. *Biomed. Chromatogr.* 2000, 14, 351.
7. Jemal M. *Biomed. Chromatogr.* 2000, 14, 422.
8. Kassel D.B. *Curr. Opin. Chem. Biol.* 2004, 8, 339.
9. Koh H.L. et al. *Drug Discov. Today* 2003, 8, 889.
10. Watt A.P., Morrison D., and Evans D.C. *Drug Discov. Today* 2000, 5, 17.
11. Berna M.J., Ackermann B.L., and Murphy A.T. *Anal. Chim. Acta* 2004, 509, 1.
12. Bhoopathy S. et al. *J. Pharm. Biomed. Anal.* 2005, 37, 739.
13. Kataoka H. *Trends Anal. Chem.* 2003, 22, 232.
14. Fang L. et al. *Rapid Commun. Mass Spectrom.* 2003, 17, 1425.
15. Kaplita P.V. et al. *J. Assoc. Lab. Auto.* 2005, 10, 140.
16. Ekins S. et al. *J. Pharmacol. Tox. Meth.* 2000, 44, 313.
17. Fung E.N. et al. *Rapid Commun. Mass Spectrom.* 2003, 17, 2147.
18. Nassar A.E.F. and Talaat R.E. *Drug Discov. Today* 2004, 9, 317.
19. Zhao S.X. et al. *J. Pharm. Sci.* 2005, 94, 38.
20. Di L. et al. *Int. J. Pharm.* 2006, 317, 54.
21. Wan H. and Rehngren M. *J. Chromatogr. A* 2006, 1102, 125
22. Yin O.Q., Tomlinson B., and Chow M.S. *Clin. Pharmacol. Ther.* 2005, 77, 35.
23. Testino S.A., Jr. and Patonay G. *J. Pharm. Biomed. Anal.* 2003, 30, 1459.
24. Herman J.L. *Int. J. Mass Spectrom.* 2004, 238, 107.
25. Jenkins K.M. et al. *J. Pharm. Biomed. Anal.* 2004, 34, 989.
26. Peng S.X., Barbone A.G., and Ritchie D.M. *Rapid Commun. Mass Spectrom.* 2003, 17, 509.
27. Hsieh Y. and Korfmacher W.A. *Curr. Drug Met.* 2006, 7, 479.
28. Jang G.R. et al. *Med. Res. Rev.* 2001, 21, 382.
29. Huang R. et al. *Int. J. Mass Spectrom.* 2004, 238, 131.
30. Sadagopan N., Pabst B., and Cohen L. *J. Chromatogr. B* 2005, 820, 59.
31. Frick L.W. et al. *Pharm. Sci. Technol. Today* 1998, 1, 12.
32. Ohkawa T. et al. *J. Pharm. Biomed. Anal.* 2003, 31, 1089.
33. Ackermann B.L. *J. Am. Soc. Mass Spectrom.* 2004, 15, 1374.
34. Berman J. et al. *J. Med. Chem.* 1997, 40, 827.
35. McLoughlin D.A., Olah T.V., and Gilbert J.D. *J. Pharm. Biomed. Anal.* 1997, 15, 1893.
36. Beaudry F. et al. *Rapid Commun. Mass Spectrom.* 1998, 12, 1216.
37. Wu J.T. et al. *Anal. Chem.* 2000, 72, 61.
38. Zhang M.Y. et al. *J. Pharm. Biomed. Anal.* 2004, 4, 359.
39. Watanabe T. et al. *Anal. Chim. Acta* 2006, 559, 37.
40. Ohkawa T. et al. *J. Pharm. Biomed. Anal.* 2003, 31, 1089.
41. Kuo B.S. et al. *J. Pharm. Biomed. Anal.* 1998, 16, 837.
42. Hop C.E. et al. *J. Pharm. Sci.* 1998, 87, 901.
43. Hsieh Y. et al. *J. Chromatogr. B* 2002, 767, 353.
44. *FDA Guidance for Industry: Bioanalytical Method Validation*, 2001.
45. Barbarin N. et al. *J. Chromatogr. B* 2003, 783, 73.
46. Borges V. et al. *J. Chromatogr. B* 2004, 804, 277.
47. Xu R.N. et al. *J. Pharm. Biomed. Anal.* 2006, 40, 728.
48. Jemal M. et al. *Rapid Commun. Mass Spectrom.* 2001, 15, 994.
49. Xu X. et al. *J. Chromatogr. B* 2005, 814, 29.
50. Mutton I.M. *Chromatographia* 1998, 47, 291.
51. Ayrton J. et al. *Rapid Commun. Mass Spectrom.* 1998, 12, 217.
52. Nguyen D.T. et al. *J. Sep. Sci.* 2006, 29, 1836.
53. Jemal M. and Xia Y. *Rapid Commun. Mass Spectrom.* 1999, 13, 97.
54. Castro-Perez J. et al. *Rapid Commun Mass Spectrom.* 2005, 19, 843.
55. Shen J.X. et al. *J. Pharm. Biomed. Anal.* 2006, 40, 689.
56. Zeng H., Deng Y., and Wu J.T. *J. Pharm. Biomed. Anal.* 2002, 27, 967.
57. Schellen A. et al. *Rapid Commun. Mass Spectrom.* 2000, 14, 230.
58. Alnouti Y. et al. *J. Chromatogr. A* 2005, 1080, 99.
59. Wang T. et al. *Rapid Commun. Mass Spectrom.* 1998, 12, 1123.
60. Fang L. et al. *Rapid Commun. Mass Spectrom.* 2002, 16, 1440.

61. Bayliss M.K. et al. *Rapid Commun. Mass Spectrom.* 2000, 14, 2039.
62. McCauley-Myers D.L. et al. *J. Pharm. Biomed. Anal.* 2000, 23, 825.
63. Jemal M., Ouyang Z., and Powell M.L. *J. Pharm. Biomed. Anal.* 2000, 23, 323.
64. Xue Y.J. et al. *J. Chromatogr. B* 2003, 795, 215.
65. Mallet C.R. et al. *Rapid Commun. Mass Spectrom.* 2003, 17, 163.
66. Yang A.Y. et al. *J. Pharm. Biomed. Anal.* 2005, 38, 521.
67. Allanson, J.P. et al. *Rapid Commun. Mass Spectrom.* 1996, 10, 811.
68. Biddlecombe R.A., Benevides C., and Pleasance S. *Rapid Commun. Mass Spectrom.* 2001, 15, 33.
69. Zhang N. et al. *J. Pharm. Biomed. Anal.* 2000, 22, 131.
70. Xue Y.J., Liu J., and Unger S. *J. Pharm. Biomed. Anal.* 2006, 41, 979.
71. Xue Y.J., Pursley J., and Arnold M.E. *J. Pharm. Biomed. Anal.* 2004, 34, 369.
72. Wang P.G. et al. *J. Chromatogr. A* 2006, 1130, 302.
73. Peng S.X., Branch T.M., and King S.L. *Anal. Chem.* 2001, 73, 708.
74. Shen Z.Z., Wang, S., and Bakhtiar R. *Rapid Commun. Mass Spectrom.* 2002, 16, 332.
75. Zhang N. et al. *J. Pharm. Biomed. Anal.* 2004, 34, 175.
76. Wang A.Q. et al. *Rapid Commun. Mass Spectrom.* 2002, 16, 975.
77. Yang L. et al. *J. Chromatogr. B* 2004, 799, 271.
78. Yin H. et al. *Xenobiotica* 2000, 30, 141.
79. Tang C. et al. *J. Chromatogr. B* 2000, 742, 303.
80. Heinig K. and Henion J. *J. Chromatogr. B* 1999, 732, 445.
81. Tanaka N. et al. *J. Chromatogr. A* 2002, 965, 35.
82. Ikegami T. and Tanaka N. *Curr. Opin. Chem. Biol.* 2004, 8, 527.
83. Leinweber F. et al. *Anal. Chem.* 2002, 74, 2470.
84. Miyabe K. et al. *Anal. Chem.* 2003, 75, 6975.
85. Vervoort N. et al. *Anal. Chem.* 2003, 75, 843.
86. Cabrera K. *J. Sep. Sci.* 2004, 27, 843.
87. Huang M.Q. et al. *Rapid Commun. Mass Spectrom.* 2006, 20, 1709.
88. Zhou S. et al. *Rapid Commun. Mass Spectrom.* 2005, 19, 2144.
89. Plumb R.S. et al. *Rapid Commun. Mass Spectrom.* 2001, 15, 986.
90. Swartz M.E. and Murphy B. *Lab. Plus Int.* 2004, 18, 6.
91. Wren S.A.C. *J. Pharm. Biomed. Anal.* 2005, 38, 337.
92. Swartz M. *J. Liq. Chromatogr. Rel. Technol.* 2005, 28, 1253.
93. Villiers A.D. et al. *J. Chromatogr. A* 2006, 1127, 60.
94. Plumb R. et al. *Rapid Commun. Mass Spectrom.* 2004, 19, 2331.
95. Li X. et al. *Anal. Chim. Acta.* in press, 2006.
96. Yu K. et al. *Rapid Commun. Mass Spectrom.* 2006, 20, 544.
97. Wang G. et al. *Rapid Commun. Mass Spectrom.* 2006, 20, 2215.
98. Kalovidouris M. et al. *Rapid Commun. Mass Spectrom.* 2006, 20, 2939.
99. O'Connor D. et al. *Rapid Commun. Mass Spectrom.* 2006, 20, 851.
100. Xu X., Veals J., and Korfmacher W.A. *Rapid Commun. Mass Spectrom.* 2003, 17, 832.
101. Tiller P.R. et al. *J. Chromatogr A* 1997, 771, 119.
102. Yang L., Wu N., and Rudewicz P.J. *J. Chromatogr. A* 2001, 926, 43.
103. Hsieh Y. et al. *Rapid Commun. Mass Spectrom.* 2001, 15, 2481.
104. King R.C. et al. *Rapid Commun. Mass Spectrom.* 2002, 16, 43.
105. Wang P.G. et al. *Biomed. Chromatogr.* 2005, 19, 663.
106. Jemal M. et al. *Rapid Commun. Mass Spectrom.* 1999, 13, 1462.
107. Zimmer D. et al. *J. Chromatogr. A* 1999, 854, 23.
108. Lim H.K. et al. *Anal. Chem.* 2001, 73, 2140.
109. Kawano S.I. et al. *Rapid Commun. Mass Spectrom.* 2005, 19, 2827.
110. Weng N. et al. *Rapid Commun. Mass Spectrom.* 2002, 16, 1965.
111. Weng N. et al. *Rapid Commun. Mass Spectrom.* 2004, 18, 2963.
112. Xia Y.Q. et al. *Rapid Commun. Mass Spectrom.* 2000, 14, 105.
113. Quinn H.M. and Takarewski J.J., Jr. *U.S. Patent 05,772,874*, 1998.
114. Wehr T. et al. *LCGC* 2000, 18, 716.
115. Van Pelt C.K. et al. *Anal. Chem.* 2001, 73, 582.
116. Karas M. and Hillenkamp F. *Anal. Chem.* 1988, 60, 2299.
117. Tanaka K. et al. *Rapid Commun. Mass Spectrom.* 1988, 2, 151.

118. Hellenkamp F. In *Biological Mass Spectrometry: Present and Future*, Matsuo T. et al., Eds., 1994, p. 101.
119. Feng W.Y., Chan K.K., and Covey J.M. *J. Pharm. Biomed. Anal.* 2002, 28, 601.
120. Wilbert S.M. et al. *Anal. Biochem.* 2000, 278, 14.
121. Rose M.J. et al. *J. Pharm. Biomed. Anal.* 2005, 38, 695.
122. Grimshaw C. et al. *J. Pharm. Biomed. Anal.,* 1997, 16, 605.
123. Miller K.J. et al. *Pharm. Res.* 2001, 18, 1373.

13 Designing High-Throughput HPLC Assays for Small and Biological Molecules

Roger K. Gilpin and Wanlong Zhou

CONTENTS

13.1 INTRODUCTION

Traditionally, the development time needed to bring a new drug candidate from initial synthesis to clinical testing and finally to the marketplace involved 10 or more years of effort. However, over the past two decades, the advent of modern combinatorial techniques and advanced analytical methods significantly reduced the timeline for developing pharmaceuticals from the preclinical to clinical phase from about 3 to 5 years.[1] Unfortunately, this time reduction is offset by more stringent regulatory requirements and the need for more rigorous clinical evaluations.

A single medicinal chemist can now prepare more than 2000 compounds annually via combinatorial chemistry technologies compared to only about 50 compounds prepared by classic synthetic approaches.[2] This dramatic increase in output has placed greater demands on the separation methodology used to isolate and purify bulk drug substances, assure their identities and quality, and evaluate their physical, chemical, and physiological properties. The current trend in analytical laboratories is to increase the use of rapid and highly automated methods that can handle new levels of sample throughput.

Although a variety of separation techniques are used daily in modern pharmaceutical development, the front-line approach continues to be high performance liquid chromatography (HPLC) because of its greater flexibility, scalability, and ruggedness.[1,3–5] Likewise, most HPLC methodology employed in laboratories (excluding separations of larger macromolecular analytes) is based on some form of reversed-phase (RP) separation.[5] This chapter focuses exclusively on optimizing RP/HPLC methodology as it relates to increasing sample throughput by (1) reducing chromatographic separation time, (2) improving the performance of and/or addressing problems with the supporting hardware needed to successfully conduct such analyses, or (3) increasing the efficiency of sample work-up.

In many instances, very rapid chromatographic separations in a few dozen seconds may be possible. However, the overall hourly throughput of the assay may be low because of lengthy and complex preseparation sample work-up steps or hardware constraints such as (1) the cycle time needed for the autosampler to perform the washing and filling steps, thus limiting the total number of injections per hour and (2) the time required for the detection system to produce enough points within a very narrow peak to adequately define its shape as the result of the number of signal co-acquisitions needed to obtain an acceptable ratio of signal to noise. A good example of the later problem occurs in coupling a high flow rate conventional HPLC separation to a time-of-flight mass spectrometer via an electrospray interface.

It is difficult to define the exact boundary where conventional HPLC ends and fast HPLC begins. However, the basic strategy of all modern high-throughput procedures is to improve resolution and reduce analysis time to a point where many dozens of samples can be assayed in the time that only a few could have been assayed in the past. It is important to stress that decreasing total analysis time— the underlying goal of high-throughput procedures—is not only dependent on the speed of chromatographic separation, but must account for all aspects of the analysis including sample work-up and injection, post-separation detection, and data analysis. Although this chapter focuses on time optimization of the separation process, time management studies (discussed below) have shown that pre- and post-separation processing are usually problematic aspects of increasing sample throughput.

The fast HPLC concept is not new; it dates back about three decades.[6–8] It is based on the concepts that (1) most separations, excluding natural product profiling, can be achieved with a few thousand plates and (2) by decreasing the band broadening contributions of diffusion-controlled mass transfer effects (i.e., reducing particle diameters and improving pore geometry), the required number of plates can be generated by short columns with relatively small backpressures. Under these conditions, very fast flow rates (high linear velocities) can be used to achieve total separation in seconds instead of minutes. An important refinement of fast HPLC arose from the use of higher operating pressures and sub-5-μm packings.[9–23] This new high efficiency separation is called ultrahigh performance liquid chromatography (UPLC).

The current state of the art in terms of commercially available instrumentation extends the operating pressure range to about 1000 bar. However custom-assembled equipment allows use of even higher operating pressures.[11] The introduction of ultrahigh pressure liquid chromatography (UHPLC) allows analysis of more complex mixtures such as plant extracts, fermentation broths, and combinatorial mixtures in times by utilizing longer columns packed with even smaller (sub-2-μm) particles. Like many of the other refinements in the field of liquid chromatography, the use of very high pressures to conduct liquid chromatographic separations is not new and dates back nearly four decades.[24] Unfortunately, UHPLC presents additional challenges related to separation design and chromatographic reproducibility that do not apply to conventional HPLC.[25,26] The physicochemical aspects of these challenges are discussed in a separate section.

13.2 DESIGN STRATEGY

Figure 13.1A shows a conventional high performance reversed-phase separation of a three-component mixture of aromatic acid esters obtained with a standard 4.6 mm × 250 mm octadecyl column and methanol:water as the eluent. From the view of chromatographic resolution and ruggedness, this is an excellent separation. However, from a practical standpoint, an assay based on this particular separation would not be satisfactory since it wastes large amounts of time between elutions of the individual components.

In designing high-throughput assays, keep in mind that unnecessary baselines between peaks waste time and should be minimized. A resolution of 1.5 between each pair of analytes in a chromatogram is ideal for quantitation. However, slightly larger resolution values may be advantageous in providing an assay with greater operational ruggedness. In developing any assay, a compromise of quantitation, speed, and ruggedness must be considered.

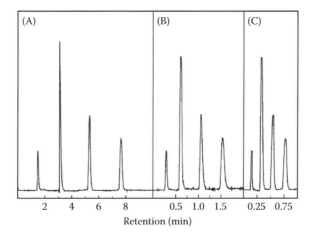

FIGURE 13.1 Reversed-phase separation of p-hydroxybenzoic acid and its methyl, ethyl, and propyl esters: (A) 4.6 mm × 250 mm ODS column at flow rate of 1.5 mL/min; (B) 4.6 mm × 50 mm ODS column at flow rate of 1.5 mL/min; (C) 4.6 mm × 50 mm ODS column at flow rate of 3.0 mL/min.

In optimizing the time requirements of a method, several strategies can be used to compress the elution profile. The first step is deciding whether the separation is to be carried out under isocratic or gradient conditions. Typically, isocratic conditions are more attractive for repetitive assays that monitor the quality of a manufactured product. Gradient conditions are more useful for nonroutine work involving separations of complex mixtures of compounds. Likewise, gradient conditions are especially helpful if a mixture contains two or more clusters of compounds exhibiting large differences in polarity and similar retention properties within each grouping. In this case, the gradient profile is adjusted to provide approximately equal spacing of all analytes or as close to equal spacing as possible. If required, analysis time may be reduced further by decreasing column length, increasing eluent flow rate, increasing column temperature, or combining these parameters with appropriate adjustments in the gradient profile to maintain at least a 1.5 resolution between each peak pair.

Another consideration in choosing a gradient or isocratic approach is the time necessary to re-establish the initial elution conditions. If the total analysis is not significantly shorter under gradient conditions and the polarity and clustering problem discussed above is not present, isocratic optimization may be a better approach for developing a high-throughput assay because it is less prone to problems such as ghost peaks, drifting baselines, etc.

Typically, the most common approach for shortening analysis time when optimizing isocratic reversed-phase separations is to increase the strength of the eluent either by decreasing the amount of water present or using an organic co-additive that has a higher eluotropic strength (e.g., using acetonitrile instead of methanol). Of these two alternatives, the use of binary mixtures of acetonitrile:water in place of methanol:water brings the added advantage of significantly lower eluent viscosity at a given elution strength and hence lower column backpressure at equivalent flow rates. Figure 13.2 shows plots of viscosity versus eluent composition for mixtures of water:methanol and water:acetonitrile.[27] Often a simple change from an eluent prepared using water:methanol to an eluent of equivalent strength using water:acetonitrile in combination with an increase in flow rate (because of lower viscosity) can significantly reduce separation time.

Figure 13.3 shows inlet pressure versus flow rate profiles calculated using relationships explained elsewhere.[28] A 4.6 mm × 250 mm column packed with particles ranging in size from 1 to 5 μm was used for each case. The solid lines in the figure represent an eluent of 100% water. The dotted lines above each solid line represent a binary mixture of 60:40 methanol:water, and the dashed lines represent binary blends of 50:50 acetonitrile:water. As an example, a column packed with 5 μm

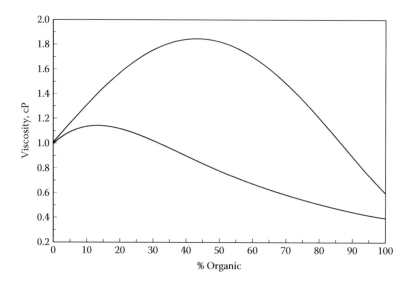

FIGURE 13.2 Plots of viscosity versus eluent composition for binary mixtures of water and methanol (top curve) and water and acetonitrile (bottom curve).

particles can be operated at a flow rate of only about 2 mL/min using 60:40 methanol:water as the eluent before the column inlet pressure is 300 bar. However, when the same separation is carried out with 50:50 acetonitrile:water that shows approximately the same elution strength as 60:40 methanol:water, the column can be operated at about 5.0 mL/min at an inlet pressure of 300 bar. These simple changes reduce separation time by a factor of 2.5, assuming the loss in plate count due to mass transfer effects will not adversely degrade the resolution below a value needed for quantitation.

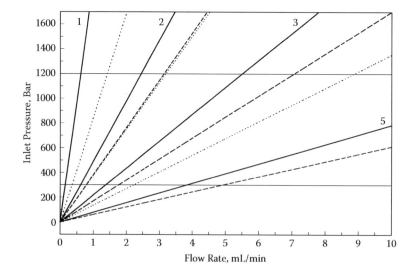

FIGURE 13.3 Comparison of relationship between column inlet pressure and eluent flow rate using 4.6 mm inner diameter × 25 cm columns packed with 1, 2, 3, and 5 μm particles. Solid lines = 100% water. Dashed lines = 50:50 acetonitrile:water. Dotted lines = 60:40 methanol:water. In the case of 1 μm particles, only 100% water is shown.

Another approach for compressing a separation profile is to reduce the hydrophobic character of the column packing. Generally changing from long chain hydrocarbon polymeric phases such as heavily loaded octadecylsilane (ODS) materials to ODS monomeric packings will decrease analyte retention. Similarly, changing from ODS packing to an octyl material or to a shorter bonded hydrocarbon phase will result in substantial reductions in retention at equivalent eluent strength. Of course, the overall elution profile of the separation can be altered dramatically and require substantial assay redesign. Although the most common HPLC phases are ODS-based, in many instances baseline separations can be obtained on less hydrophobic packings.

In the example in Figure 13.1, the analytes are nearly equally spaced under isocratic conditions, but clearly, as noted above, the amount of baseline between each peak pair is excessive and the chromatogram can be compressed by changing the eluotropic strength of the eluent or decreasing the hydrophobic character of the packing. In addition to these approaches for optimizing (reducing) the time required for separation, other possible strategies include modifications of column length, flow rate, and operating temperature, or a combination of modifications of these factors. The final goal is to eliminate as much wasted time between peaks as possible and hence develop an assay with higher sample throughput.

In contrast to commonly used 150 to 250 mm analytical columns that provide efficiencies equivalent to many thousands of plates, most rapid analysis HPLC columns are much shorter and often range in length from 20 to 100 mm. By using a column having the same retention properties (containing the same packing) as longer analytical columns but shorter in length and retaining the other eluent composition, flow rate, and temperature operating parameters, the gain in analysis speed is directly related to the reduction in length as illustrated in Figure 13.1B. In this example, the time of the chromatographic analysis was compressed from 8+ min to less than 2 min simply by using an equivalent octadecyl 4.6 mm × 50 mm column.

Under rapid analysis conditions, additional gains in speed may be achieved by using higher eluent flow rates. This is possible with shorter columns because they have significantly reduced backpressures compared to longer analytical columns. In Figure 13.1C, the flow rate of the eluent has been doubled compared to that used for the separations shown in A and B. With this increase in the flow rate, the overall column efficiency degraded from ~2300 plates (B) to ~1500 plates (C), but is still sufficient to provide enough resolving power for the three aromatic acid esters to be separated easily.

In more demanding separations that require higher plate counts, specially designed rapid analysis columns packed with very high efficiency 2 to 3 μm porous particles are available from several manufacturers. In addition, monolithic columns with improved flow-through characteristics are also commercially available. Figure 13.4 depicts a comparison of inlet pressure and flow rate for 4.6 mm inner diameter × 50, 100, and 150 mm columns packed with 5 μm particles.

Both the 2 to 3 μm porous particle-based and monolithic columns exhibited shallower van Deemter profiles at higher linear eluent velocities, resulting in smaller decreases in performance with corresponding increases in flow rate. Likewise, recently developed nonporous particles can serve as alternative media for carrying out highly efficient fast separations.[29,30] One advantage of this type of material is the elimination of band broadening due to mass transfer into and out of pores that contain stagnant mobile phase. However, the principal disadvantage of nonporous materials is significantly lower surface area per unit weight. Figure 13.5A plots surface area versus particle size for nonporous spherical materials. Significant gains in surface area are not realized until the particles reach submicrometer range. Unfortunately, very short columns in combination with extremely high pressures are needed for most separations using these types of materials.

Rapid analysis HPLC has been applied to a variety of acidic, basic, and neutral compounds and to various sample types.[12,13,31] However, a general approach works best for assays involving single components, simple combination products, and mixtures of compounds with widely and incrementally varying retention such as the separation illustrated in Figure 13.1 because these types of samples do not require the high resolving power of conventional analytical columns. Rapid analysis using shorter packed columns and conventional HPLC hardware is less suitable for complex mixtures that

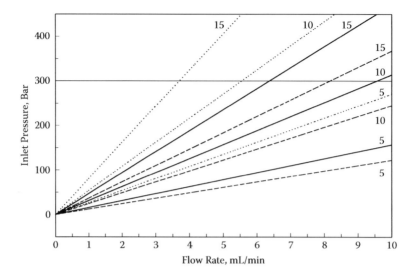

FIGURE 13.4 Comparison of relationship between column inlet pressure and eluent flow rate using 4.6 mm inner diameter columns packed with 5 μm particles at lengths of 5, 10, and 15 cm. Solid lines = 100% water. Dashed lines = 50:50 acetonitrile:water. Dotted lines = 60:40 methanol:water.

contain many analytes. Greater success has been reported for these types of samples using longer monolithic columns that can be operated at higher linear velocities compared to packed columns of equivalent efficiency. Examples of these types of separations are discussed below.

Although the basic ideas behind rapid analysis HPLC have not changed since the early work was done, one notable change is the development of better separation media (available columns) and hardware with higher pressure limits for using even smaller particles in combination with longer

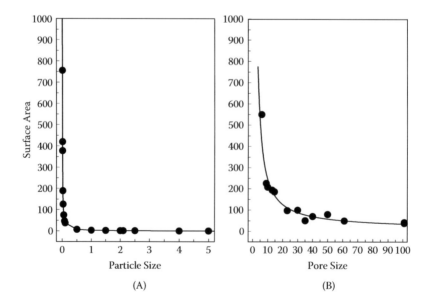

FIGURE 13.5 Surface area of solid (A) and porous particle-based (B) packings. Surface area expressed as square meters per gram, particle size as micrometers, and pore size as angstrom units.

columns. Using longer columns with very small particles allows more difficult separations that meet selectivity limits and process large numbers of components in mixtures. Higher pressure pumping systems are commercially available as are the traditional 350 to 400 bar systems that have served as standard instrumentation since the 1970s. Modern higher pressure systems can reliably deliver solvents up to 1000 bar and handle routine operations in the 500 to 800 bar range. Recent publications citing the use of higher pressures range from generic approaches for assaying a variety of different classes of drugs to more specific procedures for physiological and metabolic studies.[14–21] Additional examples appear in two articles that discuss general aspects of ultrahigh pressure LC[22,23] and a comprehensive review of pharmaceutical methodology.[5]

In addition to the above strategies, the use of higher column temperatures is another approach that may decrease analysis time and improve sample throughput. The relationship between the chromatographic retention factor, k′, and separation temperature is shown in Equation 13.1:

$$Lnk' = -\Delta H/RT + \Delta S/R + \ln \Phi \tag{13.1}$$

where R is the universal gas constant, T is the temperature in Kelvin, ΔH and ΔS are respectively the enthalpy and entropy associated with a solute transferring from the mobile phase to the stationary phase, and Φ is the phase ratio (stationary phase volume/mobile phase volume of column). All forms of chromatography follow this relationship, assuming the presence of a single invariant retention mechanism.[32–34]

In cases where elevated temperatures can be employed safely (analyte and column stability are not problems), separation can be reduced typically about 40 to 50% with an increase in column temperature of 25 to 30°C with only small changes in the relative spacing (selectivity) of analytes within the chromatogram. The reason is that van Hoff plots of many closely related compounds such as impurities and reaction by-products of a target analyte are nearly parallel (similar ΔH values). Thus, operating at a higher temperature will compress the chromatogram, but exert only minor effects on selectivity. Added advantages of increased temperature are a reduction in eluent viscosity and the ability to use significantly higher flow rates.

Assuming linear thermodynamic behavior, the two principal disadvantages of operating at higher eluent and column temperatures are decreased column lifetimes of the traditional silica-based bonded phases and the potential chemical instability of the analytes. To eliminate or at least minimize column stability problems, other nonsilica-based HPLC packings can be used. One is a zircon-based bonded phase[35–37] that has greater hydrolytic stability. These materials will operate over a wider range of eluent pH levels and higher temperatures.

In addition to these inorganic materials, organic polymer-based (polystyrene-divinyl) packings can withstand even greater extremes in pH and temperature. Unfortunately, many polymeric materials generally do not provide the same level of separation efficiency as that obtained with silica- and zircon-based packings. In a recently published high-throughput screening of drug components in rat plasma, a 3 μm, 2.0 × 50 mm C18 modified polystyrene-coated zircon column was operated at a flow rate of 1.0 mL/min at 110°C compared to only 0.2 mL/min at 30°C. This simple modification resulted in a decrease of analysis time from 3 min at 30°C to 30 sec at 110°C without significant loss of resolution. Quantitation results were equivalent to results from analysis of the same samples using standard silica-based reversed-phase conditions.[35]

Unfortunately, for many analytes, retention is not governed by a single retention mechanism (highly heterogeneous system) and the relationship between lnk′ and 1/T in K is curved. Likewise, the retention mechanism can change with temperature, usually as a result of (1) a modification of the physical organization or structure of the surface layer (bonded phase) or (2) the equilibrium and structural properties of the analyte. In both cases, plots of lnk′ versus 1/T are nonlinear. Shapes range from simple curvatures to more complex relationships. Regardless of the cause, when nonlinear retention mechanisms are present, general statements about optimization may be unreliable. Consequently, temperature adjustment is often a last resort in the overall design strategy for optimization of HPLC separation in contrast to retention manipulation in gas chromatography in which temperature is one of the most important variables.

13.3 DEVELOPMENTS IN COLUMN TECHNOLOGY

HPLC columns can be divided into three broad categories depending on the packing materials used in construction that influence the flow patterns of the eluent through the packed bed and the mass transfer mechanisms that affect the solute as it partitions between the moving and stationary phases. The three types of packings are totally nonporous, porous nonreticulated, and porous reticulated materials. The latter two categories can be divided further into two additional groupings. The two types of nonreticulated materials are totally porous and pellicular, both of which have been used since HPLC was invented. The two categories of reticulated materials are based on flow-through particles or monolithic construction.

It is interesting that some manufacturers and researchers have rediscovered pellicular construction in the form of smaller 1.5 to 2.5 μm particle-based packings with very thin porous outer layers.[38] The original pellicular work can be traced back to the late 1960s[39] and pellicular construction was the packing of choice until the introduction of completely porous 5 to 10 μm materials in 1972.[40] The reintroduction of pellicular materials in combination with longer columns and higher solvent delivery pressures allowed highly efficient separations of complex proteomic mixtures.[41] Modern porous shell packings with improved solute mass transfer kinetics are highly efficient because their outer micro-particulate layers are only 0.25 to 1.0 μm thick and contain pores in the 300 Å range.[42,43] Typically, these materials have surface areas in the 5 to 10 m^2/g range. This type of construction (Figure 13.6A) serves as a useful medium when properly chemically modified for separating larger biomolecules such as angiotensin II, insulin, lysozymes, myoglobins, enkephalins, and a number of other proteins.[42]

As noted, the totally reticulated materials are of particle-based (Figure 13.6B) technology or monolith construction (Figure 13.7). Both types of construction produce media that contain combinations of larger flow-through channels (macro paths) and nonflow areas that are highly porous (contain micro diffusion pores). From a historical perspective, the reticulated materials are the most modern separation media. They are especially useful for separating macromolecular samples such as larger peptides and proteins[44–49] and for carrying out rapid analysis separations using high flow velocities. The macro channels of particle-based technology allow the mobile phase to pass through the particles, thus reducing solute mass transfer distances within the diffusion pores. As a result of improved mass transfer, particles larger than 10 μm can be used to produce packed columns that have lower pressure drop and flow rate profiles that allow them to operate at higher flow velocities and produce faster separations.

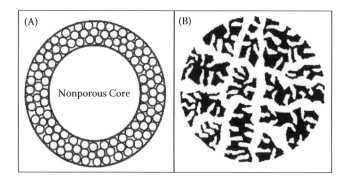

FIGURE 13.6 Particle construction: (A) nonreticular pellicular and (B) reticular.

FIGURE 13.7 Scanning electron micrograph showing flow-through channels in silica-based monolithic rods.

Unlike conventional particle-based construction, monolithic columns consist of continuous beds or rods containing through pores. Although the preparation and chromatographic use of the monolith column was first reported in 1967,[50,51] the initial device suffered from poor flow characteristics and attracted little interest as a feasible approach for producing separation media. Since the 1990s, manufacturing improvements have resulted in monoliths with better performance characteristics and utility for high-throughput assays.[52,53] Many investigators have used various types of monoliths to conduct a wide variety of pharmaceutical and biomedical analyses. Monoliths are very useful for separating larger peptides, proteins, and polycyclic compounds.[54–59]

Monoliths are constructed of several types of materials including silica, polymers, and graphitized carbon.[60] Silica monoliths are the most common and are manufactured using a sol–gel process *in situ* or in a manufacturing mold. Molded columns are removed and encapsulated in a suitable construction material such as polytetrafluoroethylene (PTFE) or poly(ether ketone) (PEEK). Figure 13.7 illustrates a group of monolithic silica rods prior to encapsulation along with an electron micrograph of the inner flow construction.[61] The flow channel sizes of these materials average about 2 μm and the diffusion pores about 120 Å in diameter. The porosity is 65 to 70% compared to packed columns with porosities in the range of 35 to 40%.[12] The result is reduced nonflow through diffusion paths that improve the solute mass transfer characteristics.

Since monoliths can be operated at much higher linear velocities and generate relatively flat van Deemter profiles, they are good choices for optimizing efficiency and speed using either constant high flow rates or a combination of gradient and flow programming. Several types of monoliths are employed to carry out high-throughput separations including those based on polystyrene, acrylate polymers, and silica. Sizes range from standard analytical to capillary dimensions. In one case,[54] a 14-fold reduction in analysis time for aprepitant and several related compounds was obtained using a 4.6 mm inner diameter × 50 mm silica-based C18 monolith compared separation with a conventional 4.6 mm × 250 mm porous particle-based C18 analytical column. In another study, a series of five proteins were separated in less than 20 sec using a 4.6 mm × 50 mm monolith operated at a flow rate of 10 mL/min.[55] Both examples illustrate significant gains in analysis speed made possible by combining shorter column lengths with improved flow-through properties.

Completely solid packings produce no mass transfer effects related to solute movement into and out of tortuous pore structures. The principal limitation of solid packings is low total surface area and consequent limited sample capacity. Even with very small particles, the overall surface area generated per unit length of the column is still low compared to the porous materials with mean pore sizes in the 60 to 300 Å range. Figures 13.5 A and B show a comparison of surface areas of

FIGURE 13.8 Scanning electron micrograph showing nonreticulated porous structures of standard spherical silica particles.

solid and porous particle-based packing materials. To overcome the sample capacity problem with solid materials, longer columns packed with very small particles are needed[11] and larger numbers of theoretical plates can be generated. However, the disadvantages of these types of columns are the need for ultrahigh pressure pumping and injection systems and the inability to handle larger sample sizes. Other problems also come into play including frictional heating and compressibility of the eluent that affect both the thermodynamics and kinetics of the separation, and are discussed in Section 13.4. In most cases, solid packings are experimentally interesting but not practical in terms of ease of use and reliability.

The second particle type is nonreticulated; the transfer of analyte into and out of the porous packing is totally diffusion-controlled. Figure 13.8 consists of scanning electron micrographs of nonreticulated porous structures of standard spherical silica particles. These are the most common HPLC packings because they provide the best compromise of efficiency, sample capacity, and mechanical stability. Over the past three decades, the sizes of standard high efficiency particles decreased from the 5 to 10 μm range in the 1980s, to the 3 to 5 μm range in the 1990s, and currently to the 1 to 3 μm range for ultrahigh efficiency separations.

13.4 ULTRAHIGH PRESSURE LIQUID CHROMATOGRAPHY

As noted, decreasing the HPLC analysis time involves a combination of improved columns with higher efficiency resulting from the use of smaller particles and increased flow velocities for the eluent for less complex mixtures, and the use of smaller particles, longer columns, and increased flow velocities for more complex mixtures. Both approaches make greater demands on the pumping system to deliver higher pressures. In addition to increased strain on the equipment, greater pressure can result in two problems not encountered in conventional HPLC: solvent heating and compressibility that can affect the quality and expected behavior of a separation.

A detailed theoretical discussion of UHPLC is complex and beyond the limited space and scope of this chapter. Nevertheless, a few comments may be helpful in providing a fundamental overview and aid the search for relevant information in the literature. Unfortunately, many papers understate the underlying concepts governing the physical processes that occur at very high pressures or they provide incomplete or even incorrect information. A number of investigators may not be fully aware of the complex nature of physicochemical changes resulting from eluent compression and decompression in pumps and columns. Two recently published articles serve as good starting points to learn about UHPLC because they present detailed discussions of the fundamental and applied aspects of the subject.[25,26]

In conventional HPLC, one assumption made to simplify the overall treatment of solute migration is the incompressibility of the eluent. For moderate pressures up to approximately 250 to 300 bar, this approximation is reasonable since about 2% or less occurs and the resulting compression and expansion heating, frictional heating, and changes in eluent structure and polarity may be ignored. For all practical purposes, retention factors measured at one flow rate (pressure) are identical to those obtained at a different flow rate (pressure). However, at higher inlet pressures, both thermal and polarity changes in an eluent become significant and the effects of compression and decompression cannot be ignored.

The first physical change to consider with UHPLC is the compression of the eluent in the piston chamber that produces thermal heating. The change in temperature related to a change in pressure (dT/dP) can be estimated per Equation 13.2 in which C_p represents eluent heat capacity, α is the thermal compression/expansion coefficient, T is the temperature in K, and V is molar volume.

$$(dT/dP)_S = \alpha T V / C_p \tag{13.2}$$

Consider the case of pure methanol for which the values of C_p and α are known. Using a specific volume of 0.00127 m^3/kg, a temperature of 25°C (298 K), and a compression pressure of 1000 bar, Equation 13.2 predicts the eluent temperature will increase approximately 15°C assuming adiabatic conditions. In actual practice, the increase in eluent temperature entering the column will be lower than this upper limit due to thermal losses in the pump, connecting tubing, and injection system, as well as entropic changes ($\Delta S \neq 0$).

It is important to note that using Equation 13.2 assumes no entropic changes occur in the system ($\Delta S = 0$) during compression. In practice this is not the case and values calculated using this approximation over-estimate the amount of heat produced (increase in temperature) during compression. In addition to producing thermal energy during compression, significant amounts of internal energy are stored via a change in ΔS. This energy is released during decompression of the eluent between the inlet and outlet of the column (i.e., conservation of energy).

The second physical change is decompression as the eluent travels through the packed bed. Although this change is present under normal HPLC conditions, ΔP is small enough to be ignored. However, this is not the case in UHPLC where the amount of the decompression can be modeled using the Joule-Thomson expansion of a liquid through a porous plug effect and by assuming the change in enthalpy for the process is zero (isenthalpic). The resulting mathematical model is shown in Equation 13.3. In solving this expression, it is important to note that the Joule-Thomson effect can be positive or negative, depending on the exact inversion point, resulting in either heating or cooling.

$$(dT/dP)_H = (\alpha T - 1) V / C_p \tag{13.3}$$

For methanol, where α is approximately 1×10^{-3} K^{-1} and temperature ranges from 30 to 60°C, decompression results in an increase of 35 to 30 K/kbar.[25] Accordingly, Equations 13.2 and 13.3 predict decompression heating in the column to be about twice that of compression heating in the pump. Again it is important to recognize that under actual UHPLC conditions (binary solvents, different compression volumes, temperatures, pressures, and nonadiabatic heat losses), the calculated values serve as only very rough estimates of what to expect. Often the measured increases in temperature are significantly less than values predicted. Likewise, with smaller bore columns and appropriate temperature control, the problem of thermal gradients in the column can be minimized.

In addition to thermal and viscosity changes, a third and perhaps the most important physical consideration in UHPLC is the influence of pressure on the solvent structure of the hydroorganic eluents and the resulting changes in the equilibrium constants that govern solute migration. Although this aspect has received far less attention in the literature, attempts have been made to model these effects.[25,26] It is important to recognize that the elution profile of a mixture of compounds observed using moderate pressures can differ significantly from the elution profile of the same mixture measured at ultrahigh pressure even if an identical eluent and surface or column are used. Furthermore, the elution profile will become less predictable as the flow rate (pressure) changes. This effect was first reported nearly four decades ago[24] and suggests that chromatographic behaviors of UHPLC are far more difficult to predict and reproduce than those obtained under conventional HPLC conditions.[25]

13.5 SAMPLE INTRODUCTION AND AUTOMATION

Except for bulk drug substances, samples can rarely be analyzed directly following simple preparation steps. In many cases, an active ingredient is found in a formulated form or in a physiological fluid or tissue. Likewise, interferences associated with the matrix and the presence of possible degradation products, metabolites, and other closely related compounds mean the target analyte is often highly diluted. Also, an analyte may be difficult to detect. Effective sample preparation steps serve up to three broad purposes: (1) eliminate and/or minimize possible interferences, (2) concentrate the sample, and (3) render the analyte of interest into a more easily detectable form.

An additional consideration for sample preparation is to ensure that the final sample solution is miscible with the HPLC eluent and will not alter or degrade the column.[62] The total time needed for sample preparation may be longer than that required to conduct chromatographic separation and therefore becomes is the rate-determining step for the analysis.[63] A survey cited by several authors indicated that on average chromatography separation accounts for about 15% of the total analysis time, sample preparation, about 60%, and data analysis and reporting, 25%.[64–66]

Sample preparation can involve several steps including collection, drying, grinding, filtration, centrifugation, precipitation, and various forms of classical and modern extraction. Several different approaches can be used to produce a final solution that can be analyzed chromatographically. Some are less cumbersome and time-consuming and more easily automated. Because of the scope of this chapter, it is not possible to discuss all of them. However, many books, chapters, reviews,[62,65–69] and special journal issues[70,71] deal with various aspects of sample preparation in reducing human involvement through the use of robotics.

Offline parallel processing using automated instruments such as accelerated solvent extractors and sample handling robotics, and inline sequential sample pretreatment using fluid switching are two basic approaches used to increase sample throughput prior to the chromatographic separation step. Two important developments are the automation and miniaturization of solid phase extraction via conventional and capillary cartridge designs. Solid-phase microextraction (SPME) was introduced about 15 years ago.[72,73] It is based on the partitioning of analytes between a thin film immobilized onto the exterior of a fiber (fused silica) or to the interior wall of a delivery capillary (syringe needle). Figure 13.9 illustrates these two types of mounting designs.[72] Typical surface coatings are immobilized polysiloxane, polystyrene, and polyacrylate polymers.[73] One advantage of capillary design is that SPME can be connected easily to LC using available instrumentation.[72,74]

Many other types of solid phase adsorbents, including those based on conventional and specialty materials like restricted access media (RAM), can increase analysis speed and improve assay performance. These types of materials, also known as internal reversed-phase packings, are especially useful for assaying target compounds in biological samples such as serum and plasma. They are chemically modified porous silicas that have hydrophilic external surfaces and restricted-access hydrophobic internal surfaces. The ratio of interior to external surface areas is large. Macromolecules such as proteins cannot enter the pores of the RAM (they are excluded from the hydrophobic internal surface) and they elute quickly through the column. However, the smaller analyte molecules that can enter the pores are retained via interactions with the hydrophobic bonded phase within

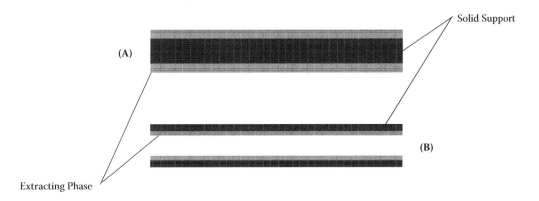

(A)

Solid Support

Extracting Phase

(B)

FIGURE 13.9 Implementation of SPME techniques: (A) coating on outer surface of fiber and (B) coating on inner wall of capillary.

the pores. Although some assays reported use of a single RAM column for both sample clean-up and analyte separation, most often RAMs are used in combination with an intermediate switching valve or serve as precolumns for the inline clean-up of biological samples prior to analytical separation.[75,76]

13.6 CONCLUSIONS

During the past 40 years, modern liquid chromatography has gone from infancy to a mature and widely employed technique. In pharmaceutical research and development laboratories, it is the most used analytical approach for studying a wide variety of sample types and topical areas. It is first used in the earliest stages of the drug discovery process, later in the clinical testing stage, and finally as a routine tool for monitoring manufactured product quality. Many changes in instrumentation and column technology have greatly improved reliability and increased the speed of LC based assays.

An important and emerging technique is the use of columns packed with smaller and smaller particles operated at higher and higher pressures. Although the use of ultrahigh pressures in liquid chromatography dates back to the very early days of modern HPLC, little work was done before the 1990s, primarily because of the lack of off-the-shelf instrumentation able to operate at higher pressures and appropriate sub-3-μm column technology designed to provide better performance. Unlike conventional HPLC separations in which fluid compressibility can be ignored, the use of higher operating pressures means that fluid compressibility can no longer be ignored and creates new operational challenges.

REFERENCES

1. Okafo, G.N. and Roberts, J.K. in *Pharmaceutical Analysis*, Lee, D.C. et al., Eds., Blackwell: Oxford, 2003, Chap. 2.
2. Lee, M.S., *LC/MS Applications in Drug Development*, John Wiley & Sons: New York, 2002, Chap. 1.
3. Dixon, S.P., Pitfield, I.D., and Perrett, D. *Biomed. Chromatogr.* 2006, *20*, 508.
4. Hanai, T. *Curr. Med. Chem.* 2005, *12*, 501.
5. Gilpin, R.K. and Gilpin, C.S. *Anal. Chem.* 2007, *79*, 4275.
6. Knox, J.H. *J. Chromatogr. Sci.* 1977, *15*, 352.
7. Gilpin, R.K. and Gaudet, M.H. *J. Chromatogr.* 1982, *248*, 160.
8. Lindsay, S. *High Performance Liquid Chromatography*, 2nd ed., John Wiley & Sons: London, 1992.
9. Jones, M.D. and Plumb, R.S. *J. Sep. Sci.* 2006, *29*, 2409.
10. Al-Dirbashi, O.Y. et al. *Anal. Bioanal. Chem.* 2006, *385*, 1439.

11. MacNair, J.E., Lewis, K.C., and Jorgenson, J.W. *Anal. Chem.* 1997, *69*, 983.
12. Wu, N.J. and Thompson, R., *J. Liq. Chromatogr. Rel. Technol.* 2006, *29*, 949.
13. Nováková, L., Matysová, L., and Solich, P. *Talanta* 2006, *68*, 908.
14. Wren, S.A.C. and Tchelitcheff, P. *J. Chromatogr. A* 2006, *1119*, 140.
15. Johnson, K.A. and Plumb, R. *J. Pharm. Biomed. Anal.* 2005, *39*, 805.
16. Mensch, J. et al. *J. Chromatogr. B* 2007, *847*, 182.
17. Kalovidouris, M. et al. *Rapid Commun. Mass Spectrom.* 2006, *20*, 2939.
18. Yu, K. et al. *Rapid Commun. Mass Spectrom.* 2006, *20*, 544.
19. Dear, G.J., James, A.D., and Sarda, S. *Rapid Commun. Mass Spectrom.* 2006, *20*, 1351.
20. King, S. et al. *LCGC N. Am.* 2005, *23 (May Suppl.)*, 36.
21. Nováková, L., Solichová, D., and Solich, P. *J. Sep. Sci.* 2006, *29*, 2433.
22. Swartz, M. E. *J. Liq. Chromatogr. Rel. Technol.* 2005, *28*, 1253.
23. Nguyen, D.T.T. et al. *J. Sep. Sci.* 2006, *29*, 1836.
24. Bidlingmeyer, B.A. et al. *Sep. Sci.* 1969, *4*, 439.
25. Martin, M. and Guiochon, G. *J. Chromatogr. A*, 2005, *1090*, 16.
26. Gritti, F. and Guiochon, G. *J. Chromatogr. A*, 2006, *1131*, 151.
27. Van der Wal, S. *Chromatographia* 1985, *20*, 274.
28. Giddings, J.C. *Unified Separation Science*, John Wiley & Sons: New York, 1991, Chap. 4.
29. Wu, N.J. et al. *J. Microcolumn Sep.* 2000, *12*, 462.
30. Xiang, Y. et al. *Chromatographia* 2002, *55*, 399.
31. Gilpin, R.K. and Dudones, L.P. in *Encyclopedia of Analytical Chemistry: Applications Instrumentation and Applications*, John Wiley & Sons: New York, 2000, p. 7143.
32. Grushka, E., Colin, H., and Guiochon, G. *J. Chromatogr.* 1982, *248*, 325.
33. Jandera, P., Blomberg, L.G., and Lundanes, E. *J. Sep. Sci.* 2004, *27*, 1402.
34. Takeuchi, T., Kumaki, M., and Ishii, D. *J. Chromatogra.* 1982, *235*, 309.
35. Hsieh, Y.S., Merkle, K., and Wang, G.F. *Rapid Commun. Mass Spectrom.* 2003, *17*, 1775.
36. Teutenberg, T. et al. *J. Chromatogr. A* 2006, *1114*, 89.
37. Li, J.W. and Carr, P.W. *Anal. Chem.* 1997, *69*, 837.
38. Majors, R.E. *Am. Lab.* 2003, *35*, 46.
39. Horvath, C.G. and Lipsky, S.R.. *J. Chromatogr. Sci.* 1969, *7*, 109.
40. Majors, R.E. *Anal. Chem.* 1972, *44*, 1722.
41. Wang, X.L., Barber, W.E., and Carr, P.W. *J. Chromatogr. A* 2006, *1107*, 139.
42. Kirkland, J.J. et al. *J. Chromatogr. A* 2000, *890*, 3.
43. Kirkland, J.J. *J. Chromatogr. Sci.* 2000, *38*, 535.
44. Afeyan, N.B. et al. *J. Chromatogr.* 1990, *519*, 1.
45. Banks, J.F. *J. Chromatogr. A* 1995, *691*, 325.
46. Fulton, S.P. et al. *J. Chromatogr.* 1991, *547*, 452.
47. Garcia, M.C., Marina, M.L., and Torre, M. *J. Chromatogr. A* 2000, *880*, 169.
48. Hofstetter, H., Hofstetter, O., and Schurig, V. *J. Chromatogr. A* 1997, *764*, 35.
49. Rodrigues, A.E. *J. Chromatogr. B* 1997, *699*, 47.
50. Kubin, M. et al. *Coll. Czechosl. Chem. Commun.* 1967, *32*, 3881.
51. Švec, F. and Tennikova, T.B. in *Monolithic Materials: Preparation, Properties, and Applications*, Švec, F. et al., Eds., Elsevier: Boston, 2003, Chap. 1.
52. Nakanishi, K. and Soga, N. *J. Am. Ceram. Soc.* 1991, *74*, 2518.
53. Minakuchi, H. et al. *Anal. Chem.* 1996, *68*, 3498.
54. Liu, Y. et al. *J. Liq. Chromatogr. Rel. Technol.* 2005, *28*, 341.
55. Xie S. et al. *J. Chromatogr. A 1999*, 865, 169.
56. Walcher, W. et al. *J. Chromatogr. A* 2004, *1053*, 107.
57. Luo, Q.Z. et al. *J. Chromatogr. A* 2001, *926*, 255.
58. Leinweber, F.C. et al. *Rapid Commun. Mass Spectrom.* 2003, *17*, 1180.
59. Rieux, L. et al. *J. Sep. Sci.* 2005, *28*, 1628.
60. Liang, C.D., Dai, S., and Guiochon, G. *Anal. Chem.* 2003, *75*, 4904.
61. Cabrera, K. *J. Sep. Sci.* 2004, *27*, 843.
62. Snyder L.R., Kirkland J.J, and Glajch J.L. *Practical HPLC Method Development*, 2nd ed., John Wiley & Sons: New York, 1997, Chap. 4.
63. Jinno, K. *Anal. Bioanal. Chem.* 2002, *373*, 1.
64. Majors, R.E. *LCGC N. Am.* 1991, *9*, 16.

65. Rouessac, F. and Rouessac, A. *Chemical Analysis: Modern Instrumentation Methods and Techniques*, John Wiley & Sons: New York, 2000, Chap. 20.
66. Smith, R.M. *J. Chromatogr. A* 2003, *1000*, 3.
67. Hyötyläinen, T. and Riekkola, M.L. *Anal. Bioanal. Chem.* 2004, *378*, 1962.
68. Theodoridis, G.A. and Papadoyannis, L.N. *Curr. Pharm. Anal.* 2006, *2*, 385.
69. Visser, N.F.C, Lingeman, H., and Irth, H. *Anal. Bioanal. Chem.* 2005, *382*, 535.
70. *Anal. Bioanal. Chem.* 2002, *373*.
71. *J. Chromatogr. A* 2000, *885*.
72. Lord H. and Pawliszyn J. *J. Chromatogr. A* 2000, *885*, 153.
73. Dietz, C., Sanz, J., and Cámara, C. *J. Chromatogr. A* 2006, *1103*, 183.
74. Kumazawa T. et al. *Anal. Chim. Acta* 2003, *492*, 49.
75. Rossi D.T. and Zhang N. *J. Chromatogr. A* 2000, *885*, 97.
76. Berna M.J., Ackermann B.L., and Murphy A.T. *Anal. Chim. Acta* 2004, *509*, 1.

14 Advances in Capillary and Nano HPLC Technology for Drug Discovery and Development

Frank J. Yang and Richard Xu

CONTENTS

14.1 INTRODUCTION

New drug discovery and development is very expensive and time-consuming process. Ten to twelve years of development may be required before a new drug is safe and effective for marketing and distribution. Many thousands of compounds must be screened to find a few lead candidates. For every 5000 new compounds evaluated, only about 5 may be safe enough to be considered for testing in healthy human volunteers. After 3 to 6 years of further clinical testing, only one of these compounds will ultimately gain approval as a drug treatment.

Accelerating drug discovery has become the goal of all pharmaceutical companies. It is important to evaluate physicochemical properties such as absorption, distribution, metabolism, elimination, and toxicity (ADMET) and conduct pharmacokinetic screening of drug candidates as early as possible to allow faster decision making and save time and money. The screening of proteins as drug interaction targets is a rapidly developing field. To deliver high quality compounds, good separations of complex cellular and plasma proteins are required. Corens[1] reviewed traditional approaches such as GC-MS, CE-MS, SFC-MS, and conventional LC-MS and introduced a parallel analysis strategy using multiple columns in a valve-switching scheme.

Drug discovery, particularly for a small molecule drug target, faces four great challenges: (1) complexities of sample matrices and drug metabolites, (2) limited numbers and amounts of biological samples, (3) trace amounts of compounds of interest that require careful sample pretreatment and concentration prior to identification by analytical techniques, and (4) wide dynamic ranges of target sample concentrations. Because of the complexities of biological fluids, sample clean-up and concentration steps are often required. Recent developments in discovery techniques (combinatorial chemistry, proteomics, ADME, and toxicology profiling) demand high-throughput separation, efficient sensitivity detection, and accurate data handling techniques to speed the discovery process. MALDI-TOF-MS offers great advantages such as speed and small sample size. It is effective for the direct analysis of complex metabolites and drug mixtures because of its high tolerance to sample matrices. Tandem MS is the cornerstone of drug metabolite identification,[2,3] and includes a variety of scanning techniques such as product ion, precursor ion, and neutral loss.

Online LC-MS is a good solution for separation, identification, and quantification because it permits the confirmation of polar and nonvolatile compounds without need for derivatization.[4] The use of LC-MS for biological sample detection and data analysis has grown rapidly during the past few years. Many reliable and easy to use LC-MS systems are commercially available and have been adapted for solving analytical problems by scientists in proteomics research, metabolic study, complex natural product separation and characterization, and drug discovery.

Recent advances in nano HPLC and nano spray mass spectrometry have significantly increased the resolution power of complex sample analyses and also improved sensitivity for trace detections of biological samples to the attamole range. In theory, nano HPLC is the best choice for high-throughput chromatographic applications. The goal of this chapter is to discuss the theoretical bases and demonstrate the advantages of capillary HPLC in high-throughput drug discovery and development.

Nano LC-MS method development consumes as little as 12 μL solvent for a 1-hr run (0.2 μL/min flow rate). A 100 mL solvent reservoir for 7/24 applications can last nearly a year. Fundamentally, the promise of chromatographic speed depends heavily on separation techniques such as the utilization of nano particles in short column lengths. In addition, online sample preparation and fully automated multidimensional technique and instrumentation greatly enhance sample throughput, reproducibility, and data quality. An ultrahigh pressure capillary-HPLC system is advantageous because it allows the use of capillary HPLC columns packed with 1.5 to 2 μm particles at high flow rates for ultrafast analysis. An ultrahigh pressure splitless HPLC system that allows operating pressures up to 15,000 psi is required for use with ultrahigh pressure capillary columns packed with 1.5 μm particles.

14.2 BASIS OF FAST HPLC

The speed of analysis in HPLC is a potential bottleneck for complex sample analysis. Various approaches such as utilizing short columns at high flow rates and the recent focus on 1.5 to 2 μm particles have been reported to increase the speed of analysis. Multidimensional chromatographic approaches have also been demonstrated to increase the throughput of HPLC. The five major parameters that may affect the speed of capillary and nano LC are discussed below.

14.2.1 COLUMN LENGTH

Theoretically, chromatographic resolution depends on the square root of the column length. Separation of small molecules may be improved 40% by doubling column length. Retention time may also

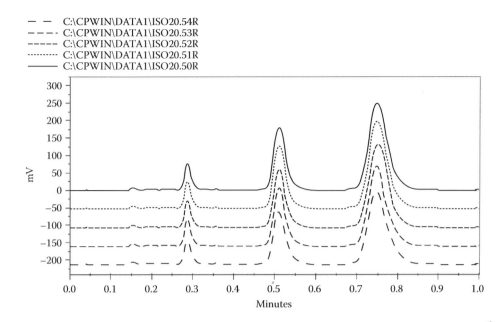

— —   C:\CPWIN\DATA1\ISO20.54R
— — ·   C:\CPWIN\DATA1\ISO20.53R
— — —   C:\CPWIN\DATA1\ISO20.52R
·········   C:\CPWIN\DATA1\ISO20.51R
————   C:\CPWIN\DATA1\ISO20.50R

FIGURE 14.1 One-minute fast elution of three small molecules from a 5 cm × 1 mm column packed with 5 μm C18 particles. Column head pressure at 180 μL/min flow rate is 2800 psi. Solvent composition is 60:40 water:acetonitrile.

be lengthened by a factor of two and a long column requires an ultrahigh pressure HPLC pumping system to deliver solvent at a high flow rate to compensate for the increased pressure drop in proportion to column length.

Conventional HPLC in general has a maximum column length of 25 cm × 4.6 mm inner diameter (ID) due to the difficulty of achieving uniform packed-bed density across a long length column of 4.6 mm ID. For fast HPLC, 3 to 5 cm short columns packed with 1.5 to 5 μm particles are normally used. Figure 14.1 shows a 1-min fast elution of three small molecules from a 5 cm × 1 mm ID, 5 μm C18 column. Reproducible retention time and area count are noted. Long capillary columns (up to 100 cm) can be well packed.[5,6] As shown in Figure 14.2, a complex base peak chromatogram of five protein digests was obtained using a 50 cm × 75 μm ID column packed with 3 μm C18 particles. The flow rate for nano LC was 0.4 μL/min and the column head pressure was 6,500 psi.[7] Recent development of 1.5 μm nonporous and porous particle columns with 120 Å pores in a length of 1 to 5 cm offer ultrafast separation of drug metabolites and small molecules.

To analyze complex biological samples such as proteins and polypeptides, a 15 cm capillary column is normally utilized to separate nearly 200 peptides in a single run. Short capillary columns packed with 1.5 to 3 μm particles can be run at higher flow rates to achieve high throughput with great mass sensitivity in sample detection operations. Figure 14.3 shows a fast analysis of several control drugs and metabolites using a 15 cm × 300 μm ID capillary column packed with 3 μm C18 particles operated at 10 μL/min.[8] Capillary column runs at high flow rates achieve excellent retention time reproducibility because the gradient regeneration is fast and small flow path leakage does not significantly affect chromatographic performance at high flow. Figure 14.4 shows chromatograms for reproducible runs in fast capillary LC using a 5 cm capillary column at 40 μL/min flow rate—eight times faster than the optimum flow rate for the column. Reproducible fast chromatography analyzed three small test probe molecules in less than 1 min.

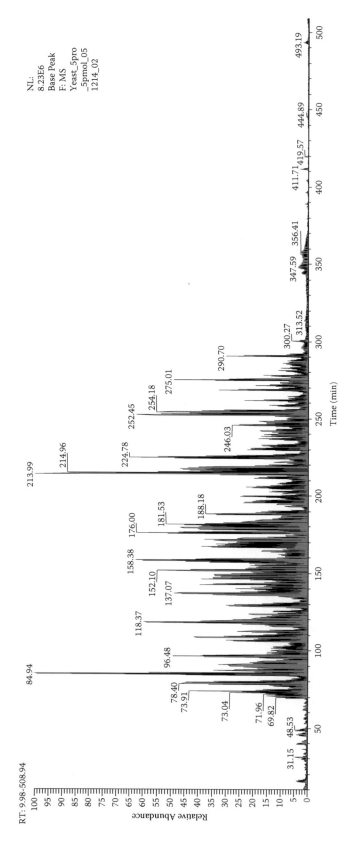

FIGURE 14.2 Base peak chromatogram for a five protein tryptic digest mixtures containing 3.75 pmol of each protein. Gradient time program: 260 min, 5 to 23.5% B; 150 min, 23.5 to 50% B; 90 min, 50 to 80% B. Nano column size: 50 cm × 75 μm inner diameter packed with Microsil 3 μm C18 particles (Micro-Tech Scientific: MC-50-C18WSS-75-EU). Column head pressure: 6500 psi at 0.4 μL/min splitless gradient flow rate. Solvent A: water with 0.1% formic acid; solvent B: acetonitrile with 0.1% formic acid. Ultrahigh pressure splitless nano LC system: XTS binary UHPLC from Micro-Tech Scientific, Vista, California. LTQ MS: Thermal-Finnigan with nano spray interface, San Jose, California.

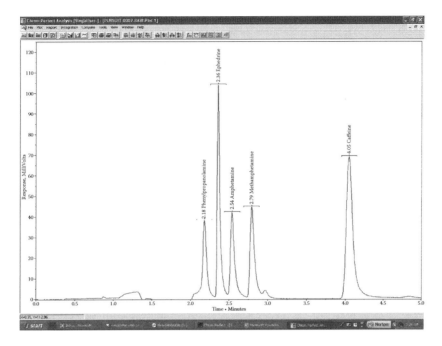

FIGURE 14.3 Fast analysis of control drugs and metabolites using a 15 cm × 300 μm inner diameter capillary column packed with 3 μm C18 particles (Micro-Tech Scientific: MC-15-C18SS-320-EU) operated at 10 μL/min gradient flow rate. UV at 278 nm. (*Source:* Drug Enforcement Administration, Southwest Laboratory, Vista, California and S. DiPari.)

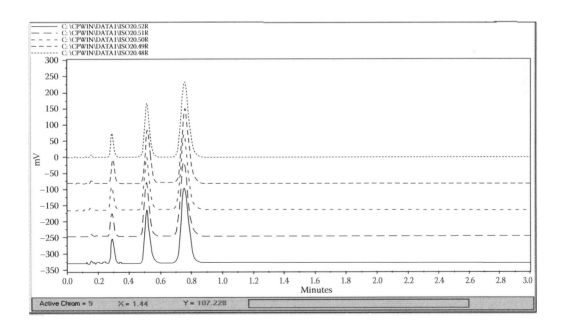

FIGURE 14.4 Chromatograms of high speed isocratic capillary LC elution of three components. Column: 15 cm × 320 μm inner diameter, 5 μm C18 particles. Column head pressure: 6800 psi at 48 μL/min flow rate. System: XTS two-dimensional splitless ultrahigh pressure nano UHPLC, Micro-Tech Scientific, Vista, California.

14.2.2 Column Diameter

Column diameter is an important parameter to consider in life science applications in which sample amounts are very limited and the components of interest may not be abundant. Researchers have reviewed micro HPLC instrumentation and its advantages.[9,10] Nano LC-MS offers 1000- to 34,000-time reductions in the dilution of a sample molecular zone eluted from nano LC columns of 25 to 150 μm IDs in comparison to a 4.6 mm ID column. This represents a large enhancement of ion counts in comparison to counts obtained for the same amount of sample injected into a conventional 4.6 mm column. Solvent consumption for an analysis run or sample amount required for injection in a nano LC application may be reduced 1000 to 34,000 times compared to amounts required by an analytical column operated at a 1 mL/min flow rate.

When nano LC is combined with mass spectrometer detection, attamole detection can be achieved for low abundance components in biological fluids, drug metabolites, and natural products such as Chinese herb medicines. Nano LC-MS-MS has become an essential tool for complex biological and drug metabolite studies. Nano LC-MS presents two significant differences from conventional analytical HPLC: (1) large enhancement factor for sample detection and (2) direct interface to MS without flow splitting. The enhancement in MS ion counts relative to a conventional 4.6 mm ID column is proportional to the ratio of the square of the column diameter:

$$\text{MS detection enhancement factor} = [\text{conventional HPLC}$$
$$\text{column diameter (4.6 mm)/(nano LC column diameter in mm)}]^2 \qquad (14.1)$$

For a 75 μm ID nano LC column as an example, the MS detection enhancement factor (ion count) in comparison to a 4.6 mm column is much higher than $(4.6/0.075)^2 = 3761$ because of the reduction in sample molecular zone dilution and because a nano LC solvent flow rate at 0.02 to 2 μL/min can be 100% directly sprayed into the MS ion source. No post-column flow splitting is required for nano-LC-MS as that required when 1 mL/min is used in a 4.6 mm ID column. This large enhancement of MS detection and the ability to directly interface with MS presents nano LC-MS as the best tool for life science research.

Nano LC columns may be operated at column flow rates much higher than optimum without significant losses of resolution of sample components. High-throughput nano LC can be achieved by increasing mobile solvent flow rates. Because of very low LC solvent consumption at high throughput, nano LC is the best choice for moderate solvent consumption. Table 14.1 compares solvent consumption for column diameters from 50 μm to 4.6 mm.

Table 14.1 shows clearly that capillary and nano LC save solvent cost by allowing use of the same solvent for several thousand analyses and thus generating reproducible chromatographic data.

TABLE 14.1

Comparison of Solvent Consumption for LC Columns

Column Diameter (μm)	Solvent Consumption per Run (μL)[a]	Number of Analyses per 100 mL Solvent
50	2.4	1,667
75	6.2	16,129
150	25	4,000
320	100	1,000
500	500	400
1,000	1,000	100
2,000	4,000	25
4,600	20,000	5

[a] Assume 2-min run time at 10 times optimum flow rate.

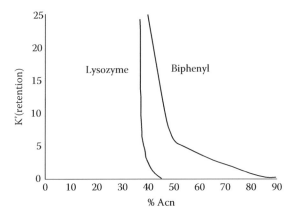

FIGURE 14.5 Comparison of partition factors for a PAH (biphenyl) and polypeptide (lysozyme) versus percent acetonitrile concentration.

For nano LC using a 75 μm ID column, 100 ml of solvent can handle more than 16,000 injections (2-min analysis time per injection). No solvent waste collects because it is evaporated. Conventional LC will produce more than 320,000 ml of solvent waste for the same number of analyses.

14.2.3 MOBILE PHASE COMPOSITION (Z NUMBER)

The elution of biological molecules such as proteins and polypeptides requires precise control of the composition of the organic modifier in the mobile solvent. According to Geng and Regnier,[11] the desorption of polypeptides from a reversed-phase column requires a critical number (Z Number) of an organic molecule. They define the Z number as a very precise and narrow range of organic modifier concentration that causes desorption of a polypeptide from the stationary phase surface.

As shown in Figure 14.5 adapted from Geng and Regnier,[11] an increase in retention of small biphenyl molecules will be detected when the percent of organic modifier concentration is reduced. However, for a polypeptide biological sample, the lysozyme is not retained when the organic solvent composition exceeds 45%. At 35%, organic, lysozyme is completely retained.

The narrow range of organic modifiers required to elute and desorb polypeptides from the reversed-phase column packing material accounts for the separation of polypeptides from a short C18 capillary column. Because polypeptide elution and separation depend on the accuracy of solvent composition in gradient nano LC, it is very important to use a system that can precisely control the LC modifier concentration even at low percents of organic modifiers in chromatographic elution compositions.

14.2.4 IONIC PROPERTIES OF ANALYTES

For high-throughput applications, it is important to consider the pH of the mobile solvent in relationship to the pKa values of acidic or basic analytes. The retention of biological and drug molecules depends on the ionic states of the analytes. Figure 14.6 shows the retention factors of acidic molecules at different mobile phase pH values against pKa values. Figure 14.7 shows the retention factors of basic molecules at different mobile phase pH values against pKa values. A general relationship[12] of the retention factor k for an analyte with pKa and pH values of the mobile phase is shown in Equation 14.2.

$$k = [k_0 + ki \exp (2.3 [pKa - pH])/(1 + \exp (2.3 [pKa - pH])] \qquad (14.2)$$

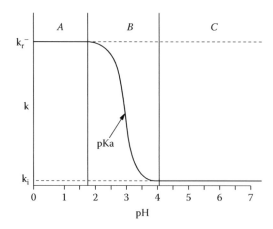

FIGURE 14.6 Retention of acidic analytes with respect to mobile phase pH. The inflection point of the curve corresponds to the pKa of the acidic analyte.

where k is current retention factor at given pH, ki is the retention factor of an ionized analyte (protonated form for basic analytes and anionic form for acids), k_0 is the retention factor of neutral molecules, and pKa is the ionization constant of the analyte. As shown in Figure 14.7, region A indicates relatively low retention for basic analytes because the analytes are in protonated forms. Figure 14.6 shows very strong retention for acidic molecules that are in neutral form in the A region. In region C of Figure 14.6, however, basic analytes show great retention. Acidic analytes show low retention in region C of Figure 14.6. In region B, a small change in pH and mobile phase composition may cause significant changes in the selectivity and retention for both acidic and basic analytes. Region B is not a good area for chromatography because the peak is usually broad and may have a split shape. Region A for basic and region C for acidic components show very low retention changes with variations in mobile phase pH. Therefore, the pH regions A and C are generally employed for basic and acidic analytes, respectively.

14.2.5 Particle Size

Recent advances in column stationary phases are remarkable. High performance silica-based reversed-phase 3 to 5 μm packing materials have been developed for biological sample separations

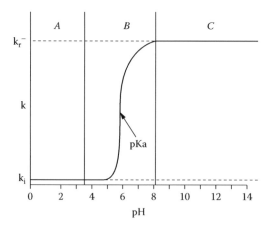

FIGURE 14.7 Retention of basic analytes at different mobile phase pH values against pKa for basic molecules.

with formic acid for MS detection. Recently, 1.5 to 2 μm diameter C18 particles became available for ultrahigh speed chromatographic separations of polypeptides and drug metabolites. Particle diameter is a critical parameter for improving separation efficiency and speed. Theoretically, the separation power is inversely proportional to the diameter of the particle size dp, and the speed of analysis is proportional to dp^2. The time required (Tr) to generate a theoretical plate number (N) to obtain a baseline resolution (Rs = 1.0) of a component[13] can be expressed as Equation 14.3.

$$Tr = N\,h\,dp^2\,(1 + k')/v\,Dm \qquad (14.3)$$

where h is the reduced plate height; v is the reduced linear velocity; Dm is the analyte molecular diffusivity in the mobile phase; and k' is the partition ratio. Assuming h = 2, v = 1, and Dm = 1.5 × 10^{-5} cm^2/sec, a mixture of two components that has a partition ratio of k' = 5 can be baseline separated in 90 sec using a 3 cm column that has N = 10,000 theoretical plates packed with 1.5 μm particles.

The relationship between column length and particle size is L = NH = Nhd$_p$ = 10,000 × 2 × 1.5 × 10^{-4} cm = 3 cm. Assuming the column has a reduced plate number of 2 at its optimum flow velocity, $v = V_{opt} = 1$, then a 3 cm column could produce 10,000 plates when packed with 1.5 μm particles.

The smaller the particle size, the faster the rate of generating theoretical plate (HETP) per unit of time. Figure 14.8 shows a plot of HETP versus linear carrier velocity u for small particles. It indicates that the smaller the particle size, the lower the HETP. It is also important to note that small particles provide nearly the same HETP over a wider range of flow rate.[14]

For high-throughput applications, it is advantageous to operate a column packed with small particles at a high flow rate, particularly fused silica nano columns in which the friction heat effect is negligible even at 10 times the optimum flow rate for the column. High performance 1.5 to 3 μm reversed-phase particle nano columns (<150 μm ID), capillary columns (320 to 500 μm ID), microbore columns (1 mm ID), and narrow bore columns (2 mm ID) are commercially available for ultrahigh-throughput applications. The speed of analysis for a short column packed with 1.5 to 3 μm particles depends upon the pressure limitations of the HPLC system. A commercially available splitless ultrahigh pressure nano LC system allows operation pressures to 15,000 psi. Figure 14.9 shows that retention time reproducibility can be demonstrated using a 3 cm × 150 μm ID column packed with

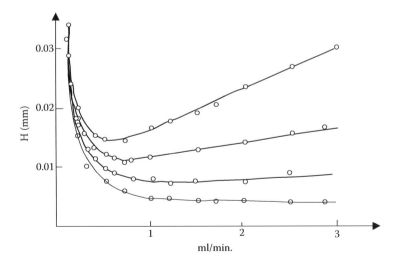

FIGURE 14.8 Relationship of HETP and mobile phase linear velocity for column packing materials of 2, 3, 5, and 8 μm ROSIL C18 particle size. Mobile phase was 75:25 acetonitrile:water. Sample test probe was pyrene at k' = 6.

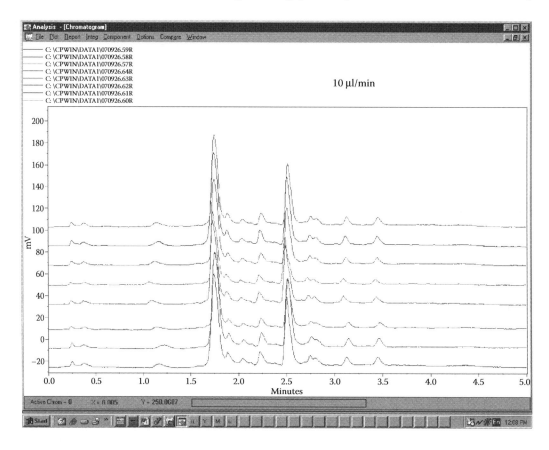

FIGURE 14.9 Performance of short column ultrahigh pressure nano LC system at 10 μL/min for the separation of tranylcypromine sulfate, perphenazine, and their impurities. Nano LC may be operated at 10 times optimum column flow rate and achieve ultrahigh throughput and reproducibility. Short column (3 cm × 150 μm inner diameter) was packed with 1.8 μm C18 particles. Solvent A was water with 0.4% ammonia; solvent B was acetonitrile with 0.4% ammonia. Gradient: 0 to 1 min, 3 to 10% B; 1 to 1.3 min, 10 to 35% B, 1.3 to 3.5 min, 35 to 90% B; held at 90% B through 4.9 min and then returned to 3% B. Column head pressure was 7200 psi.

1.8 μm C18 particles to analyze tranylcypromine sulfate, perphenazine, and their impurities. Excellent retention time reproducibility can be obtained using high speed gradient nano LC at 6900 psi.

14.3 ULTRAHIGH PRESSURE NANO LC FOR HIGH-THROUGHPUT APPLICATIONS

According to Giddings' equation,[15] the pressure drop across a packed column ($\blacktriangle$P/L) may be expressed as:

$$\blacktriangle P/L = 2\ \Phi\ \eta\ u\ /\ d_p^2 \qquad (14.4)$$

where Φ is the overall flow resistance factor; Φ ranges from 300 to 600 for nonporous to porous particles; η is the viscosity of the solvent; and u is the mean fluidic linear velocity through a column cross section. The relationship between pressure drop and particle size for the column of same dimensions is:

$$\blacktriangle P_1/\blacktriangle P_2 = d_{p2}^2 / d_{p1}^2 \qquad (14.5)$$

A comparison of pressure drops across a typical 15 cm nano LC column packed with particles of various sizes appears below.

Particle Size (μm)	Pressure Drop (psi)
5	1,000
3	2,775
2	6,250
1.5	11,111
1.0	25,000

Based on the table, utilizing a column packed with 1.0 μm particles requires the fluidic delivery system to have an operation pressure exceeding 25,000 psi. A 10,000 psi pressure limit pumping system can be used for a maximum 10 cm long column packed with 1.5 μm particles. Using a 15,000 psi pumping system, the applicable length of a column packed with 1.5 μm particles is 15 cm. An ultrahigh pressure pumping system equipped with an ultrahigh pressure injector, switching valve, piston seals and connection tubing, nuts, and ferrules must be leak-free at ultrahigh pressures.

It is particularly difficult when the gradient nano LC flow rate is below 1 μL/min at which any small plumbing leakage could significantly affect retention time and gradient reproducibility. A 3 cm capillary column packed with 1.5 to 3 μm particles can be run at a high flow rate to reach 15,000 psi and perform ultrafast separations. An ultrafast separation of four antidepressant drugs shown in Figure 14.10 compares retention times of a 3 cm $\times$ 75 μm ID nano LC column packed with 1.8 μm C18 particles. Trace A was obtained from running the column at a 1.2 μL/min flow rate (u_{opt} = optimum linear velocity). Trace B shows operation at 10,000 psi. The resolution of benzene, naphthalene, and biphenyl was maintained, but the analysis time was reduced from 5.68 to 0.70 min.

Chromatographic reproducibility for nano LC packed with 1.8 μm C18 particles is demonstrated in Figure 14.11. A 3 cm $\times$ 150 μm ID fused silica column packed with 1.8 μm C18 particles was run at both 1.25 and 10 μL/min. Gradient time programming for the 1.25 μL/min runs was 5 to 45% acetonitrile in 8 min; 45 to 80% in 24 min; held at 80% for 10 min; and then returned to 5%. For the fast splitless nano LC, the gradient was 3 to 10% acetonitrile in 1.0 min; 10 to 35% in 1.3 min; 35 to 90% in 3.5 min; held at 90% to 4.9 min; then returned to 3%. Figure 14.11 shows that nano LC is very reproducible. High speed gradient nano LC can also achieve reproducible performance. Nano columns can be operated at very high flow rates and greatly increase analysis speed even in gradient mode in which the time required for gradient regeneration is reduced 7.5 times from 15 to 2 min using flow rates of 1.25 and 10 μL/min, respectively.

14.4 MULTIDIMENSIONAL HPLC

HPLC separation is the rate-limiting step for high-throughput biological sample analysis. Online SPE-LC-MS-MS has been widely used for such analyses in the pharmaceutical industry. Column switching systems are easily built and controlled. Numerous commercial systems and SPE cartridges are readily available for online sample pretreatment. As shown in Section 14.3, the need for ultrafast LC-MS-MS analysis may be met by using short columns (1 to 5 cm) packed with 1.5 to 2 μm particles. However, the resolving power of a short column may not be sufficient to separate complex biological samples.

The future of LC-MS in polypeptide drug discovery lies in the increasing use of automated online sample clean-up and use of nano LC-MS, both of which greatly increase speed, sensitivity,

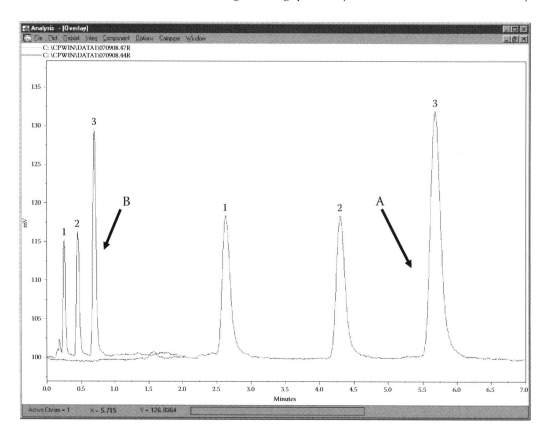

FIGURE 14.10 Comparison of benzene (1), naphthalene (2), and biphenyl (3) test probes eluted from a 3 cm × 75 μm inner diameter column packed with 1.8 μm C18 particles. Trace A was obtained at 1.2 μL/min flow rate and trace B was obtained at 10,000 psi column head pressure (10 μL/min flow rate). Mobile phase was 60:40 acetonitrile:water.

and accuracy. Cheng et al.[16] demonstrated a simple LC-MS-MS for the rapid analysis of acids, neutral, and basic pharmaceutical compounds. Wu et al.[17] described a high-throughput multi-channel LC-MS capable of analyzing 1152 plasma samples in 10 hr. This system performed sample extraction, separation, and detection in a four-channel parallel platform that achieved a throughput of approximately 30 sec per sample.

Using nano LC-MS at submicroliter per minute flow rates requires special attention to plumbing, system dead volume, valve switching, large volume sample injection, precolumn methodology, automation, online sample clean–up, and multichannel parallel operation of a single MS. The techniques discussed below are particularly useful for nano LC-MS-MS applications.

14.4.1 Fast Injection of Large Sample Volumes

Drug metabolites are often present at very low levels of concentration in complex biological fluids. Hence, an online trapping column that allows large sample volume injection, trapping, concentrating, desalting, and subcellular membrane filtering is useful. Figure 14.12 shows a fully automated capillary LC with four pumps, two trap columns, an analytical column, and an autosampler for rapid and effective analysis of biological fluids. Methods 1 and 2 allow alternative injection and trapping

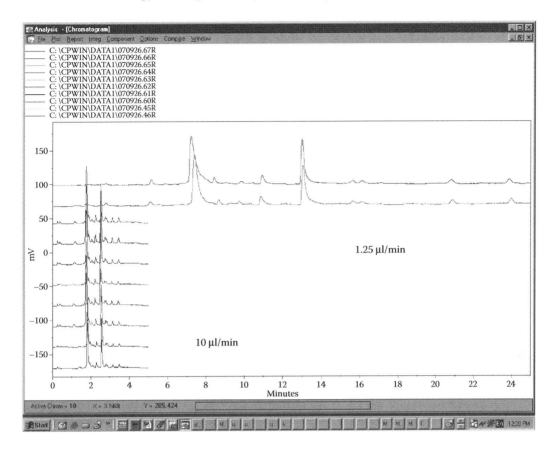

FIGURE 14.11 Comparison of chromatography performance obtained from gradient nano LC at 1.25 μL/min (900 psi) and 10 μL/min (7200 psi) flow rates. Column was 3 cm $\times$ 150 μm, packed with 1.8 μm C18 particles. Reproducible retention time and peak areas for two antidepressant drugs and their impurities are shown.

of sample in trap column 1 while trap column 2 elutes sample into the analytical column and to the MS. The switching valve position for both methods 1 and 2 is given below:

14.4.1.1 Method 1: Trap Column 1 (Figure 14.12)

Step 1: Condition autosampler sample loop and liquid transfer line with water from pump 1. Condition trap column 1 and analytical column with 95 to 100% water delivered from pumps 3 and 4 during equilibration. Switching valve A = off; switching valve B = off.

Step 2: Sample injection from autosampler into trap column 1 for 3 to 5 min using 100% pump 1. Sample is focused, filtered, and concentrated in trap column 1. Condition pumps 3 and 4 to 95% water in composition. Switching valve A = on; switching valve B = on.

Step 3: Run chromatography and elute sample from trap column 1 to analytical column to MS for detection using pumps 3 and 4. Clean autosampler sample loop with 90% organic from pumps 1 and 2 and then condition with 95 to 100% water from pumps 1 and 2. Switching valve A = off; switching valve B = off.

Step 4: Clean autosampler sample loop and trap column 2 using 90% organic solvent from pumps 1 and 2 and then condition trap column 2 and sample loop with 95 to 100% water from pump 1. Continue to run gradient elution of sample from trap column 1 and analytical column using pumps 3 and 4 until the end. Switching valve A = on; switching valve B = off.

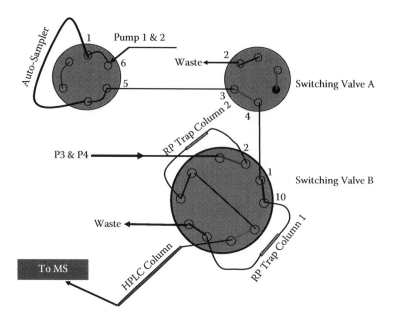

FIGURE 14.12 Fully automated nano HPLC-MS with autosampler injection of large sample volume using two parallel online sample trapping, concentrating, desalting, and filtering columns. Pumps 1 and 2 transfer sample from loop of the autosampler into trap column 1 while pumps 3 and 4 elute sample in trap column 2 into the analytical column then to MS.

14.4.1.2   Method 2: Trap Column 2 (Figure 14.13)

Step 1: Condition autosampler sample loop and liquid transfer line with water from pump 1. Condition trap column 1 and analytical column with 100% water delivered from pump 3 during equilibration. Switching valve A = off; switching valve B = off.

Step 2: Sample injection from autosampler into trap column 2 for 3 to 5 min using 100% pump 1. Sample is focused, filtered, and concentrated in trap column 2. Condition pumps 3 and 4 to 95% water in composition. Switching valve A = on; switching valve B = off.

Step 3: Run chromatography and elute sample from trap column 2 to analytical column to MS for detection using pumps 3 and 4. Clean autosampler sample loop with 90% organic from pumps 1 and 2 and then condition with 95 to 100% water from pumps 1 and 2. Switching valve A = off; switching valve B = on.

Step 4: Clean autosampler sample loop and trap column 1 using 90% organic solvent from pumps 1 and 2 and then condition trap column 1 and sample loop with 100% water from pump 1. Continue to run gradient elution of sample from trap column 2 and analytical column using pumps 3 and 4 until the end. Switching valve A = on; switching valve B = on.

14.4.2   Two-Dimensional Nano LC-MS

The complexity of biological samples presents a great challenge for analytical scientists. Drug metabolites are complicated as a result of drug metabolism and can involve multiple enzymatic pathways. As many as 30 metabolites of various concentrations may be detectable from one drug.[18] A cellular protein sample may contain several hundred to several thousand proteins. The concentration of a cellular protein sample component may range from highly abundant to a trace amount.

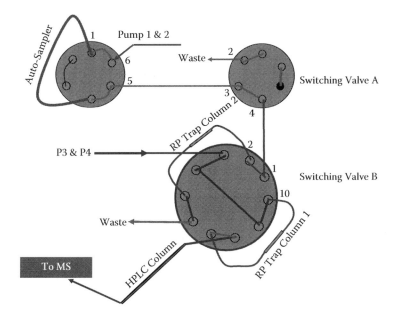

FIGURE 14.13 Fully automated nano-HPLC-MS with autosampler injection of large sample volume using two parallel online sample trapping, concentrating, desalting, and filtering columns. Pumps 1 and 2 transfer sample from loop of the autosampler into trap column 2 while pumps 3 and 4 elute sample in trap column 1 into the analytical column then to MS.

Two-dimensional polyacrylamide gel electrophoresis (2-D PAGE) is the primary method for resolving complex protein mixtures.[19] 2-D PAGE offers extremely high resolution of more than 2000 proteins in a single run. The characterization of protein is achieved by excision of protein spots and sequence analysis using Edman degradation and reversed-phase chromatography of the degradation products. The recent advances in MS allow rapid protein characterization by coupling 2-D PAGE with MS. Using a combination of 2-D PAGE and capillary LC-MS-MS to confirm low abundance proteins offers significant advantages in terms of accuracy and sensitivity. Figure 14.14 is a flow diagram of a two-dimensional capillary LC-MS-MS system. A strong cation exchanger column (SCX) is installed between port 4 of the six-port valve and port 1 of the 10-port switching valve. The selection of the SCX column depends on the amount of sample. Its column diameter is usually about twice the diameter of the analytical column. A 5 μm SCX packing material with 300 Å pore size is normally used for polypeptides and 9 μm particles of 900 Å pore size are common for proteins. The separation power of SCX is not critical in this application. However, in practice, the SCX should have sufficient peak capacity for all components in the sample. The trap column should have sufficient peak retention capacity to prevent sample break-through resulting in sample loss. A 2.5 cm guard column packed with the same C18 particles contained in the analytical column was tested and produced no break-through for polypeptides. A nano LC-MS-MS analytical column normally has a 75 to 150 μm ID and is packed with 3 or 5 μm C18 particles.

2-D nano LC-MS-MS involves fast injection of a large sample volume, i.e., 40 μL from a 2-D PAGE protein tryptic digest into an SCX column. It is important to activate the ion exchange sites by pumping a 10-column volume of the aqueous solvent containing strong acidic H+ ions prior to sample injection. In the SCX and capillary LC method, the first fraction of polypeptide sample is eluted from the SCX column into the C18 trapping column 1 using a 25mM ammonia acetate salt buffer. Note that after salt elution of sample from the SCX into the C18 reverse column, a

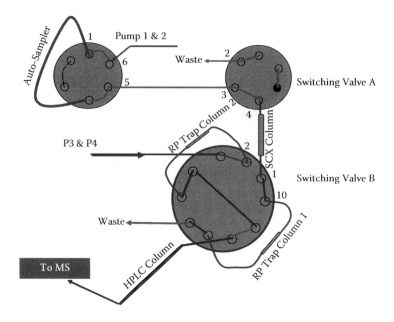

FIGURE 14.14 Fully automated two-dimensional nano HPLC-MS with autosampler injection of large sample volume from autosampler loop via pumps 1 and 2 into the SCX column and then to online sample trapping column 1. Pumps 3 and 4 elute sample from trapping column 2 into the analytical column and to MS.

desalting step for trapping column 1 should follow to prevent salt-out of the ammonium acetate when in contact with organic solvent that could result in the plugging of the column.

The 10-port valve is then switched to allow the elution of polypeptides from trapping column 1 into the analytical column (Figure 14.15) then to MS using pumps 3 and 4. At the same time, pumps 1 and 2 elute the second fraction of polypeptides with $50mM$ ammonia acetate into trapping column 2, followed by desalting and conditioning with 100% water using pump 1. After complete elution of the sample from trapping column 1, both trapping column 1 and the analytical column are conditioned with 95 to 100% water. The 10-port switching valve is switched back to allow pumps 3 and 4 to elute sample from trapping column 2 (Figure 14.15). The SCX column is now connected to trapping column 1 and the third polypeptide fraction from the SCX column is now eluted into trapping column 1. At the same time, trapping column 2 and the analytical column are connected to elute sample to MS using pumps 3 and 4 in gradient mode. At the completion of elution from trapping column 2 and conditioning of the analytical column, the 10-port valve is switched again to elute sample from trapping column 1 using pumps 3 and 4. This process continues with increasing salt concentration in each step for eluting sample from the SCX column until the sample is completely eluted into the MS.

2-D nano LC-MS-MS can identify approximately 100 components per salt step within 100 min. Ten-step 2-D SCX capillary LC-MS can separate approximately 1000 polypeptides in about 17 hr.

Due to the limited peak capacity of the 15 cm analytical column utilized in 2-D nano LC-MS, several elution steps are required to achieve the required separation. The 15 cm analtical column can be replaced with a 100 cm nano LC column to increase the resolution of sample in each step. As shown by Yang,[20] a 100 cm column allows the one-step separation of more than 2000 polypeptides from trypsin digest of mouse brain lysate, P2 fraction using XtremeSimple ultrahigh pressure nano LC (Micro-Tech Scientific, Vista, California) and LTQ MS (Thermo Electron, San Jose, California) in 6 hr (Figure 14.16). In addition to the improvement of resolving power with a 100 cm column, it

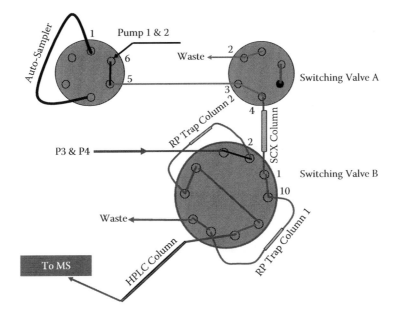

FIGURE 14.15 Fully automated two-dimensional nano HPLC-MS with autosampler. Pumps 1 and 2 elute sample from SCX column into trapping column 2 while pumps 3 and 4 elute sample from trapping column 1 into the analytical column and to MS.

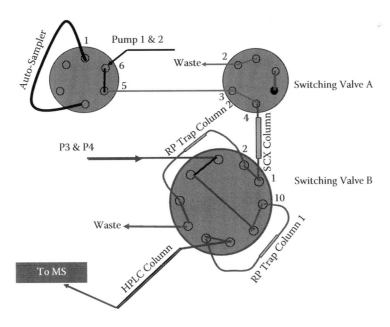

FIGURE 14.16 Fully automated two-dimensional nano HPLC-MS with autosampler. Pumps 1 and 2 elute sample from SCX column into trapping column 1 while pumps 3 and 4 elute sample from trapping column 2 into the analytical column and to MS.

is also possible to reduce the number of salt elution steps to improve the accuracy of peak identification and quantification by eliminating the duplication of polypeptides detected in two sequential steps.

14.5 ONLINE SAMPLE PREPARATION

Because of the short liquid flow path of nano LC and the small orifice spray tip of the MS interface, column and flow path plugging is a common problem with nano LC-MS. Sample clean-up is critical for ensuring reliable daily operation and generation of quality data. Online desalting and particle filtering are particularly important steps. Four online sample clean-up factors should be considered with nano LC:

1. Dead volume could cost long gradient delay; bypass with valve switching should be employed after sample clean-up.
2. Selecting proper sample solvent and stationary phase selectivity can prevent sample break-through.
3. Stationary phase, mobile phase, and column length should yield sufficient sample and peak capacity.
4. Selection of stationary phase should provide 100% sample recovery.

Many precolumns and trap cartridges for sample clean-up are commercially available. In our experience, a 2 to 3 cm short column with twice the analytical column inner diameter and packed with the same particles performs satisfactorily. An antibody affinity column for selective removal of highly abundant proteins from human serum samples provides better sensitivity for the discovery of low abundance protein markers that may represent revolutionary therapeutic diagnosis and monitoring.

The Agilent multiple affinity removal system utilizes the specificity of antibody–antigen recognition for 14 highly abundant proteins from human serum samples. The affinity column achieves reproducible and specific depletion from human serum and plasma to eliminate 94% of interfering proteins. It allows identification of proteins down to nanograms per milliliter level as reported by Agilent.

Techniques for sample clean-up also include SPE and turbulent flow chromatography on large particle size packing. Turbulent flow chromatography offers high-throughput analysis of drugs in biological fluids. It allows direct injection of crude plasma samples onto the column and achieves very high sample throughput.[21] A capillary column packed with large particles in turbulent flow chromatography enhances detection sensitivity for fast analysis of drugs in biological samples. Capillary turbulent flow chromatography at higher flow rates is very attractive for pharmacokinetics studies; antibody affinity chromatography is for abundant cellular protein elimination. It offers ultrafast analysis (1 to 2 min) along with enhanced detection sensitivity and reproducibility.

14.6 ADVANCES IN INSTRUMENTATION

A conventional HPLC pumping system was designed to deliver LC solvent at high flow rates. For capillary LC, a conventional reciprocating pumping system cannot deliver gradient flow at rates of 5 μL/min or less without flow splitting. In 1983, van der Wal and Yang[22] introduced a split flow gradient HPLC based on a Varian 5000 chromatograph for capillary LC applications. As shown in Figure 14.17 a flow rate of 5 μL/min was achieved by splitting the pump flow rate from 300 to 1000 μL/min to 1 to 5 μL/min before the injection valve using a bypass tee. Excellent retention time reproducibility at 0.7% variation was reported when the column temperature was controlled at 4°C.

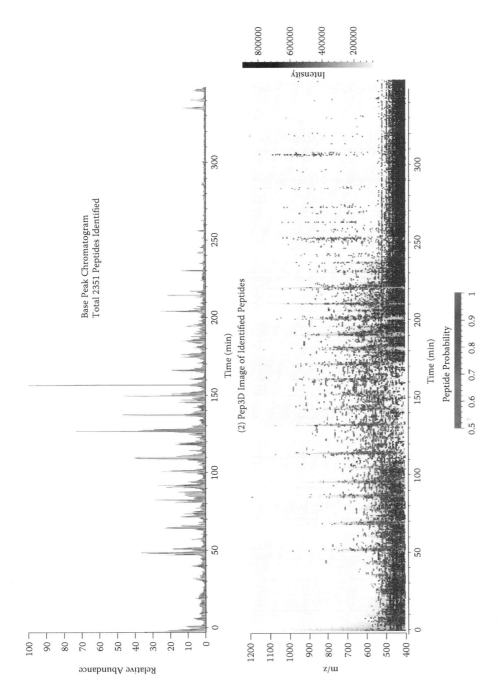

FIGURE 14.17 Ultrahigh resolution nano LC-MS separation of base peak chromatogram of 2351 peptides identified in trypsin digest of mouse brain lysate P2 fraction using Micro-Tech XtremeSimple nano-LC and Thermo Electron LTQ. Column: 100 cm × 75 μm C18 column, 3 μm, 8000 psi column head pressure. Solvent composition time: 350 min gradient, 5 to 35% B. Solvent A: 2% acetonitrile, 0.1% formic acid. Solvent B: 95% acetonitrile, 0.1% formic acid. Data analysis: Sequest, PeptideProphet, and Protein Prophet.

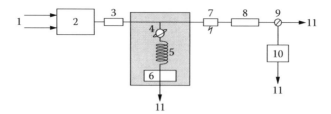

FIGURE 14.18 Flow diagram of split flow capillary LC system. 1. Solvent reservoirs. 2. Model 5000 syringe pump (Varian, Walnut Creek, California). 3. Static mixer. 4. Injection port. 5. Column. 6. Detector. 7. Pressure transducer. 8. Pulse dampener. 9. Purge valve. 10. U-flow controlling device. 11. Waste.

The split flow technique (Figure 14.18) based on constant pressure at the splitter tee does not work well if capillary column flow restriction is changed due to sample contaminants, solvent composition changes during gradient elution, temperature variation, restrictor tubing plugs, and other conditions. Recent advances in nano flow controllers based on thermal conductivity detection allow online nano LC flow rate monitoring and control.

Many commercial split flow capillary LC systems incorporate a nano flow sensor mounted online to the capillary channel. The split flow system can be easily modified from a conventional system and performs satisfactorily for capillary LC applications. However, the split flow system may require thermal control and the LC solvent requires continuous degassing. In addition, the system may not work reliably at a high flow split ratios and at pressures above 6000 psi due to technical limitations of the fused silica thermal conductivity flow sensor. The split flow system based on conventional check valve design may not be compatible with splitless nano LC applications. The conventional ball-and-seat check valve is not capable of delivering nano flow rates and is not reliable for 7/24 operation at low flow.

Schwartz and Brownlee introduced a syringe pump for micro LC applications in early 1984.[23] It exhibits a number of advantages over a conventional reciprocating pump:

1. Delivery of LC solvent in one stroke without pressure or flow pulsation induced by piston refill strokes
2. Precise control of constant flow rate as low as 0.1 μL/min
3. High pressure mixing to allow minimum gradient delay to the column
4. Absence of pump head inlet and outlet check valves to ensure reliable 7/24 operation without check valve leakage and breakage
5. Easy calibration of direct control lead screw piston drive with flow calibration software

The Brownlee syringe pump was designed for micro LC at a 50 μL/min flow rate and has a piston volume of 10 or 20 ml. For nano LC applications, the required gradient total flow rate could be as low as 0.1 μL/min. The LC system should have zero solvent leakage throughout the entire liquid path. The syringe pump design is the most reliable approach to achieve ultrahigh pressure for nano flow rate delivery without solvent leakage.

Yang introduced an ultrahigh pressure splitless nano flow syringe pumping system for multidimensional nano LC in 2005. The design incorporates the patented closed-loop digital motion control technology[24–26] with a 0.0625″ zirconium oxide piston driven by a 1 mm pitch lead screw controlled by a 4072 count optical encoder for a subnanoliter per minute flow delivery at 0.49 nL/digital count. The syringe pumping system performs reproducibly at 0.2 μL/min splitless gradient nano LC to 15,000 psi operation pressure. It can deliver 20 nL/min splitless flow rate for isocratic nano LC applications to 15,000 psi and is commercially available.

14.7 FUTURE DEVELOPMENTS AND CONCLUSIONS

Drug discovery requires fast LC separation because high-throughput drug candidate lead generation and identification impact speed and cost. Integration of online sample clean-up, sample pre-treatment, cellular protein digestion, sample concentration, desalting, subcellular particle filtering, sample component separation, and MS detection is particularly important for biological sample analysis and disease marker identification. Advances in antibody affinity column technology that can selectively and reproducibly remove high abundance proteins from human serum represent major breakthroughs in improving sensitivity in the detection of low abundance protein disease markers from human serum. Size-selective column packing that can fractionate cellular proteins based on their size differences is also highly desirable. The coupling of a capillary affinity column or size-selective column with ultrahigh speed nano LC-MS-MS, data mining, and availability of bio-information library for drug metabolite and cellular protein analysis can also expedite drug discovery.

Advances in nano particles for micro and capillary LC promise ultrahigh speed drug analysis to shorten the drug discovery process. Long capillary columns maximize resolving power for complex biological samples. Developments in nano LC-MS technology and related techniques continues in the direction of fully automated online sample preparation, complex component separation, MS detection, and computer data searching.

The integration of online multidimensional sample preparation and separation techniques including size exclusion chromatography, isoelectrical focusing column chromatography, ion exchange chromatography, affinity chromatography, online enzymatic digestion, and reversed-phase chromatography in a fully automated capillary format offers hands-free operation for highly reproducible and accurate identification of low level disease markers in human serum and other body fluids. Multiparallel nano LC instrumentation with a single MS should generate multiple sets of data outputs. Complex metabolite, ADME, and toxicity profiling studies in parallel with drug candidate identification using online nano LC-MS-MS require the development of suitable 1.5 to 2 μm particles and ultrahigh pressure HPLC systems for high-throughput analysis. In addition, high-throughput LC-MS generates great amounts of data, thus requiring an information management system. The future bio-information and drug metabolite library should allow sharing of data mining tools to enable scientists to interpret analytical data obtained via different techniques. Spectral data interpretation, processing, and visualization tools merged in a common software platform will aid decision making related to identities and purities of compounds and greatly accelerate the drug discovery process.

REFERENCES

1. D. Corens, *Recent Appl LC-MS*, November, 2002.
2. R.A. Yost and C.G. Enke, *J Am Chem Soc* 100, 2274, 1978.
3. K.L. Busch, G.L. Glish, and S.A. McLuckey, *Mass Spectrometry/Mass Spectrometry: Techniques and Applications*, VCH: New York, 1988.
4. J.F. Van Bocxlaer et al., *Mass Spectrom Rev* 19, 165, 2000.
5. F.J. Yang, *J. Chromatogr* 236, 265 1982.
6. F.J. Yang, U.S. Patent 4,483,773, 1984.
7. Greenebaum Cancer Center, Department of Anatomy and Neurobiology, University of Maryland, Baltimore.
8. Drug Enforcement Administration, Southwest Laboratory, Vista, CA, and S. DiPari.
9. *Microbore Column Chromatography: A Unified Approach to Chromatography*, J.F. Yang, Ed., Marcel Dekker: New York, 1984.
10. F.J. Yang, *HRCCC* 6, 348, 1983.
11. X. Geng and F.E. Regnier, *J Chrom* 296, 15, 1984.
12. H.M. McNair, HPLC Training Course.
13. F.J. Yang, *J Chrom Sci* 120, 241, 1982.

14. M. Verzele and C. Dewale, *Microbore Column Chromatography: A Unified Approach to Chromatography*, J.F. Yang, Ed., Marcel Dekker: New York, 1984, p. 37.

15. J.C. Giddings, *Dynamics of Chromatography, Part I: Principles and Theory*, Marcel Dekker, New York, 1965.

16. Y. Cheng, Z. Lu, and U. Neue, *Rapid Commun Mass Spectrom* 15, 141, 2001.

17. J.T. Wu et al., *Rapid Commun Mass Spectrom* 15, 1113, 2001.

18. N.J. Clarke et al., *Anal Chem* 430, 439, 2001.

19. P.H. O'Farrell, *J Biol Chem* 250, 4007, 1975.

20. A. Yang, Greenebaum Cancer Center, Department of Anatomy and Neurobiology, University of Maryland, Baltimore. Private communication.

21. J. Ayrton et al., *Rapid Commun Mass Spectrom* 13, 1657, 1999.

22. S.J. van der Wal and F.J. Yang, *HRCCC* 6, 216, 1983.

23. H. Schwartz and R.G. Brownlee, *Am Lab* 1610, 43, 1984.

24. F.J. Yang et al. U. S. Patent 5,253,981, 1993.

25. F.J. Yang, U.S. Patent 5,630,706, 1997.

26. F.J. Yang, U.S. Patent 5,664,938, 1997.

15 High-Throughput Analysis of Complex Protein Mixtures by Mass Spectrometry

Kojo S. J. Elenitoba-Johnson

CONTENTS

15.1 INTRODUCTION

Proteomics is defined as the study of the proteome—the total complement of proteins present in a complex, an organelle, a cell, a tissue, or an organism. It encompasses studies of protein expression, interaction, post-translational modification, and function at the cellular level. Mass spectrometry (MS) offers significant opportunities for the analysis of single proteins and unbiased large-scale analyses of proteins in complex mixtures. The opportunity to conduct large-scale unbiased investigations of proteins is advantageous for biological discovery and is relevant for acquiring novel biological insights into physiology and disease. MS is considered a vital technology for identifying key proteins involved in disease detection and treatment. This chapter provides brief synopses of the techniques employed in MS-based proteomics and the opportunities these technologies offer in biological discovery.

15.2 BASIC PRINCIPLES AND TOOLS OF PROTEOMICS

15.2.1 GENERAL STRATEGY FOR MASS SPECTROMETRY-BASED PROTEOMICS

Proteomic analysis via MS can be categorized as a top-down or bottom-up approach. In top-down proteomics, intact proteins are analyzed and data obtained by fragmentation of the intact protein. In bottom-up proteomics, the sample is initially digested using a proteolytic enzyme such as trypsin. The resulting peptides are separated by chromatography and then analyzed by MS. The source proteins are identified by matching the experimental tandem mass spectra with those from theoretical tandem mass spectra of translated genomic databases subjected to *in silico* cleavage using specific enzymes.

15.2.2 BIOLOGICAL SAMPLES FOR PROTEOMIC ANALYSIS

15.2.2.1 Protein Isolation

Proteomic studies generally entail isolation of a protein or proteins of interest prior to analysis by MS. Cellular proteins must be extracted from material containing other biological molecules including carbohydrates, lipids, and nucleic acids. Thus protein extraction protocols involve the homogenization of cells and tissues with subsequent application of detergents such as 3-(dimethylammonio])-1 propane sulfate (CHAPS).[1] Tween and sodium dodecyl sulfate (SDS) facilitate solubilization of the proteins and separate them from the lipid components, reducing agents such as dithiothreitol (DTT), denaturants such as urea that disrupt the bonds leading to formation of secondary and tertiary conformational structures, and enzymes that degrade nucleic acids such as DNAses and RNAses.

The sample materials from which proteins for proteomics studies may be extracted include fresh or snap-frozen cells from varied sources such as biological fluids, (serum, urine, plasma) and solid tissues such as biopsy specimens. Moreover, proteins isolated from ethanol-fixed paraffin-embedded tissues can be utilized for MS analysis.[2] Protocols for the identification of proteins from formalin-fixed paraffin-embedded (FFPE) tissues have been recently developed.[3,4] FFPE materials are the most common forms of biopsy archives utilized worldwide, and represent an important advancement for the large-scale interrogation of proteins in archival patient-derived materials. Finally, laser capture microdissected tissues have been successfully used for MS analysis.[4,5]

15.2.3 ENZYMES FOR PROTEOMIC STUDIES

The enzymes used for bottom-up proteomic studies can be classified as those with specific cleavage specificity and those with nonspecific proteolytic activity.

15.2.3.1 Enzymes with Specific Cleavage Activities

Trypsin is the most frequently used enzyme in proteomic analyses. It specifically cleaves proteins at the carboxy terminal ends of lysine (K) and arginine (R) residues except when a proline residue is located C-terminal of the K or R residue. In general, trypsin yields tryptic fragments from 9 to 30 amino acid residues in length which is suitable for the mass range of analysis for most MS instruments. Trypsin cleavage of a 50 kD protein is estimated to yield up to 30 tryptic peptides from the protein. By comparison Glu C or V8 protease demonstrates cleavage carboxy terminal to glutamic acid residues. If sodium phosphate is the buffer, then Glu C may exhibit aspartic acid residue cleavage. Utilization of enzymes with different cleavage specificities is advantageous because the different enzymes can provide overlapping information that is complementary and facilitates identification of proteins and their post-translational modifications.

15.2.3.2 Enzymes with Nonspecific Cleavage Activities

The utilization of non-specific proteases including proteinease K, pronase, and elastase among others may be very useful in proteomic studies.[6–8] These enzymes generate multiple overlapping peptides, thereby increasing the coverage on any individual protein. Parallel analyses using enzymes with specific cleavage activities provide higher confidence identifications and are used to map amino acid substitutions arising from genetic point mutations, oncogenic chimeric fusions encoded by chromosomal translocations, and post-translational modifications.[6–8]

15.2.3.3 Preanalytical Sample Simplification

Analytical separation of proteins into simpler complexes is critical for optimal proteomics analysis. With one-dimensional gel electrophoresis (1D-GE), proteins are resolved by their migration characteristics on polyacrylamide gels based on their molecular weights. The proteins in two-dimensional gel electrophoresis (2D-GE) are separated based on their isoelectric points (pIs) on one axis, and then by molecular weights in the second dimension. High performance liquid chromatography (HPLC), ion exchange, and different types of affinity chromatography have been used very successfully and integrated seamlessly with MS.[9,10] In particular, ion exchange liquid chromatography (LC) combined with reversed-phase (RP) HPLC is an efficient approach for resolving complex peptide mixtures.[8,11–13]

This multidimensional protein identification technology (MudPIT) specifically incorporates a strong cationic exchange (SCX) column in tandem with an RP column to achieve maximal resolution and exquisite sensitivity. MudPIT is effective for studying complex proteomes such as mammalian cellular samples. It has been applied to large-scale protein characterization with identification of up to 1484 proteins from yeast in a single experiment.[12]

15.2.4 MASS SPECTROMETERS

Biological MS has largely been aided by technological approaches that permit the ionization of larger biological molecules such as proteins without extensive fragmentation of the molecules (so-called soft ionization).[14–16] This development in combination with the production of sensitive mass analyzers with mass range characteristics suitable for handling larger biological molecules such as proteins has brought MS to prominence as a powerful technique for large-scale analysis of proteins.

MS instruments measure the mass-to-charge ratio (m/z) values of the smallest of molecules very accurately. In addition, the development of translated genomic databases and specialized software algorithms that rapidly search MS data against theoretical spectra of known or predicted proteins within databases is an important component that greatly facilitated the emergence of mass spectrometry-based proteomics as a key approach for large-scale proteomic analysis.[15]

The basic components of an MS instrument are an ionization source, a mass analyzer, and a detector. The ionization source generates ions from the sample to be analyzed. The mass analyzer

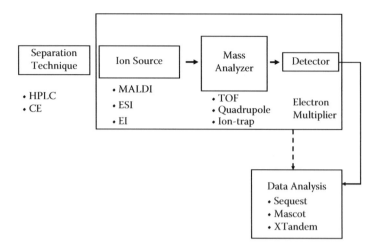

FIGURE 15.1 Overview of configuration of MS and MS-based proteomic analysis. Proteins are extracted from biologic samples and fractionated by a variety of separation methods including gel separation, HPLC, and capillary electrophoresis. The common ion sources and mass analyzers used are indicated.

resolves the ions by their *m/z* ratios. The detector determines the masses of the ions. The most common ionization sources and mass analyzers are discussed below. Figure 15.1 shows the basic set-up of MS equipment.

15.2.5 IONIZATION SOURCES

15.2.5.1 Matrix-Assisted Laser Desorption/Ionization (MALDI)

The development of soft (low energy) ionization techniques, in particular, MALDI[17,18] and electrospray[14] dramatically enhanced the feasibility of proteomic analysis by MS. In MALDI, the sample to be analyzed is incorporated into a chemical matrix containing crystallized molecules of compounds such as 3,5-dimethoxy-4-hydroxycinnamic acid (sinapinic acid), α-cyano-4 hydroxyninnamic acid (alpha-cyano or alpha-matrix) and 2,5-dihydroxybenzoic acid (DHB). Ionization is achieved by laser activation of a target, leading to the release of peptide and protein ions into a gas phase (Figure 15.2).

MALDI-generated peptide ions are characteristically singly charged; multiply charged species are infrequent. In a design similar to MALDI, surface-enhanced laser desorption/ionization (SELDI)[19,20] is embodied in the Ciphergen Chip.™ This system includes a variety of chip matrices with preferential affinities for different proteins including hydrophobic proteins and an immobilized metallic ion chromatography chip with selectivity for phosphorylated peptides among others.

15.2.5.2 Electrospray Ionization (ESI)

ESI involves the generation of ions from macromolecules in aqueous solution. The solution aerosolizes and the ions transition into a gas phase by passage of the solution through a needle subjected to high voltage[14] (Figure 15.2). The solution stream is ejected from the needle orifice as a spray of droplets. An inert carrier gas such as nitrogen can be utilized to nebulize the solvent. Because solutions with acidic pH favor protonation of N-terminal amines and histidine, nitrogens and peptide fragmentation are favored when peptide ions are positively charged. ESI protocols commonly include acidification steps prior to peptide ion analysis in the mass analyzer.

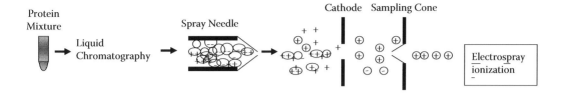

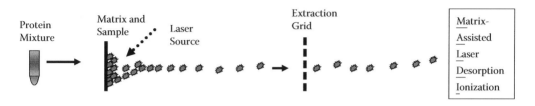

FIGURE 15.2 Common protein ionization methods used for MS-based proteomics. Two common ionization technologies are currently available for protein analysis. Top: ESI volatilizes and ionizes peptides and proteins in solution. Bottom: MALDI uses analytes that are co-crystallized in a matrix composed of organic acid on a solid support. A pulse of ultraviolet laser evaporates the matrix and analyte into gas phase, resulting in generation of single charge ions.

15.2.6 Mass Analyzers

Mass analyzers interrogate and resolve ions produced by an ion source based on their m/z ratios. Several types of mass analyzers are utilized for proteomic analysis including time-of-flight (TOF) quadrupoles, ion traps, and Fourier transform ion cyclotron resonance (FTICR). Mass analyzers may be assembled in hybrid configurations. MS instruments such as quadrupole TOF and quadrupole ion trap-FTICR facilitate diversified applications and achieved great success.

MS equipment is evaluated on several performance metrics. Mass accuracy, mass resolution, and mass range are standard parameters frequently assessed to determine the suitability of an instrument. Mass accuracy is defined as the extent to which a mass analyzer reflects "true" m/z values and is measured in atomic mass units (amu), parts per million (ppm), or percent accuracy.

Mass resolution describes the capability of an MS to distinguish ions with different m/z values. It is defined by the $M/\Delta M$ equation in which M is the m/z ratio of a mass peak and ΔM is the full width of a peak at half its maximum height. The mass resolution of an instrument often correlates with its accuracy. Mass range indicates the m/z range at which the mass analyzer best functions. For example, quadrupole mass analyzers exhibit a mass range of up to 4000 m/z, while the mass ranges of TOF extend up to 100,000. The operating principles of common MS instruments are discussed below.

15.2.6.1 Time of Flight (TOF) Analysis

With TOF, the ions from an ion source are accelerated linearly down a chamber containing an electrical field. The flight chamber is at very low pressure that facilitates the flight of the peptide ions with minimal collisions with other molecules. The ions travel in a linear trajectory until they impact a detector at the other end of the tube. The heavier ions travel more slowly than the lower molecular weight ions and reach the detector later. Hence, TOF analyzers derive their name from the concept that the "time of flight" of an ion is related to its m/z ratio and velocity within a fixed distance. Linear mode TOF analyzers contain single chambers and are not favored for proteomics applications because of their lower mass accuracy.

The mass accuracy and resolution of TOF analyzers was improved by the *reflectron* design in which traveling ions are reflected by an ion mirror and are turned around in their flight paths.

Having traveled a greater distance, they reach the detector later. TOF instruments have very wide mass ranges because ion detection is not limited by the mass range of the analyzer. This property of TOF analyzers is advantageous in the analysis of larger biological molecules.

TOF analyzers are especially compatible with MALDI ion sources and hence are frequently coupled in a MALDI-TOF configuration. Nevertheless, many commercial mass spectrometers combine ESI with TOF with great success. For proteomics applications, the quadrupole TOF (QqTOF) hybrid instruments with their superior mass accuracy, mass range, and mass resolution are of much greater utility than simple TOF instruments.[21,22] Moreover, TOF instruments feature high sensitivity because they can generate full scan data without the necessity for scanning that causes ion loss and decreased sensitivity. Linear mode TOF instruments cannot perform tandem mass spectrometry. This problem is addressed by hybrid instruments that incorporate analyzers with mass selective capability (e.g., QqTOF) in front of a TOF instrument.

15.2.6.2 Quadrupoles

A quadrupole consists of two pairs of charged poles that separate ions generated from the ion source based on their m/z values. Direct and alternating radiofrequency voltages applied between each pair of poles deflect the ions in the direction of one or the other rod in a pair. The rapid oscillation of the polarity of the electrical field around the quadrupoles causes the ions to travel in a spiral trajectory. The specific oscillation frequency determines which ions (based on their m/z ratios) can pass through, collide with the poles with loss of charge, or are ejected from the quadrupole. The mass range and resolution achievable with a quadrupole assembly are determined mainly by the lengths and diameters of the poles.

ESI is highly compatible with quadrupole mass analyzers because quadrupole mass analyzers are fairly tolerant of the high pressures generated by ESI. Additionally, ESI generates multiple charged ions, and since mass analyzers measure m/z, the higher charged states result in m/z values within the mass range that can be measured by the mass analyzer. In proteomics applications, the triple quadrupole configuration (three quadrupole analyzers aligned in tandem) has been successful for several proteomics applications including product ion, precursor ion, and neutral loss scanning. These features, particularly the neutral loss scanning ability are useful for analyzing post-translational modifications.[21,23–25] Triple quadrupoles are effective for quantitative analysis. The sensitivity of these instruments is further improved by their ability to perform single and multiple reaction monitoring analysis. However, their sensitivity may be limited by losses in ion beam transmission. In addition, discrepancies in the elution rates of analytes from fast HPLC runs and the slower scanning speeds of many triple quadrupoles may limit their ability to analyze co-eluting molecules.

15.2.6.3 Ion Trap MS

The ion trap mass analyzer is similar to the quadrupole but with the important distinction that it can isolate and trap ions in an electrical field. Notably, the ion trap differs significantly from quadrupoles in design and operation in that triple quadrupoles perform tandem mass analysis on ions as they pass through an analyzer; ion traps are capable of isolating and retaining specific ions for fragmentation upon collision with an inert gas in the same cell. An ion trap is about the size of a tennis ball and consists of a donut-shaped electrode and two perforated disk-like end-cap electrodes.

The mass-selective instability mode of operation permits the selection and trapping of all ions created over a specified period with subsequent ejection to the detector.[26] Ions with different m/z values can be confined within the ion trap and scanned singly by application of voltages that destabilize the orbits of the ions and eject them to the detector. Ion trap instruments interface readily with liquid chromatography, ESI,[15] and MALDI.[27] The motions of the ions and the dampening gas (e.g., helium) concentrate around the middle of the ion trap, thereby diminishing ion loss through collisions with electrodes.

Ion traps are favored for proteomics studies because of their ability to perform multistage mass analysis (MSn), thereby increasing the structural information obtained from molecules. Ion traps, however, do not provide information for ions that have lower mass-to-charge values (the one-third rule). Additionally, the sensitivity of ion traps can also be limiting because only about 50% of the ions within a trap are ejected to the detector. Ion traps are also subject to a space charging phenomenon that may occur when the concentration of ions in the trap is high and produces ion repulsion within the trap. Nevertheless, the versatility and robustness of ion trap MS underlies its popularity for several proteomics-related applications.

15.2.6.4 Fourier Transform Ion Cyclotron Resistance (FTICR) MS

This is a very powerful MS analyzer that determines the m/z ratios of ions based on their cyclotron frequencies in a fixed magnetic field. The distinctive property of FTICR is that the ion trapping cell is surrounded by a magnetic field within which entrapped ions can resonate at their cyclotron frequencies.[28] Ions in FTICR can be focused into coherent packets and can then be excited to a larger cyclotron radius by an oscillating electrical field. An image current is generated by the ions as they pass near a pair of plates as they cyclotron. The resulting free induction decay signal is composed of a complex profile of sine waves.

The m/z values of peptide ions are mathematically derived from the sine wave profile by the performance of a fast Fourier transform operation. Thus, the detection of ions by FTICR is distinct from results from other MS approaches because the peptide ions are detected by their oscillation near the detection plate rather than by collision with a detector. Consequently, masses are resolved only by cyclotron frequency and not in space (sector instruments) or time (TOF analyzers). The magnetic field strength measured in Tesla correlates with the performance properties of FTICR. The instruments are very powerful and provide exquisitely high mass accuracy, mass resolution, and sensitivity—desirable properties in the analysis of complex protein mixtures. FTICR instruments are especially compatible with ESI[29] but may also be used with MALDI as an ionization source.[30] FTICR requires sophisticated expertise. Nevertheless, this technique is increasingly employed successfully in proteomics studies.

15.2.7 TANDEM MASS ANALYSIS

Tandem MS entails multiple rounds of mass analysis. Thus tandem MS instruments are capable of ion isolation and fragmentation and mass analysis of fragment ions. Fragmentation can be achieved in several ways including collision-induced dissociation (CID), electron capture dissociation (ECD), and electron transfer dissociation (ETD). Fragmentation of the parent ion generates daughter ions that can be used to determine the amino acid sequences of peptides, and characterize post-translational modifications of proteins. The most common fragmentation method used in proteomics is CID. Peptide ions analyzed in an initial mass analyzer are directed into a collision cell where individual ions can be isolated and fragmented by collision with an inert gas (helium). Fragment ions exit the collision cell and are separated in a second mass analyzer before scanning out to the detector. The fragmentation pattern of peptide ions by CID occurs along the peptide backbone and yields a fairly predictable array of ions. The resulting m/z data from the original precursor ion and the fragmentation data from product ions can be used to determine the peptide sequence.

15.2.8 PROTEIN IDENTIFICATION STRATEGIES

15.2.8.1 Peptide Mass Fingerprinting (PMF)

Spectra to be analyzed via PMF are derived from a protein sample been treated with an enzyme (e.g., trypsin) or other chemical (cyanogen bromide) with specific cleavage activity. The experimental m/z values for each peptide are converted into peptide masses and compared with the theoretical mass spectra of proteins in a sequence database. The proteins with the highest numbers of experimental and theoretical peptide mass matches are ranked highest in probability of identification. The significance of each matching peak is calculated and all matches cumulatively computed to produce a score. The scores for all possible matches are ranked in descending order and the protein with the highest score is ranked as the putative identification.

Clearly, PMF is most effective when applied for MS analysis of proteins from species whose complete genome sequences are available. An important caveat is that PMF algorithms may assign higher scores to larger proteins that contain more peptides, thereby leading to an increased propensity for incorrect assignment of peptide matches that could result in misidentifications. Several software algorithms that facilitate peptide mass mapping are available and include PeptIdent/MultiIdent and ProFound.[31,32]

PMF is generally used to identify proteins that have been previously separated by 2-D GE so that additional information including the molecular weights and isoelectric points can be used to supplement PMF identification. PMF is not well suited for searching expressed sequence tag (EST) databases that contain incomplete gene coding information for particular ESTs and it is not adequate for the analysis of complex protein mixtures in solution.

15.2.8.2 Tandem MS Peptide Sequencing

Peptide ions in tandem mass spectrometry (MS/MS) undergo fragmentation upon collision with neutral gas atoms in a collision chamber of triple quadrupoles. The collision-induced dissociation occurs along the peptide backbone. The most frequently observed cleavage site is at the amide bond between the amide nitrogen and the carbonyl oxygen. This leads to the generation of *b*- and *y*-ion series that constitute the primary data for peptide sequencing. When a positive charge is retained on the carboxy terminus of an original peptide, the result is a *y*-ion. If the positive charge is retained on the amino-terminal fragment of the original peptide, the fragment is a *b*-ion.

The experimental MS/MS spectra are matched against theoretical spectra and cross correlation scores are calculated based on the extent to which the predicted and experimental spectra overlap.[8] The higher cross correlation scores reflect a high level of matching of the experimental and predicted MS/MS spectra, and vice versa. The difference between a normalized cross correlation score and the next best match is reported as the (ΔCn) and indicates the quality of the top match in comparison to the next ranked sequences in the database.[8]

Several algorithms are available for the analysis of MS/MS spectra including SEQUEST, MASCOT, and X!Tandem among others. Note that additional secondary quality control of assessment of MS/MS data has recently been implemented to assess identification probabilities and false positivity rates. The MS/MS spectra from an experiment can be interrogated against a concatenated forward and reverse database and an assessment of the intrinsic error rate of the data set can be made. Other approaches for secondary analysis of matching scores for peptide sequencing data include XCorr score normalization routines that are independent of peptide and database size.[33]

Algorithm suites that employ statistical modeling strategies to assign confidence values for large scale datasets such as PeptideProphet, INTERACT, and ProteinProphet[34–37] are also useful for the quality control of protein identifications by MS/MS. One useful aspect of ProteinProphet is that it allows the determination of false positive rates in specific datasets. Because of sequence conservation among similar or related proteins, it is important to identify a unique peptide that distinguishes a specific protein from related family members. Accordingly, a single tandem mass spectrum alone should not be construed as providing a unique identification of a particular protein. Rather, multiple peptide "hits" corresponding to a specific protein sequence may provide unequivocal evidence of identification of that protein.

Tandem MS has emerged as a definitive approach for identifying proteins from multiple sources including complex mixtures. In comparison to PMF, MS/MS permits more definitive identification of proteins. Matching of multiple MS/MS spectra to peptide sequences within the same protein increases the confidence in identification. MS/MS-based protein identification is applicable to EST databases with reliable matches.

15.3 APPLICATIONS OF MASS SPECTROMETRY-BASED PROTEOMICS

Several categories of proteomics techniques apply to the studies of human disease. Expression proteomics involves the large-scale identification of proteins from biological materials of interest such as subcellular organelles, enriched cell populations, tissues, or entire organs. Expression proteomics also includes the comprehensive identification of proteins in body fluids such as saliva and urine. Quantitative proteomics encompasses global proteomic studies that monitor the proteome-wide changes occurring in different biological states. The quantitative analysis of the protein expression profiles of tissues or body fluids from specific disease conditions in comparison to normal states holds promise for identifying disease biomarkers. Accordingly, the protein expression "signatures" consisting of discriminative expression patterns may facilitate recognition of deregulated protein pathways in specific disease contexts.

Functional proteomics encompasses the study of proteins in their functional environments and the biological consequences of perturbations of normal functional proteins, including the analysis of protein interactions with other proteins, interactions with DNA or RNA, and post-translational modifications such as phosphorylation and glycosylation. These studies allow investigators to obtain information regarding protein functions such as identifying networks of signaling pathways characteristic of physiologic and pathologic states.

15.3.1 PROTEIN EXPRESSION PROFILING

Global proteomic profiling by MS is gaining significant attention as a tool for discovering disease biomarkers. Two basic approaches have been explored. With the first, MS analysis is performed with a material from a specific disease condition and the mass spectra are compared to those of normal individuals or related disease conditions. SELDI-TOF MS gained popularity in this area because of its simplicity and the requirement for only small amounts of samples.[38–43] In MS-pattern based disease categorization, the mass spectral patterns are considered reflective of the proteins present in samples from distinct clinical conditions.

Bioinformatics analyses can reveal the distinctive mass spectral signatures of a disease condition of interest. Discriminating mass spectral signatures have been reported for a number of neoplastic conditions including ovarian,[41] breast,[44] prostate,[45] and liver cancers.[46] Because these studies rely predominantly on mass spectral profiles without tandem MS or identification of the peptides or proteins involved, the utilization of mass spectral signatures to discriminate specific disease conditions requires extensive pre- and post-acquisition procedures including mass calibration, baseline correction, and noise subtraction to aid in identifying bona fide features that are robust discriminators of normal, benign, and malignant states.[47,48] Proteomic patterns of nipple aspirate fluids,[49] cytologic specimens,[50] and tissue biopsies[51] by SELDI-TOF have also shown potential for discovery of novel biomarker profiles that aid in diagnosis.

15.3.2 IMAGING MASS SPECTROMETRY

Imaging mass spectrometry involves MS performed on tissue sections mounted on a MALDI plate. The mass spectra generate images and an *in situ* protein expression profile of the specimen is analyzed. Specifically, the frozen tissue sections applied to a MALDI plate are subjected to laser interrogation and analyzed at regular spatial intervals. The mass spectral data obtained at different intervals are compared to generate a spatial distribution of masses (proteins) across the tissue section.

Analyses using this approach have revealed more than 1500 protein peaks from histologically selected 1 mm diameter regions of single frozen sections.[52] Imaging MS allows investigators to distinguish glial neoplasms from benign brain tissues and differentiate tumors of different histological grades.[53] However, imaging MS has been successful only with frozen tissue sections.

15.3.2.1 MS of Laser Capture Microdissected Tissues

An alternative approach to assessing tissue-specific expression at the proteomic level can be achieved by MS of laser capture microdissected tissues.[4] An important development in this arena is the ability to perform LCM and MS/MS on formalin-fixed paraffin-embedded tissues.

15.3.3 QUANTITATIVE PROTEOMICS ANALYSIS

Quantitative studies comparing the relative abundances of proteins in different cellular states may be performed with MS. Methods such as 2D-GE have been utilized extensively with great success and differentially represent spots excised and then subjected to MS/MS for final identification of the differentially expressed proteins. 2D-GE requires approximately 50 µg of starting material and is limited by its bias toward high abundance proteins and propensity to detect proteins with extreme pI values. Furthermore, proteins at both extremes of molecular weight and those associated with membrane fractions are not well represented by 2D-GE.[10]

Despite these limitations, investigators have successfully used 2D-GE followed by MALDI-TOF to determine differential expression of protein profiles in many comparisons of normal and tumor tissues.[54] Cellular responses to stimulating and differentiating agents such as LPS[55] and Fas[56] have been studied using this approach. Finally, the proteomic consequences of exposure to cytotoxic agents such as butyrate[57] have been studied with 2D-GE and MS. 2D-GE can reveal as many as 50 differentially expressed proteins per experiment, depending on the complexities of the proteomes compared.

15.3.3.1 Stable Isotope Labeling in Cell Culture (SILAC)

Another useful strategy that utilizes different isotopes of the same element for proteomic quantification is known as stable isotope labeling with amino acids in culture (SILAC).[58,59] This technique entails culturing of cells from two distinct biologic conditions in parallel culture media that lack natural amino acids and are supplemented with an synthetic amino acid that contains only one distinct isotope of an element (12C, 13C; 14N, 15N, respectively). The two cell populations metabolically incorporate the corresponding light or heavy isotopes in the synthesis of their respective cellular proteins during propagation in culture. Proteins from each sample can thus be isolated, mixed at a 1:1 ratio, and subjected to enzymatic digestion and MS analysis.

Corresponding peptides from each sample co-elute during liquid chromatography and relative quantification of a particular peptide represented in both samples are performed by measuring the ratios of the peptide mass peak intensities from matching isotopic pairs. The peptide sequence is determined by MS/MS and database interrogation identifies the differentially expressed protein. This method is readily adaptable for comparing different conditions and can be used to reliably interrogate signaling pathways and post-translational modifications. However, it is effective only with viable and metabolically active cells, so biopsy tissues cannot be utilized.

15.3.3.2 Isotope-Coded Affinity Tagging (ICAT)

The isotope-coded affinity tag approach utilizes chemical labeling that allows quantitation when combined with mass spectrometry. ICAT™ is desirable because the chemical labeling takes advantage of the mass defects of monoisotopic stable isotopes. ICAT uses an ICAT™ reagent to differentially label protein samples on their cysteine residues. ICAT™ is advantageous because it permits the evaluation of low-abundance proteins and proteins at both extremes of molecular weights and isoelectric points.[60]

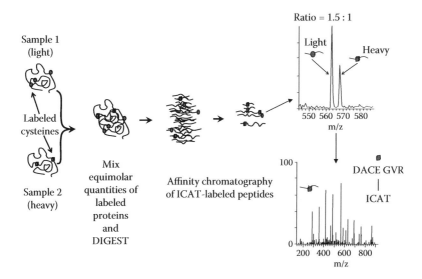

FIGURE 15.3 Outline of experimental protocol used for ICAT differential protein expression profiling. Protein mixtures from two cell populations are labeled with light or heavy isotopic versions of a cleavable ICAT reagent. Labeled proteins are combined, subject to multidimensional separation by SCX, RP, and avidin affinity chromatography, then analyzed by tandem MS for peptide and protein identification. Based on the relative ratio of the two isotopically labeled peptides, a relative abundance of protein expression can be determined.

The ICAT™ reagent is composed of (1) a biotin tag that enables affinity isolation and detection of peptides labeled with heavy or light versions of the ICAT™ reagent, (2) a thiol-reactive iodoacetamide group that reacts with the cysteine residues, and (3) a linker incorporating stable isotopes. One sample is labeled with a tag containing a light isotope and the other (comparison) sample is labeled with a heavy isotope. The two samples are combined, enzymatically digested, and analyzed by MS (Figure 15.3).

Because ICAT™-labeled peptides co-elute as pairs from an HPLC column, calculating the ratio of the areas under the curve for identical peptide peaks labeled with light and heavy ICAT™ reagent allows determination of the relative abundance of that peptide in each sample. ICAT™'s advantages include internal quantitation, automation, and reduced complexity of the peptide mixture. The commercially available cleavable (c)ICAT™ reagent contains nine 13Cs in the heavy version of the linker and nine 12Cs in the light version. The resultant database search is constrained by the requirement for a cysteinyl group. ICAT is compatible with analysis of low abundance proteins and may be performed with proteins from snap-frozen archival tissues.

The iTRAQ labeling technique employs isobaric tags and involves labeling at the peptide level.[61] iTRAQ is beneficial because it permits analysis of up to eight different samples, allowing multiple comparisons that are desirable in time-course experiments. It also yields highly reproducible results. Quantitative proteomic analysis may also be performed using endoproteinase-catalyzed incorporation of stable isotopes of oxygen (^{16}O and ^{18}O). Labeling is performed at the peptide level and the approach is simple and inexpensive.

15.3.4 IDENTIFICATION OF PROTEIN–PROTEIN INTERACTIONS

MS can perform large-scale analyses of protein interactions. Interacting partners of proteins in complexes can be purified by several strategies including affinity chromatography and immunoprecipitation with antibodies specific to the bait protein. The purified components can then be subjected to LC-MS/MS and proteins within the complex identified. Using a variety of approaches including protein complex purification, immunoprecipitation, affinity chromatography[62] followed by HPLC and ESI-MS/MS, the interacting proteins of CD4 receptor complex, PCNA,[63] nonmuscle myosin heavy chain II,[64] and protein kinase Cε signaling complex[65] have been identified.

The advantage of such co-purification protocols is that the fully processed protein serving as the bait can allow interactions in a native environment and cellular location to allow isolation of multicomponent complexes. One limitation with this approach is the necessity for an antibody with specific immunoreactivity and immunoprecipitative capability for the bait protein. This drawback can be addressed by expression of the protein with an epitope tag. Excellent antibodies to a variety of epitope tags are available and can be utilized for immunoaffinity purification. Tags such as 6-histidine and GST allow purification using affinity characteristics to nickel and GSH beads, respectively.

Expression of a recombinant protein using an inducible vector system would permit expression at endogenous levels to simulate physiologic levels of expression of a protein of interest. Tandem affinity purification strategies have recently been employed and facilitate the analyses of highly interactive proteins when the bait protein is expressed at endogenous levels. Immunoaffinity or immunoprecipitation followed by LC-MS/MS does not readily permit determination of the stoichiometry of interacting partners. Additionally, when compared to yeast hybrid experiments, it is difficult to determine whether interactions are binary when identified in complexes by MS/MS.

We used immunoprecipitation and LC-MS/MS to identify the interacting partners of the NPM/ALK oncogenic chimeric fusion kinase. NPM/ALK results from the t(2;5)(p23;q35) chromosomal aberration characteristic of a subtype of T-cell lymphoma known as anaplastic large cell lymphoma.[66–68] The interaction partners of the deregulated ALK kinase play important roles in transducing the aberrant signals of oncogenic tyrosine kinase to mediate its downstream cellular effects. The elucidation of novel interacting partners provides opportunities to identify previously uncharacterized interaction partners and mechanisms by which ALK overexpression affects cellular homeostasis.

Identification of the interacting partners also provides opportunities for elucidation of therapeutic targets that may be useful in treating ALK-deregulated neoplasia. Our experiments identified a total of 46 proteins unique to the ALK immunocomplex. Many previously reported proteins in the ALK signal pathway were identified including PI3-K, Jak2, Jak3, Stat3, Grb2, IRS, and PLCγ1. More importantly, many proteins previously not recognized as associated with NPM-ALK but having potential NPM-ALK interacting protein domains were identified. Proteins identified by MS were confirmed by western blotting and reciprocal immunoprecipitation and show the potential of MS for identification of novel proteins in a well studied signaling pathway.[69]

15.3.5 MS-Based Analysis of Post-Translational Modifications (PTMs)

PTMs are important for the regulation of protein function and the maintenance of cellular hemostasis. There are 300 or more reported PTMs of proteins. PTMs may involve the addition of functional groups such as acetyls in acetylation, hydroxylation, amidation, and oxidation or the addition of peptides or proteins such as ubiquitination, SUMOylation (addition of small ubiquitin-like modifier), and ISGylation (addition of interferon-stimulated gene[15]).

Other PTMs may involve changes in the chemical nature of amino acids (e.g., citrullination or deimination). Because many of these modifications result in mass changes that are measurable by MS, they are amenable to detection by MS-based approaches. A number of emerging MS-based strategies allow the identification of PTMs. Several MS-based methods to determine the types and sites of protein phosphorylation and ubiquitination have been developed. Phosphorylation occurs mainly on serine, threonine, and tyrosine residues at a frequency ratio of 1800:200:1 in vertebrates.[70] Although the phosphorylation of tyrosine residues occurs less frequently in the proteome, it has been extensively studied.

Attempts have been made to define the phosphorylation status of protein on a global scale.[71] Most approaches involve the use of phospho-specific antibodies to enrich for proteins with phosphorylated residues. Due to the availability of excellent antibodies that react with phosphotyrosines, studies focused on tyrosine phosphoproteins far outnumber those analyzing serine and threonine phosphoproteins. Using phospho-specific antibodies, serine- and threonine-phosphorylated proteins have been enriched by immunoprecipitation with subsequent identification by MS.

Using this approach, a novel protein demonstrated to be a substrate of protein kinase A was identified.[72] More large-scale studies of phosphoproteins have used commercially available immobilized metal ion affinity chromatography (IMAC) that allows enrichment of phosphopeptides[73] and identification of several phosphorylation sites on single proteins.[74] Similar enrichment strategies have been applied to analyses of ubiquitinated proteins. Peng et al. identified over 1000 ubiquitinated proteins in yeast.[9] Their ability to identify the sites of ubiquitination of over 100 cases validates their high-throughput proteomic-based methodology for ubiquitination site mapping.

An important emerging strategy is the simultaneous utilization of SILAC-based approaches for the quantitative identification of legitimate PTMs. This approach has been used successfully in the interrogation of phosphorylation for the analysis of phosphoproteomic signaling.[75,76]

15.3.6   PROTEOMICS OF SPECIFIC SUBCELLULAR COMPARTMENTS

Enriched subcellular compartments can be analyzed by MS/MS to determine their constituent proteins. One advantage of analysis of different cellular fractions is pre-analytical simplification that offers rewarding yields in dealing with the proteins identified in large-scale MS experiments. One of the major initiatives of the Human Proteome Organization (HUPO) is the comprehensive characterization of the complete subproteome of each cell type.

Defining the global fingerprints of proteins expressed in a certain cell type will aid in identifying deregulated proteins that are characteristic of certain disease states and also in diagnosis and prognostication. Similarly, it is possible to analyze proteins from body fluids and proteins secreted from different cell types. Martin et al.[77] used a combination of ICAT and tandem MS to identify and quantitate more than 500 proteins secreted from a neoplastic prostate cancer cell line (LNCaP) in the presence or absence of androgen receptor stimulation. Similar studies identified numerous secreted proteins during differentiation of 3T3-L1 preadipocytes to adipocytes.[78] Nuclear, cytosolic, and mitochondrial subproteomes may be analyzed following the utilization of appropriate enrichment protocols. Strategies for the MS analysis of membrane proteins have shown promising results based on utilization of the proteinase K non-specific proteolytic cleavage enzyme under appropriate buffer conditions.[79]

15.4   CONCLUSIONS

The large-scale study of protein expression, interactions, and post-translational modifications at organellar, cellular, tissue, organismal levels is known as proteomics. Because proteins are the functional effectors of cellular processes, analysis of aberrations at the proteomic level promises to yield novel insights into the cellular consequences of alterations of proteins and their networks in physiological and disease processes. The advent of MS-based proteomics techniques provides tremendous opportunities to leverage the advantages of this powerful technology in biological discovery research. Future challenges include the archiving and integration of large amounts of proteomic data into meaningful knowledge of biological processes. MS clearly provides significant opportunities to utilize novel unanticipated findings in biological research and aid the discovery of disease biomarkers and therapeutic targets in specific clinical contexts.

REFERENCES

1. Wubbolts R et al. 2003. *J Biol Chem* 278: 10963.
2. Ahram M et al. 2003. *Proteomics* 3: 413.
3. Crockett DK et al. 2005. *Lab Invest* 85: 1405.

4. Hood BL et al. 2005. *Mol Cell Proteomics* 4: 1741.
5. Li C et al. 2004. *Mol Cell Proteomics* 3: 399.
6. Elenitoba-Johnson KS et al. 2006. *Proc Natl Acad Sci USA* 103: 7402.
7. Gatlin CL et al. *Anal Chem* 72: 757.
8. MacCoss MJ et al. 2002. *Proc Natl Acad Sci USA* 99: 7900.
9. Peng J et al. 2003. *Nat Biotechnol* 21: 921.
10. Gygi SP et al. 2000. *Proc Natl Acad Sci USA* 97: 9390.
11. Link AJ et al. 1999. *Nat Biotechnol* 17: 676.
12. Washburn MP, Wolters D, and Yates JR, 3rd. 2001. *Nat Biotechnol* 19: 242.
13. Washburn MP et al. 2002. *Anal Chem* 74: 1650.
14. Fenn JB et al. 1989. *Science* 246: 64.
15. Yates JR, 3rd. 1998. *J Mass Spectrom* 33: 1.
16. Hillenkamp F et al. 1991. *Anal Chem* 63: 1193A.
17. Tanaka K et al. 1998. *Rapid Commun Mass Spec* 2: 151.
18. Karas M and Hillenkamp F. 1988. *Anal Chem* 60: 2301.
19. Kuwata H et al. 1998. *Biochem Biophys Res Commun* 245: 764.
20. Merchant M and Weinberger SR. 2000. *Electrophoresis* 21: 1164.
21. Yost RA and Boyd RK. 1990. *Methods Enzymol* 193: 154.
22. Chernushevich IV, Loboda AV, and Thomson BA. 2001. *J Mass Spectrom* 36: 849.
23. Li J et al. 2000. *Anal Chem* 72: 599.
24. Dainese P et al. 1997. *Electrophoresis* 18: 432.
25. Steen H, Kuster B, and Mann M. 2001. *J Mass Spectrom* 36: 782.
26. Stafford G et al. 1984. *Int J Mass Spectrom Ion Processes* 60: 85.
27. Krutchinsky AN, Kalkum M, Chait BT. 2001. *Anal Chem* 73: 5066.
28. Marshall AG, Hendrickson CL, and Jackson GS. 1998. *Mass Spectrom Rev* 17: 1.
29. Bruce JE et al. 1999. *Anal Chem* 71: 2595.
30. Solouki T et al. 2001. *J Am Soc Mass Spectrom* 12: 1272.
31. Wilkins MR et al. 1998. *Electrophoresis* 19: 3199.
32. Zhang W and Chait BT. 2000. *Anal Chem* 72: 2482.
33. MacCoss MJ, Wu CC, and Yates JR, 3rd. 2002. *Anal Chem* 74: 5593.
34. Nesvizhskii AI et al. 2003. *Anal Chem* 75: 4646.
35. Von Haller PD et al. 2003. *Mol Cell Proteomics* 2: 426.
36. Von Haller PD et al. 2003. *Mol Cell Proteomics* 2: 428.
37. Keller A et al. 2002. *Anal Chem* 74: 5383.
38. Petricoin EF et al. 2002. *Nat Rev Drug Discov* 1: 683.
39. Petricoin EF, 3rd et al. 2002. *J Natl Cancer Inst* 94: 1576.
40. Petricoin EF, 3rd et al. 2002. *Nat Genet* 32 Suppl: 474.
41. Petricoin EF, 3rd et al. 2002. *Lancet* 359: 572.
42. Rosenblatt KP et al. 2004. *Annu Rev Med* 55: 97.
43. Rui Z et al. 2003. *Proteomics* 3: 433.
44. Li J et al. 2002. *Clin Chem* 48: 1296.
45. Adam BL et al. 2002. *Cancer Res* 62: 3609.
46. Poon TC et al. 2003. *Clin Chem* 49: 752.
47. Coombes KR et al. 2003. *Clin Chem* 49: 1615.
48. Baggerly KA et al. 2004. *Bioinformatics* 20: 777.
49. Paweletz CP et al. 2001. *Dis Markers* 17: 301.
50. Fetsch PA et al. 2002. *Am J Clin Pathol* 118: 870.
51. Lin Z et al. 2004. *Mod Pathol* 17: 670.
52. Yanagisawa K et al. 2003. *Lancet* 362: 433.
53. Schwartz SA et al. 2004. *Clin Cancer Res* 10: 981.
54. Friedman DB et al. 2004. *Proteomics* 4: 793.
55. Lian Z et al. 2001. *Blood* 98: 513.
56. Gerner C et al. 2000. *J Biol Chem* 275: 39018.
57. Tan S et al. 2002. *Int J Cancer* 98: 523.
58. Ong SE et al. 2002. *Mol Cell Proteomics* 1: 376.
59. Ong SE, Kratchmarova I, and Mann M. 2003. *J Proteome Res* 2: 173.
60. Gygi SP et al. 1999. *Nat Biotechnol* 17: 994.

61. Ross PL et al. 2004. *Mol Cell Proteomics* 3: 1154.
62. Kumar A and Snyder M. 2002. *Nature* 415: 123.
63. Ohta S et al. 2002. *J Biol Chem* 277: 40362.
64. Rey M et al. 2002. *J Immunol* 169: 5410.
65. Edmondson RD et al. 2002. *Mol Cell Proteomics* 1: 421.
66. Morris SW et al. 1994. *Science* 263: 1281.
67. Shiota M et al. 1994. *Oncogene* 9: 1567.
68. Shiota M et al. 1994. *Blood* 84: 3648.
69. Crockett DK et al. 2004. *Oncogene* 23: 2617.
70. Hunter T. 1998. *Philos Trans R Soc Lond B Biol Sci* 353: 583.
71. Mann M et al. 2002. *Trends Biotechnol* 20: 261.
72. Gronborg M et al. 2002. *Mol Cell Proteomics* 1: 517.
73. Ficarro SB et al. 2002. *Nat Biotechnol* 20: 301.
74. Fuglsang AT et al. 1999. *J Biol Chem* 274: 36774.
75. Blagoev B et al. 2003. *Nat Biotechnol* 21: 315.
76. Blagoev B and Mann M. 2006. *Methods* 40: 243.
77. Martin DB et al. 2004. *Cancer Res* 64: 347.
78. Kratchmarova I et al. 2002. *Mol Cell Proteomics* 1: 213.
79. Wu CC et al. 2003. *Nat Biotechnol* 21: 532.

Index

Bold locators indicate material in figures and tables.